ance AD and DA Converters,
Scaled Technologies,
main Signal Processing

Pieter Harpe · Andrea Baschirotto
Kofi A. A. Makinwa
Editors

High-Performance AD and DA Converters, IC Design in Scaled Technologies, and Time-Domain Signal Processing

Advances in Analog Circuit Design 2014

Springer

Editors
Pieter Harpe
Department of Electrical Engineering
Eindhoven University of Technology
Eindhoven
The Netherlands

Kofi A. A. Makinwa
Delft University of Technology
Delft
The Netherlands

Andrea Baschirotto
Department of Physics
University of Milan-Bicocca
Milan
Italy

ISBN 978-3-319-07937-0 ISBN 978-3-319-07938-7 (eBook)
DOI 10.1007/978-3-319-07938-7
Springer Cham Heidelberg New York Dordrecht London

Library of Congress Control Number: 2014945140

Printed on acid-free paper

Springer is part of Springer Science+Business Media (www.springer.com)

Preface

This book is part of the *Analog Circuit Design* series and contains contributions of all 18 speakers of the 23rd workshop on Advances in Analog Circuit Design (AACD). The local chairs were Jorge Guilherme (from Instituto Politécnico de Tomar, Tomar) and Nuno Horta (from Instituto Superior Técnico, Lisbon). The sponsors of the workshop this year have been SYNOPSYS, FCT—Fundação para a Ciência e a Tecnologia, IST—Instituto Superior Técnico, IPT—Instituto Politécnico de Tomar, IT—Instituto de Telecomunicações, and Turismo de Lisboa. The workshop was held at Instituto Superior Técnico in Lisbon, Portugal, in April 8–10, 2014.

The book comprises three parts, covering advanced analog and mixed-signal circuit design fields that are considered highly important by the circuit design community:

- High-Performance AD and DA Converters
- IC Design in Scaled Technologies
- Time-Domain Signal Processing

Each part contains six chapters from experts in the field.

The aim of the AACD workshop is to bring together a group of expert designers to discuss new developments and future options. Each workshop is followed by the publication of a book by Springer in their successful series of *Analog Circuit Design*. This book is the 23rd in this series. The book series can be seen as a reference for all people involved in analog and mixed-signal design. The full list of the previous books and topics in the series is given next.

We are confident that this book, like its predecessors, proves to be a valuable contribution to our analog and mixed-signal circuit design community.

Eindhoven, The Netherlands — Pieter Harpe
Milan, Italy — Andrea Baschirotto
Delft, The Netherlands — Kofi A. A. Makinwa

Contents

The Topics Covered Before in this Series

2013	Grenoble (France)	Frequency References Power Management for SoC Smart Wireless Interface
2012	Valkenburg (The Netherlands)	Nyquist A/D Converters Capacitive Sensor Interfaces Beyond Analog Circuit Design
2011	Leuven (Belgium)	Low-Voltage Low-Power Data Converters Short-Range Wireless Front-Ends Power Management and DC–DC
2010	Graz (Austria)	Robust Design Sigma Delta Converters RFID
2009	Lund (Sweden)	Smart Data Converters Filters on Chip Multimode Transmitters
2008	Pavia (Italy)	High-Speed Clock and Data Recovery High-Performance Amplifiers Power Management
2007	Oostende (Belgium)	Sensors, Actuators and Power Drivers for the Automotive and Industrial Environment Integrated PAs from Wireline to RF Very High Frequency Front Ends
2006	Maastricht (The Netherlands)	High-Speed AD Converters Automotive Electronics: EMC issues Ultra Low Power Wireless
2005	Limerick (Ireland)	RF Circuits: Wide Band, Front-Ends, DACs Design Methodology and Verification of RF and Mixed-Signal Systems Low Power and Low Voltage
2004	Montreux (Swiss)	Sensor and Actuator Interface Electronics Integrated High-Voltage Electronics and Power Management Low-Power and High-Resolution ADCs
2003	Graz (Austria)	Fractional-N Synthesizers Design for Robustness Line and Bus Drivers

(continued)

(continued)

2002	Spa (Belgium)	Structured Mixed-Mode Design Multi-bit Sigma-Delta Converters Short-Range RF Circuits
2001	Noordwijk (The Netherlands)	Scalable Analog Circuits High-Speed D/A Converters RF Power Amplifiers
2000	Munich (Germany)	High-Speed A/D Converters Mixed-Signal Design PLLs and Synthesizers
1999	Nice (France)	XDSL and Other Communication Systems RF-MOST Models and Behavioural Modelling Integrated Filters and Oscillators
1998	Copenhagen (Denmark)	1-Volt Electronics Mixed-Mode Systems LNAs and RF Power Amps for Telecom
1997	Como (Italy)	RF A/D Converters Sensor and Actuator Interfaces Low-Noise Oscillators, PLLs and Synthesizers
1996	Lausanne (Swiss)	RF CMOS Circuit Design Bandpass Sigma Delta and Other Data Converters Translinear Circuits
1995	Villach (Austria)	Low-Noise/Power/Voltage Mixed-Mode with CAD Tools Voltage, Current and Time References
1994	Eindhoven (The Netherlands)	Low-Power Low-Voltage Integrated Filters Smart Power
1993	Leuven (Belgium)	Mixed-Mode A/D Design Sensor Interfaces Communication Circuits
1992	Scheveningen (The Netherlands)	OpAmps ADC Analog CAD

Part I
High-Performance AD and DA Converters

Pieter Harpe

The first part of this book is dedicated to "High-Performance AD and DA Converters." It discusses recent developments, challenges, and future trends in converter design. A variety of architectures is covered, such as pipeline, SAR, Sigma-Delta, and time-interleaved ADCs as well as current-steering DACs. Further, it addresses both system innovations, novel analog circuits, digital methods to enhance analog performance, or even software tools to assist converter design.

Low-Power, High-Speed and High-Effective Resolution Pipeline Analog-to-Digital Converters in Deep Nanoscale CMOS by João Goes (Universidade Nova de Lisboa) gives an overview of techniques that can be used to improve pipeline ADCs, especially in view of future technologies. The main focus is on novel amplification solutions that enhance overall performance while being able to operate at low supply voltages.

In Digitally Assisted Analog to Digital Converters, Bob Verbruggen (imec) discusses how digital assistance can be leveraged to alleviate performance limitations in AD converters. Besides a general overview, chip implementations and measurement results of a SAR ADC and a pipelined SAR ADC are also discussed.

Lukas Kull et al. (IBM Research – Zurich) describe the design of a 90 GS/s time-interleaved SAR ADC in Energy-Efficient High-Speed SAR ADCs in CMOS. The architectural and circuit innovations to achieve such an extremely high speed are discussed in detail and the implementation in an advanced 32nm SOI CMOS process is described.

In Automated Design of High-Speed CT ΣΔ Modulators Employing Compensation and Correction of Non-idealities, Timon Brückner and Maurits Ortmanns (University of Ulm) present an automated tool to develop Continuous-Time Sigma-Delta Modulators on architectural level while including circuit-level non-idealities. With smart acceleration techniques, this allows quick design optimization. The tool is applied to a practical design case with measurement results.

Pedro Figueiredo (Synopsys) presents a unified treatment of ADC architectures in Recent Advances and Trends in High-Performance Embedded Data Converters to provide a deeper understanding of fundamental limitations and trade-offs. Further, the evolution of embedded ADC architectures in SoC applications is discussed and various implementations are described.

In High-Performance DACs: Unifying 16-Bit Dynamic Range with GS/s Data-Rates of Part I, Joost Briaire et al. (Integrated Device Technology) describe the trends in high-performance DAC design. The two main approaches, based on either intrinsic design or mismatch calibration, are compared. The development of a state-of-the-art 16-bit multi-GS/s current-steering DAC is further elaborated.

Low-Power, High-Speed and High-Effective Resolution Pipeline Analog-to-Digital Converters in Deep Nanoscale CMOS

João Goes and Nuno Pereira

Abstract This chapter reviews recent advances in low-power design techniques for high-speed and high-effective resolution pipeline Analog-to-Digital Converters (ADCs). The advantages of replacing, in a pipeline ADC architecture, the traditional local low-resolution parallel (flash) quantizers by low-resolution successive-approximation register (SAR) ADCs are shown through a set of selected works. Some of the most promising energy-efficient residue amplification techniques are reviewed, spanning from open-loop and closed-loop amplifierless approaches to innovative and highly-scalable amplifier-based topologies.

1 Introduction

More and more signal processing is being transferred to the digital domain to profit from the advancements of digital circuits. While technology scaling enhances the capabilities of digital circuits, it degrades the performance of analog circuits. However, it is important to note that the impact of technology scaling on digital circuits is becoming smaller, which means that, with nanotechnologies, to enhance energy and area efficiency, we cannot simply depend on scaling. Although part of the efficiency increase can be obtained from technology, new circuit architectures and techniques have to be developed to really push the limits of efficiency.

In analog-to-digital converters (ADCs), a fundamental decision can be made: either investigate energy and area efficient analog circuit techniques and architectures that cope with technological scaling issues, or design algorithms that use

J. Goes (✉) · N. Pereira
Department of Electrical Engineering (DEE), Faculty of Sciences and Technology, Centre of Technologies and Systems (CTS/UNINOVA), Universidade Nova de Lisboa (FCT/UNL), Campus FCT/UNL, 2829-516 Caparica, Portugal
e-mail: jg@uninova.pt

P. Harpe et al. (eds.), *High-Performance AD and DA Converters, IC Design in Scaled Technologies, and Time-Domain Signal Processing*,
DOI: 10.1007/978-3-319-07938-7_1,

digital circuitry to assist the poor analog technological performance. The former option is behind the techniques described in this chapter.

High-effective resolution (i.e., effective-number-of-bits, ENOB > 10.5 bits) with sampling rates higher than 160 MS/s are required in high-quality imaging and modern digital communication systems. When very high conversion rates are envisaged, the ADCs usually employ pipelining to relax the speed requirements of the analog components. A cascade of stages is used, each consisting of a low resolution quantizer and a low resolution multiplying digital-to-analog (DAC) converter (MDAC), which computes and amplifies the residue to be quantized by the following stages.

The accuracy of a pipeline ADC is mainly limited by the gain-error and static linearity errors in the front-end stages, caused by capacitor mismatches in the DACs and by the finite DC gain and linearity of the residue amplifiers. The accuracy required for the first stage is that of the overall ADC and it is progressively relaxed in the following stages. Without using digital factory trimming [1], self-configuring schemes [2] or self-calibration techniques, the overall resolution of these ADCs is limited to 10 bits by the capacitor's matching accuracy provided by most CMOS processes available today. This accuracy is limited by the lithography and subsequent processing steps [3].

As factory trimming can add to the manufacturing costs, self-calibration, applied in the field, must be considered for extending the resolution beyond 10.5 bits. Since digital circuitry shrinks with the evolution of CMOS technology, it is a good strategy to reduce analog circuit complexity at the expense of increased digital complexity. Hence, several purely digital and mixed-signal self-calibration techniques have been intensively investigated over the past years. However, it is out of the scope of this chapter to cover this lengthy subject. A partial survey of this matter can be found elsewhere [4].

This chapter is organized as follows. In section II the SAR-assisted pipeline A/D architectures are described. The advantages of replacing, in a pipeline ADC architecture, the traditional local low-resolution parallel (flash) quantizers by low-resolution successive-approximation register (SAR) ADCs are shown through a set of selected works. Sections 3 and 4 describe energy-efficient residue amplification techniques. On the one hand, Sect. 3 describes efficient techniques for implementing residue amplification without using operational transconductance amplifiers (OTAs) and, on the other hand, Sect. 4 shows some of the state-of-the-art of amplifier-based residue amplification techniques. Finally, Sect. 5 draws the main conclusions.

2 SAR-Assisted Pipeline ADC Architectures

With the scaling down of CMOS technology, the achievable precision for a limited power budget has become an issue, as it is increasingly difficult to design reliable high-gain and high-speed amplifiers operating at low supply voltages. For moderate

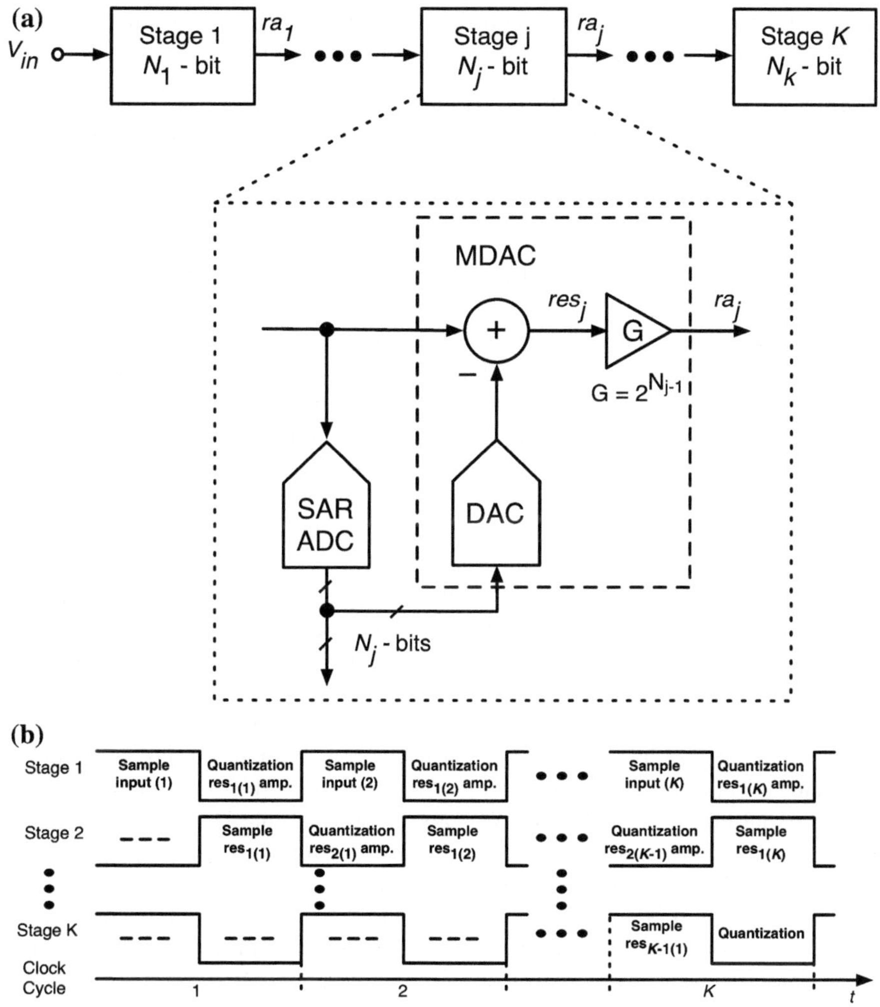

Fig. 1 **a** Block diagram of a conventional pipeline stage comprising a local quantizer and an N_j-bit MDAC circuit, in which, the parallel-type (flash) ADC has been replaced by a SAR-ADC; **b** Timing diagrams

resolutions, SAR ADCs have surpassed pipeline converters in terms of energy efficiency. However, SAR-ADCs provide excellent energy-efficiency for large-signals but, when design become "noise-limited", SAR-ADCs need more energy to extract the LSBs since no amplification is used to bring-down the noise level. In fact, pipeline ADCs can have better energy efficiency for noise-limited designs.

Recent "SAR-assisted" pipeline ADCs aim to capture the best of both worlds, by replacing the traditional flash quantizers used in the pipelined stages by SAR ADCs, as shown in Fig. 1. In the particular case of using only two pipelined stages, the two local quantizers can be replaced by two energy-efficient (6-to-8 bit) SAR-ADCs. In this case, the advantages are the following:

- only a residue amplifier will be required;
- the first DAC can be, itself, embedded into the front-end SAR-ADC (stage 1);
- multiple comparators can be used in the SAR-ADCs to make them faster (providing multiple output bits per clock-cycle);
- the residue amplifier can be switched OFF when is not being used.

Some of the most recent and more efficient techniques/architectures are discussed in this section.

Several structures have been proposed to implement capacitor- based DACs for SAR-assisted ADCs for both radix-2 and non-radix-2 architectures [5]. In radix-2 capacitive-array DACs, the digital circuit is simple, yet the matching of capacitors in the array is essential. On the other hand, in non-radix-2 architectures, the matching requirement in the capacitive-array can be more relaxed but the digital circuit has a larger complexity. Several methods have been proposed in literature for implementation of radix-2 capacitive-array DACs. All of these radix-2 architectures are based on three fundamental structures, i.e., conventional binary-weighted (CBW) capacitive-array, binary-weighted capacitive-array with attenuation capacitor (BWA) and split binary-weighted (SBW) capacitive-array DAC [5]. The major details of the "SAR-assisted" pipeline ADCs described in the following paragraphs are summarized in Table 1.

In [6], a two-stage 10-bit 204 MS/s "SAR-assisted" pipeline ADC that employs a dual-channel interleaved structure with an asynchronous timing technique is presented. While the DAC/quantizer of the 1st stage (4-bit) is implemented as a conventional binary-weighted (CBW) capacitive array, the DAC of the 2nd stage (7-bit) SAR ADC uses a binary-weighted capacitor-array relying on a BWA architecture. No frontend sample-and-hold (S/H) block is required, since the sampling operation is already embedded in the 1st stage DAC/quantizer. The use of an attenuation capacitor divider, splits the N-bit capacitor-array of the 2nd stage into a K-bit main-DAC and a M-bit sub-DAC (where $K + M = N$). This allows for the total capacitance value to be reduced, thus optimizing both die area and power dissipation. The residue amplifier is designed based on an amplifier-based approach, with a closed-loop gain of 8 and its amplification time is increased due to the use of an asynchronous timing technique. This technique uses a 4-phase clock and the period of the input sampling clock is $T_s/2$. However, the amplification time is allowed to vary, unlike in traditional pipeline ADCs, thereby improving operation speed.

To further reduce power and area, the residue amplifier is shared between channels thanks to the non-overlapping residue amplification periods of the dual-channel structure. Nevertheless, since the amplifier requires high gain-bandwidth-product (GBW), it dominates the power dissipation. A reference voltage error correction circuit reduces the static current of the internal reference buffer by means of a dynamic current control, thereby reducing the settling time of reference

Table 1 Details of some published works on SAR-assisted pipeline ADCs [6–9]

	[6]	[7]	[8]	[9]
Architecture	10-bit, two-stage pipeline ADC with 2 SAR-based quantizers of 4b and 7b	12-bit, two-stage pipeline ADC with 2 SAR-based quantizers of 6b and 7b	10-bit, two-stage pipeline ADC with 2 SAR-based quantizers of 6b and 7b	11-bit, two-stage pipeline ADC with 2 SAR-based quantizers of 6b and 7b
Residue amplification technique	Amplifier-based, closed-loop (8x)	Amplifier-based, closed-loop (16x) and reference voltages for 2nd.-stage scaled by 2x	Amplifier-based, closed-loop (8x) and reference voltages for 2nd.-stage scaled by 4x	Amplifier-based, open-loop, dynamic charge-steering; reference voltages for 2nd.-stage scaled by 4x
1st stage DAC/quantizer architecture	CBW and a single comparator	CBW and a single comparator	SBW and a single comparator	CBW with a series sampling capacitor, front-end sampling switches and multiple comparators
2nd stage DAC/quantizer architecture	BWA and a single comparator	CBW and a single comparator	BWA and a single comparator	CBW with front-end sampling switches and multiple comparators
Gain calibration	No	No	No (only offset correction)	Yes, calibration of the residue gain error
ENOB (bits)	8.9	10.7	8.9	9.5
Energy-efficiency (fJ/conv.-step.)	95.4	52	50	13

CBW conventional binary-weighed capacitive-array; *BWA* binary-weighted capacitive-array with attenuation capacitor; *SBW* split binary-weighted capacitive-array

voltages. The fabricated two-stage 204 MS/s prototype ADC reached a peak ENOB of 8.9 bits and an energy-efficiency of 95.4 fJ/conv.-step.

In [7] a two-stage 12-bit 50 MS/s "SAR-assisted" pipeline ADC with a high-resolution front-end stage (6b SAR ADC) is presented. The backend (2nd) stage is a 7-bit SAR ADC. Similarly to the previous example, the first stage DAC/quantizer is based on a CBW structure and, hence, a frontend S/H block is not required. The difference here is that the gain of the residue amplifier is reduced by a factor of two, when compared with conventional architectures. Therefore, the amplifier specifications are relaxed, as the closed-loop bandwidth nearly doubles and the amplifier's output-swing requirement also decreases by a factor of two. This translates into lower power dissipation and improved linearity.

The amplifier has been implemented based on an NMOS input triple-cascode topology. By scaling down the gain of the residue amplifier, the reference voltages of the subsequent stage must be adjusted accordingly (i.e., reduced by the same factor). The solution employed in [7] is to split all the capacitors of the 2nd stage SAR, where the DAC/quantizer is built using a CBW structure, into two halves. This way, during the decision phase, only half of the capacitors are connected to a full reference voltage, while the other half is grounded. The correct reference voltage is thereby provided. Such implementation requires doubling the number of unit capacitors. Thanks to the large front-end gain, the noise contribution of the 2nd stage can be almost eliminated. The fabricated 2-stage ADC reached a peak ENOB of 10.7 bits and an energy-efficiency of 52 fJ/conv.-step.

In [8] a two-stage, 10-bit, 160 MS/s, two-channel interleaved "SAR-assisted" pipeline ADC with self-embedded offset cancelation is presented. Here, the frontend stage uses a 6-bit SAR ADC (in which the DAC relies on a SBW topology) while the backend stage relies on a 7-bit SAR ADC (with a BWA DAC). Like the example described in [6], the residue amplifier is shared between channels. To improve energy-efficiency, the amplifier is designed to have only a quarter of the typical required gain (i.e., eight instead of thirty-two) of that required for a conventional architecture. The DAC/quantizer of the first-stage is composed of two binary-weighted arrays of capacitors with 64 unity elements each. To reach the desired residue amplification, 16 elements of one of the binary-weighted capacitor-arrays are flipped-around the residue amplifier ($128/16 = 8$). The use of the flipped-around method further reduces power consumption (since the feedback-factor is higher) and the total number of capacitors, as the feedback element only uses a fraction of the capacitor-array.

To adjust the subsequent stage's reference voltage (a quarter of the reference voltage of the first stage is used), the SAR ADC quantizer of the 2nd-stage uses a BWA topology. Here, the first split array scales down the reference voltage at the comparator's input by a factor of 8, behaving as a voltage divider. Moreover, the use of this architecture leads to a sufficiently low output equivalent capacitance, improving the speed of the amplifier. An offset cancelation technique is developed and made possible through the use of two extra bits of redundancy in the 2nd-stage SAR ADC. The fabricated ADC prototypes reached a measured peak ENOB of 8.9 bits and an energy-efficiency of 50 fJ/conv.-step, without any gain calibration.

In [9] a two-stage, 11-bit, 250 MS/s, two-channel interleaved "SAR-assisted" pipeline ADC with fully dynamic residue amplification is presented. Two identical ADC channels operating at half the sampling frequency are used. Each is composed of a frontend stage 6-bit SAR ADC, a dynamic residue amplifier, and a 2nd-stage 7-bit SAR ADC (both employ CBW capacitive-arrays together with front-end sampling switches and multiple comparators). The main advantage here is that fully dynamic amplifiers can deliver very low-power amplification and low noise. To make up for gain uncertainty and reduced linearity, inherent to this technique, gain calibration and limited gain implementation, respectively, are used. Here, an open-loop charge-steering residue amplifier with a gain of four is used. Both stages use SAR-ADCs with multiple comparators, thereby avoiding the need for a digital controller.

To reduce the input capacitance by a factor of 2, a series sampling capacitor is added between the input and reference. The fabricated 2-stage ADC reached a measured peak ENOB of 9.5 bits and an energy-efficiency of 13 fJ/conv.-step, with self-calibration of the open-loop gain of the residue amplifier.

3 Amplifierless Residue Amplification

In pipeline ADC research, the focus has been on techniques to reduce the power dissipation of the MDAC, which is typically the most power-hungry block in an ADC. In the vast majority of pipeline ADCs, the MDACs are implemented using an amplifier-based approach. In recent years, energy efficiency has been increased in amplifier-based topologies, by replacing the amplifier by more efficient amplifying circuitry. Some of these innovative techniques are described in this section, either using open-loop approaches or, with closed-loop, by employing comparator-based or zero-crossing-detector-based circuits.

3.1 Open-Loop Techniques

3.1.1 Capacitive Charge-Pump

In a classical capacitive charge-pump, voltage gain is achieved by sampling an input voltage on multiple capacitors, and subsequently, connecting the capacitors in series to yield a total voltage, which is the sum of the individual capacitor voltages. An output buffer/source-follower is normally used to prevent charge sharing between sampling and load capacitors. Although there are clear power and noise advantages in performing residue amplification through the use of charge-pumps, there are important limitations, such as imprecise gain, due to parasitic capacitors and poor common-mode (CM) rejection. To deal with such problems, a differential charge-pump residue amplifier, shown in Fig. 2, has been proposed in [10].

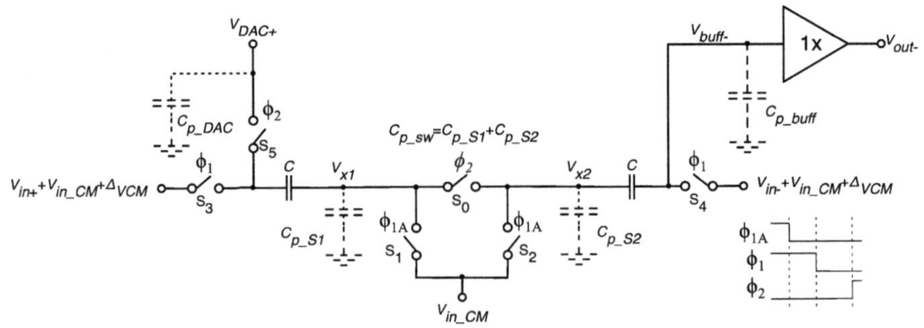

Fig. 2 Differential capacitive charge-pump 1.5-bit MDAC (half-circuit) [10]

In this circuit, the input sampling network is re-arranged so that the differential input signal is sampled into the sampling capacitors, during Φ_1, using a bridge switch configuration. Due to this AC-coupling structure, CM variations in the differential input are, therefore, rejected during Φ_2. Switch S0 is included to isolate nodes V_{X1} and V_{X2} during Φ_1, and thus ensure that S1 and S2 act as bottom-plate sampling switches (minimizing charge-injection effects).

As stated in [10], to deal with the nonlinearities affecting the residue amplification gain due to the parasitic capacitors, self-calibration had to be employed in some of the front-end stages. After calibration, the measured 12-bit 50 MS/s IC prototypes reached a peak ENOB of 9.4 bits and an energy-efficiency of 300 fJ/conv.-step. With appropriate self-calibration and few additional bits of resolution (for extra redundancy), the effective resolution can be extended above the 11 bits.

3.1.2 Dynamic Source-Follower Residue Amplification

Figure 3 shows the conceptual operations of a dynamic source-follower amplifier [11]. During the sampling-phase, Φ_1, the MOS transistor is biased in depletion with the gate at a constant DC voltage V_{BIAS} and the drain, source and bulk tied to the input signal. During the residue amplification phase, Φ_2, the drain is connected to V_{DD} and the gate is left floating while the source and the bulk are tied to the circuit's output node, V_{out}. The device now acts as a source-follower and the amplifier charges the load capacitance until V_{GS} approximately equals the threshold voltage, V_{Tn}. Notice that the parasitic capacitances C_{gd}, C_{gs} and C_{gb} in phase Φ_1, are different from the values of C_{gd}' C_{gs}' and C_{gb}' in phase Φ_2. As demonstrated in [11], since there is charge conservation at the gate terminal when switching from Φ_1 to Φ_2, the residue amplification gain is given by

$$G_{\text{DynAmp}} \cong -\left(C_{gs} + C_{gd} + C_{gb}\right)/C_{gd}' \tag{1}$$

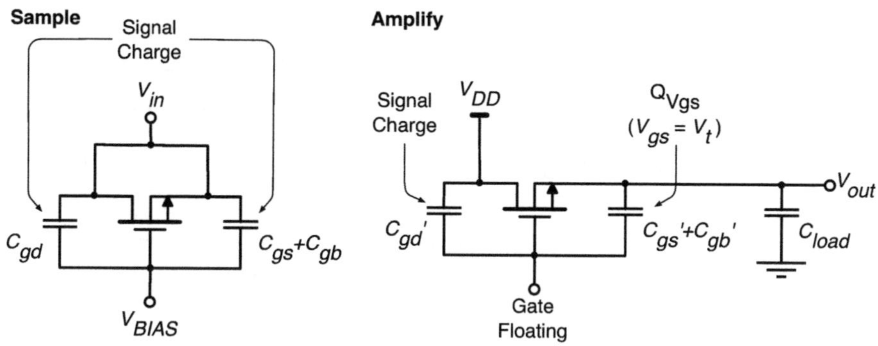

Fig. 3 Dynamic source-follower residue amplifier [11]

To provide an additional degree of freedom and have a better gain control, an extrinsic $C_{gs,ext}$ is added in parallel to the intrinsic parasitic C_{gs}. Adding this explicit capacitor, the gain becomes

$$G_{\text{DynAmp}} \cong -\left(C_{gs,ext} + C_{gs} + C_{gd} + C_{gb}\right)/C_{gd}' \quad (2)$$

Moreover, to avoid the MOS device entering in sub-threshold regime when V_{GS} approaches V_{Tn}, a small bleed current-source, I_{bleed}, is added to the circuit. A single-ended schematic of the circuit is shown in Fig. 4. Capacitors $C_{dac,pos}$ and $C_{dac,neg}$ pre-charge the reference voltages, $V_{refdac,pos}$ and $V_{refdac,neg}$ during the sampling phase and, based on the output of the local quantizer, they are properly connected to implement, during phase Φ_2, the required DC level-shifting to obtain the desired stage transfer function. Since the dynamic source-follower has stronger pull-up than pull-down capabilities, any possible change in the decision of the comparator will make that the amplifier will not have enough time to settle properly. To overcome this problem, a 4-phase clocking scheme has to be used.

Similarly to [10], after employing a digital self-calibration technique, the fabricated 14-stage 50 MS/s prototype ADC reached a peak ENOB of 9.4 bits and an energy-efficiency of 119 fJ/conv.-step.

3.1.3 Parametric Amplification

Low-gain voltage amplification, e.g., to implement a 1.5-bit MDAC, can be achieved by using the parametric MOS structure described in [13], where a discrete-time (DT) MOS parametric amplifier (MPA) was experimentally evaluated for the first time. In this amplifier, the gain is set by varying the total equivalent gate capacitance of a single MOSCAP device, while maintaining the total gate charge between the sampling phase Φ_1 and the amplification phase Φ_2. As explained in [13], the capacitance reduction of a MOSCAP can be achieved by

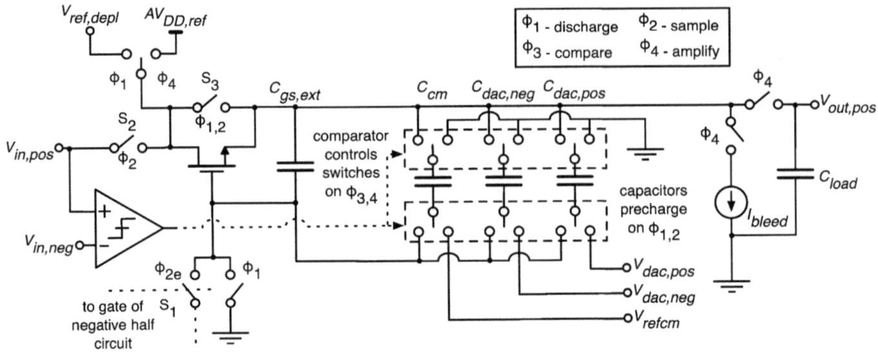

Fig. 4 Pseudo-differential dynamic source-follower 1.5-bit pipeline stage [11]

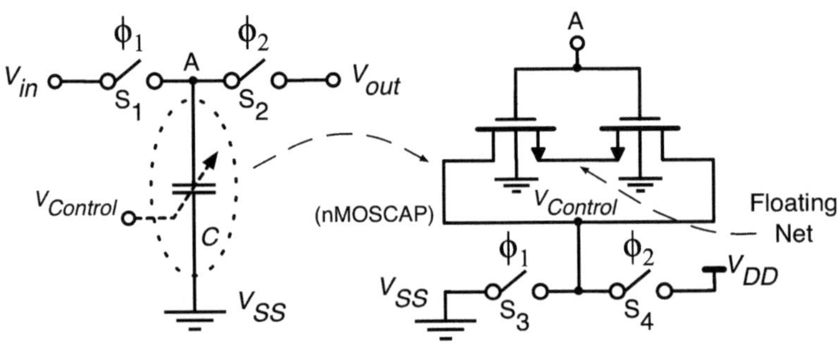

Fig. 5 MOS parametric amplifier using two separated devices with a floating terminal [12]

changing its state from inversion into depletion, as result of changing the control voltage, $V_{control}$, applied to the drain, as shown in Fig. 5.

The first difference of the MOSCAP structure proposed in [12] from the original structure in [13] is that two half-sized MOSCAPs are used in parallel rather than a single MOSCAP and with one pair of terminals left floating. The division into more than two devices can be used for higher unit capacitance values. Care must be taken in using neither large L (to avoid speed limitations) nor minimum L (to avoid short-length effects). With this modified structure, it becomes possible to decrease the effect of the extrinsic (overlap) gate capacitances during the amplification phase (Φ_2). Hence, amplification gains above 2 (considering the loading effect) can now be easier to achieve with a nMOS-type MOSCAP (nMOSCAP). By properly adjusting the load, it becomes possible to design MBTA circuits with gain accuracy above the 6-bit level, without calibration. Using a suitable self-calibration technique, the accuracy can reach the 11-bit level.

The amplifier gain in Fig. 6 was determined using the intrinsic MOS gate capacitance values obtained from electrical simulations of the circuit shown in Fig. 5. Figure 6 also shows that, besides the load capacitance, C_L, the maximum

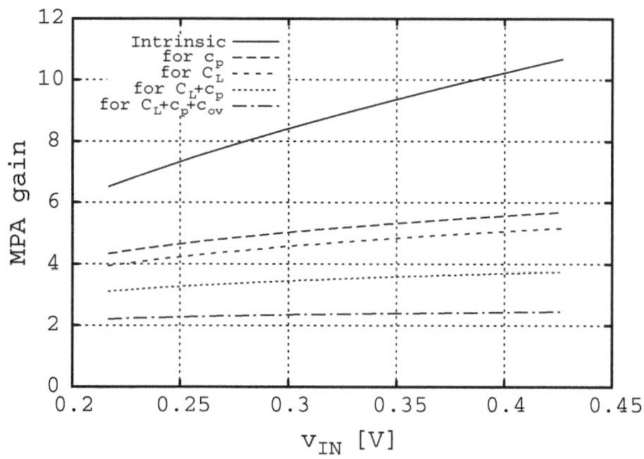

Fig. 6 Parametric amplifier gain as a function of the input DC level for an nMOSCAP sized with W = 14 μm and L = 2 μm in 130 nm CMOS [12]

achievable gain also depends on the CM level of the input voltage, reflecting how well the MOS device is biased in the inversion region during the sampling phase. Therefore, an appropriate DC level has to be carefully chosen.

The schematic of a 1.5-bit MDAC circuit is shown in Fig. 7. MOSCAPs C_X, C_Y, and C_Z, operating in the inversion region, perform the DAC function. C_Y is required to provide the same loading, although it does not add any charge (differentially). Only one of these capacitors is selected in each clock cycle for charge redistribution by the *X*, *Y*, and *Z* codes provided by the local quantizer. Due to charge conservation, the input signal applied to the MDAC block is amplified by the MPA principle through C_I. This process adds a positive DC level shifting. As stated above, to prevent this shift from exceeding V_{DD}, a MOSCAP C_2 (pMOS) is used to control the DC (CM) level. By optimizing the size of all five MOSCAPs, the MDAC function is implemented at the output of a source-follower (SF), i.e., $v_{outd} = 2 \cdot v_{ind} - X \cdot V_{REFD} + Z \cdot V_{REFD}$ (where *X* and *Z* are the usual digital control signals provided by the local quantizer). During Φ_1, the sampling capacitance C_I is 312 fF (in inversion). This value drops by a factor of roughly 4-to-6 during Φ_2 when C_I goes into depletion.

3.2 Comparator-Based and Zero-Crossing-Based Techniques

A comparator-based switched-capacitor (CBSC) circuit is shown in the simplified schematic of Fig. 8a [14]. Observe that the amplifier is replaced by a comparator and a current source. Similarly to the amplifier-based implementation, when Φ_1 is high during the sampling phase, the input voltage v_I is sampled onto C_I and C_2.

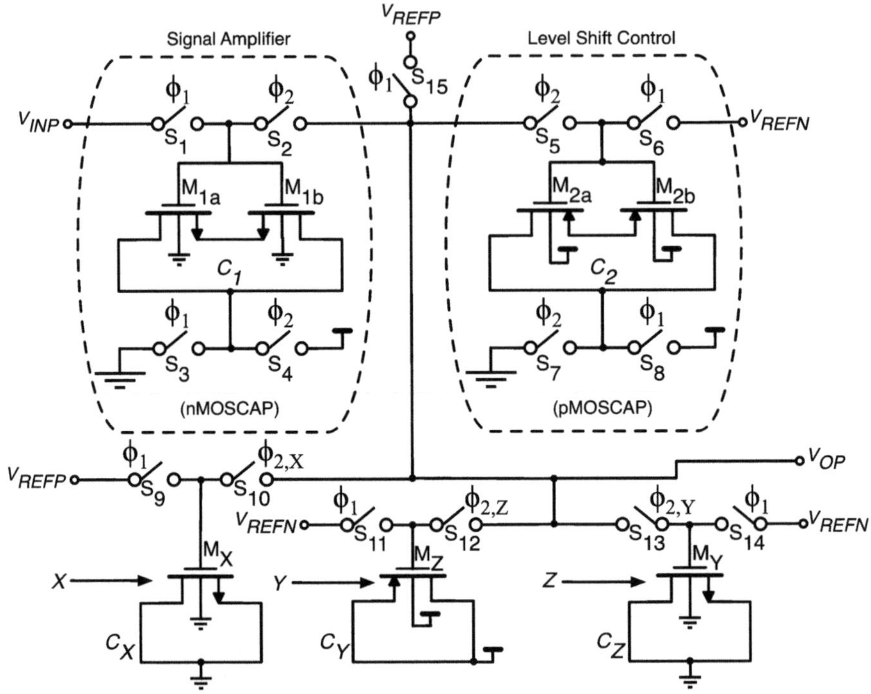

Fig. 7 Example of a 1.5-bit MDAC employing MPA [12]

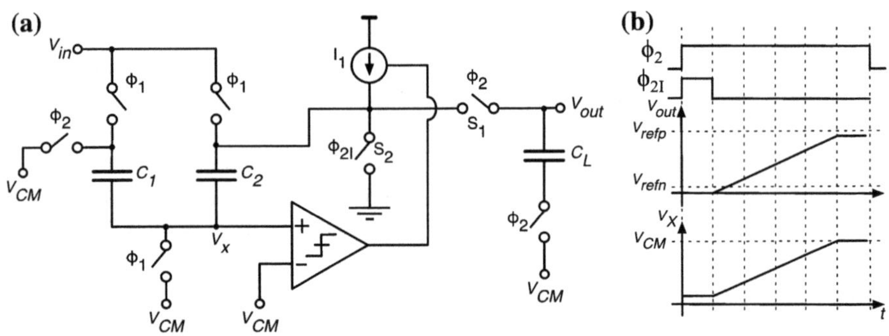

Fig. 8 **a** Comparator-based switched-capacitor (CBSC) gain stage; **b** Transient-response of a CBSC switched-capacitor gain stage [14]

When Φ_2 goes high to enter the transfer phase, a short pulse Φ_{2I} is used to initialize the charge transfer by closing switch S_2 to discharge the output voltage v_O to ground. Following this pulse, S_2 opens, and the current source I_1 charges the capacitors to generate a constant voltage ramp at the output v_O. This causes the virtual ground voltage v_X to ramp with it via the capacitor divider consisting of C_1 and C_2. As the voltage ramp proceeds, the comparator will *detect* when the both of its inputs become equal and then, it will cut off the current source. In this way, it is

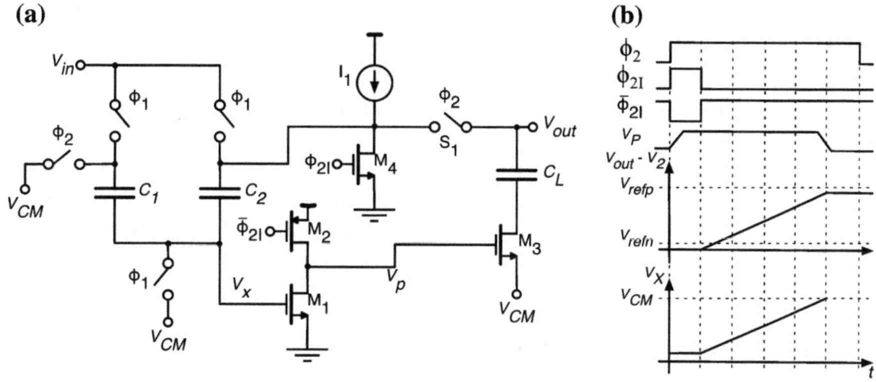

Fig. 9 **a** Zero-crossing-based switched-capacitor gain stage; **b** Transient response of a ZCBC switched-capacitor gain stage [15]

realized the same charge transfer as an amplifier-based implementation. The resulting transient response for voltages v_O and v_X is shown in Fig. 8b. Whereas an amplifier-based implementation forces the virtual ground condition, the CBSC implementation sweeps the output voltage and searches for the condition in which, the voltages at both inputs of the comparator become equal. Both, however, realize the same charge transfer despite their quite different transient responses.

Generally, a comparator must be accurate enough to distinguish between two input voltage waveforms. The input into the comparator of a CBSC circuit, however, is not arbitrary. In fact, it is a constant slope voltage ramp, so the comparator actually performs a zero-crossing detection. Therefore, a general-purpose comparator is not strictly necessary, and it can be replaced by a zero-crossing-detector (ZCD). This allows the simplification of CBSC circuits into zero-crossing-based circuits (ZCBC) [15].

Figure 9a shows a simplified implementation of a ZCBC. The comparator of the CBSC implementation shown in Fig. 8a has been replaced by a dynamic zero-crossing detector (DZCD), which consists of devices M_1 and M_2. During the sampling phase, Φ_1, the input voltage is sampled onto C_1 and C_2. Then, as shown in the timing diagram of Fig. 9b, Φ_2 and Φ_{2I} go high to start the transfer phase. Enabling phase Φ_{2I} will turn on M_4 to discharge the output voltage to ground. This pushes the virtual ground node voltage v_X down to turn off M_1. Simultaneously, M_2 will be turned on to pre-charge the voltage v_P to a high level and turn on the sampling switch M_3.

When Φ_{2I} is disabled, node v_P is left floating at a high level to keep the sampling switch on, and the output voltage v_O begins to ramp since the current source pulls it up. Node voltage v_X will ramp with it according the capacitor divider established by C_1 and C_2. As v_X ramps up, it will at some point give M_1 sufficient gate overdrive to start pulling down the floating v_P node. When v_P is pulled down sufficiently to turn off the sampling switch M_3, the voltage on the load capacitor C_L is sampled and the charge transfer is complete. Opening M_3 to define the sampling instant minimizes signal dependent charge injection by performing bottom plate sampling. The DZCD consisting of M_1 and M_2 is not suitable as a general purpose comparator, since it can not detect differences in two arbitrary voltages. However, it is suitable as a zero-crossing detector in this architecture because the constant slope voltage ramp created by the current source I_I ensures consistent switching at the same voltage.

Efficient practical realizations of 12-bit pipeline ADCs have been reported [16, 17]. Energy-efficiencies slightly above 50 fJ/conv.-step have been reached, at moderate conversion-rates, and with effective-resolution of about 10.5 bits.

4 Amplifier-Based Residue Amplifiers

As stated above, the effects of technology scaling have made the task of performing amplification in switched-capacitor (SC) circuits increasingly difficult to implement. A multitude of innovative SC-based techniques have emerged, either to reduce the finite-gain errors of the amplifiers without introducing long transient time, or to improve energy efficiency, speed and noise performance. Classic topologies of amplifiers, which are ill suited to scaling-down in modern technologies, have been replaced with novel amplifier structures. The former are described in this section as the lossless bottom-plate sampling (LBPS) techniques and, the latter, are the based on the charge steering and ring amplifiers.

From the different amplifier-based techniques presented in this section, it is worth to mention that, in general, they are able to achieve higher effective-resolutions, with minimum calibration requirements, than the amplifierless techniques described in the previous section.

4.1 Lossless Bottom-Plate Sampling (LBPS) Techniques

Recent LBPS techniques are based on the multiple feedback networks patented in 1956 by McMillan [18], as illustrated in Fig. 10, for the particular case of a two-channel (i.e., dual path) amplifying system. The principle behind the application to a SC residue amplifier is relatively straightforward. Blocks μ_1 and β_1 constitute, respectively, the main OTA and the SC network of a given MDAC stage. The virtual-ground of the main OTA can be seen as a summing point (18). Amplifier μ_2 and feedback network β_2 constitute the auxiliary system for finite-gain compensation (FGC).

As it is demonstrated in [18], as long as $\mu_1\beta_1 >> 1$, $\mu_2\beta_2 >> 1$ and $\Delta\beta/\beta << 1$ (where $\Delta\beta = (\beta_1 - \beta_2)$, $\beta_1 = (\beta + \Delta\beta/2)$ and $\beta_2 = (\beta - \Delta\beta/2)$), the effective gain becomes

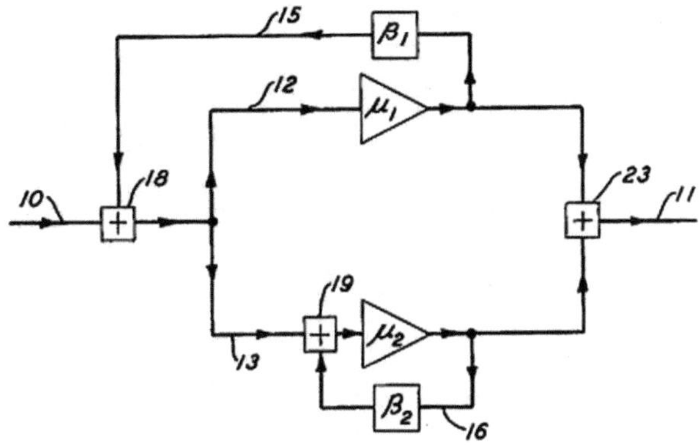

Fig. 10 A double-feedback, two-channel (dual-path), amplifying system [18]

$$\text{Effective Gain} = (\mu_1\mu_2)\,/\,(1/\beta + \mu_2\Delta\beta/\beta) \quad (3)$$

In the analog approach proposed in [19], the OTA summing-node error is treated as a separate signal, and subtracted from the residue output. A discrete-time, closed-loop SC circuit is used as the auxiliary path as shown in Fig. 11. It is worth noticing that capacitor C_L represents the sampling capacitors of the next pipelined stage. Since the error term is present at the output of the auxiliary circuitry (V_{o2}), it will be subtracted at the bottom-plate of C_L.

The experimental evaluation results of the integrated IC prototypes reported in [19] show a peak signal-to-noise-plus-distortion ratio (SNDR) of about 56 dB, corresponding to an ENOB of about 9 bits.

Unlike the previous scheme depicted in Fig. 11, the continuous-time solution proposed in [20] and shown in Fig. 12a is much more effective and it even cancels the thermal noise of the main OTA during the residue amplification phase. In this improved version of the dual-path amplifying system represented in Fig. 10, the MDAC output is sampled after its OTA-induced error is cancelled so that no residual error can be left in the digital output. The auxiliary error amplifier relies on a single-stage amplifier with an adjustable open-loop gain around the value of $1/f' \approx 1/\beta_2$. A differential-pair is used as a programmable gain element. If the top plate of the next-stage sampling capacitor is initialized with the error of the previous stage, the sampled residue is free from the previous-stage amplifier error.

The gain of the error path is adjusted using a zero-forcing least-mean-square (LMS) feedback algorithm, as illustrated in Fig. 12b. Measurements of fabricated IC prototypes indicate a peak SNDR of about 73.3 dB.

4.2 Charge-Steering Dynamic Amplifiers

A conventional two-stage amplifier can be modified into a charge-steering topology as shown in Fig. 13a [21, 22]. Interestingly, dynamic charge-steering amplifiers have been used either in open-loop [9] or in closed-loop [21]. As explained in [21],

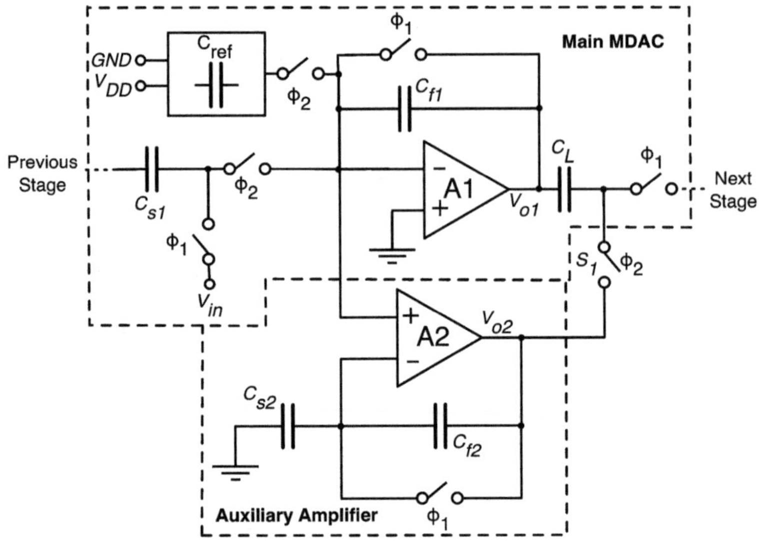

Fig. 11 Schematic of the FGC MDAC proposed in [19]

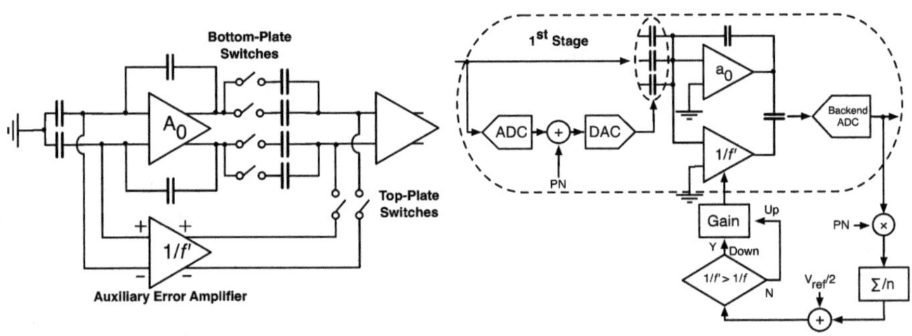

Fig. 12 **a** Schematic of the proposed open-loop FGC scheme proposed in [20]; **b** Illustrating the LMS-algorithm used to adjust $1/\beta_2 = 1/f'$

when *CLK* is low, capacitors C_1–C_4 are pre-charged to V_{DD} and the tail nodes of the two differential-pairs are left floating. When *CLK* signal is enabled, the output nodes are released and the tail nodes are connected to ground, causing differential and common-mode (CM) currents to be drawn by pairs M_1–M_2 and M_3–M_4 from their respective loads. As the CM level at nodes *X* and *Y* falls, transistors M_3 and M_4 will eventually turn off. At a given time-instant, the cascade of the two stages provides a certain differential voltage gain.

This type of amplification basically works based on integration, since the differential voltage gain is defined by: Gain $= gm_{1,2}{*}T_{INT}/C_{1,2}$. The integration time, T_{INT}, needs to be fixed and 'jitter clean'. Notice that the voltage gain does not

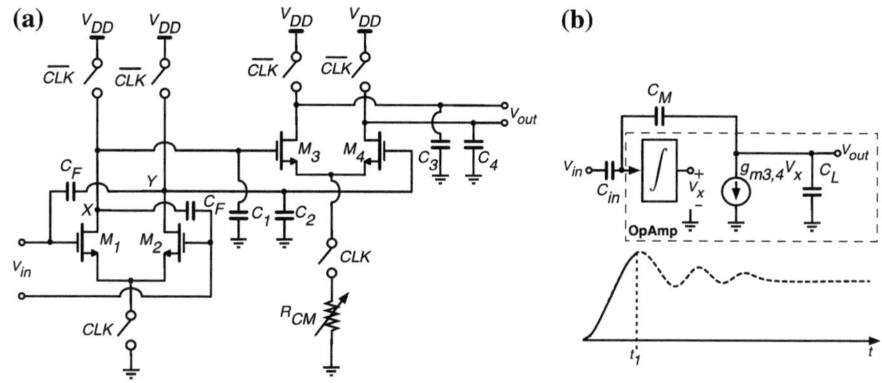

Fig. 13 **a** Possible implementation of a charge-steering amplifier; **b** Closed-loop model [21]

really depend on the capacitance value of $C_{1,2}$, since for smaller capacitance value, T_{INT} can be set smaller as well. The major drawback of these charge-steering dynamic amplifiers relies on the high gain sensitivity to output common-mode (CM) variations, which will require the use of self-calibration.

Note that the capacitance values C_1 and C_2 are primarily determined by the first pipelined stage thermal noise constraints, while C_3 and C_4 represent the input capacitances of the next pipelined stage. The basic charge-steering amplifier faces three issues: (i) the open-loop voltage gain resulting from the two stages is quite limited. Therefore, as suggested in [21], capacitors C_F can be added to introduce positive-feedback around the first stage, thus raising the gain; (ii) if utilized in a MDAC, this topology limits the closed-loop linearity to about 7 bits. Hence, as stated before, the use of self-calibration will be mandatory; (iii) the open-loop voltage gain and the nonlinearity of the circuit in Fig. 13a are very sensitive to the input CM level. Proper CM feedback-loops have to be used.

It can be demonstrated theoretically that, if designed for the same power dissipation, voltage gain, and load capacitance, charge-steering amplifiers exhibit half input noise power and a quarter settling time with respect to a conventional two-stage amplifier. Moreover, the closed-loop behavior of charge-steering amplifiers presents interesting and useful properties. As evident from the simplified model shown in Fig. 13b [21], the overall circuit resembles a loop containing two integrators, which become lossy-integrators if the output resistances are included. The step response would thus exhibit an overshoot, but the output stage turns off around time-instant t_1. In other words, the closed-loop voltage gain can exceed the ratio defined by C_{in}/C_M.

The SAR-assisted 11-bit, 250 MS/s pipeline ADC reported in [9] used a dynamic charge-steering amplifier in open-loop. Measured performance achieved a peak ENOB of 10-bits and an energy-efficiency at Nyquist-rate of 13 fJ/conv.-step.

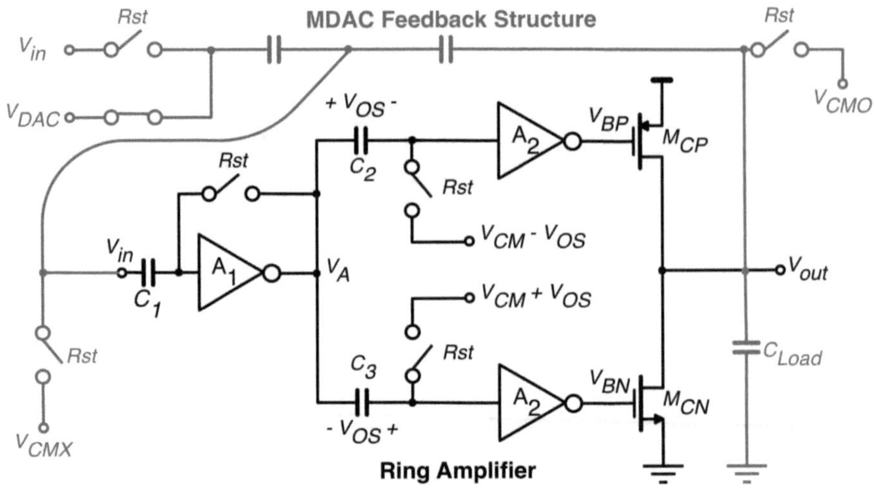

Fig. 14 Example of a 1.5-bit MDAC employing a RAMP [22]

4.3 Ring Amplifiers

Fundamentally, a ring amplifier (RAMP) is a ring oscillator that has been split into two (or more) separate signal paths [22]. A different offset is embedded into each signal path in order to create a range of input values for which neither output transistor M_{CN} nor M_{CP} of Fig. 14 will fully conduct. If this non-conduction "dead-zone" is sufficiently large, the ring amplifier will operate by slewing-to, stabilizing, and then locking into the dead-zone region.

As illustrated in Fig. 14, in which the RAMP is embedded in an SC MDAC feedback structure, capacitor C_1 is used to cancel the difference between the MDAC virtual-node sampling reference, V_{CMX}, and the trip-point of the first stage inverter. This ensures that the ideal settled value for V_{IN} will always be V_{CMX}, independent of the actual inverter threshold. Any offset that is generated after the first stage inverter will not be removed by C_1, but the typical input-referred value of such offsets will be negligibly small.

The dead-zone of the RAMP in Fig. 14 is embedded prior to the second stage inverters by storing a voltage offset across capacitors C_2 and C_3. Any value for V_{IN} within the dead-zone region is a viable steady-state solution for the ring amplifier, and the input-referred value of the dead-zone will determine the overall accuracy of the amplifier for most practical cases. In other words, the error at V_{IN} when the RAMP has stabilized and locked will be [22]

$$-\left|V_{DZ}/\left(2A_1\right)\right| \leq \varepsilon v_{IN} \leq \left|V_{DZ}/\left(2A_1\right)\right| \tag{4}$$

where $V_{DZ} = 2V_{OS}$, A_I is the final settled small-signal gain of the first-stage inverter, and it is assumed that the finite gain effects of the latter stages are ignored.

After the initial charging ramp, the ring amplifier will begin to oscillate around the settled value, with certain amplitude. With no dead-zone, the structure is functionally identical to a three-inverter ring oscillator, and will continue to oscillate indefinitely. However, as the size of the dead-zone is increased, the RAMP will eventually reach an operating condition where it is able to self-stabilize. If the dead-zone size is increased still further, the time required to stabilize, decreases substantially, and for most practical designs, a RAMP will stabilize in only one or two periods of oscillation.

In [23] a 15-bit, 20 MS/s pipeline ADC has been presented, in which, only RAMPs have been employed (a composite "coarse" and "fine" structure has been proposed). Measured results achieved 12.3 bits of ENOB (for a peak SNDR of 75.9 dB) together with an energy-efficiency as good as 29 fJ/conv.-step.

4.4 Amplification with Unity Feedback Factor

A 1.5-bit MDAC is shown in Fig. 15a, which implements the desired function $V_{od} = 2 \cdot V_{id} - X \cdot V_{REFD} + Z \cdot V_{REFD}$. The circuit operates in two clock phases [24, 25]. During Φ_1, the differential input voltage ($V_{id} = V_{ip} - V_{in}$) is sampled in capacitors C_{1j}, C_{2j}, and C_{3j}, $j = \{1,2\}$. During Φ_2, capacitors C_{1j} and C_{2j} are associated in series around the amplifier's feedback loop and the gain of two is obtained by voltage sum, instead of charge redistribution, as occurs in the conventional 1.5-bit flipped-around MDAC. This association of capacitors has two benefits: unity feedback-factor (β) and insensitivity to capacitor mismatch.

Reference shifting occurs (during Φ_2) when current sources, I_P and I_N, are turned ON (see Fig. 15b). These current sources sink/source current through the series associated capacitors changing the output voltage by an amount proportional to the respective current, feedback capacitance and duration of Φ_2. By the end of Φ_2 the output voltage should have changed by an amount equal to $\pm V_{REF}$ (differentially), for X and Z modes. Regarding the output waveforms, in $Y = \overline{X + Z}$ operation mode it is exponential and for X and Z it has a ramped integrating characteristic until the end of Φ_2. Following this approach for obtaining a capacitor mismatch insensitive gain of two, the circuit becomes sensitive to parasitic capacitors C_{p2j} andC_{p3j}, at the nodes connecting C_{2j} and C_{1j} (see Fig. 15a). To attenuate this sensitivity, capacitors C_{3j} are employed. Notice, however, that these capacitors are shorted during Φ_2 because only the charge stored in their parasitic capacitances (C_{p6j}) is used to compensate the charge stored in C_{p2j} and C_{p3j}.

Notice that there is another interesting 1.5-bit MDAC topology reported in the literature that achieves a unity feedback-factor; it is described in [26]. However it requires either two single-ended amplifiers or a four-input differential amplifier (possibly doubling the static power dissipation of the MDAC).

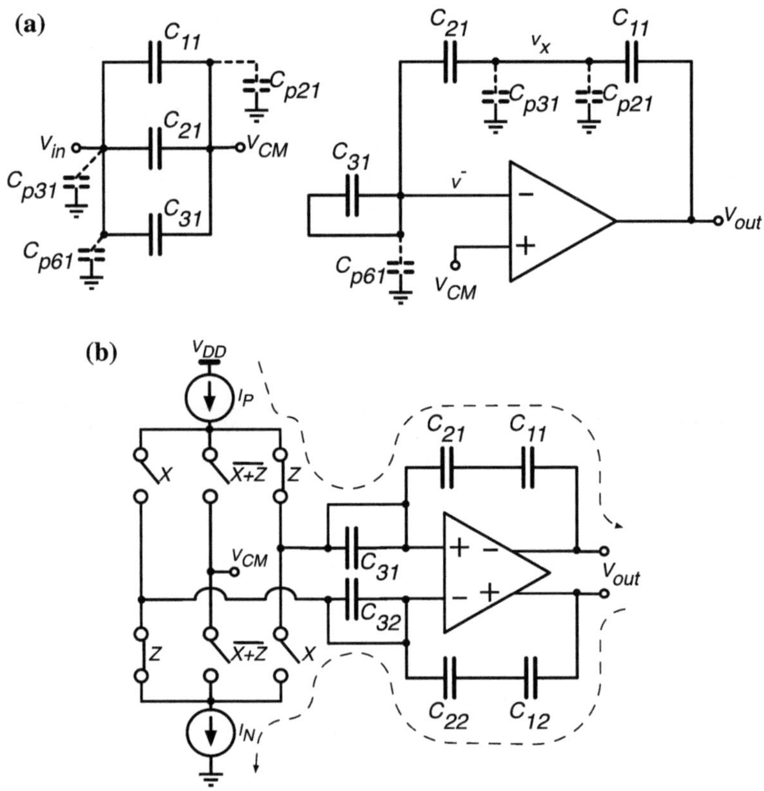

Fig. 15 **a** 1.5-bit mismatch-insensitive MDAC with current-mode reference shifting; single-ended circuit during Φ_1 and Φ_2; **b** Equivalent circuit to demonstrate the current-mode reference shifting. Situation shown for Z = 1 [25]

5 Conclusions

This chapter described some of the most promising of the recent techniques for enhancing the power and area efficiency of high-speed and high-effective resolution pipeline ADCs. First, SAR-assisted pipeline ADC architectures have been reviewed and, then, some possible solutions for performing energy-efficient residue amplification were pointed out. Several other interesting techniques have also been reported in the literature, such as, time-and-capacitor sharing [27] and hybrid time-based pipeline architectures [28]. However, it is not the aim of the present chapter to cover all recent techniques but, rather, to select and describe some of the most suited for designing low-power, high-speed and high-resolution pipeline ADCs.

Acknowledgments This work was supported by national funds through FCT—Portuguese Foundation for Science and Technology under projects PEst-OE/EEI/UI0066/2011 and DISRUPTIVE (EXCL/EEI-ELC/0261/2012). The authors would like to thank Prof. Manuel Medeiros-Silva for the many suggestions that improved the quality of this chapter.

References

1. S. Devarajan, *et. al.*, "A 16-bit, 125 MS/s, 385 mW, 78.7 dB SNR CMOS Pipeline ADC", *IEEE J. Solid State Circuits,* vol. 44, 3305–3313, Dec. 2009.
2. S. Ray, B.-S. Song, "A 13-bit Linear, 20-MS/s Pipeline ADC With Self-Configured Capacitor Matching", *IEEE J. Solid State Circuits,* vol. 42, 463–474, Mar. 2007.
3. R. Aparicio, A. Hajimiri, "Capacity Limits and Matching Properties of Integrated Capacitors", *IEEE J. Solid State Circuits,* vol. 37, 384–394, Mar. 2002.
4. J. Goes *et. al.*, "Purely-Digital *versus* Mixed-Signal Self-Calibration Techniques in High-Resolution Pipeline ADCs", *Proc. NORCHIP'2010*, pp. 1–8, Finland, Nov. 2010.
5. M. Saberi, R. Lotfi, k. Mafinezhad, W. A. Serdijn, "Analysis of Power Consumption and Linearity in Capacitive Digital-to-Analog Converters Used in Successive Approximation ADCs", IEEE *Transactions on Circuits and Systems – I,* vo. 58, n°. 8, pp. 1736–1748, Aug. 2011.
6. Y.-D. Jeon et al., "A 9.15mW 0.22 mm^2 10b 204MS/s pipelined SAR ADC in 65 nm CMOS", in 2010 IEEE *Proc. CICC*, pp. 1–4, 2010.
7. C. C. Lee and M. P. Flynn, "A SAR-Assisted Two-Stage Pipeline ADC", IEEE *Journal of Solid-State Circuits*, vol. 46, n. 4, pp. 859–869, April 2011.
8. Y. Zhu et al., "A 50-fJ 10-b 160-MS/s Pipelined-SAR ADC Decoupled Flip-Around MDAC and Self-Embedded Offset Cancellation", IEEE *Journal of Solid-State Circuits*, vol. 47, n. 11, pp. 2614–2626, Nov. 2012.
9. B. Verbruggen, M. Iriguchi, and J. Craninckx, "A 1.7 mW 11b 250 MS/s 2-Times Interleaved Fully Dynamic Pipelined SAR ADC in 40 nm Digital CMOS", IEEE *Journal of Solid-State Circuits*, vol. 47, n. 12, pp. 2880–2887, Dec. 2012.
10. I. Ahmed, J. Mulder, D. A Johns, "A Low-Power Capacitive Charge Pump Based Pipelined ADC", IEEE *Journal of Solid-State Circuits*, vol. 45, n. 5, pp. 1016–1027, May 2010.
11. J. Hu, N. Dolev, B. Murmann, "A 9.4-bit, 50-MS/s, 1.44-mW Pipelined ADC Using Dynamic Source Follower Residue Amplification", IEEE *Journal of Solid-State Circuits*, vol. 44, n. 4, pp. 1057–1066, April 2009.
12. J. Oliveira, J. Goes, *et. al.*, "An 8-bit 120 MS/s Interleaved CMOS Pipeline ADC Based on MOS Parametric Amplification", IEEE *Transactions on Circuits and Systems – II,* pp. 105–109, Feb. 2010.
13. S. Ranganathan and Y. Tsividis, "Discrete-time parametric amplification based on a three-terminal MOS varactor: Analysis and experimental results", *IEEE J. Solid-State Circuits*, vol. 38, no. 12, pp. 2087–2093, Dec. 2003.
14. J. K. Fiorenza *et. al.*, "Comparator-Based Switched-Capacitor Circuits for Scaled CMOS Technologies", IEEE *J. Solid-State Circuits*, vol. 41, no. 12, pp. 2658–2668, Dec. 2006.
15. L. Books and H.-S. Lee, "A Zero-Crossing-Based 8-bit 200 MS/s Pipelined ADC", IEEE *J. Solid-State Circuits*, vol. 42, no. 12, pp. 2677–2687, Dec. 2007.
16. J. Chu and H.-S. Lee, "A Zero-Crossing Based 12b 100MS/s Pipelined ADC with Decision Boundary Gap Estimation Calibration", IEEE *Proc. VLSI Circuits,* pp. 237–238, 2010.
17. J. Kuppambatti and P. R. Kinget, "A Low-Power Zero-Crossing Pipeline-SAR ADC with On-Chip Dynamically Loaded Pre-Charged Reference", IEEE *Proc. ESSCIRC'13,* pp. 113–116, 2013.
18. B. McMillan, "Multiple-Feedback Systems", US patent 2,748,201, May 1956.

19. S.-W. Sin, S.-P. U, R. P. Martins, "Generalized Circuit Techniques for Low-Voltage High-Speed Reset- and Switched-Opamps", IEEE *Transactions on Circuits and Systems* – I, pp. 2188–2201, Sep. 2008.
20. Y. Miyahara, *et. al.*, "Adaptive Cancellation of Gain Nonlinearity Errors in Pipelined ADCs", IEEE *Proc. ISSCC'13*, pp. 282–283, 2013.
21. S.-H. W. Chiang, H. Sun, Behzad Razavi, "A 10-Bit 800-MHz 19-mW CMOS ADC", IEEE *Proc. VLSI Circuits*, pp. C100–C101, 2013.
22. B. Hershberg, *et. al.*, "Ring Amplifiers for Switched-Capacitor Circuits", IEEE *J. Solid-State Circuits*, vol. 47, no. 12, pp. 2928–2942, Dec. 2012.
23. B. Hershberg and Un-Ku Moon, "A 75.9 dB-SNDR 2.96mW 29fJ/conv-step Ringamp-Only Pipelined ADC", IEEE *Proc. VLSI Circuits*, pp. C94–C95, 2013.
24. M. Figueiredo, G. Evans, J. Goes, "Reference-Free High-Speed CMOS Pipeline Analog-to-Digital Converters", Springer, ISBN 978-1-4614-3466-5, 2013.
25. M. Figueiredo, G. Evan, J. Goes, "A Reference-Free 7-bit 500 MS/s Pipeline ADC Using Current-Mode Reference Shifting and Built-in Threshold Quantizers", *Analog Integrated Circuits and Signal Processing Journal* (AICSP, Springer), vol. 75. no. 1, pp. 53–65, 2013.
26. P. Quinn and M. Pribytko, "Capacitor Matching Insensitive 12-bit 3.3 MS/s Algorithmic ADC in 0.25 μm CMOS", IEEE Proc. CICC'03, pp. 425–426, 2003.
27. Y.-C. Huang and T.-C. Lee, "A 10-bit 100-MS/s 4.5-mW Pipelined ADC With a Time-Sharing Technique", IEEE *Transactions on Circuits and Systems* – I, vol. 58, pp. 1157–1166, June 2011.
28. T. Oh, H. Venkatram and U.-K. Moon, "A 70MS/s 69.3 dB SNDR 38.2fJ/conversion-step Time-Based Pipelined ADC", IEEE *Proc. VLSI Circuits,* pp. C96–C97, 2013.

Digitally Assisted Analog to Digital Converters

Bob Verbruggen

Abstract This chapter discusses how digital assistance can be leveraged in the design of analog to digital converters. Different types of digital assistance are defined, and a few of the possible applications selected for detailed discussion. Finally, an example of an ADC implementation heavily leveraging digital assistance is presented.

1 Introduction

Digital assistance in ADCs can be generally defined as using digital techniques to relax requirements for analog non-idealities. This is not a particularly new concept: as early as 1981 digital correction was used to overcome accuracy limitations in component matching [1]. Since then, technology scaling has massively reduced the power and area cost of digital logic, and the prevalence of digital assistance has increased to match. This has happened to such a degree that in some cases the line between the pure analog architectures and digitally assisted architectures has blurred.

Let us consider for example a conventional 1.5b/stage pipelined converter. Strictly speaking even this architecture is digitally assisted [2]: analog comparator errors are compensated by digitally combining different stage outputs. However, the underlying redundancy technique is so common nowadays that few people would consider this architecture to be overtly "digitally assisted". The scope of this paper will be limited to digital assistance of Nyquist converters that involve clearly defined observation and correction steps for analog non-idealities. A more general overview of digital assistance in data converters is presented in [2].

B. Verbruggen (✉)
Imec, Kapeldreef 75, Leuven, Belgium
e-mail: bob.verbruggen@imec.be

P. Harpe et al. (eds.), *High-Performance AD and DA Converters, IC Design in Scaled Technologies, and Time-Domain Signal Processing*,
DOI: 10.1007/978-3-319-07938-7_2, © Springer International Publishing Switzerland 2015

Using this definition and the data from [3] the Nyquist converters published at ISSCC from 1997 to 2013 can be divided into two categories, based on whether they use any kind of digital assistance. Figure 1 shows the Walden FoM versus clock speed for these two populations and reveals two interesting trends. First: 40 % of the Nyquist converters published at ISSCC over this time period include digital assistance, which indicates that these techniques are quite common. Second: for clock speeds above 1 MS/s, all of the designs with best Walden FoM leverage digital assistance. Obviously ADC performance cannot be reduced to just the Walden FoM, and many designs achieve excellent performance without any digital assistance, but this trend does lend credence to the idea that digital assistance can be used to lower power consumption in analog to digital converters.

In the remainder of this paper, we will try to show why this is the case. In Sect. 2, we will first briefly define different types of observation and correction. Next, it will be explained how digital assistance can change analog design trade-offs and how this could benefit power consumption in comparators, amplifiers and DACs. In Sect. 4, these ideas are illustrated in a detailed look at the design of a pipelined SAR ADC heavily leveraging digital assistance. Finally, in Sect. 5 conclusions will be drawn.

2 Types of Digital Assistance

A first important classification of digital assistance is based on how analog non-idealities are corrected. In this chapter, digital compensation will refer to a digital mapping of raw output bits to a corrected, final set of output bits, as illustrated in Fig. 2a. On the other hand, digital calibration refers to the use of digital calibration settings to adjust analog circuit parameters and thus directly correct output values of the analog core as shown in Fig. 2b. The relative merits of compensation and calibration depends on the effect to be corrected, and both can be used in conjunction.

Compensation can be done using a simple look-up table to correct for purely static effects [1], or using elaborate digital signal processing to correct for interleaving artefacts, linearity or even self-heating gradients [4]. Digital calibration is applied most commonly to comparator offsets [5], amplifier gains [6] or capacitor values [7], and in all cases involves some digital programmability for said non-idealities.

The choice between compensation and calibration is not a simple one. From the point of view of area and power consumption, calibration is a preferred solution if analog tuning can be implemented with negligible overhead: it avoids the constant digital activity required for compensation logic. However, from the perspective of design-time the flexibility of compensation may be a compelling argument. Some effects, such as DNL degradation, are trivial to calibrate but are extremely difficult to compensate, because they directly affect the quantization of the ADC core. Higher level non-idealities, such as non-linear distortion, which may have many

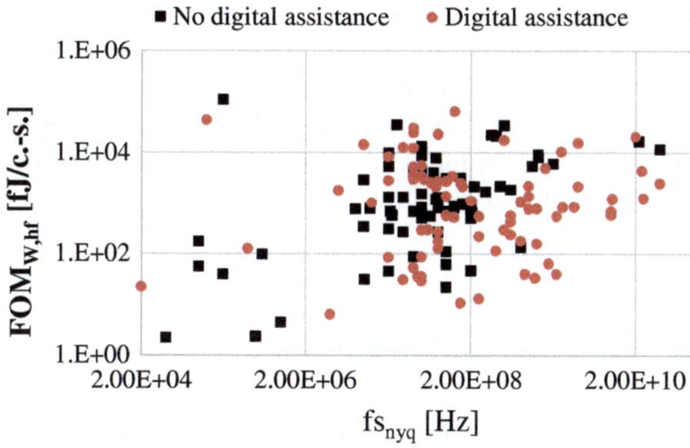

Fig. 1 Nyquist converters published at ISSCC between 1997 and 2013, divided based on use of digital assistance

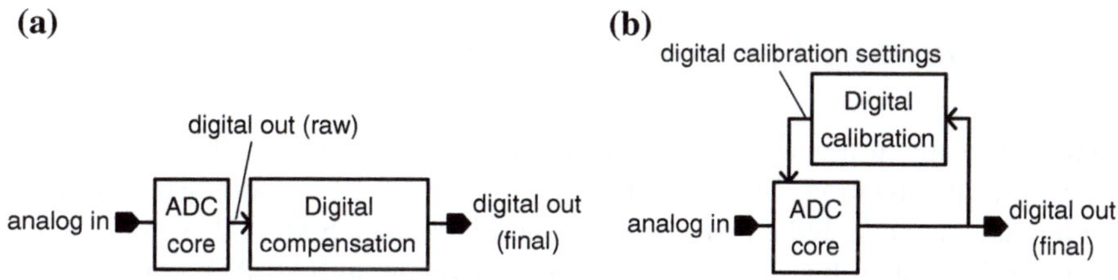

Fig. 2 High-level block diagrams comparing digital compensation (**a**) with digital calibration (**b**)

different analog contributions, are usually difficult to calibrate but comparatively easy to compensate for.

A second important distinction is based on when the ADC core output is observed. In this paper, we will refer to foreground observation if this occurs using a well-known input signal while the ADC operation is interrupted. On the other hand, if observation occurs while the ADC is quantizing an unknown input we will refer to background observation.

The ability to observe non-idealities during continuous ADC operation is a significant advantage to background techniques but this advantage is not without its cost. Since background observation is subject to statistics of the ADC input, it will typically take significantly longer to measure non-idealities with a given confidence level than using foreground techniques. In addition, background observation nearly always requires some assumptions about the ADC input statistics to guarantee convergence. On the other hand, foreground observation is vulnerable to changes of non-idealities after the observation phase, while the ADC is quantizing an unknown input. In many cases, this results in the need to periodically repeat the foreground observation, and thus periods when the ADC is not quantizing its input. In burst-mode applications, this might be a valid choice, while in others this is unacceptable.

These general methods for observation and correction provide a huge design-space when implementing an ADC with digital assistance. Proper choices in this design-space often require some knowledge about the system integration, in addition to ADC design know-how.

3 Designing with Digital Assistance

In this section we will discuss a few common applications for digital assistance in recent ADCs. We will consider how digital assistance changes design trade-offs for comparators, amplifiers and DACs. We will not discuss digital assistance for interleaved channels, despite its usefulness and popularity.

3.1 Calibration of Comparator Thresholds

Comparators are at the heart of nearly all ADC architectures, as they are required to perform quantization. Their accuracy is limited by offset and noise, but not all architectures are affected equally by these non-idealities. SAR ADCs, for example are generally quite robust to offset since in a conventional SAR ADC a comparator offset does not introduce any non-linearity. However, they are highly sensitive to comparator noise, as this noise is added directly to the input. Redundancy, which is nearly omnipresent in pipelined ADCs and gaining popularity fast in SAR ADCs, can reduce sensitivity to both offset and noise. In flash-based architectures, however, comparator accuracy is still key, and calibration is an extremely useful technique.

Indeed, the stringent requirements on offset can be all but completely avoided by adding threshold calibration. This is illustrated in Fig. 3 for comparators with a dynamic input pair followed by a latch. These comparators are activated on a falling clock edge and generate a rising slope at the drains of the input pair, depending on the input voltage. This slope is then resolved into digital levels by a latch in different configurations [8, 9]. Threshold calibration can be added by controlling the capacitance to the drain of the input pair [10], by controlling the body terminals of the input pair [11] or by controlling the voltage on the gates of a redundant input pair [12], among others [13, 14]. These calibration schemes have different advantages and disadvantages: using load capacitance is low-noise but slightly degrades speed, using body voltage tracks environmental changes reasonably well but requires a separate N-well and using a redundant input pair is probably the highest speed solution but slightly degrades noise.

Assuming comparator offset is calibrated, the final comparator accuracy is limited by noise. In nearly all cases, for a given power consumption and speed

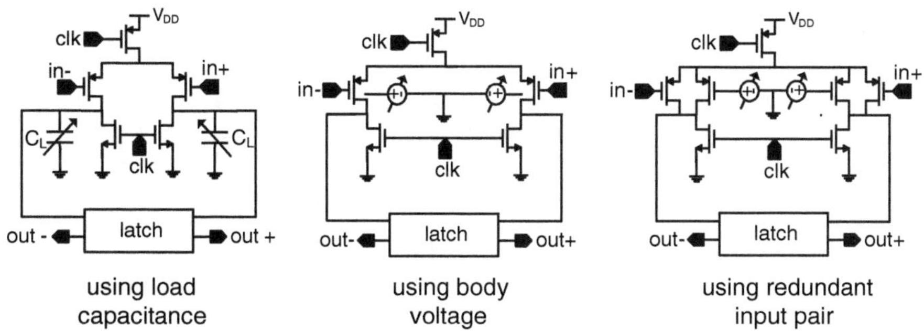

Fig. 3 Popular methods for comparator calibration

comparator noise spread is significantly lower than comparator offset spread, which implies a significant reduction in power consumption. In addition to these power savings, offset calibration typically significantly reduces comparator input capacitance. This is especially useful in flash-based architectures, where many parallel comparators are required, and target speeds are typically high.

Comparator calibration also enables a few architectural changes. One example is the noise-tolerant SAR ADC shown in Fig. 4 [15]. This design uses two offset-calibrated fully dynamic comparators in parallel: one of these is designed for low power and consequently has fairly high noise (HN), whereas the other is optimized for low noise (LN) at a power penalty. In [15] the HN comparator is activated for the first eight SAR cycles, after which the LN comparator resolves two final cycles. By giving the final cycles 1b redundancy with respect to the first eight, the ADC can tolerate a fairly large r.m.s. noise in the HN comparator without an SNR penalty. Since only two low noise comparisons are required instead of nine, comparator power can be significantly reduced. Without offset calibration, this scheme would need far more redundancy to compensate for differences in offset between the two comparators.

Another example is the comparator-controlled SAR ADC proposed in [16] and illustrated for a 4b example in Fig. 5. In this example, the input is tracked on two pseudo-differential 3b DACs and the input of 4 dynamic comparators. At a rising edge of the clock, the tracking switch opens and the first comparator is activated. When this comparator decides, either the positive or negative DAC MSB is discharged and a ready signal is generated. This ready signal is delayed to allow time for DAC settling, and then activates the second comparator. This comparator in turn generates feedback and asynchronously activates the next comparator in line. When all comparators have decided, the outputs of the 4 comparators can simply be latched to obtain the final output code. This arrangement thus implements a conventional SAR algorithm without the need for a controller, but in the absence of comparator calibration, this configuration would be crippled by offset.

Regardless of how comparator offset calibration is implemented and leveraged, a method for controlling the calibration codes is required. The most straightforward choice is foreground calibration: applying the desired threshold at the

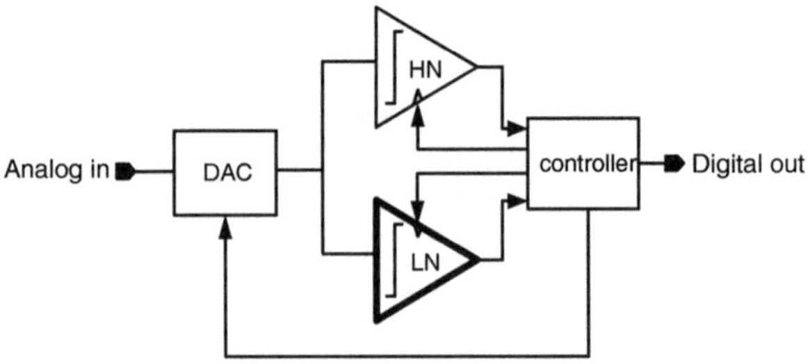

Fig. 4 Simplified block diagram of noise-tolerant SAR ADC

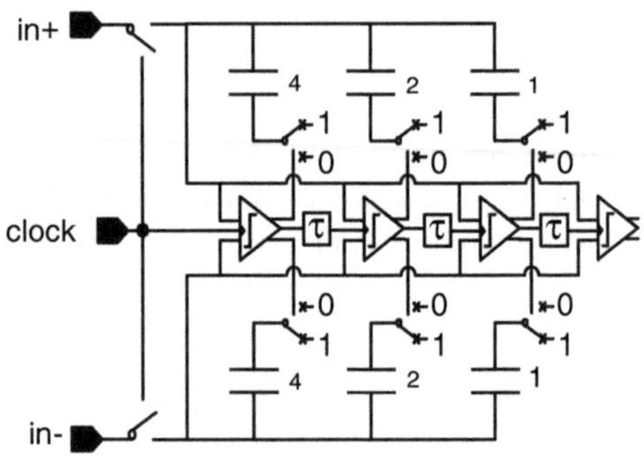

Fig. 5 Simplified block diagram of comparator-controlled SAR

comparator input and changing the calibration code until the comparator outputs as many zeroes as ones [5], or some variation on this scheme [14]. Background schemes also exist but vary wildly depending on architecture: adding a redundant ADC channel and averaging a rotating set of comparator pairs in a flash architecture [17], observing output code density [18, 19] or using redundancy to detect erroneous decisions [18, 19].

In conclusion, comparator threshold calibration can be used to shift the critical accuracy constraint from offset to noise, and thus reduce comparator power consumption and input capacitance. In addition, this calibration enables some architectural changes that are potentially useful. The disadvantage is that most offset calibration methods are sensitive to changes of temperature, common-mode input voltage or supply voltage. This sensitivity thus needs to be handled either through recurring foreground calibration or a background calibration loop for threshold calibrated flash converters to be applicable in a real-world environment.

3.2 Digital Assistance for Amplifiers

Conventional pipeline converters rely on closed-loop high-gain amplifiers to perform residue amplification. Feedback serves to linearize the amplifier and to stabilize the residue gain, but it requires high gain to achieve good accuracy, which is increasingly difficult in deeply scaled CMOS. Alternatives to high gain amplifiers have been explored: comparator-based switched capacitor circuits [20] or ring amplifiers [21] also offer the benefits of closed-loop settling without some of the drawbacks.

Digital assistance offers a different alternative: in [22] a digital loop tracks gain and distortion of an open-loop single stage amplifier. This implementation uses digital compensation: the digital logic which processes the stage outputs uses adjustable coefficients for linear combination of the stage outputs, as well as adjustable coefficients to correct dominant harmonic distortion components from some of the stage amplifiers.

Digital assistance significantly changes the requirements for residue amplifiers. Indeed, if gain stability and intrinsic linearity are not required, residue amplifier requirements can be reduced to:

- some fairly low amount of gain, for example approximately 12 dB in [6]
- distortion at a level that can be corrected in the digital domain, for example no significant non-third-order distortion components in [22]
- sufficiently low input referred noise
- sufficiently fast amplification.

While the above requirements for speed, gain and linearity can be further relaxed through architectural choices, the input referred noise is a fundamental limitation which currently dictates the lower power limit for amplifiers. To explore this power limit, a number of unconventional amplification paradigms has been proposed: charge-domain pipelines [23, 24], dynamic source follower-based amplifiers [25] or charge-steering amplifiers [6, 26].

While some of these would be applicable at low resolutions without digital assistance, most use digital compensation for gain and distortion. The exception is [5] which uses calibration for gain and avoids the need for distortion correction by limiting input and output range in the architecture choice. Using calibration to reduce distortion is probably impractical: it is not straightforward to implement analog controls guaranteed to fully linearize a circuit.

Background algorithms to observe residue gain and distortion are readily available. For example, by adding pseudo-random dither to the input of the residue amplifier and correlating the ADC back-end output with said dither sequence in the digital domain gain and distortion can be estimated [22]. Methods based on output histogram also exist [27]: these do not require any analog dither injection, but typically do impose more stringent assumptions about the ADC input signal for convergence. Even running at fairly low sample rates these algorithms are certainly fast enough to track most temperature variations, but to obtain an accurate measurement of gain and/or distortion at start-up, a foreground observation is often faster and more practical.

In conclusion, digital assistance can shift the critical requirements for residue amplifiers from gain accuracy and distortion to noise. While these relaxed requirements are certainly beneficial in their own right, they have also enabled exploration of a number of non-conventional amplification paradigms in the quest to find a more beneficial noise-power trade-off.

3.3 Digital Assistance for DACs

SAR and many pipeline architectures rely on matched capacitor or resistor arrays, specifically in feedback DACs. Given their current popularity, we will focus the rest of this discussion on capacitor DACs, but some of the conclusions apply equally to resistive or current DACs.

Assuming a careful common-centroid layout to eliminate systematic mismatch, DAC matching is limited by random effects determined by the well-known Pelgrom model [28]:

$$\sigma\left(\frac{\Delta C}{C}\right) = \frac{A_c}{\sqrt{W \cdot L}} \tag{1}$$

While this formula predicts a fairly straightforward 4× area scaling to improve linearity by a factor 2, in some cases this scaling is anything but straightforward. Indeed, as the size of an array increases, it becomes more sensitive to systematic effects, even while tolerance for such effects decreases. This is especially noticeable in high resolution SAR ADCs: in addition to very accurate matching these also need an extremely large number of units, which further increases area due to unit and routing overhead. In state of the art, there seems to be an intrinsic matching plateau around 10.5b resolution [29]: higher resolution SAR ADCs tend to use techniques to either relax matching requirements [1] or reduce unit count [30].

Both compensation and calibration can be applied to capacitor DACs. Compensation simply involves adjusting digital bit weights to match analog feedback weights, but it requires redundancy to ensure a sufficiently fine quantization grid in the presence of mismatch. In the absence of this redundancy, or to simplify digital logic, calibration [7] is also an option.

A common misconception is that avoiding the matching limit allows sizing of the DAC for kT/C noise, with obvious speed and power consumption benefits. This is true from a certain point of view, but it does assume a fixed capacitance density. In practice the matching constraint of (1) enforces a minimum DAC area, but does not place a constraint on DAC capacitance. Indeed, in vertical MOM capacitors the finger distance can usually be more or less freely chosen. Choosing a greater finger distance thus in theory reduces capacitance without affecting matching. The real benefit of avoiding the matching limits is thus in terms of area, and not in terms of capacitance. Theoretical considerations aside, for high resolution arrays reducing size is often very beneficial as it lowers sensitivity to systematic effects. In addition, the theory indicating that A_c should be the same regardless of capacitance density is not always verified during device modeling. As a result, relying on a theoretical extrapolation of device matching might be somewhat risky. In addition, digital assistance allows switching schemes that are very difficult to design for intrinsic linearity [1].

Background calibration algorithms to detect capacitor mismatch have been extensively applied in pipelined ADCs [27], and could be generalized to cover SAR ADCs. However, foreground calibration might in this case be sufficient, since the primary goal is often to compensate for capacitor mismatch which changes only very slightly as a function of environmental conditions. In many cases, since drift is negligible over the chip lifetime, mismatch is even measured at manufacturing and fused into a ROM [1].

In conclusion, DACs can be compensated if redundancy is available, and can be calibrated in general. In theory, the benefit from doing so is in terms of array area, not in terms of power or speed, which should be taken into account when deciding whether to go for intrinsic matching or not, especially at medium resolution. It can be done in background, but in many cases a foreground-only approach is also quite valid, since the non-ideality to be corrected is often quite stable over environmental conditions.

4 Pipelined Dynamic SAR

In this section we will discuss the design of an analog to digital converter for a software defined receiver. Power consumption will be aggressively minimized using digital assistance. The basic architecture is reused from the dynamic pipelined SAR architecture of [6], which scales favorably to the 0.9 V supply 28 nm technology. Comparator offsets and residue amplifier gain are automatically calibrated on-chip, in either a foreground or a background mode. For these non-idealities calibration is preferred over compensation because of the relative ease of implementing analog controls. Channel offset is observed and compensated off-chip, as compensation simply requires a half-rate digital addition. Channel gain mismatch is observed off-chip but calibrated on-chip to avoid fairly power-hungry digital multipliers that would be required for channel gain compensation. The ADC prototype achieves a peak SNDR of 59.8 dB at 410 MS/s for 2.1 mW of power [19].

4.1 Architecture

The implemented ADC consists of two interleaved channels with a full-rate front-end sampling switch as shown in Fig. 6. The channels each consist of a 6b coarse SAR stage, a dynamic residue amplifier, a 7b fine SAR stage and are accompanied by a calibration engine. The calibration detects and corrects comparator offsets and residue amplifier gain either in the foreground or in the background. Since each channel is observed individually, the on-chip calibration engine does not correct interleaving effects such as offset or gain mismatch. Both of these are observed externally, and offset is compensated externally while channel gain is calibrated.

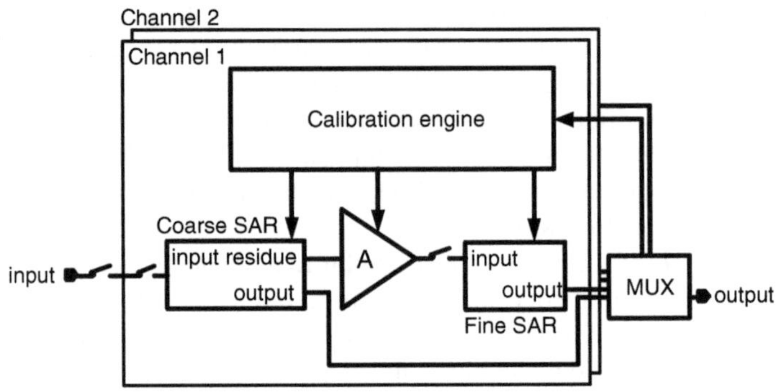

Fig. 6 High-level block diagram of ADC

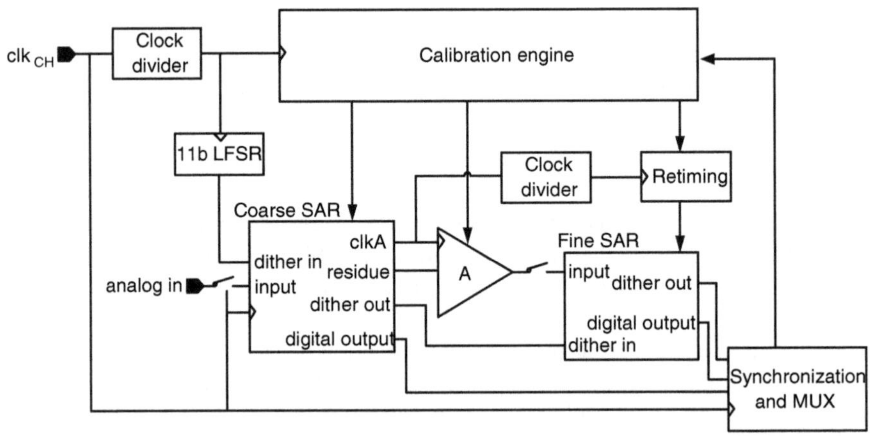

Fig. 7 Detailed block diagram of channel

A detailed view of one of the interleaved channels is shown in Fig. 7. Each channel operates with one synchronous main channel clock used for sampling and the start of the coarse SAR operation. After the coarse SAR has finished its conversion, the residue amplifier is asynchronously activated using clkA and in turn activates the fine SAR. Since the actual calibration engine operates on a single clock, its outputs are retimed using clkA to ensure calibration inputs change only when appropriate. Both the main calibration engine clock and the retiming clock are divided by a programmable factor ranging from 1 to 1,024 in binary scaled steps which allows a trade-off between calibration settling time and digital power consumption. An 11b linear feedback shift register (LFSR) is used to generate a 2,047 length PRBS dither sequence that is synchronized in each SAR stage and finally sent to the calibration engine for use in the residue amplifier gain calibration as will be explained in the section on the calibration algorithms.

4.2 Core ADC Design

The coarse SAR itself is implemented similar to [6, 16] using a comparator controlled step-down DAC. There are three potential benefits to this arrangement. First, the power consumption and delay in the SAR controller is avoided, which admittedly is a fairly minor advantage in deeply scaled CMOS. Second, the above arrangement avoids constraints on common-mode dependence of comparator offsets [31]. Indeed, in the step-down DAC used here the DAC common-mode changes every cycle. This results in a different comparator input common-mode in every cycle, and since in most comparator designs the comparator offset is common-mode dependent, this changing common-mode could introduce comparator errors. While these errors can be mitigated using redundancy or a modified comparator design [31], having multiple comparators ensures that each comparator operates at a specific common-mode, and is thus insensitive to common-mode dependent offset. Finally, the above arrangement trivially allows cycle-specific choices for comparator sizes and DAC settling delays. For example, since MSB settling is typically critical in a SAR ADC, one could easily use a larger delay value for the delay line between the first and second comparator than for the later cycles.

There are also two disadvantages. First, the need to implement multiple comparators rather than a single one: while this an energy-neutral operation using dynamic comparators, it does impose a small area penalty. The more significant disadvantage is the fact that each of the 6 comparators in the coarse SAR have potentially different offsets, which need to be calibrated to ensure accuracy. This can be done in foreground or background, but definitely requires digital logic, with a further area penalty. In practice, the area increase due to the calibration logic, especially combined with the area for multiple comparators, is larger than whatever area is saved by eliminating the need for a controller in the first place, so this is not a beneficial choice from an area perspective. In addition, the digital calibration logic requires some energy, which should be lower than the energy typically required by the SAR controller. For reasonable speed ADCs this is usually the case: whereas a SAR controller is active and thus consumes energy in every ADC cycle, calibration logic only needs to be activated often enough to track environmental changes. Since in most environments, kilocycle/second calibration speeds are sufficient, a comparator-controlled SAR results in energy savings even for an ADC running at a fairly modest 10 MS/s.

Due to a 1 bit redundancy between the coarse and fine stage, the coarse stage is robust to any comparator error up to ± 0.5 LSB_{coarse}. Since comparator offset is calibrated, this redundancy can be used almost completely for errors due to DAC settling or comparator noise. The coarse SAR also adds ± 0.5 LSB_{coarse} dither to the residue for use in the amplifier gain calibration as will be discussed later. The coarse SAR is designed for intrinsic matching, since the 1 pF required at maximum density is not yet prohibitive in terms of power consumption or speed.

The residue amplifier must amplify the residue so that it can be sampled by the fine SAR converter. Similar to [6] a dynamic amplifier is chosen since it combines low, purely dynamic power consumption with fast settling. Linearity issues are avoided by using a limited output range and the gain uncertainty is compensated using a calibration. Power supply rejection is not an issue, since the amplifier will be powered from the 0.9 V reference, which must be extremely stable anyway. Common-mode rejection is a potential issue: this approach assumes a fairly stable common-mode, which changes at a speed that can be tracked by background calibration.

The fine SAR architecture is similar to that of the coarse SAR except in the fact that dither is subtracted prior to the first comparator in the fine stage, whereas it is injected after the last comparison in the coarse stage. The top plate of a step-down DAC is connected to the input of 5 high noise, and 2 low noise comparators directly controlling aforementioned DAC. The DAC is implemented using 2.2 fF units, not limited by the relaxed 6b matching requirements. Extra capacitance is added to the DAC top plate to reduce the range of the DAC, thereby reducing the required output range of the residue amplifier. One bit of redundancy between the 5th and 6th cycles of this SAR renders the converter robust to comparator noise errors in the first 5 cycles. This allows the use of a relatively high noise comparator in these cycles whereas the last two comparisons are noise-critical and therefore done in comparators with lower noise and higher power consumption.

As each comparison in both coarse and fine SAR ADCs is done with a different comparator, comparator offset errors potentially affect the converter accuracy. The comparators can be sized for noise if comparator offset is calibrated as will be discussed in the next section. This lowers the power consumption of these comparators to the same level required in a conventional, single-comparator, SAR converter, as offset calibration can be implemented with an extremely small power penalty. All comparators are implemented as in [9] with their offset calibrated using digitally controllable MOS-capacitors.

4.3 Calibration Engine

The calibration engine supports both foreground and background calibration. It is clocked at the start of tracking, which ensures that all its outputs change during the track-time while the coarse SAR and amplifier are reset. The fine SAR calibration inputs are retimed to change at the start of the fine SAR reset, which ensures they also only change while the circuits they control are in reset mode.

4.3.1 Foreground Calibration

Foreground calibration assumes a zero differential input signal at the ADC common-mode. Comparator offsets are calibrated first, followed by amplifier gain calibration. Since all coarse SAR comparators are activated at a different common-mode during normal operation, this common-mode is replicated during calibration.

This is done by shifting the common-mode of the zero differential sampled signal using direct control of the internal DACs before activation of the comparator in question. The output of said comparator is then accumulated during a programmable number of cycles and a binary search drives the comparator threshold to obtain an equal number of zeros and ones at the comparator output.

To calibrate the fine SAR comparator offsets, the coarse SAR DAC is programmed to shift the common-mode amplifier input voltage to the appropriate value. The amplifier is then activated to ensure a zero differential voltage of the appropriate common-mode at the amplifier output. This voltage is then shifted by the fine SAR DAC to the correct common-mode value for each fine SAR comparator. The average comparator output is again driven to 0.5 using a binary threshold search.

In our implementation, all 13 comparators are calibrated sequentially for simplicity, since the direct DAC control allows only one common-mode shift per ADC cycle. Using slightly more complex direct DAC control, multiple common-mode steps per cycle are possible, which could allow calibration of all 13 comparator thresholds to happen concurrently.

In the last step of the foreground calibration the amplifier gain is tuned. The zero differential input voltage is now transformed by the coarse DAC into ± 0.5 LSB_{coarse} amplifier inputs at the correct common-mode for the residue amplifier. The fine SAR stage is then operated normally and its output is accumulated for a programmable number of cycles with positive and negative input each. If the average difference between the two obtained second stage outputs is smaller than 32, the gain of the residue amplifier is increased, if it is larger than 32, the gain is decreased.

4.3.2 Background Calibration

Background calibration is implemented using three different algorithms running concurrently. The first algorithm calibrates the six coarse SAR and first 5 fine SAR comparator offsets using the redundancy available in the ADC [18]. The second algorithm calibrates the final two comparators of the fine SAR based on their average outputs. A final algorithm is used to calibrate the residue gain using dither injection similar to [32].

The first algorithm is illustrated in Fig. 8 for calibration of offsets in the first stage using a one bit redundant second stage, with V_{in1} representing the first stage input and V_{in2} representing second stage input. Since at design-time the ideal residue transfer characteristic is well-known, the nominal upper and lower limits of this characteristic are also known. The implemented algorithm assumes the second stage is ideal, in which case the quantized output residue can only exceed these nominal limits when an error occurs in one of the first stage comparators. Based on the second stage output the sign of the error can be determined, and based on the digital output of the first stage, the specific comparator responsible for the error can be identified.

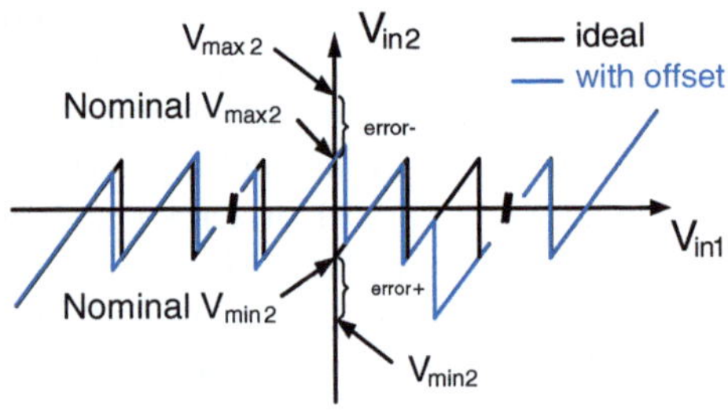

Fig. 8 Illustration of first calibration algorithm

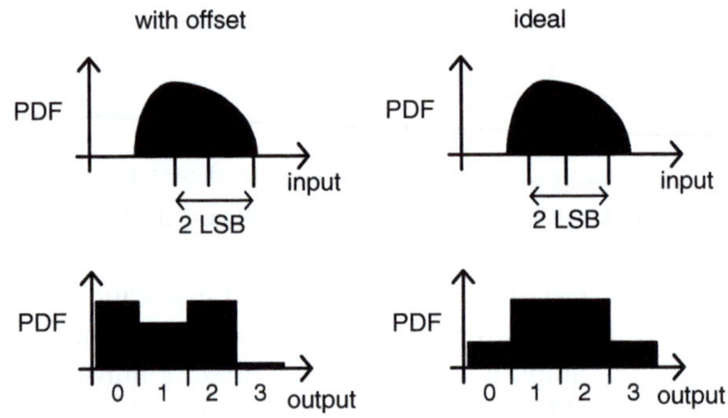

Fig. 9 Illustration of second calibration algorithm

In our implementation the difference between the number of incorrect positive and the number of incorrect negative decisions for each comparator is counted, and when this counter exceeds a programmable limit the comparator offset is adjusted to reduce the probability of the more common type of error. By comparing the relative probabilities of positive and negative errors the impact of comparator noise, incomplete DAC settling and the fact that the fine SAR is not actually ideal are reduced.

Since the last two comparators of the fine SAR do not have a redundant comparison available, a second calibration algorithm uses output statistics to calibrate these comparators as illustrated in Fig. 9. Because of the redundancy in our implementation, the PDF at fine SAR comparator input after the first five bits of the fine SAR would ideally be exactly 2 LSBs wide, but because of non-idealities in the preceding residue generation this PDF will be somewhat wider. Assuming that this PDF is more or less symmetric, the optimal placement of the comparator thresholds is such that a symmetric output histogram is obtained. This corresponds uniquely to a 50 % distribution for both comparator outputs. The final two comparator thresholds are thus calibrated by accumulating the comparator output during a programmable number of cycles and adjusting the threshold if the number of positive decisions is larger than 75 % or smaller than 25 %.

The third and final algorithm is used to calibrate the residue amplifier gain as illustrated in Fig. 10, slightly modified from [22]. A pseudo-random signal PN is added to the residue amplifier input using the coarse stage DAC, and an ideally

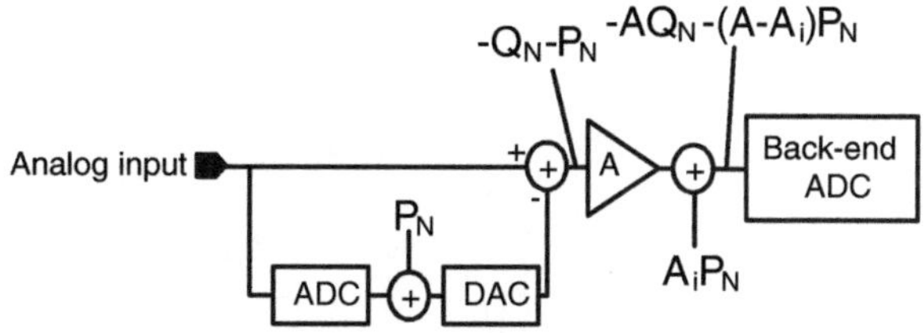

Fig. 10 Illustration of amplifier gain calibration

scaled version of this dither is subtracted at the output of the amplifier by the second stage DAC, before the fine SAR quantization. The fine SAR output is accumulated in one of two registers based on the sign of the dither and when the difference between these two accumulations exceeds a certain value the residue amplifier gain is adjusted. In our implementation we use ± 0.5 LSB_{coarse} and ± 0.5 MSB_{fine} dither amplitude and a 2047 length pseudo-random signal generated using an 11b LFSR. To reduce the complexity only the 4 MSBs of the fine SAR are accumulated. Complexity is further reduced by comparing only the 7 MSBs of the accumulators, instead of calculating the difference between the two accumulator outputs and comparing this difference to a certain threshold. If the 7 MSBs are not the same for both accumulators, gain is adjusted. While in the simplified scheme the gain might be adjusted based on a very small actual difference between the two accumulators, the amplifier gain step is sufficiently small that a few unrequired changes of the amplifier gain do not significantly affect performance.

The calibration engine is described in VHDL and synthesized with a total gate count of 6,080. Block area after place and route is approximately 60 μm by 250 μm for each calibration engine. Area and power consumption could be optimized further by omitting some debugging and testing options.

4.4 Measurements

The ADC prototype has been manufactured in an 1P9 M 28 nm CMOS process with a core chip area of 0.11 mm^2 including the calibration engines but not decoupling (Fig. 11). Individual supply domains with approximately 180 pF decoupling each are used for channel 1 and channel 2; both calibration engines are implemented on a third supply domain with another 120 pF decoupling. The chip area including this decoupling capacitance is 0.25 mm^2, which could be somewhat optimized by using MOS capacitors for decoupling instead of only MOM. Gain and offset mismatch between the two channels are measured by applying non-zero inputs and observing both channel outputs. Gain mismatch is calibrated by changing the top plate DAC capacitance in the coarse SAR, offset is compensated digitally.

INL and DNL performance is measured at 20 MS/s after a 3,000 cycle on-chip foreground calibration and shown in Fig. 11. While the DNL is unremarkable, the INL shows a significant and consistent pattern in both channels and across different dies. This has been identified to be due to unintended front-end layout

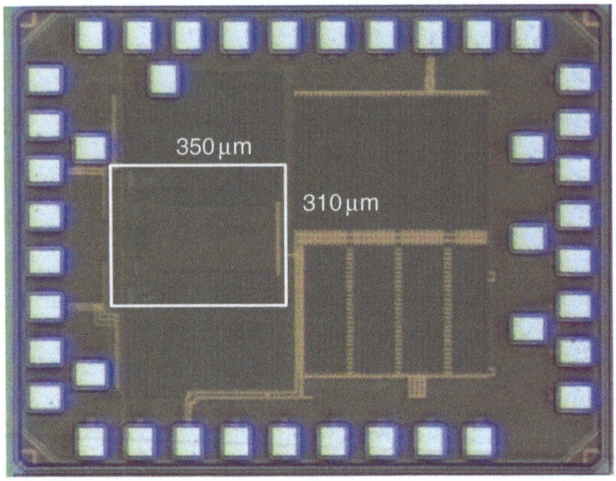

Fig. 11 Chip micrograph with core area indicated

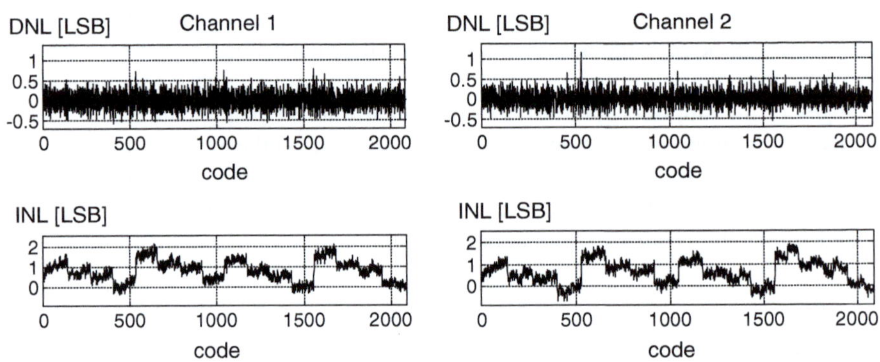

Fig. 12 INL and DNL for channel 1 (*left*) and channel 2 (*right*)

dummies (active and poly) in the first stage DAC. Since redundancy is available, this INL pattern is digitally compensated by adjusting the first stage bit weights for the remaining measurements (Fig. 12).

The settling behavior of the background calibration is measured by programming the chip in default, center-range settings and applying a full-scale sine wave input. The SNDR calculated in 16 k windows with variable start point is shown in the left plot of Fig. 13 for different divider ratios for the calibration engine clock: the background calibration converges within approximately 50 k calibration cycles per channel. This settling behavior is dominated by the amplifier calibration, which needs at least 2047 cycles per step. Presetting the amplifier gain to a nearly ideal value, settling to steady-state is significantly faster, as shown in the right plot of Fig. 13.

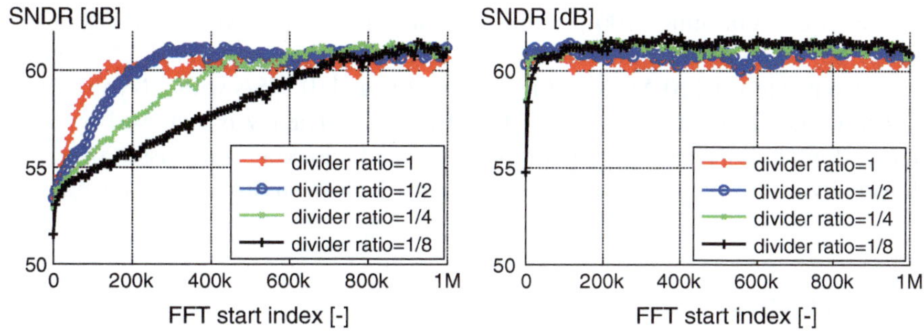

Fig. 13 SNDR versus FFT start index after programming default settings (*left*) and after programming default comparator settings and correct amplifier gain (*right*)

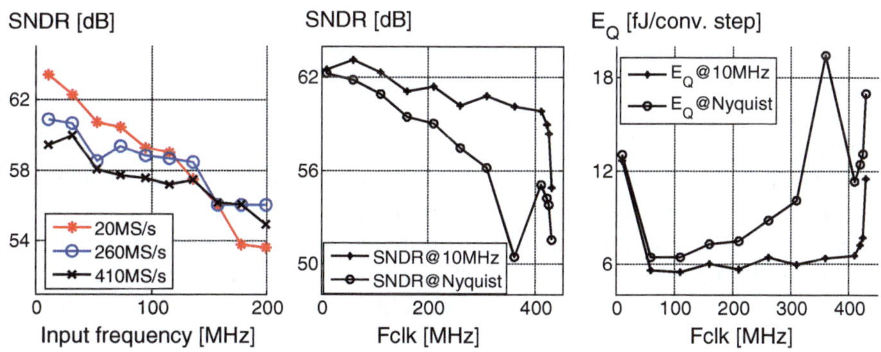

Fig. 14 SNDR versus input frequency (*left*), SNDR versus. clock frequency (*middle*) and energy per conversion step versus clock frequency (*right*)

Dynamic performance is measured with background calibration clocked every 1,024 cycles. The SNDR versus input frequency is shown in the left plot of Fig. 14 for different clock frequencies: at 20 MS/s the ERBW is limited by external clock phase noise, at 260 and 410 MS/s the ERBW exceeds 130 MHz. The SNDR versus clock frequency is shown in the middle plot of Fig. 14 for low and near-Nyquist input frequency. The low frequency SNDR degrades fairly gradually from 62.5 dB at 10 MS/s to 59.8 dB at 410 MS/s. The high frequency behavior degrades gradually at moderate clock frequencies, shows significant degradation around 360 MS/s and then increases somewhat again. The high input frequency degradation at 360 MS/s is due to time skew induced by supply coupling: the clock generation is powered by the channel 1 supply, and at 360 MS/s generation of the sampling clock occurs during DAC reset, which generates some supply bounce. This supply bounce modulates the sampling clock delay and because this only occurs every other sample, a 4.5 ps time skew occurs.

The circuit consumes 100 μA leakage current, 5 pJ per ADC clock cycle and 3.3 pJ per calibration engine clock cycle from a 0.9 V supply. Peak efficiency is 5.5 fJ/conv. step at 110 MS/s and less than 12 fJ at 410 MS/s with a Nyquist input signal as shown in the right plot of Fig. 14. This efficiency is achieved by leveraging digital assistance wherever practical and clocking said assistance at a drastically reduced clock rate, sufficient to track slow environmental changes.

5 Conclusions

It has been shown that digital assistance significantly changes the design tradeoffs in analog to digital converters. Comparator calibration shifts accuracy constraints from offset to noise which significantly reduces comparator power consumption and input capacitance. In addition, comparator calibration allows some alternative architectures that potentially further increase power efficiency. In amplifiers, digital assistance relaxes requirements on linearity and gain precision and shifts the critical requirement to input referred noise. In addition to lowering power consumption in conventional amplifiers, this opens up many alternative amplifier topologies. In DACs, digital assistance can be used to reduce required area for a matching level, which is potentially useful at high resolutions.

Some of these concepts have been illustrated in the design of an interleaved 410 MS/s 11 bit pipelined SAR ADC enabled by digital assistance. A SAR architecture leveraging comparator offset calibration, residue amplifier gain calibration, channel gain calibration and channel offset compensation are assumed during the design phase to relax analog requirements. In addition, due to a design mistake, DAC linearity compensation is used. The measured prototype obtains a peak SNDR of 63.3 dB at 20 MS/s and up to 59.8 dB at 410 MS/s with energy per conversion step below 12 fJ. To date, similar performance has not been achieved with similar energy consumption without digital assistance.

Acknowledgments The author would like to thank Masao Iriguchi, Manuel de la Guia Solaz, Guy Glorieux, Kazuaki Deguchi, Badr Malki, Ewout Martens and Jan Craninckx for their contributions to this paper.

References

1. Z. G. Boyacigiller, B. Weir, P. D. Bradshaw, "An Error-Correcting 14b/20μs CMOS A/D Converter," IEEE ISSCC 1981
2. B. Murmann, "Digitally assisted data converter design," IEEE ESSCIRC, September 2013, pp. 24–31
3. B. Murmann, "ADC Performance Survey 1997-2013," [Online]. Available: http://www.stanford.edu/~murmann/adcsurvey.html

4. B. Setterberg, et al., "A 14b 2.5GS/s 8-way-interleaved pipelined ADC with background calibration and digital dynamic linearity correction," IEEE ISSCC 2013
5. G. Van der Plas, et al., "A 0.16pJ/Conversion-Step 2.5mW 1.25GS/s 4b ADC in a 90 nm Digital CMOS Process," IEEE ISSCC 2006
6. B. Verbruggen, et al., "A 1.7mW 11b 250MS/s 2× interleaved fully dynamic pipelined SAR ADC in 40 nm digital CMOS," IEEE JSSC, Vol. 47, No. 12, December 2012, pp. 2880–2887
7. A. Shikata, et al., "A 0.5 V 1.1 MS/sec 6.3 fJ/Conversion-Step SAR-ADC With Tri-Level Comparator in 40 nm CMOS," IEEE JSSC, Vol. 47, No. 4, April 2012, pp. 1022–1030
8. T. Kobayashi, et al., "A current-controlled latch sense amplifier and a static power-saving input buffer for low-power architecture," IEEE JSSC, Vol. 28, No. 4, April 1993, pp. 523–527
9. M. Miyahara, et al., "A low-offset latched comparator using zero-static power dynamic offset cancellation technique," IEEE ASSCC 2009
10. M.-J. E. Lee, et al., "Low-power area-efficient high-speed I/O circuit techniques," IEEE JSSC, Vol. 35, No. 11, November 2000, pp. 1591–1599
11. F. H. Gebara, et al., "A body-driven offset cancellation technique in PD-SOI," International Conference on Microelectronics, May 2004, Vol. 2, pp. 567–570
12. B. Verbruggen, et al., "A 2.2 mW 1.75 GS/s 5 Bit Folding Flash ADC in 90 nm Digital CMOS," IEEE JSSC, Vol. 44, No. 3, March 2009, pp. 874–882
13. Y.-S. Shu, "A 6b 3GS/s 11mW Fully Dynamic Flash ADC in 40 nm CMOS with Reduced Number of Comparators," Symposium on VLSI Circuits, June 2012, pp. 26–27
14. V. H.-C. Chen, et al., "An 8.5mW 5GS/s 6b flash ADC with dynamic offset calibration in 32 nm CMOS SOI," Symposium on VLSI Circuits, June 2013, pp. 264–265
15. V. Giannini, et al., "An 820 μW 9b 40MS/s Noise-Tolerant Dynamic-SAR ADC in 90 nm Digital CMOS," IEEE ISSCC 2008
16. G. Van der Plas, et al., "A 150 MS/s 133 μW 7 bit ADC in 90 nm Digital CMOS," IEEE JSSC, Vol. 43, No. 12, December 2008, pp. 2631–2640
17. Y. Nakajima, et al., "A Background Self-Calibrated 6b 2.7 GS/s ADC With Cascade-Calibrated Folding-Interpolating Architecture," IEEE JSSC, Vol. 45, No. 4, April 2010, pp. 707–718
18. Z. Gu, et al., "A novel self-calibrating scheme for video-rate 2-step flash analog-to-digital converter," IEEE International Symposium on Circuits and Systems, May 1992, Vol. 2, pp. 601–604
19. B. Verbruggen, et al., "A 2.1 mW 11b 410 MS/s dynamic pipelined SAR ADC with background calibration in 28 nm digital CMOS," Symposium on VLSI Circuits, June 2013, pp. 268–269
20. J. K Fiorenza, et al., "Comparator-Based Switched-Capacitor Circuits for Scaled CMOS Technologies," IEEE JSSC, Vol. 41, No. 12, December 2006, pp. 2658–2668
21. B. Hershberg, et al., "Ring Amplifiers for Switched Capacitor Circuits," IEEE JSSC, Vol. 47, No. 12, December 2012, pp. 2928–2942
22. B. Murmann, et al., "A 12-bit 75-MS/s pipelined ADC using open-loop residue amplification," IEEE JSSC, Vol. 38, No. 12, December 2003, pp. 2040–2050
23. M. Anthony, et al., "A process-scalable low-power charge-domain 13-bit pipeline ADC," Symposium on VLSI Circuits, June 2008, pp. 222–223
24. N. Dolev, et al., "A 12-bit, 200-MS/s, 11.5-mW pipeline ADC using a pulsed bucket brigade front-end," Symposium on VLSI Circuits, June 2013, pp. 98–99
25. J. Hu, et al., "A 9.4-bit, 50-MS/s, 1.44-mW Pipelined ADC Using Dynamic Source Follower Residue Amplification," IEEE JSSC, Vol. 44, No. 4, April 2009, pp. 1057–1066
26. S.-H. W. Chiang, et al., "A 10-Bit 800-MHz 19-mW CMOS ADC," Symposium on VLSI Circuits, June 2013, pp. 100–101
27. L. Brooks, et al., "Background Calibration of Pipelined ADCs Via Decision Boundary Gap Estimation," IEEE TCAS I, Vol. 55, No. 10, November 2008, pp. 2969–2979
28. M. J. M. Pelgrom, et al., "Matching properties of MOS transistors," IEEE JSSC, Vol. 24, No. 5, October 1989, pp. 1433–1439

29. P. Harpe, et al., "A 2.2/2.7fJ/conversion-step 10/12b 40kS/s SAR ADC with Data-Driven Noise Reduction," IEEE ISSCC 2013
30. N. Verma, et al., "A 25/spl mu/W 100kS/s 12b ADC for wireless micro-sensor applications," IEEE ISSCC 2006
31. C.-C. Liu, et al., "A 10-bit 50-MS/s SAR ADC With a Monotonic Capacitor Switching Procedure," IEEE JSSC, Vol. 45, No. 4, April 2010, pp. 731–740
32. E. J. Siragusa, et al., "Gain error correction technique for pipelined analogue-to-digital converters," Electronics Letters, vol. 36, no. 7, March 2000, pp. 617–618

Energy-Efficient High-Speed SAR ADCs in CMOS

Lukas Kull, Thomas Toifl, Martin Schmatz, Pier Andrea Francese, Christian Menolfi, Matthias Braendli, Marcel Kossel, Thomas Morf, Toke Meyer Andersen and Yusuf Leblebici

Abstract An ADC featuring a new architecture for an 8 b 64× interleaved CMOS ADC running at up to 100 GHz sampling frequency is presented. The ADC fulfills all specifications for 100 Gb/s ITU-OTU4 communication over long-distance optical fiber channels. It is based on a SAR ADC, known for its superior energy efficiency and suitability for deep-submicron digital CMOS processes, as the comparator is the only true analog element. Several improvements to existing SAR ADC architectures are presented. Alternate comparators are used to increase the sampling speed at no power and area penalty, and dynamic memory is used to reduce latency in the CDAC feedback. A deep-trench capacitor-based reference buffer significantly reduces power at low output impedance, and a differential CDAC with constant common mode and fractional reference voltages optimizes comparator performance and silicon area. The 64× interleaved ADC consists of a dedicated sampling and interleaving block and 64 SAR ADCs. Four interleaved passive samplers based on a sampling switch with in-line 1:4 demultiplexer provide an initial 1:16 interleaving with high linearity and more than 20 GHz input bandwidth while using only a single supply voltage.

1 Introduction

As internet traffic continues to grow exponentially, future networking standards have to support higher data rates. Next-generation optical communication links require fast, but energy-efficient ADCs to enable complex digital equalization of long-distance fiber channels.

L. Kull (✉) · T. Toifl · M. Schmatz · P. A. Francese · C. Menolfi · M. Braendli · M. Kossel · T. Morf · T. M. Andersen
IBM Research—Zurich, Rüschlikon, Switzerland
e-mail: lku@zurich.ibm.com

Y. Leblebici
EPFL, Lausanne, Switzerland

P. Harpe et al. (eds.), *High-Performance AD and DA Converters, IC Design in Scaled Technologies, and Time-Domain Signal Processing*,
DOI: 10.1007/978-3-319-07938-7_3,

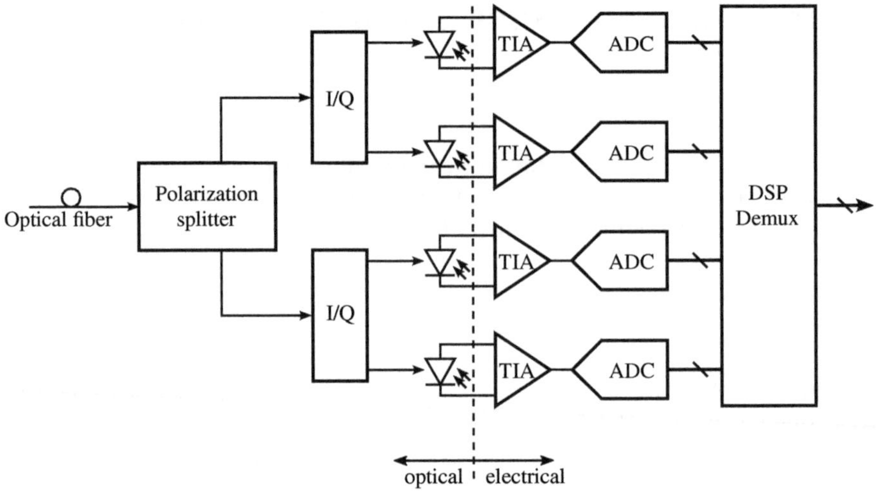

Fig. 1 Receiver of an OIF scalable SERDES framer interface (SFI-S) [1]

Starting with an outline of the target application, the first part of this contribution describes the sub-ADC, where high speed at low area and power is crucial to reach the performance of the interleaved design. The second part describes the architecture of the interleaver, which defines the speed and bandwidth performance of the interleaved ADC.

1.1 ITU-OTU4

The optical communications standard ITU-OTU4 defines 100 Gb/s over a single fiber for long haul, i.e., up to a few thousand kilometers. A receiver implementation is shown in Fig. 1. The OTU4 channel consists of four sub-channels, an I/Q pair each for two orthogonal polarizations, which implements dual polarization quadrature phase shift keying (DP-QPSK). QPSK combines two orthogonal non-return to zero (NRZ) symbols in the complex domain (in-phase and quadrature phase). DP-QPSK with coherent detection, i.e., with known phase and amplitude, is implemented. The baud rate is specified from 28 to 32 GS/s, with four bits per symbol from DP-QPSK and some margin for forward error correction (FEC).

It is generally suggested to run the ADC at 2× the baud rate for coherent detection, resulting in a sampling rate requirement of 56–64 GS/s [2]. Chromatic dispersion and polarization mode dispersion impact the signal integrity of long optical fibers. Both effects have to be compensated by the DSP shown in Fig. 1. Taking into account the channel imperfection and signal constellation with the bit-error rate requirement of $<10^{-12}$, about 5.5–6 ENOB at DC and 5–5.5 ENOB at 15–20 GHz input frequency are required with a 3 dB bandwidth >15 GHz [3].

The 64× interleaved 8 b ADC primarily targets ITU-OTU4. A CMOS-only implementation provides significant benefits for power savings, in particular if the ADC is implemented in a modern CMOS process, where the large DSP can be implemented on the same chip, thus benefiting from the power efficiency of modern CMOS processes.

2 High Speed SAR ADC

Newer technology nodes inherently result in better SAR ADC performance, but the benefit obtained from switching to the most recent nodes in deep sub-micron CMOS shrinks with every newer node. Innovative design approaches can still be found and provide significant improvements in SAR ADCs. The SAR ADC described herein makes use of asynchronous timing, a redundant capacitive DAC (CDAC) and the Set-and-Down principle [4, 5].

The ADC extends these known principles with a constant common-mode CDAC, a low-power reference voltage buffer and two alternate comparators. The two alternate comparators increase speed while relaxing the electro-migration issue at no extra complexity. In Table 1, the features implemented, and described below, are classified based on their impact on performance.

The resulting 8 b SAR ADC runs at 1.2 GS/s on a single-channel in 32 nm CMOS, and achieves 39.3 dB SNDR and a Figure-of-Merit (FoM) of 34 fJ/conv.-step. The ADC consumes 3.1 mW from a 1 V supply and occupies 0.0015 mm^2.

2.1 Architecture

As shown in Fig. 2, the ADC is based on relatively simple building blocks. A target size of 128 fF on nodes V_{cp} and V_{cn} was chosen to achieve low-power CDAC switching, a high input-bandwidth and a reasonably large unit capacitance for good matching.

No in-line switches are placed between the CDAC and the comparators to minimize delays. For maximum speed, the comparator is operated without a pre-amplifier. Therefore more focus is put on the design of the comparator, as it is largely responsible for achieving low noise at high conversion speed.

The logic of an asynchronous SAR ADC can be divided into two parts, a clock-generation block and a DAC control block. The clock-generation block senses the completion of a decision from one of the two comparators and generates the subsequent clock edges, as illustrated by ck_1 and ck_2 in Fig. 2. The DAC control block receives the output of the comparators and controls the CDAC. It stores the output of the comparators and keeps track of the state of the SAR ADC.

To eliminate inter-symbol interference (ISI) in the subsequent input-sampling phase, the CDAC is reset prior to the sampling phase. The auto-zero phase of the

Table 1 Benefits of high-performance design features implemented in this SAR ADC

Feature	↑ Speed	↑ Precision	↓ Power	↓ Area
Alternate comparators	+	o	o	o
Asynchronous internal timing	+	o	+	o
Set-and-up/down conversion	+	+	+	o
Redundant DAC	+	o	o	o
CDAC and reference capacitor stacking	+	o	o	+
Fractional reference voltages	o	o	o	+
Clocked reference buffer	o	o	+	o
Dynamic logic	+	o	o	o
Memory with state detection	+	o	+	+

+ beneficial
o mostly unchanged

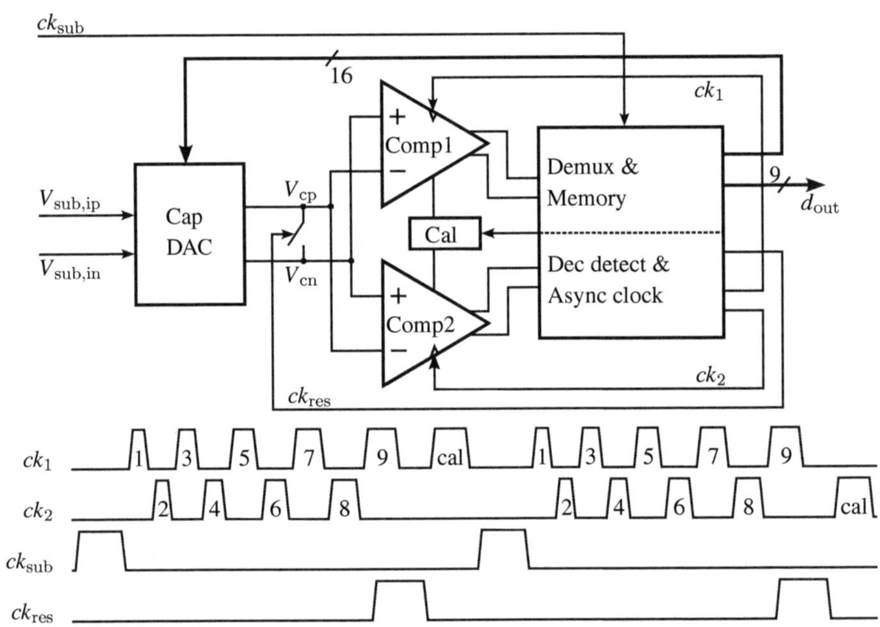

Fig. 2 Asynchronous SAR ADC overview with alternate comparators

comparators is also located at the end of the conversion and requires a zero input to the comparators. Therefore the CDAC is reset as soon as possible after the last decision of the SAR has been initiated. The time required for calibration of the comparator is simultaneously used to compensate any charge lost on the reference capacitor by triggering the clocked reference buffer.

The input to the comparator is statistically largest for the MSBs and reduces statistically towards the LSBs. As higher amplification for smaller input signals at the comparator requires more time, the conversion of the bits inside a SAR cycle is statistically fastest for the MSBs and becomes longer for the LSBs, as seen from ck_1 and ck_2 in Fig. 2.

2.2 Alternate Comparators

The comparator is important for fast conversion of a SAR ADC. It has two phases, a decision phase and a reset phase. The reset phase is not of importance for many designs, as storing the comparator output and settling of the CDAC requires significantly more time than the reset phase of the comparator. In our design, the memory of the DAC control block is optimized with a signal by-pass, that directly connects to the CDAC. The settling time of the CDAC is kept short by using a redundant switching scheme, a low-impedance capacitive reference voltage and short connections between the reference capacitor and the capacitors inside the DAC. Therefore the reset time of the comparator becomes the limiting factor for the overall conversion speed. Large reset switches inside the comparator could reduce the reset time at the price of higher power and slower decision speed because of the additional parasitics.

A solution to this problem based on two comparators is suggested. The two comparators are operated in an alternating fashion, as shown in Fig. 2. Decision 1 is taken by comparator 1, while comparator 2 resides in reset mode. After comparator 1 has finished its decision, it triggers its reset with minimum delay and initiates the decision phase of comparator 2. Using two comparators provides much more time for the reset phase and thus eliminates the reset time from the critical path.

2.3 Timing

A detailed picture of the timing sequence is given in Fig. 3. The initial sampling phase is triggered by the rising edge of ck_{sub}, during which both comparators are in reset state (R). The falling edge of ck_{sub} triggers the asynchronous clock logic of the SAR and initiates the first comparison with comparator 1 (D). No capacitive switching is required prior to the first decision with Set-and-Down switching. The end of the decision phase of comparator 1 is sensed by a decision-detect block to reset comparator 1 (R) prior to activating comparator 2 (D).

As can be seen from Fig. 3, at the end of the conversion cycles one of the comparators is calibrated to zero offset (C). Alternating calibration is sufficient to track offset drifts from history effects and $1/f$ noise. Simultaneously the clocked reference buffer supplies charge to the reference capacitor (RC).

The total duration of the SAR conversion is not pre-determined, because the comparator decision time depends on the input signal at the time of the decision.

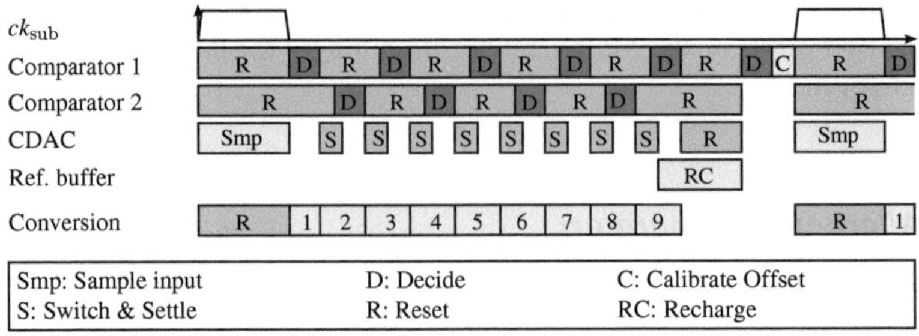

Fig. 3 Asynchronous timing sequence of the SAR ADC with two alternate comparators. The corresponding schematic is found in Fig. 2

In a synchronous design, the time to resolve metastability in the comparator has to be added to each cycle, resulting in a significant loss of conversion speed. In an asynchronous design, this buffer time can be added once, because not every conversion can become metastable. In a binary ADC with a very low-noise comparator only one comparison can be metastable. The buffer time is shifted to the end of the conversion and used to calibrate one of the comparators and resupply charge to the reference capacitor. If, for example, comparator 1 requires more time in the fifth decision owing to an almost zero input, the calibration is skipped. As this happens only rarely, it has no impact on the precision of the ADC. If even more time is required owing to metastability, which is even less likely, charge would not be resupplied to the reference capacitor and the CDAC not reset. The resulting ISI is marginal as the input bandwidth of the sampling switch is sufficiently high. Cases in which metastability requires more time than used for CDAC reset and comparator calibration are very rare, with a probability $<10^{-12}$, but still do not impact the functionality of the ADC. Even if the last one or two conversions are skipped, the ADC is still reset correctly by the rising edge of the sampling clock, ck_{sub}, and correct outputs of all completed conversion are stored. Unfinished decisions result in a zero output. Unlike previously published asynchronous designs, this design does not require special circuitry to handle metastability.

2.4 Building Blocks

2.4.1 Capacitive DAC

No bottom plate sampling is required for the 8 b converter. Thus the CDAC can be based on the Set-and-Down principle described in [6]. Set-and-Down is based on top-plate sampling and allows a direct comparison after sampling with no capacitive switching. Two additional features to the Set-and-Down CDAC are implemented; constant common mode during CDAC switching and redundancy. A constant common mode as shown in Fig. 4 requires twice the number of

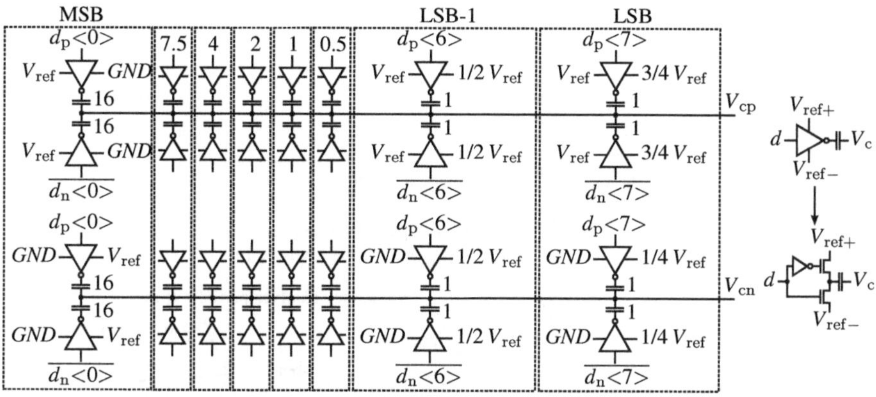

Fig. 4 CDAC design with constant common mode and fractional reference voltages to reduce the number of unit capacitors

capacitors, but provides a constant common mode to the comparator. As the comparator performance with respect to noise, speed and offset calibration highly depends on the common mode, the effort for a constant common mode CDAC can be justified. Each output of the SAR memory, with d_p<0:7> corresponding to the positive output of the comparators and d_n<0:7> corresponding to the negative output of the comparators, connects to both sides of the CDAC. Assuming d_p<0> switches from 0 to 1 and increases the voltage at V_{cn}, the voltage at V_{cp} is decreased by the same amount to keep the common mode constant. The unit capacitors of the CDAC are custom designed to achieve best area efficiency and compatibility with underlying reference capacitors. Each unit capacitor connects to two switches: One is connected to a positive reference voltage and the other to a negative reference voltage. Both reference voltages are below $V_{DD}/2$, and NMOS switches are used.

Capacitors in the CDAC are not sized in a binary way to achieve redundancy, as shown in Fig. 4. Therefore one additional comparison cycle is introduced in decisions 2–6, as seen from Fig. 5. The first decision range equals full-scale and no redundancy is required in the first decision, as the CDAC voltage remains constant after the end of the sampling phase. Therefore the decision range for the second decision is reduced by the ideal factor of 2. By the time the second decision is taken, the CDAC voltage is not fully settled. Therefore the decision range for the third decision is larger than the 50 % from a binary DAC to cover the incomplete settling. The settling time in Fig. 5 is indicated in multiples of *RC* time constants. The settling time in a binary DAC is indicated by t_{bin} and the settling time from the redundant DAC with capacitor sizing according to Fig. 4 is indicated by t_{red}. These numbers assume that each unit capacitor in the DAC has its own buffer and that the *RC* time constant is independent of the size of the capacitor.

The relaxed settling achieved from the redundancy of 93 % introduced, compared with 99.2 % (allowing for <0.5 LSB error) without redundancy, significantly reduces the settling time by almost a factor of 2. Assuming 93 % settling in

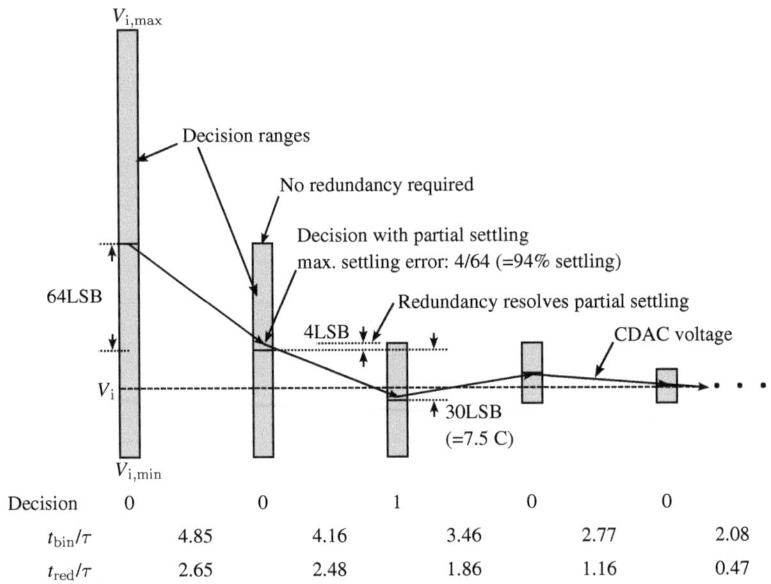

Fig. 5 Details of the redundancy switching scheme with time comparison for redundant and non-redundant switching. Values are given in multiples of time constants ($\tau = RC$) for a settling error <0.5 LSB

the non-redundant LSBs results in a maximum error of 0.15 LSB and is therefore negligible.

Introducing a constant common mode in the CDAC requires twice the number of capacitors. To reduce the total number of unit capacitors in the CDAC and optimize area, a new approach with fractional reference voltages is used, as seen in Fig. 4: Instead of switching fractions of capacitors by the full difference of the reference voltages, one capacitor is switched by a fraction of the reference voltage. For example, switching one capacitor between $V_{ref}/2$ and GND is equivalent to switching half a capacitor between V_{ref} and GND. The impact on the CDAC output is the same except for some minor additional parasitic capacitance associated with switching by fractional reference voltages. The parasitic capacitance results in a small gain error, which is compensated by adjusting V_{ref}.

If the capacitors located at V_{cn} and V_{cp} are not switched between the same reference voltages to achieve a voltage difference of $V_{ref}/2$, the precision requirement of $V_{ref}/2$ is greatly relaxed. To switch half a unit capacitor, as shown for the capacitor connected to $d_{p/n}$<6> in Fig. 4, the capacitor at node V_{cp} is switched between $V_{ref}/2$ and V_{ref} and the capacitor at node V_{cn} is switched between GND and $V_{ref}/2$. The resulting impact of an imprecise $V_{ref}/2$ is zero on the output voltage difference $V_{cp} - V_{cn}$. An imprecise $V_{ref}/2$ only affects the common mode, with negligible impact as only few capacitors are switched to or from $V_{ref}/2$.

The precision of $V_{ref}/2$ becomes more important if smaller fractions of V_{ref} are used, as in this CDAC with $V_{ref}/4$. The quarter-fraction capacitor is also balanced, but not necessarily immune to variations in $V_{ref}/2$. As only the LSB capacitor uses $V_{ref}/4$, the impact on the total ADC performance is negligible though.

Introducing $V_{ref}/2$ and $V_{ref}/4$ results in unit capacitors of 2 fF with a total capacitance of only 128 fF on each side, which is well above the kT/C noise limit, but provides good matching. The 2 fF consist of 1.3 fF which can be switched, and 0.7 fF from the parasitic capacitances associated with the neighboring metals of the unit capacitor, the MOS parasitic capacitance of the unit capacitor switches, the track-and-hold switches, the input transistors of the comparator, and the reset switch.

To achieve fast switching in the CDAC with a low time constant ($\tau = RC$), also the resistance between the reference and the CDAC has to be optimized. The in-line resistance consists of the CDAC switch resistance and the resistance of the V_{ref} reference buffer. A large deep-trench (DT) capacitor [7] of 80 pF buffers V_{ref}. As the fractional reference voltages are less loaded, their associated DT capacitors are sized smaller. The large capacitors reduce the reference impedance significantly. By placing the CDAC on top of the voltage-reference DT capacitors, area is saved and the in-line resistance between the switched capacitors and the reference voltage is further reduced. Stacking is possible, because DT capacitors, which are equivalent to DRAM capacitors, only occupy space in the substrate and are connected on the first metal level, thus leaving the upper metals for the CDAC capacitors.

2.4.2 Comparator

The comparator is the key component of this design and determines the accuracy and, to a large extent, the speed of the ADC. As it accounts for a significant part of the total power consumption, special care has to be taken for the comparator design.

A sense-amp latch, also known as Strong-ARM latch, is chosen as a comparator because of its superior speed enabled by the single-stage design, as shown in Fig. 6 [8]. Dynamic history based on the SOI body-effect affects the offset and requires frequent offset cancellation to ensure that both comparators have zero offset. The second differential pair, M_3 and M_4, is sized $5\times$ smaller and used to cancel the offset. A smaller size of the second differential pair is beneficial for the noise performance, because the internal nodes are discharged mainly from the main differential pair connected to the input. As there is no pre-amplifier, the activation of a comparator results in direct kickback onto the CDAC nodes.

Differential kickback from input transistors M_1 and M_2 is subject to mismatch. Therefore there is only one comparator active at a time. For the same reason the calibration for zero offset of the comparators is performed with a zero value on the CDAC and a released CDAC reset switch to achieve the same dynamic behavior as during conversion.

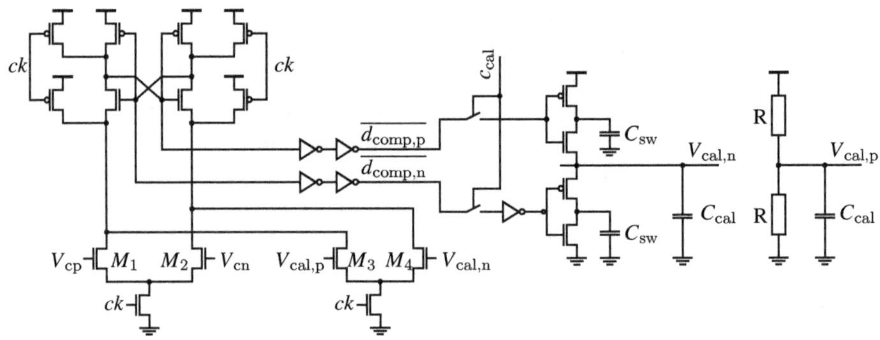

Fig. 6 Sense-Amp comparator with second differential pair for offset calibration

The calibration circuit for the auto-zero of the comparator is shown in Fig. 6 and is similar to that described in [9]. It is a clocked circuit, in which charge is added to or subtracted from C_{cal} depending on the result of the comparison at zero input of the comparator. The size of C_{sw} determines the tracking speed of the offset voltage. In general, smaller values for C_{sw} result in less noise on C_{cal}. The voltage $V_{cal,n}$ on C_{cal} is used directly as one of the differential inputs of the comparator. The other input voltage of the calibration differential pair is set to a constant voltage. Therefore the common mode is not constant on the calibration differential pair, but as the differential pair is sized much smaller, this is acceptable.

As the comparator limits the ENOB of the ADC, performance was optimized by a careful analysis based on [10, 11] and extensive simulations. The performance has to be balanced between noise and speed by choosing the common mode correctly [5].

Metastability is resolved by comparator gain, which means that the comparator must have a short time constant in the regenerative gain stage τ_{cmp}. Because of the exponential gain of the comparator, even very small input voltages can be resolved with acceptable delay [5]. In a binary SAR ADC with little noise, only one comparison can be metastable. The redundancy shown in Fig. 4 slightly increases the chance for metastability, but not significantly. Therefore there is a probability of approximately 10^{-12} for a comparator input voltage to be $<10^{-12}\text{LSB}/2$, which is 1 fV. The comparator decision time t_{dec} can be derived as

$$t_{dec} = t_0 + \tau_{cmp} \ln(V_i). \tag{1}$$

This means that a worst-case metastability event with a probability of 10^{-12} will take approximately $12\ln(10)\,\tau_{cmp}$, which is about 128 ps in this design. This is still less than the CDAC reset and comparator calibration require. There is no impact on precision if the comparator calibration is skipped sometimes. Moreover, reset of the CDAC and the last bit decisions are not affected when events up to 10^{-12} are considered.

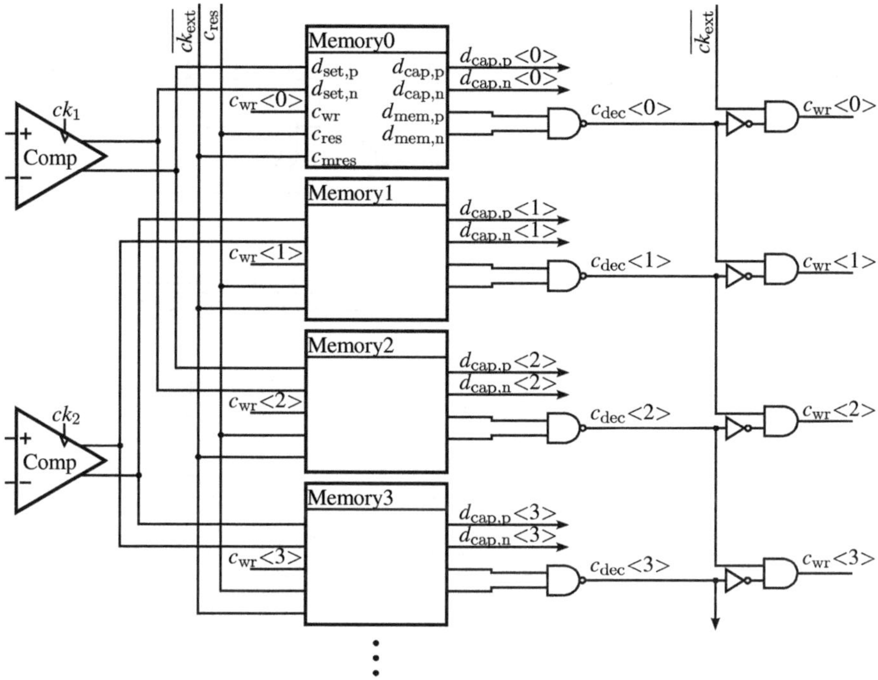

Fig. 7 Memory logic with state detection based on combinatorial logic

2.4.3 SAR Logic

The logic block can be divided into two major parts: the DAC control block and the clock-generation block. Both are described in more detail below.

The DAC control block consists of a memory block with attached state logic, as shown in Fig. 7. The decision output of each comparator is directly connected to a number of memory cells. The signal c_{wr} activates one pair of memory cells, i.e., one memory cell for each output of the comparator. The memory outputs $d_{cap,p}$ and $d_{cap,n}$ are connected to the CDAC at minimum delay.

The high conversion speed of the ADC requires a state machine with more than 10 GHz, which is hard to implement. Therefore the state machine consisting of state memory and state transition logic is replaced by a state detection block consisting of simple combinatorial logic attached to the memory cells. A successful decision of a comparator is detected by the signal c_{dec}<*n*>, leading to a falling c_{wr}<*n*> signal on the memory cell just written and a rising c_{wr}<*n*+1> signal on that memory cell to prepare the next memory cell for the next comparator decision. The presence of a second comparator does not increase the complexity of the memory and state detection logic. On the contrary, it relaxes the timing and saves power.

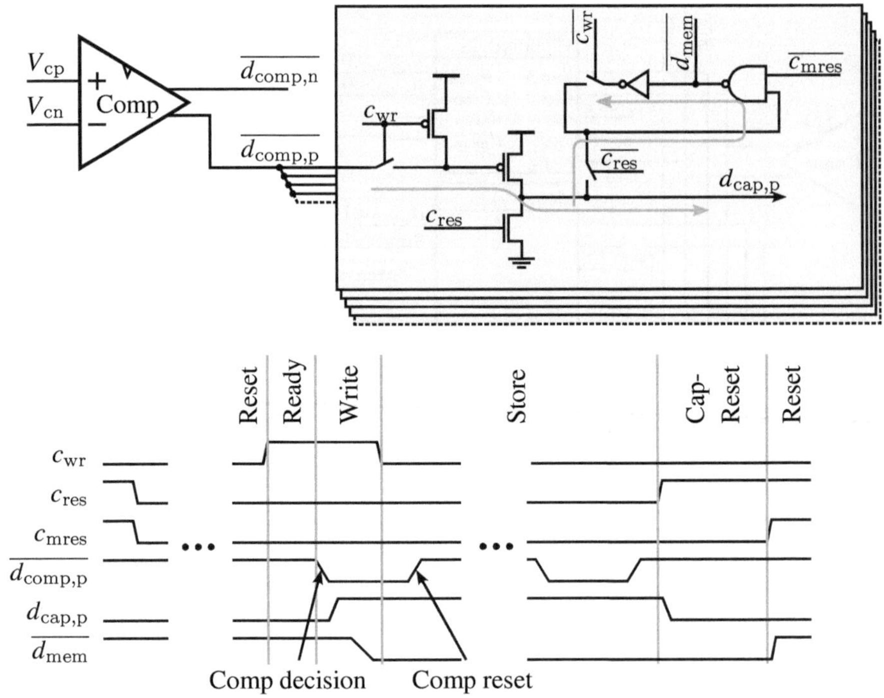

Fig. 8 Memory cell with dynamic buffer and optimized signal transition path

To achieve the highest conversion speed, the memory cell shown in Fig. 8 is optimized for a fast input-to-output transition to enable early CDAC settling. The lower half of Fig. 8 displays a typical write cycle from a comparator output to the memory cell. When c_{wr} is high, node $d_{\mathrm{cap,p}}$ is floating for increased speed. The transition from the memory cell input to the output passes only one transistor to buffer the signal. A slower side loop stores the conversion decision and holds the memory output. The signal c_{wr} is disabled after the slower memory loop has settled. The outputs of the memory cells $d_{\mathrm{cap,p/n}}$ are buffered and/or inverted and connected to the inputs $d_{\mathrm{p/n}}$ of the CDAC.

Given that the data path is highly optimized for maximum speed, the clock path has to be optimized to the same degree. The sequence of combinatorial blocks is reduced, as shown in Fig. 9. The decision detect block consists of a simple NAND gate. When a successful decision is detected, e.g., from comparator 1, ck_1 is disabled with minimum delay, and ck_2 is triggered immediately with a pulse on p_1 to enable comparator 2. The non-overlapping clocks of ck_1 and ck_2 ensure that the kickback from the comparators will never impact the precision of the neighboring comparator. The signals en_1 and en_2 are used to activate the SAR operation at the start of the conversion and to initiate the calibration cycle. The signals dis_1 and

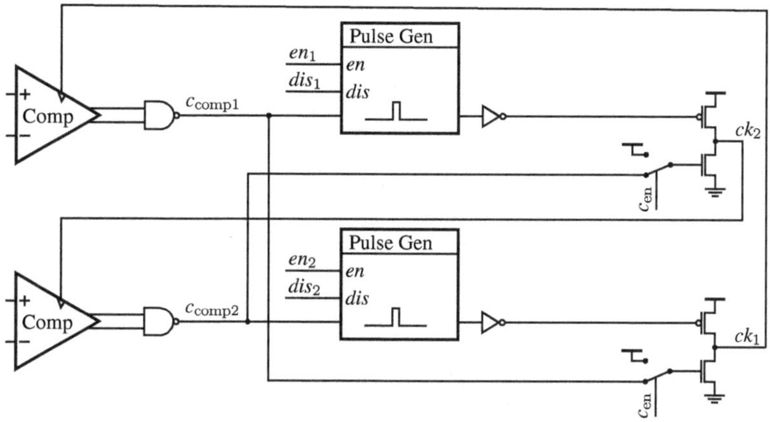

Fig. 9 Asynchronous clock logic with decision detect and dynamic clock driver

dis_2 are used to interrupt the SAR operation for DAC reset and at the end of the conversion cycle. Pre-charged/dynamic logic is used to drive the signals ck_1 and ck_2 for increased speed.

3 64× Interleaved SAR ADC

Based on the SAR ADC, an interleaved ADC architecture is presented that achieves the specifications required for the optical communication standard ITU-OTU4 [12]. The ADC achieves up to 100 GS/s at less than 1 W. First the ADC architecture is highlighted before the core component of the 64× interleaved ADC, the interleaver, is described in detail. Some measurement results conclude the chapter.

3.1 Architecture

The top-level overview of the ADC is shown in Fig. 10. The differential clock is connected at half-rate to a CML divider and CML-to-CMOS stages to generate four high-precision clock phases for the sampling switches. Skew is compensated prior to the sampling switch with the signal d_{skew}. The input is protected with reduced-size ESD diodes equivalent to the secondary CDM diodes and terminated by $2 \times 50\,\Omega$. It is unbuffered and directly connected to the hot sampling switches.

Each sampling switch is part of a 1:16 interleaver. The output of these four interleaver blocks is connected to 64 SAR ADCs. The aggregated output of the SAR ADCs is captured in an on-chip memory block which stores 8192 digitized samples.

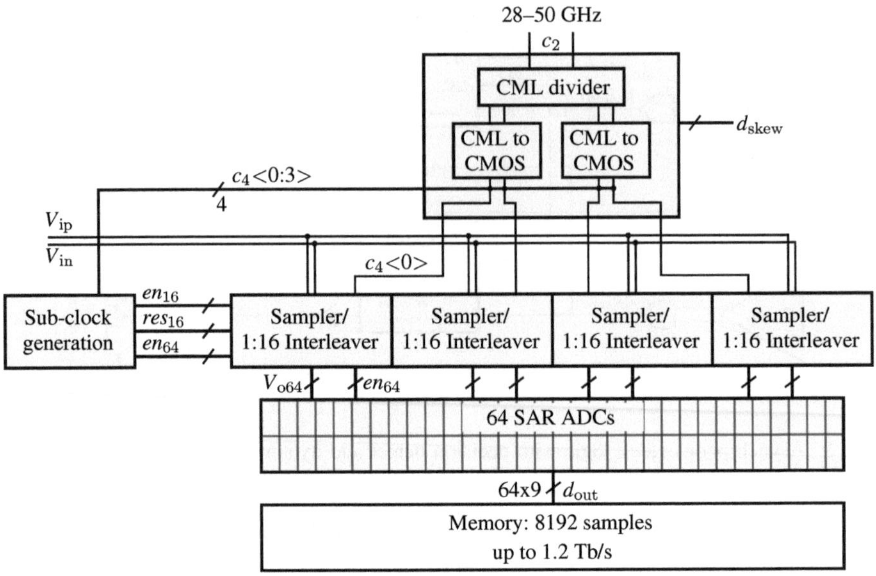

Fig. 10 Overview of the 64× interleaved ADC

A sub-clock generation block provides all non-timing critical clocks to the interleaver and the SAR ADCs.

The SAR ADC described earlier in this chapter is used and connected to the interleaver. As the SAR ADC speed exceeds 1 GS/s, 64 slices are sufficient to achieve more than 64 GS/s conversion speed.

3.2 Sampler and Interleaver

The sampler and interleaver represent the most critical part of the ADC and define the achievable sampling speed, the bandwidth and potentially also the precision. The architecture is shown in Fig. 11 and implements voltage mode sampling.

There are four sampling switches, each connected to a demux stage to increase the hold time on the capacitors. Careful optimization of the number of sampling switches and the demux ratio while limiting the number of sampling switches results in four sampling switches and a 1:4 demux for highest input bandwidth. The resulting hold time on the sampling capacitor is sufficient to sub-sample the voltage on the capacitor by buffering the signal and connecting it with a second set of demux switches to the SAR ADCs. Single NMOS switches are used to sample the signal because boot-strapped switches are not fast enough and not required to boost the linearity of the converter. The common mode of the input stage was carefully chosen to pass corners and different operating temperatures up to 100 °C.

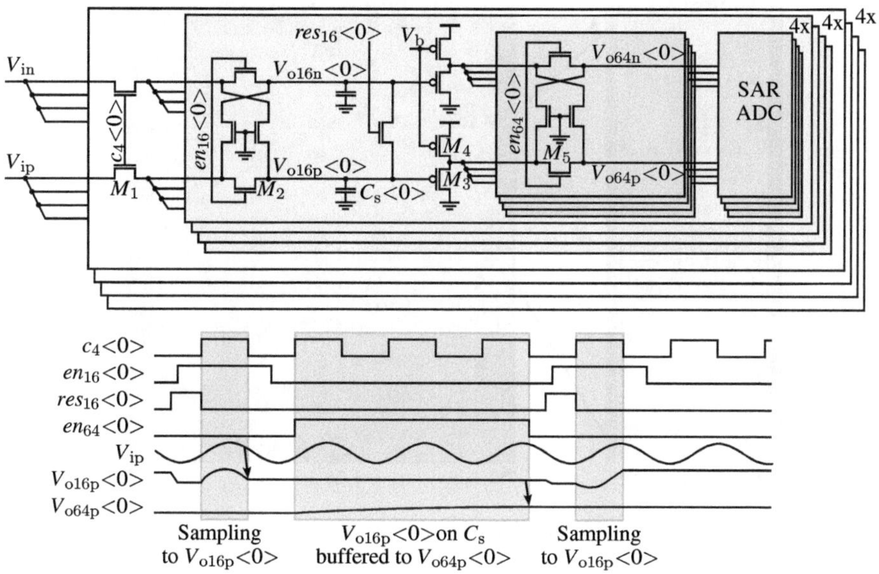

Fig. 11 Schematic details of the differentially implemented 1:64 interleaver

The in-line demux switches following the sampling switch feature feed-through compensation to prevent distortion of the sampled voltage on C_s from the sampling switch.

One of the four demux switches is activated by en_{16} prior to the rising edge of c_4 and disabled after the falling edge of c_4 to eliminate any influences of en_{16} on the sampling window. Thus there are only four c_4 clocks defining the sampling window. A reset switch driven by res_{16} shorts the sampled voltage prior to the sampling window to eliminate ISI.

The sampled voltage is buffered with a PMOS source follower formed by M_3 and M_4. A source follower is optimal for robustness and high linearity when driven with an output common-mode near half of the supply, which fits the input common-mode of the SAR ADC. The output of the source follower is connected to the SAR ADCs by a second demux with feed-through compensation.

3.3 Measurement Results

The ADC layout with a detailed view of the SAR ADC is shown in Fig. 12. The clock divider is placed closely to the interleaver to reduce parasitics in the clock path and optimize for low jitter. The size of the interleaver is optimized to reduce mismatch between the sampling paths as much as possible. The ADC occupies an area of $470 \times 960\ \mu m^2$, of which about 50 % is occupied by the SAR ADCs. Routing to the

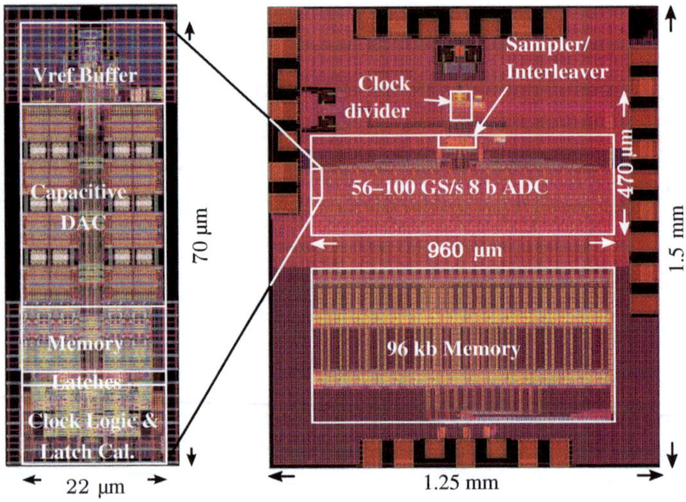

Fig. 12 Chip micrograph and layout. The 96 kb memory block stores 8192 samples with Hamming encoding

sub-ADCs is long and affects the input signal swing of the sub-ADCs because of the different wiring lengths and associated series resistance.

At 90 GS/s the ADC is run at 1.2 V supply on the interleaver, the clock input path (V_{DI}) and on the SAR ADCs (V_{DA}). The high supply voltage on the interleaver results in a higher overdrive of the sampling switches and compensates the shorter tracking time for high sampling frequencies. For lower sampling rates of 70 GS/s and below, the supply on the SAR ADCs can be decreased substantially to 1 V to save power. At 1.2 V and 90 GS/s, the measured total power consumption of 667 mW consists of 56 mW for the clock input path (consisting of CML clock divider and CML-to-CMOS stages), 112 mW for the interleaver/sampler and subclock generation, and 499 mW for the SAR ADCs.

Offset inside the SAR ADCs is eliminated by an auto-zero step because the two comparators need to have the same offset for correct operation. Offset between the channels is compensated off-chip. Gain equalization between the SAR channels is achieved by adjusting the reference voltage. The phase mismatch on the four high-speed input clock signals is adjusted prior to the sampler with the signal d_{skew} in Fig. 10. Offset, gain and skew compensation are based on off-line sine-fit data and adjusted once, at 2.1 GHz for offset and gain and at 19.9 GHz for skew, to achieve good sensitivity. Bandwidth mismatch in the input sampling path cannot be calibrated by design. The layout of the sampling switches and capacitors including the clock routing is done with highest symmetry and step repetition to eliminate mismatch as much as possible. Because the input sampling transistors, the demux transistors and the sampling capacitors are large compared with the minimum feature size, random mismatch is kept low.

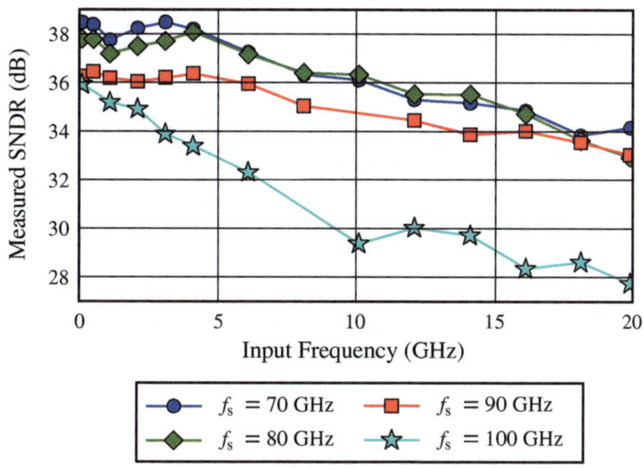

Fig. 13 Measured SNDR versus input sine frequency

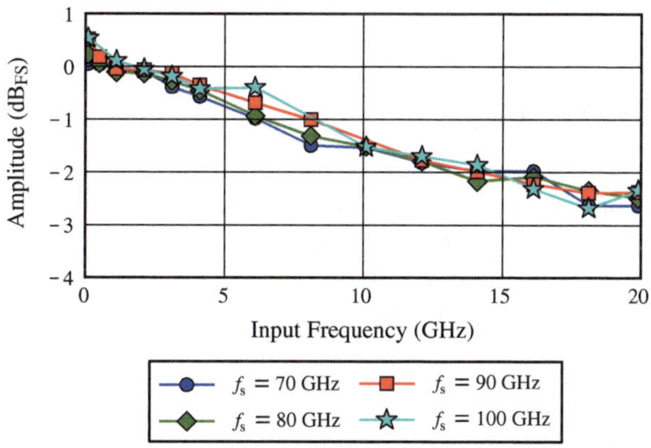

Fig. 14 Measured amplitude versus input sine frequency for different sampling frequencies referred to 2.1 GHz input frequency

The SNDR and amplitude versus input frequency are shown in Figs. 13 and 14. At 90 GS/s sampling speed, more than 36.0 dB is achieved up to 6.1 GHz and more than 33.0 dB up to 19.9 GHz. It is important to measure SNDR with a near-full-scale or full-scale signal to capture the full nonlinearity and the impact of jitter, skew and bandwidth degradation at high frequencies. The measurement series at 100 GS/s was acquired with skew calibration disabled; therefore the SNDR at higher input frequencies is limited by phase mismatch of the four input clock phases. This series displays the necessity of skew calibration to achieve best performance at high input frequencies. With skew calibrated, lower SNDR at higher input frequencies mainly stem from the reduced amplitude.

The SNDR versus sampling frequency is shown in Fig. 15 with a similar SNDR roll-off as in the single SAR ADC. The SAR ADC is designed in such a way that it remains functional even if the conversion is terminated too early. The more SAR

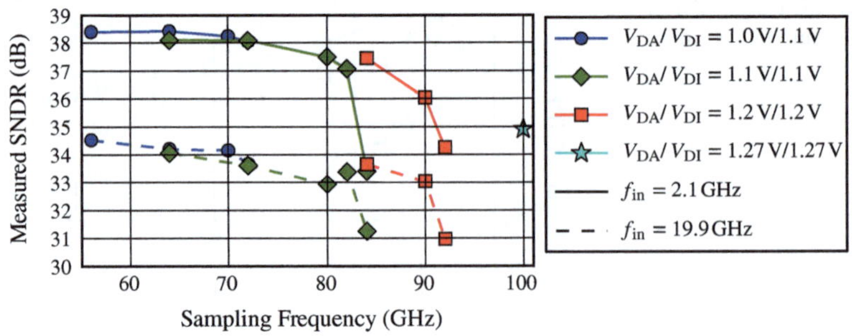

Fig. 15 SNDR versus sampling frequency

ADCs are terminated too often too early, the more significant the drop in SNDR will be. Therefore there is no hard limit to an upper sampling frequency; it is a trade-off between sampling frequency and SNDR.

4 Conclusion

This interleaved ADC design exhibits the highest sampling frequency at lowest FoM among previously reported 6 b+, >20 GS/s CMOS ADCs [13] and first reports a sampling frequency of 100 GHz. The FoM is 203 fJ/conv.-step at 19.9 GHz input frequency, 90 GHz sampling frequency and 1.2 V supply voltage.

References

1. Optical Interconnect Forum, "OIF-SFI-S-01.0. scalable SERDES framer interface (SFI-S): Implementation agreement of interfaces beyond 40G for physical layer devices," http://www.oiforum.com/public/documents/OIF_SFI-S_01.0_IA.pdf, 2008.
2. C. Malouin, B. Zhang, A. Wagner, S. Khatana, E. Ibragimov, H. Jiang, and T. Schmidt, "Sub-rate sampling in 100 Gb/s coherent optical receivers," OFC/NFOEC, pp. 1–3, Mar. 2010.
3. Y. Greshishchev, Nyquist AD Converters, Sensor Interfaces, and Robustness. Springer, 2012, ch. 6 CMOS ADCs for Optical Communications.
4. L. Kull, T. Toifl, M. Schmatz, P. A. Francese, C. Menolfi, M. Brändli, M. Kossel, T. Morf, T. M. Andersen, and Y. Leblebici, "A 3.1mW 8b 1.2GS/s single-channel asynchronous SAR ADC with alternate comparators for enhanced speed in 32 nm digital SOI CMOS," in ISSCC Dig. Tech. Papers, Feb. 2013, pp. 468–469.
5. L. Kull, T. Toifl, M. Schmatz, P. A. Francese, C. Menolfi, M. Brändli, M. Kossel, T. Morf, T. M. Andersen, and Y. Leblebici, "A 3.1mW 8b 1.2GS/s single-channel asynchronous SAR ADC with alternate comparators for enhanced speed in 32 nm digital SOI CMOS," IEEE J. Solid-State Circuits, vol. 48, no. 12, pp. 3049–3058, Dec. 2013.

6. C.-C. Liu, Y.-T. Huang, G.-Y. Huang, S.-J. Chang, C.-M. Huang, and C.-H. Huang, "A 6-bit 220-MS/s time-interleaving SAR ADC in 0.18-m digital CMOS process," in Proc. International Symposium on VLSI Design, Automation and Test, Apr. 2009, pp. 215–218.
7. B. Jayaraman, S. Gupta, Y. Zhang, P. Goyal, H. Ho, R. Krishnan, S. Fang, S. Lee, D. Daley, K. McStay, B. Wunder, J. Barth, S. Deshpande, P. Parries, R. Malik, P. Agnello, S. Stiffler, and S. S. Iyer, "Performance analysis and modeling of deep trench decoupling capacitor for 32 nm high-performance SOI processors and beyond," in Proc. IEEE International Conference on IC Design & Technology (ICICDT), May 2012, pp. 1–4.
8. T. Kobayashi, K. Nogami, T. Shirotori, and Y. Fujimoto, "A current-controlled latch sense amplifier and a static power-saving input buffer for low-power architecture," *IEEE J. Solid-State Circuits*, vol. 28, no. 4, pp. 523–527, Apr. 1993.
9. M. Miyahara, Y. Asada, D. Paik, and A. Matsuzawa, "A low-noise self-calibrating dynamic comparator for high-speed ADCs," in IEEE Proc. Asian Solid-State Circuit Conf., November 2008, pp. 269–272.
10. P. Nuzzo, F. D. Bernardinis, P. Terreni, and G. V. der Plas, "Noise analysis of regenerative comparators for reconfigurable ADC architectures," IEEE Trans. on Circuits and Systems II, vol. 55, no. 6, pp. 1441–1454, July 2008.
11. B. Wicht, T. Nirschl, and D. Schmitt-Landsiedel, "Yield and speed optimization of a latch-type voltage sense amplifier," *IEEE J. Solid-State Circuits*, vol. 39, no. 7, pp. 1148–1158, July 2004.
12. L. Kull, T. Toifl, M. Schmatz, P. A. Francese, C. Menolfi, M. Brändli, M. Kossel, T. Morf, T. M. Andersen, and Y. Leblebici, "A 90GS/s 8b 667mW 64 interleaved SAR ADC in 32 nm digital SOI CMOS," in ISSCC Dig. Tech. Papers, Feb. 2014, pp. 378–379.
13. B. Murmann, "ADC performance survey 1997-2013," http://www.stanford.edu/~murmann/adcsurvey.html, 2013.

Automated Design of High-Speed CT ΣΔ Modulators Employing Compensation and Correction of Non-idealities

Timon Brückner and Maurits Ortmanns

Abstract In this contribution, we discuss our recent advances in the automated design of continuous-time sigma-delta modulators. In the state of the art, the architectural and circuit level design of these modulators, especially if high-speed, low power or high resolution is targeted, either requires broad experience in the loop filter design, or relies on mathematical transformations which does not include the influence of non-idealities nor gives control about the signal transfer function, or the design is based on a very time consuming, massive simulation approach. Our solution is based on a detailed high level modeling, which allows to include many of the common non-idealities into the architectural design. By using an exact DT simulation of the CT modulator, and by employing a GPU-based heuristic search, a quasi real-time design optimization is obtained. The adoption of this tool, which is online available via a web page, is shown for the design of a 50 MHz, 10 bit sigma-delta modulator.

1 Introduction: Common Styles of Designing ΣΔ Modulators

Within the state of the art of design methods for ΣΔ modulators (e.g., in [1–3]), it turns out that typically three phases can be distinguished during the design of continuous-time (CT) modulators: The first phase consists of the determination of the most important modulator characteristics, such as oversampling ratio (OSR), filter order N, and internal quantizer bit width B. This can e.g., be done with the help of an analytical formula for the ideal quantization noise power inside the band of interest (e.g., from [4]):

T. Brückner · M. Ortmanns (✉)
Institute of Microelectronics, University of Ulm, Albert-Einstein-Allee 43, 89081 Ulm, Germany
e-mail: maurits.ortmanns@uni-ulm.de

P. Harpe et al. (eds.), *High-Performance AD and DA Converters, IC Design in Scaled Technologies, and Time-Domain Signal Processing*,
DOI: 10.1007/978-3-319-07938-7_4,

$$P_{\mathrm{IBN}} \approx \frac{\mathrm{FS}}{12(2^B-1)^2} \frac{\pi^{2N}}{2N+1} \frac{1}{k_q^2 \prod_{i=1}^{N} \gamma_i^2} \frac{1}{\mathrm{OSR}^{2N+1}}. \tag{1}$$

Here, FS is the quantizer full scale level, which is mostly a design constraint within the available technology. The effective quantizer gain k_q, and the loop filter scaling coefficients γ_i of stage i are not known a priori, but depend on the quantizer resolution and the loop filter order and architecture. For multi-bit ΣΔ modulators with filter orders less than 4, and single-bit modulators with filter orders less than 3, these coefficients can be set to one for an initial estimate, while for higher orders the scaling of the loop filter due to stability will significantly reduce the obtainable resolution in (1).

Based on these parameters, in the second phase, the high-level filter structure can be determined in more detail. The most common techniques are:

- Reusing or adapting existing CT ΣΔ modulator loop filters.
- Using the Delta-Sigma Toolbox [5] to obtain a standardized modulator filter.
- Designing a DT modulator and transforming it into CT domain.
- Scaling a chosen, fixed filter architecture based on experience, or simulations, or by using a heuristic search algorithm.

This phase is probably the one with the highest diversity in terms of personal flavor in the design process, which strongly depends on the designers experience, the available tools and algorithms, as well as the resulting modulator characteristics. Hence, most of the published tools, which incorporated the previously mentioned design methods, address this phase, as e.g., [3, 5, 6].

The last phase in designing CT ΣΔ modulators is the implementation of the analog circuit on transistor-level. Here, the high-level description, performances and high-level simulation including non-idealities, which are obtained from and after the second phase, serve as specification. There exist a few heuristic search approaches for this phase, as, e.g., [3], which map the high-level implementation to a certain degree into a transistor level specification; but usually, the transistor level implementation including non-ideal behavior is a manual task. A common approach is to realize the ideally modeled high-level filter blocks with low-level models including non-idealities, and then—bit by bit—replacing these by transistor level circuits, which show non-idealities not excessively deteriorating the ideal performance.

Including the non-idealities directly into the design flow, thus allowing the circuits to be non-ideal while still achieving almost optimal performance, is unfortunately not supported by most approaches. Needed for this is a simulation based tool, which is detailed enough to include non-idealities, allows performance and stability analysis of the highly non-linear system by transient simulation, and at the same time is fast in execution.

For such transient analysis, the computational effort of course depends on the time-domain of the modulator, i.e., DT or CT, as well as the level of detail in the modeled building blocks. The low-level simulation approach used for both DT and CT ΣΔ modulators presented in [3] is extremely time-consuming. Due to the continuous nature of the input signal, the integrator and the DAC outputs, simulations of CT modulators require considerably longer simulation times, since also the inter-cycle operation is of interest. A prominent example for this is excess-loop-delay (ELD), which significantly increases the quantizer input swing between the sampling points, even though its effect at the sampling instant can be perfectly compensated for. Thus, not looking to the inter-cycle waveform could lead to undesired clipping.

In contrast, for DT modulators it is mostly sufficient to calculate the steady state at the sampling instants, if the effect of non-idealities on this response is sufficiently modeled. Therefore, heuristic searches with hundreds up to millions of simulations often lead to acceptable simulation times in the DT case, while becoming—depending on the initial guess and the desired quality of the results—very time-consuming or even impossible in the CT case [7]. Thus, the execution time of a single simulation is typically the bottleneck concerning the speed for a single CT ΣΔ modulator design cycle, when the design tools rely on simulation based algorithms. This corresponds with the published CT design tools [3, 8], whose run times range from several 10 min to hours for several hundreds of simulation iterations. Especially, if the designer intends to iteratively improve the results by redefining the optimizer constraints, this obviously becomes impractical, making the design of CT ΣΔ modulators still a job for experts.

As a compromise between deep, transistor level modeled simulation and heuristic search at architectural level without any non-idealities, we aimed at a partially automated design method for CT ΣΔ modulators, which addresses all three previously mentioned design phases; this is achieved by utilizing a heuristic search for the high-level synthesis, while incorporating non-idealities to a certain extent such as to compensate for them within the design process. All non-idealities, which can not be tackled on architectural level, are then compensated for or corrected by dedicated circuitry.

The hereby utilized heuristic search can be controlled via a freely accessible web-interface, http://www.sigma-delta.de. It mitigates the bottleneck of lengthy CT simulations by the usage of Lifting [9], a method for the exact DT simulation of CT ΣΔ modulators, which computes only the design relevant CT information. Within this heuristic search many modulators are investigated in parallel, which results in single-instruction-multiple-data executions; these are perfectly suited for execution on a graphics card processors (GPU). Thus, by using Lifting and the computational power of a high-end consumer GPU, extensive heuristic searches with hundreds of thousands of simulations are feasible within seconds of response time, allowing an interactive design process. The result is a CT ΣΔ modulator architecture, which compensates some of the most prominent non-idealities, while the design environment can be handled by non-experts.

The contribution is organized as follows. In Sect. 2, a short review of the most common CT ΣΔ modulator non-idealities is presented, since these are to be considered and, if possible, compensated for or corrected during high-level or later on low-level design. Sect. 3 describes the operation of the heuristic search on the GPU and possible choices for optimization goals on the available web-interface. A design example is presented in Sect. 4, which starts at the specification, uses the tool for a non-ideality compensated modulator, and briefly reviews the circuit level design. A short conclusion is given in Sect. 5.

2 Most Common Non-idealities in CT ΣΔ Modulators

Since the intention of the half-automated design of CT ΣΔ modulators is the ability to include non-idealities into the design process and—if possible—to compensate for them, in the following we shortly review the most common sources of non-ideal behavior and some possible compensation techniques. Figure 1 illustrates the location and kind of non-idealities for an exemplary 3rd order modulator: the total loop-delay τ_D from the quantizer sampling instant to the appearance of the feedback signal at the DAC output commonly referred to as ELD, integrator gain errors (GE_i), finite DC gain and finite gain-bandwidth (GBW) of the differential amplifiers within the filters, non-linearity of the integrators, additional parasitic poles further dropping the loop-filter phase, clock jitter, DAC non-linearity, limited amplifier output swing, quantizer random offset. Of course the list can be arbitrarily extended; but most importantly, we want to split the non-idealities into two groups:

- Those which can be compensated for on the architectural level.
- Those non-idealities, which can be either minimized by design, or corrected via dedicated circuitry.

In the following, we will shortly revise those non-idealities and reference the most common approaches for their compensation.

2.1 Non-idealities with Loop Filter Compensation

2.1.1 Compensation for Excess-Loop-Delay

In modulators operating at a high sampling frequency f_s, ELD becomes an increasingly important issue and needs to be compensated for. Thereby, the quantizer is first given enough time to decide, which reduces the problem of metastability [10], and then the decision must be forwarded to the DAC. A common choice is to clock the quantizer on one and the DAC on the other clock edge, thus realizing $\tau_D = 50\ \%T_s$, where T_s is the sampling period. Over the last two decades, many techniques have been proposed, all of which implement a fast proportional feedback around the quantizer [11]. Therewith, the loop-filter impulse response can be made identical at the sampling instants for a large range of ELD.

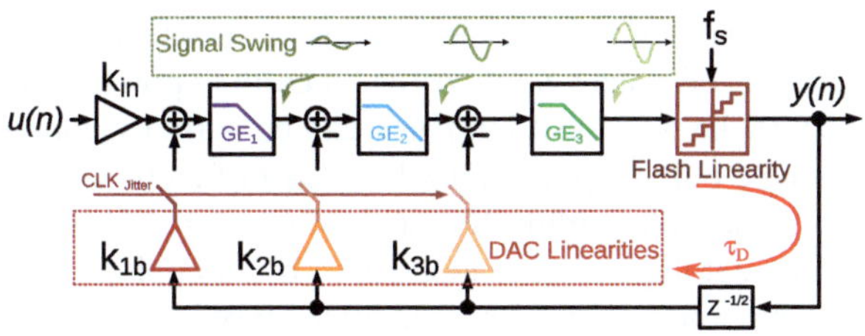

Fig. 1 Common loop filter non-idealities

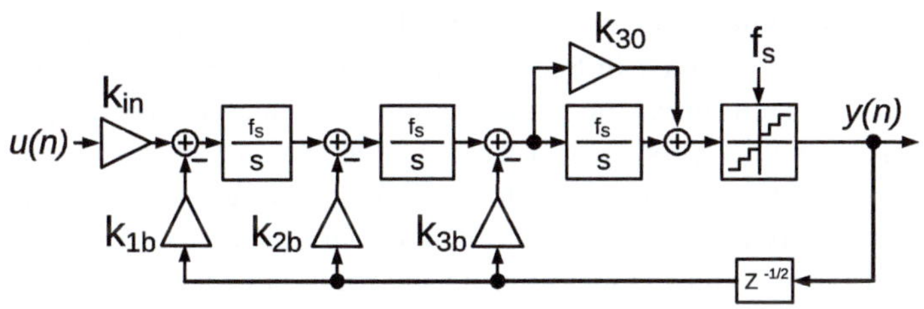

Fig. 2 PI-based ELD compensation [15]

The architectural implementation ranges from the straightforward adder, adder embedded into the quantizer [12, 13], analog differentiator and reusing the integrator summing node [14], incorporation of the proportional path into the last integrator [15], and fully digital implementation [16].

In Fig. 2 the exemplary ELD compensation using the proportional path within the last integrator [15] is illustrated, which will be called PI-integrator further on. Even though this implementation does not require additional active circuitry and therefore appears to be attractive for power reasons, closer investigations reveal that this technique tightens the speed requirements of the last integrator due to forcing the integrator now having to provide not only the integrated output but an additional proportional step-response. Moreover, the PI compensation introduces a feed-forward into the loop filter, which yields additional peaking of the signal transfer function (STF).

Thus, those compensation methods should be included into an architectural design optimization, allowing the compensation for ELD while revealing their individual requirements to the loop-filter at an early design stage.

2.1.2 Operational Amplifier Finite DC Gain and GBW

Integrator dynamics, foremost influenced by finite DC gain and finite gain-bandwidth product (GBW) of the operational amplifiers used within the integrator, are another prominent non-ideality. This is first of all since finite GBW influences loop stability, robustness and modulator performance, secondly since overdesigning integrator dynamics is obviously very costly concerning power, and finally, since it has been shown that the non-ideal influence can be partially compensated for [4]. When adding a single pole amplifier into the transfer function of an RC integrator, this yields in a second pole within the integrator transfer function, which deteriorates the overall stability of the loop-filter by dropping the phase margin. Additionally, the integrator gain is lowered by a GBW induced gain error, thus the linear modulator performance drops.

In [4] it was demonstrated that both effects can be compensated for. Therefore, the scaling coefficients were adopted such as to counteract the GBW induced gain error. Additionally, the GBW induced phase shift in the integrator was compensated for by modeling it as an additional loop-delay; loop-delay can again be compensated for as outlined in Sect. 2.1.1. While in [4] an iterative process was used for this compensation, the same method was analytically adapted in [17]: there, a DT loop filter prototype was converted into a CT loop filter using the impulse invariant transform, while finite GBW of the employed integrators and ELD was implicitly canceled. Still problematic in this approach is that whenever the achieved GBW values change during the transistor level design, e.g., due to parasitics, the whole architectural process needs to be repeated.

Additionally, as in all DT-CT transformation approaches, the designer has no a priori control on the resulting STF.

2.1.3 Internal Signal Swings

Insufficient voltage range of the transistor level circuit yields in clipping, distortion, and thus severe loss in performance. Shown in Fig. 3 is an example output spectrum simulating a 3rd order, 4-bit, mixed FB/FF compensated CT $\Sigma\Delta$ modulator [17] with and without clipping of the integrator outputs. When clipping occurs, the overall modulator SNDR is degraded to 35.9 dB comparing to an ideal SNR of 72.8 dB.

Internal swings can be easily avoided by careful scaling of the loop filter. Thereby, the internal signal swings are influenced by the choice of the architecture (e.g., FB vs. FF compensation), splitting and moving the scaling coefficients among the filter stages, or by reducing the reference levels. In the most popular design tool [5], internal swing can be equalized for DT modulators. In contrast, for CT $\Sigma\Delta$ modulators the state of the art does not provide a straightforward way to do that, nor an assistive tool. This becomes especially challenging, when it comes to compromising STF peaking (prominent in FF compensated modulators) against internal swing (which is large in FB compensated modulators), or when compromising small scaling coefficients for low swing against large RC values to realize those coefficients. Then, it is again an iterative decision process of an experienced designer, which makes the difference between a good or bad architectural choice.

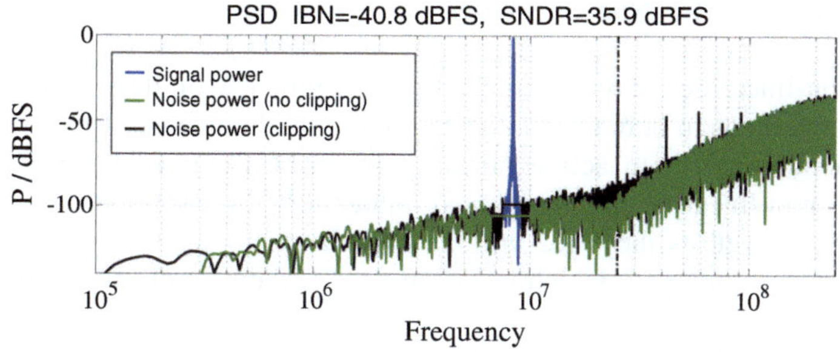

Fig. 3 Spectral distortion created by internal signal clipping [17]

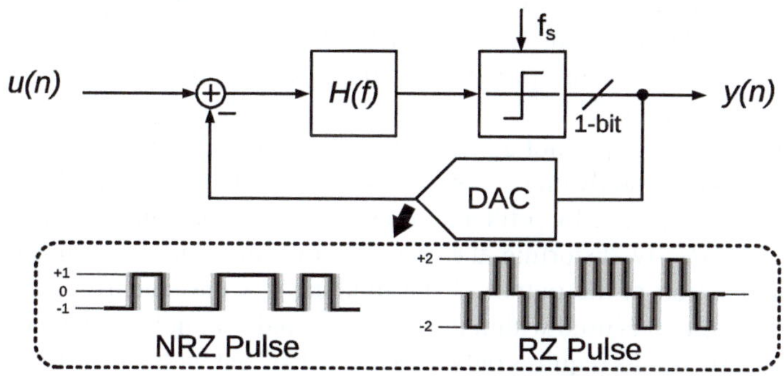

Fig. 4 ADC clock jitter due to modulation of charge in a 1-bit NRZ and RZ feedback DAC

2.2 Non-idealities with Explicit Correction

2.2.1 Clock Jitter in CT ΣΔ Modulators

As a first non-ideality with limited influence from the architectural choice, we can discuss clock jitter, indicated by the blurred edges of the feedback DAC pulses in Fig. 4. Over the years, CT ΣΔ modulators with single-bit internal quantizer and feedback DAC's have been discussed to have a tremendously increased sensitivity to clock jitter than their DT counterparts [18].

In order to reduce this sensitivity, over two decades mostly the implementation of other than single-bit rectangular DAC pulses has been proposed, such as multi-bit NRZ, or exponentially decaying ones. More recently, the influence of the NTF on the jitter sensitivity was discussed [19], while the effect was minor. In a very recent high-speed, single-bit CT ΣΔ modulator [20], the implementation of a FIR filter in the feedback path was used to reduce the jitter induced noise.

Nonetheless, in contrast to the above mentioned non-idealities, apart from the choice of multi-bit against single-bit loop filters, jitter is considered an effect which has not to be included into the architectural design phase.

2.2.2 CT Loop Filter Non-linearities

Circuit non-linearities directly affect the performance of data converters, and correction methods in both the analog and/or digital domain are usually employed. In ΣΔ modulators, three active blocks have to be designed against producing excessive non-linearity: the employed integrators (most importantly the first integrator), as well as the DAC and the quantizer; the latter if internal multi-bit quantization is employed, which has been used for almost all high-speed CT ΣΔ implementations over the last years.

Integrator non-linearities have to be kept below the intended distortion level of the modulator, especially when considering the influence of the first stage in the loop filter. The non-linearity is of course directly influenced by the required signal swing, as explained above. Apart from that, achieving the required linearity is firstly a job for the transistor level designer, even though there have been several assistive techniques being proposed, among others the assistive operational amplifier technique in [21].

The outermost DAC_1, if not designed for a ΣΔ modulator with single-bit internal quantization, is commonly one of the main sources of non-linearities [22, 23]. Since DAC_1 feeds into the same loop-filter node as the input signal, this results in any DAC non-linearity directly distorting the input of the modulator which is then seen without attenuation at the output of the modulator. Consequently, if e.g., a 4-bit internal quantization is used within the loop filter and a modulator inband resolution of 12-bit is desired, this requires only a 16 level DAC but with above 12-bit linearity and thus unit element matching. In focusing at DAC_2 and DAC_3, their non-linearities are suppressed by the preceding loop filter relaxing their linearity requirements against DAC_1. Nevertheless, the non-linearity suppression is strongly dependent on the oversampling ratio OSR [4]. Thus, especially for high speed ΣΔ modulators, where OSR is typically below 20, this suppression can easily be lower than 20 dB.

In order to overcome the aforementioned DAC non-linearities in multi-bit ΣΔ modulators, linearization techniques have been proposed over the last decades which mainly fall into two categories: error shaping and error correction. The easiest error shaping technique is dynamic element matching (DEM), where the output data of the quantizer is randomly interchanged between unit elements within the feedback DACs, which spectrally whitens the DAC non-linearities caused by the mismatching unit elements. Additionally, DEM can be improved utilizing data weighted averaging (DWA) where the whitened mismatch error is noise-shaped to out-of-band frequencies [24]. Even though these techniques have been used effectively for decades, they rely on fast oversampling, and hence, become less attractive for linearization in low OSR and thus high speed implementations.

In using error correction techniques, which do not rely on OSR, feedback DAC mismatches can either be corrected in the analog or the digital domain. In correcting DAC mismatches within a ΣΔ modulator, beyond many other implementations dynamic calibration has been successfully used where a reference unit cell of the DAC dynamically calibrates all other unit sources [25], capacitive tuning was used to linearize SC DACs, and auxiliary DACs have been used to compensate the DAC integral non-linearity. While compensation of non-linearities in the digital domain is possible [26], first the non-linearities need to be determined in a prior step before correction. One way for this estimation has been presented in [27], which has little circuit overhead and allows to digitally estimate the source in its original location, allowing to estimate both the static but also partially the dynamic mismatch [28]; the latter becoming important in high-speed implementations.

The above summary shows that there are clearly non-idealities, which can be absorbed into the architectural design process, since there are compensation techniques available and also since there are cross-dependencies, as e.g., in the case of peaking STF, ELD compensation and GBW of the last integrator, or e.g., in the case of choice of CIFF/CIFB and swing of the integrators and GBW. Those non-idealities, where these cross-dependencies can not be tackled on the architectural side, are then explicitly corrected for on the circuit level. Both together give reason for the chosen design framework and style, which is presented in the following.

3 High-Level CT ΣΔ Modulator Synthesis Tool

After choosing the key features of the intended ΣΔ modulator via (1), an architectural design of the loop filter with the corresponding coefficient scaling is started. In our approach, this is done via a web-based synthesis tool [29], which is briefly reviewed in this chapter. Thereby, the focus is set on:

- high-level modeling of low-level circuit behavior,
- utilized simulation and evaluation techniques,
- the usage of the web front-end.

3.1 High-Level Modeling of Low-Level Circuitry

Although aiming for a high-level filter scaling, a synthesis tool has to be capable of reproducing various sources of non-ideal behavior in order to produce meaningful results, and—as outlined in the previous chapter—be able to absorb the non-idealities into the design process. In order to give an example of how this low-level behavior is reproduced, the simulation model for an integrator with additional proportional path used for ELD compensation, as used in Fig. 2, is investigated in detail. The corresponding circuit is shown in Fig. 5.

To obtain a high-level model including non-idealities, first the transfer behavior is computed. Since the model shall at least include finite DC gain A_0 and gain-bandwidth of the amplifier, a one pole open-loop transfer characteristic is used.

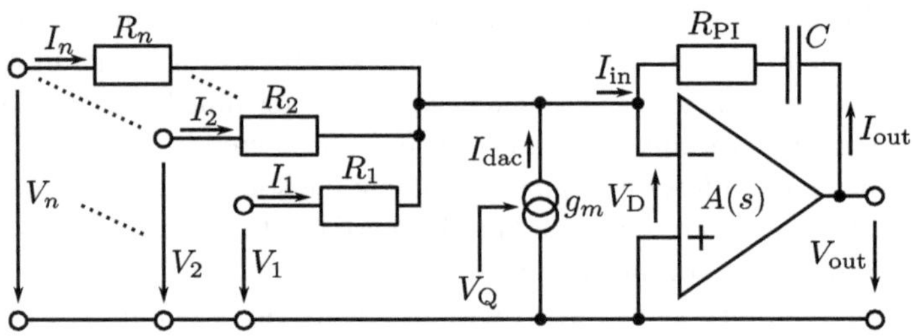

Fig. 5 Non-ideal integrator with multiple input paths and proportional feed-through behavior

Obviously, higher order non-idealities such as phase margin or even non-linearity can be modeled [30]. Nevertheless, it is sufficient to consider only first order non-ideal behavior to give the designer a good estimation on the desired amplifier requirements. Moreover, since finite GBW can be compensated for within the loop filter, this can be achieved automatically during filter design, if modeled properly. The output voltage of the circuit in Fig. 5 becomes

$$V_{\text{out}} = \frac{-\sum_{i=1}^{n}(f_s k_i + s \cdot p_i)V_i - \left(f_s k_{\text{dac}} + s \cdot p_{i,\text{dac}}\right)V_Q}{s^2 \cdot \frac{\left(1+\sum_{i=1}^{n} p_i\right)}{\omega_{\text{GBW}}} + s \cdot \left(\frac{1}{G_{\text{E}}} + \sum_{i=1}^{n} \frac{p_i}{A_0}\right) + \sum_{i=1}^{n} \frac{f_s k_i}{A_0}}, \tag{2}$$

where k and p are scaling coefficients, ω_{GBW} the GBW in [rad/s], and with the GBW induced gain error G_{E}

$$G_{\text{E}} = \frac{\omega_{\text{GBW}}}{\omega_{\text{GBW}} \cdot \left(1 + \frac{1}{A_0}\right) + \sum_{i=1}^{n} f_s k_i}. \tag{3}$$

Thereby, the circuit parameters from Fig. 5 have been substituted by high-level loop-filter coefficients according to

$$\frac{1}{R_i C} = f_s k_i, \quad \frac{R_{\text{PI}}}{R_i} = p_i, \quad \frac{g_m}{C} = f_s k_{\text{dac}}, \quad g_m R_{\text{PI}} = p_{i,\text{dac}}. \tag{4}$$

Here, the coefficients k_i model the resistive inputs with their corresponding proportional feed-forwards p_i, and k_{dac} refers to the (current source) feedback with its corresponding feed forward path $p_{i,\text{dac}}$.

Figure 6 shows one possible architectural model which is capable of reproducing the behavior from (2).

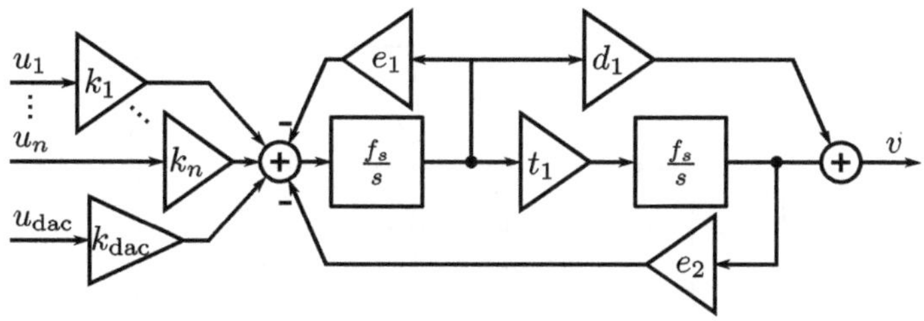

Fig. 6 Model of a non-ideal integrator with proportional feed-through, finite GBW, multiple resistive input paths, and current source feedback DAC

The corresponding coefficients calculate to

$$t_1 = \frac{\omega_{\text{GBW}}}{f_s\left(1+\sum_{i=1}^{n} p_i\right)}, \quad e_1 = \frac{\omega_{\text{GBW}}}{f_s G_{\text{E}}} \frac{1+\frac{G_{\text{E}}}{A_0}\sum_{i=1}^{n} p_i}{1+\sum_{i=1}^{n} p_i}, \quad e_2 = \sum_{i=1}^{n} \frac{k_i}{A_0}, \tag{5}$$

and

$$d_1 = \frac{\omega_{\text{GBW}}}{f_s \sum_{i=1}^{n} k_i} \frac{\sum_{i=1}^{n} p_i}{1+\sum_{i=1}^{n} p_i}. \tag{6}$$

Obviously, having resistive, current source, or capacitive input paths requires changing the model, as does e.g., RC versus GmC integrators, and many more alternatives [29]. But once the modeling is done, these can replace the filter stages in a generic ΣΔ modulator, such as exemplarily shown for a third order system in Fig. 7. This allows a broad range of architectural choices including models for a manifold of circuit implementations. The obtained high-level ΣΔ modulator architecture is then used for transient simulations, for the calculation of transfer characteristics such as STF or NTF, and is the basis for a heuristic search as shown in [29].

The method of modeling circuit non-idealities is intentionally restricted to linear-time-invariant (LTI) behavior in order to allow for a fast simulation technique, which is briefly described in the following subsection. Hence, all LTI circuit non-idealities can be covered and be absorbed into the optimization process. Non-linear non-idealities, even though they could be partially included into the simulator [30], are covered by design or explicitly via compensation or correction techniques as discussed in Sect. 2.

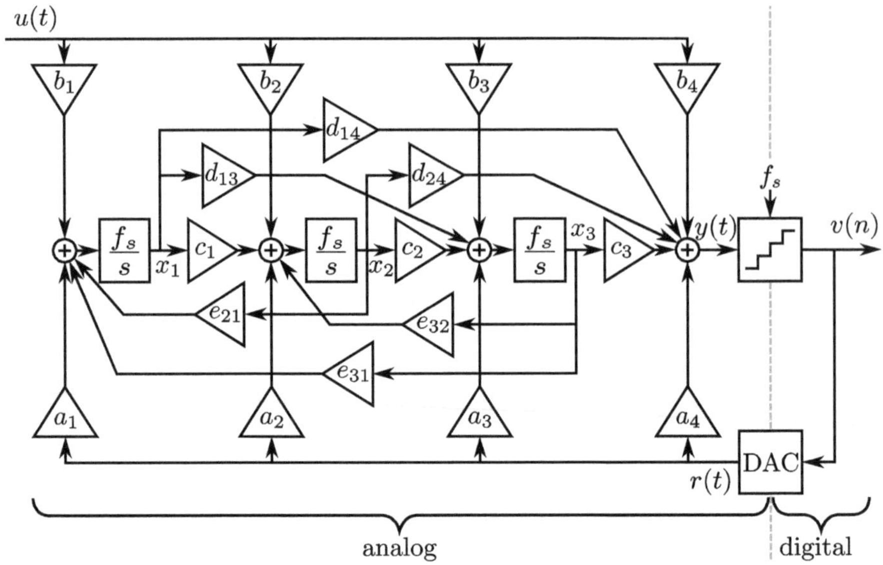

Fig. 7 Utilized generic high-level CT ΣΔ modulator model

3.2 Simulation, Evaluation, and Optimization Technique

The presented synthesis tool incorporates a heuristic search, i.e. a genetic algorithm, in order to find an application specific filter scaling. Figure 8 shows the underlying routine for the automatic filter sizing. After choosing the basic parameters like filter order, OSR, and internal quantizer bit width, the population is randomly initialized. Then, various breakup criteria are checked. In case of a breakup, the population is assumed final and the best individual, i.e. the best performing modulator scaling is returned; in case of no breakup, an iteration of the optimization loop consisting of the creation, simulation, and evaluation of a new generation is executed. Individuals, which outperform their ancestors inside the population, replace these, and another check for breakup is performed.

Using a genetic algorithm implies the requirement for a huge amount, i.e., thousands up to millions of simulations; while, simultaneously, in order to be practicably useful for a design, a short response time is necessary. These highly contrary requirements demand for extremely fast transient simulations of the CT ΣΔ modulator. As a numeric example: a genetic algorithm that executes 10^6 simulations for one optimization, and needs 1 s per simulation, which is an optimistic value for a CT ΣΔ modulator in Simulink, requires an unacceptable 11.6 days to obtain just one architecture scaling. Obviously, an iterative adoption of the result to the designers needs is not possible.

Therefore, a simulation technique called Lifting [9] is adopted, which allows for exact discrete-time simulations of CT ΣΔ modulators; this is done by adding

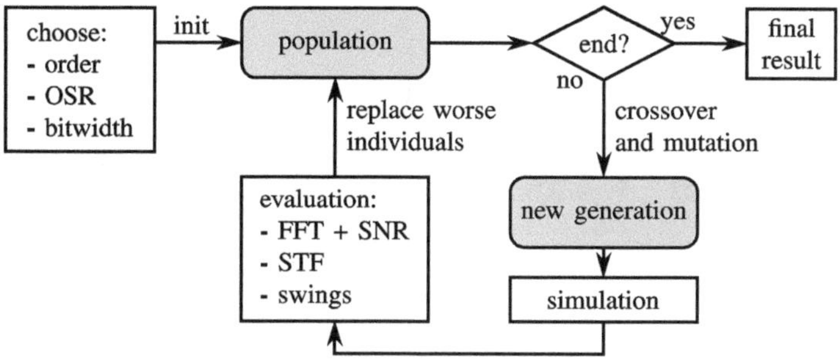

Fig. 8 Automatic sizing routine with genetic algorithm

correction values at all internal state variables of the loop-filter, which account for the changes due to the CT nature of the input and feedback signal. Additionally, hardware acceleration via a high-end consumer graphics card is employed. Both methods speed up a single simulation to an average of only ~10 μs. Thus, one architectural optimization via the genetic algorithm with one million simulations results in only ~10 s, which is short enough to allow an interactive design process, despite the utilization of a heuristic search.

3.3 Synthesis Tool Web Front-End: http://www.sigma-delta.de

In order to provide broad access to the heuristic search in contrast to many priorly published proprietary tools, a web-based front-end has been setup, which is accessible under http://www.sigma-delta.de. On this website the designer can specify all kinds of constraints and optimization goals, which define the allowed search space and the desired convergence direction upon the designers needs.

At first, the basic features of the modulator estimated from (1), such as the filter order, OSR, and internal quantizer bit-width, can be chosen. Additional constraints are e.g., a special filter architecture, ELD, coefficient limits, or also non-idealities such as finite GBW, etc. Moreover, optimization goals can be set such as e.g., maximized SNDR, desired internal signal swing, or the shape of the STF in order to obtain a desired out of band suppression e.g., for interferer suppression.

An example view of the tool is shown in Fig. 9a, where the selected third order mixed feedback/feed-forward structure with multi-bit quantizer can be seen together with finite GBW integrators and a PI-integrator to compensate for ELD. The additional unused paths of the generic structure are masked in gray. On the bottom, an interactive status bar provides information about the made selections, here showing e.g., the integrator settings. Figure 9b shows an exemplary definition of an STF constraint by an inband and out-of-band gain mask denoted by the gray boxes; an exemplary STF obtained after an optimization is shown in red.

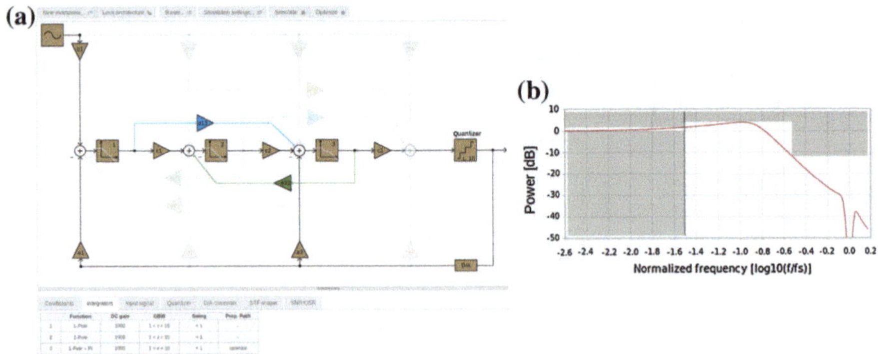

Fig. 9 Web-based CT ΣΔ modulator synthesis tool: http://www.sigma-delta.de. **a** Front-end of the tool, **b** STF specification

4 Design Strategy Example Including the Presented Tool

In this section it is briefly explained how this model based design automation is incorporated into the overall design strategy in order to obtain a high-level modulator scaling, a low-level design and finally the chip implementation. The shown example was the basis of a fabricated and published high-speed ΣΔ modulator [31]. The primary design goal of the genetic algorithm is to maximize the achievable SNR while fulfilling given specifications and constraints concerning structure, device parameters, and integrator characteristics. These specifications were concluded from our previous designs [17, 28] advancing them in terms of sampling and inband frequency:

- $f_s = 1$ GHz @ OSR $= 10$,
- 3rd order FF/FB structure (cf. Fig. 10),
- 4 bit quantizer,
- $\tau_{\text{ELD}} = 0.5\,T_s$ and compensation using proportional integrator (cf. Fig. 5),
- OAs: $GBW_{1,2,3} = 2.0/0.8/1.2$ for $\text{OA}_{1/2/3}$,
- All coefficient values in similar range: 0.5...2.0,
- Input coefficient $k_{\text{in}} = k_{1\text{b}} \geq 0.95$.

The intended speed is basically doubled compared to [28]. The architecture, bit width, etc. is chosen the same. The amplifier GBW values are chosen slightly larger compared to [17] in order to enable less excessive STF peaking, which was almost 20 dB in [17] and [28]. Similar scaling coefficients allow for similar sizing of the passives and thus better matching. The constraint for the input coefficients k_{in} and $k_{1\text{b}}$ originates from a transformation of an RC product into high-level domain, given in (4). R_{in} is a main cause for input referred thermal noise. In our previous designs with $f_B = 25$ MHz bandwidth the input resistor was chosen to be $R_{\text{in}} = 700\,\Omega$; thus, with twice the signal bandwidth, we specify $R_{\text{in}} \simeq 300\,\Omega$.

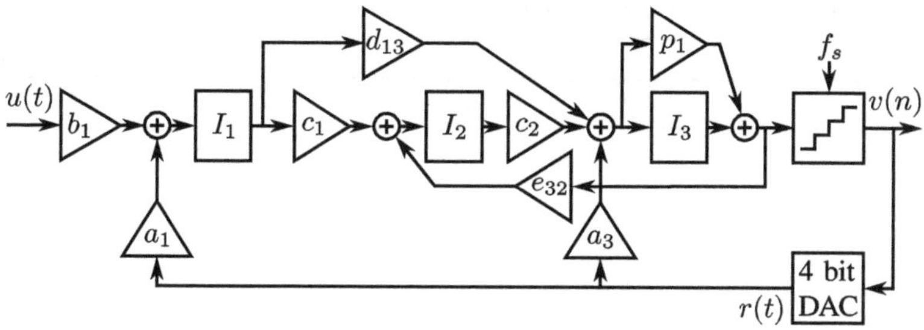

Fig. 10 3rd order, 4b, FF/FB structure with PI ELD compensation

Furthermore, if we want to limit the first integration capacitor to $C_1 \leq 3.5\,\text{pF}$ for power and area reasons, this leads to the above constraint of $k_{\text{in}} = k_{1\text{b}} \geq 0.95$.

The architecture as shown in Fig. 10 was inserted into the web-front end. After the definition of these constraints, optimization goals are inserted:

- maximize SNR,
- limit the output swing of integrator 1/2/3 to 0.4/1.0/1.0 FS,
- STF (lower limit in inband): $\text{STF} \geq -0.1\,\text{dB}$,
- STF (upper limit in dB): $\text{STF}(f) \leq \begin{cases} 7.5: & 0 \leq f/f_s < 0.2 \\ 4.0: & 0.2 \leq f/f_s < 0.3 \\ 0.0: & 0.3 \leq f/f_s < 0.4 \\ -10.0: & 0.4 \leq f/f_s < \infty \end{cases}$.

While the first design goal is always chosen by the genetic algorithm, the others are user specific: considering the low-level design, aiming for low dynamics supports a power efficient realization, good linearity especially in the first integrator, and avoids clipping. Concerning the STF, peaking restrictions can enhance the robustness of the modulator against instability. Aiming for filter roll-off and low or no peaking, is also advantageous in case of large out-of-band interferers, which can be implicitly filtered in a CT ΣΔ modulator, or also if a certain AAF is required. Therefore, in contrast to our previously realized designs with significantly higher STF peaking [17, 28], the specified design goal is now to stay below 7.5 dB peaking and reduce the STF gain in moderate steps at higher frequencies.

When providing these constraints and design goals via the web-interface to the genetic algorithm, a successful coefficient scaling is obtained within several seconds. Different scaling coefficients with similar performance values can result from the same set of constraints and design goals due to the heuristic nature of the utilized search algorithm. After studying the resulting modulator performance, the designer can change or add specifications and constraints until, in an iterative fashion, the required modulator behavior is seen. The coefficient scaling obtained and chosen for the subsequent circuit realization is shown in Table 1.

Table 1 Result of exemplary optimization

$a_1 = b_1$	a_3	c_1	c_2	d_{13}	e_{32}	p_1
0.95	1.85	0.5	1.15	1.02	0.08	0.66

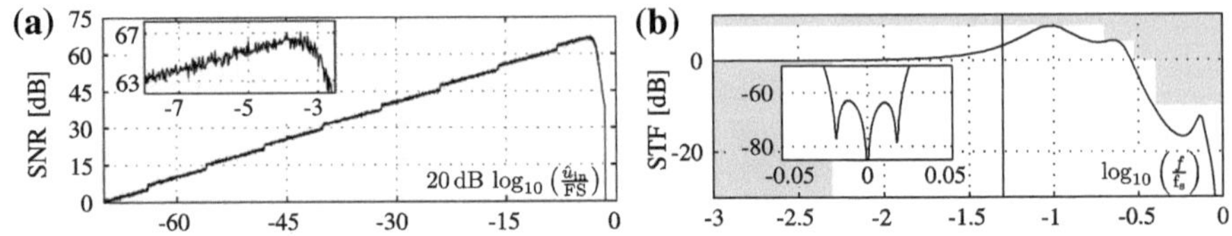

Fig. 11 SNR/SNDR and STF of the modulator found by the online tool [29]. **a** SNR result, **b** STF specification and result

The performance evaluation of the obtained modulator scaling, which is automatically generated after the genetic algorithm outputs the proposed scaling, is shown in Fig. 11. By fulfilling all constraints as, e.g., the STF shape, and using the setup of specifications, e.g., for GBW, ELD, etc., an optimal SNR $\simeq$ 67 dB is achieved.

Figure 12a shows the schematic block diagram of the manually chosen implementation. Here, all R, C and Gm values are directly calculated from the scaling coefficients obtained via the design tool. Part by part, the circuit is then realized on transistor level. In case, the parameters changed slightly, e.g., reduced GBW of the amplifiers due to adding parasitic capacitances due to DAC loading, the tool is used to fine-tune the coefficients in order to compensate for the "new" amplifier specification, rather than changing the just achieved transistor level implementation.

Since DAC non-linearities can not be compensated for in the design process, a linearization must be implemented. Due to the low OSR, dynamic element matching only would yield insufficient linearization. Thus, an explicit background digital linearization has been used for DAC_1 in Fig. 12a. Thereby, all unit element mismatches of DAC_1 are estimated. This is done by feeding a slow, 1-bit PRN test signal $e_t(n)$ into one selected feedback current source at a time; cross-correlation of the output $y(n)$ with the PRN sequence reveals the actual gain of that current source. This is subsequently done for all current sources. By obtaining the individual mismatches, those can either be tuned in the analog domain [28], or corrected for in the digital domain [17]. DAC_2 is explicitly avoided by the mixed FB/FF architecture, since it would otherwise also require linearization, since the low oversampling would not sufficiently suppress its non-linearity.

The measured SNR in Fig. 13a shows that the ideal performance–expected from the tool–only deviates by half a bit from the realized modulator, which is due to circuit noise. The chip photo and measured performance summary are given in Figs. 12b and 13b.

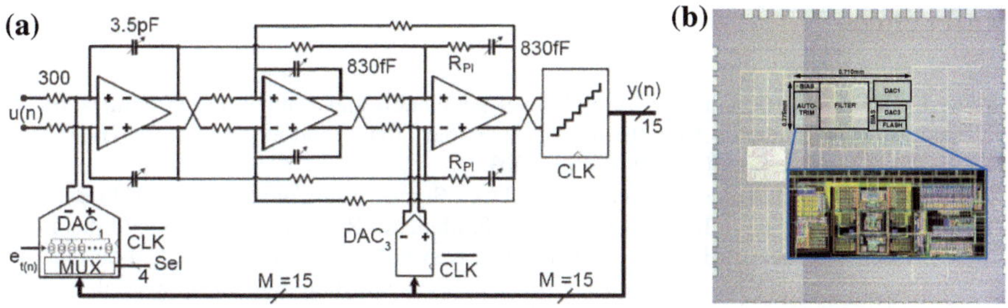

Fig. 12 Realized ΣΔ modulator [31]. **a** Circuit schematic, **b** Chip micrograph

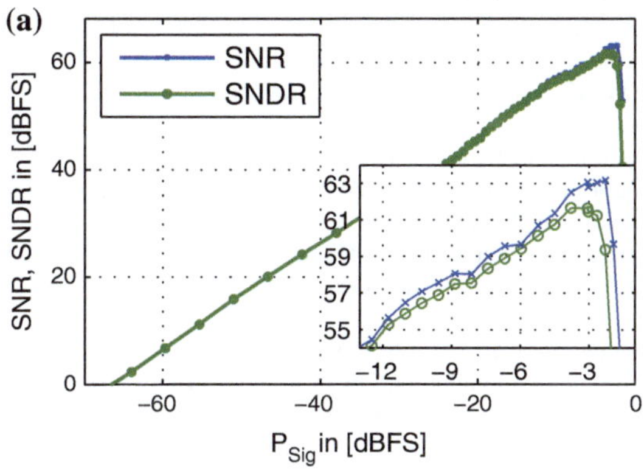

(b)

Process	1.2V 90nm TSMC CMOS
f_S/BW	1GHz/50MHz
SNR	62.75dB
SNDR	61.7dB
SFDR	72.24dB
DR	67dB
P_{Diss}	20.6mW
FOM	207fJ/conv.

Fig. 13 Measured SNR and SNDR, and performance summary [31]

5 Conclusions

A heuristic search based environment for the architectural design of CT ΣΔ modulators has been implemented. Thereby, circuit non-idealities, which can be compensated for by the architectural selection and its scaling, are modeled and the transient simulations are executed on a high-end GPU. The design environment is publicly available under http://www.sigma-delta.de. The almost real-time response of the genetic algorithm, based on hundreds of thousands of individual simulations, allows an iterative design, and even to go back to scaling during transistor level design in order to achieve fine tuning after parasitic extraction. Several successful implementations have been enabled by this design approach, underlining its practical applicability.

References

1. K. Francken and G. Gielen, "A high-level simulation and synthesis environment for ΣΔ modulators," Computer-Aided Design of Integrated Circuits and Systems, IEEE Transactions on, vol. 22, no. 8, pp. 1049–1061, aug. 2003.
2. R. Schreier and G. Temes, Understanding Delta-Sigma Data Converters. IEEE Press, 2005.
3. J. Ruiz-Amaya, J. de la Rosa, F. Fernandez, F. Medeiro, R. del Rio, B. Perez-Verdu, and A. Rodriguez-Vazquez, "High-level synthesis of switched-capacitor, switched-current and continuous-time ΣΔ modulators using SIMULINK-based time-domain behavioral models," Circuits and Systems I: Regular Papers, IEEE Transactions on, vol. 52, no. 9, pp. 1795–1810, sep 2005.
4. M. Ortmanns and F. Gefers, Continuous-Time Sigma-Delta A/D Conversion. Springer Berlin, 2006, vol. 21.
5. R. Schreier, "The sigma-delta toolbox." [Online]. Available: "http://www.mathworks.com/matlabcentral/fileexchange/19-delta-sigma-toolbox"
6. A. Buhmann, M. Keller, M. Maurer, M. Ortmanns, and Y. Manoli, "DISCO - a toolbox for the discrete-time simulation of continuous-time sigma-delta modulators using MATLAB®," in Circuits and Systems, 2007. MWSCAS 2007. 50th Midwest Symposium on, aug. 2007, pp. 1082–1085.
7. J. L. A. de Melo, B. Nowacki, N. Paulino, and J. Goes, "Design methodology for sigma-delta modulators based on a genetic algorithm using hybrid cost functions," in Circuits and Systems (ISCAS), 2012 IEEE International Symposium on, may 2012, pp. 301–304.
8. F. Medeiro, B. Perez-Verdu, A. Rodriguez-Vazquez, and J. Huertas, "A vertically integrated tool for automated design of ΣΔ modulators," Solid-State Circuits, IEEE Journal of, vol. 30, no. 7, pp. 762–772, jul 1995.
9. M. Keller, A. Buhmann, M. Ortmanns, and Y. Manoli, "A method for the discrete-time simulation of continuous-time sigma-delta modulators," in Circuits and Systems, 2007. ISCAS 2007. IEEE International Symposium on, may 2007, pp. 241–244.
10. J. Cherry and W. M. Snelgrove, "Excess loop delay in continuous-time delta-sigma modulators," IEEE Trans. Circuits Syst. II, vol. 46, no. 4, pp. 376–389, April 1999.
11. M. Keller, A. Buhmann, J. Sauerbrey, M. Ortmanns, and Y. Manoli, "A comparative study on Excess-Loop-Delay compensation techniques for Continuous-Time Sigma–Delta modulators," Circuits and Systems I: Regular Papers, IEEE Transactions on, vol. 55, no. 11, pp. 3480–3487, 2008.
12. S. Paton, A. D. Giandomenico, L. Hernandez, A. Wiesbauer, T. Potscher, and M. Clara, "A 70-mW 300-MHz CMOS continuous-time ΣΔ ADC with 15-MHz bandwidth and 11 bits of resolution," IEEE J. Solid-State Circuits, vol. 39, no. 7, pp. 1056–1063, 2004.
13. P. Crombez, G. V. der Plas, M. Steyaert, and J. Craninckx, "A Single-Bit 500 kHz-10 MHz multimode Power-Performance scalable 83-to-67 dB DR CT ΣΔ for SDR in 90 nm digital CMOS," IEEE J. Solid-State Circuits, vol. 45, no. 6, pp. 1159–1171, 2010.
14. G. Mitteregger, C. Ebner, S. Mechnig, T. Blon, C. Holuigue, E. Romani, A. Melodia, and V. Melini, "A 14b 20mW 640 MHz CMOS CT ΣΔ ADC with 20 MHz signal bandwidth and 12b ENOB," in IEEE Int. Solid-State Circuits Conf. ISSCC, 2006, pp. 131–140.
15. M. Vadipour, C. Chen, A. Yazdi, M. Nariman, T. Li, P. Kilcoyne, and H. Darabi, "A 2.1mW/3.2mW delay-compensated GSM/WCDMA ΣΔ analog-digital converter," in VLSI Circuits, 2008 IEEE Symposium on, jun 2008, pp. 180–181.
16. P. Fontaine, A. Mohieldin, and A. Bellaouar, "A low-noise low-voltage ct delta; sigma; modulator with digital compensation of excess loop delay," in Solid-State Circuits Conference, 2005. Digest of Technical Papers. ISSCC. 2005 IEEE International, Feb 2005, pp. 498–613 Vol. 1.
17. J. Kauffman, P. Witte, J. Becker, and M. Ortmanns, "An 8.5 mW continuous-time ΣΔ modulator with 25 MHz bandwidth using digital background DAC linearization to achieve

63.5 dB SNDR and 81 dB SFDR," Solid-State Circuits, IEEE Journal of, vol. 46, no. 12, pp. 2869–2881, dec. 2011.
18. J. Cherry and W. Snelgrove, "Clock jitter and quantizer metastability in continuous-time delta-sigma modulators," Circuits and Systems II: Analog and Digital Signal Processing, IEEE Transactions on, vol. 46, no. 6, pp. 661–676, jun 1999.
19. H. Shamsi and M. Yavari, "On the design of a less jitter sensitive ntf for nrz multi-bit continuous-time ΣΔ modulators," in Circuits and Systems, 2009. ISCAS 2009. IEEE International Symposium on, May 2009, pp. 1553–1556.
20. P. Shettigar and S. Pavan, "A 15mW 3.6GS/s CT delta sigma ADC with 36 MHz bandwidth and 83 dB DR in 90 nm CMOS," in Solid-State Circuits Conference Digest of Technical Papers (ISSCC), 2012 IEEE International, feb. 2012, pp. 156–158.
21. S. Pavan and P. Sankar, "Power reduction in continuous-time delta-sigma modulators using the assisted opamp technique," IEEE J. Solid-State Circuits, vol. 45, no. 7, pp. 1365–1379, 2010.
22. M. Ortmanns and F. Gerfers, Continuous-Time Sigma-Delta A/D Conversion: Fundamentals, Performance Limits and Robust Implementations, 1st ed. 1em plus 0.5em minus 0.4em Springer, Berlin, Dec. 2005.
23. R. Schreier and G. C. Temes, Understanding Delta-Sigma Data Converters. 1em plus 0.5em minus 0.4em Wiley-IEEE Press, Nov. 2004.
24. L. Carley, "A noise-shaping coder topology for 15 + bit converters," Solid-State Circuits, IEEE Journal of, vol. 24, no. 2, pp. 267–273, 1989.
25. S. Yan and E. Sanchez-Sinencio, "A continuous-time sigma-delta modulator with 88-dB dynamic range and 1.1-MHz signal bandwidth," IEEE J. Solid-State Circuits, vol. 39, no. 1, pp. 75–86, 2004.
26. X. Wang, U. Moon, M. Liu, and G. Temes, "Digital correlation technique for the estimation and correction of DAC errors in multibit mash ΣΔ ADCs," in IEEE Int. Symp. on Circuits Syst., vol. 4, 2002.
27. P. Witte and M. Ortmanns, "Background DAC error estimation using a pseudo random noise based correlation technique for Sigma-Delta Analog-to-Digital converters," IEEE Trans. Circuits Syst. I, Reg. Papers, vol. 57, no. 7, pp. 1500–1512, 2010.
28. P. Witte, J. Kauffman, J. Becker, Y. Manoli, and M. Ortmanns, "A 72 dB-DR ΔΣ CT modulator using digitally estimated auxiliary DAC linearization achieving 88fJ/conv in a 25 MHz BW," in Solid-State Circuits Conference Digest of Technical Papers (ISSCC), 2012 IEEE International, feb. 2012, pp. 154–156.
29. T. Brückner, C. Zorn, J. Anders, J. Becker, W. Mathis, and M. Ortmanns, "A GPU-accelerated web-based synthesis tool for CT sigma-delta modulators," Circuits and Systems I: Regular Papers, IEEE Transactions on, vol. PP, no. 99, pp. 1–13, 2014.
30. S. Pavan, "Efficient simulation of weak nonlinearities in continuous-time oversampling converters," Circuits and Systems I: Regular Papers, IEEE Transactions on, vol. 57, no. 8, pp. 1925–1934, Aug 2010.
31. J. Kauffman, C. Chu, J. Becker, and M. Ortmanns, "A 67 dB DR 50 MHz BW CT Delta Sigma Modulator Achieving 207 fJ/conv," in Solid State Circuits Conference (A-SSCC), 2013 IEEE Asian, 2013, p. 401–404.

Recent Advances and Trends in High-Performance Embedded Data Converters

Pedro Figueiredo

Abstract This chapter discusses the architectures of Nyquist rate Analog-to-Digital Converters (ADCs), that are embedded in systems-on-chip (SoCs) implementing some of the most widespread mobile communication, wireless and wireline connectivity standards. The typical requirements for these ADCs are presented, and the main conversion architectures are described in terms of the fundamental operations realized inside them. This constitutes a unified treatment that relates all architectures, thus providing a deeper understanding of their fundamental limitations and trade-offs. We then discuss some of the solutions recently published in the literature to improve ADC energy efficiency. Finally we disclose implementation details from two 12 bit digitally calibrated, high-speed ADCs, using the pipeline and SAR architectures.

1 Introduction

Electronic systems evolve quickly to support the ever-increasing consumer demand for more functionality, higher speed, network access anywhere and longer battery life. This evolution is supported by the continuous scaling of CMOS processes according to Moore's law, where every new process node brings higher speed, lower power consumption, and higher density which leads to reduced transistor cost [1].

To follow these trends, advanced technology nodes are used to implement the systems-on-chip (SoCs) that enable mobile communications and/or wireless connectivity in a diversity of electronic equipment (smartphones, tablets, laptop/desktop PCs, video game consoles, digital cameras, e-book readers, residential

P. Figueiredo (✉)
Synopsys, Lagoas Park – Edifício 4, 2 Piso, 2740-267 Porto-Salvo, Portugal
e-mail: pmff@synopsys.com

P. Harpe et al. (eds.), *High-Performance AD and DA Converters, IC Design in Scaled Technologies, and Time-Domain Signal Processing*,
DOI: 10.1007/978-3-319-07938-7_5,

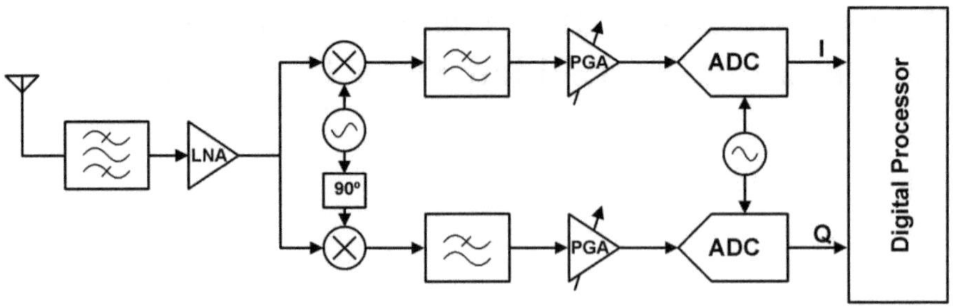

Fig. 1 Block diagram of a direct conversion receiver (architecture most frequently used in WLAN radios [3])

gateways and home entertainment systems, etc.), and wireline communications as well as multimedia applications at home (e.g. set-top-boxes). Note that communication systems in general, and wireless networking in particular, are a strong driver of the integrated circuit industry—for example, more than 2 billion WiFi enabled devices are being shipped annually [2].

Communication SoCs process modulated analog signals, which must be accurately digitized to enable their processing in the digital domain (see Fig. 1). This is done by analog-to-digital converters (ADCs), which are thus an essential component in these SoCs. Naturally, on the transmit path digital-to-analog converters (DACs) produce the signal that after proper filtering, frequency translation and amplification, is transmitted by the antenna.

This chapter discusses the architectures of Nyquist rate ADCs[1] that are currently being embedded in SoCs implementing some of the most widespread mobile communication, wireless and wireline connectivity standards.

More specifically, this chapter focus on embedded ADCs, which differ from standalone ADCs in several aspects [4, 5], but particularly in the choice of technology used to implement them. In standalone ADCs, cost and technical considerations such as the technology maturity, available process options, analog properties of the transistors, supply voltage levels and expertize of the design team, typically lead to choosing a technology that is a few generations behind the state of the art [4]. Typically, in SoCs the vast majority of the area is occupied by digital processing blocks and so more advanced technology nodes are selected [5], as they offer increased speed and lower occupied silicon area (hence lower cost). Embedded data converters must, thus, be designed in those less mature technologies, which typically feature lower supply voltages, less process options, higher variability, more layout dependent effects, higher leakage currents, and worse analog device properties [5–8].

[1] Nyquist rate architectures are those theoretically capable of digitizing signals with a bandwidth up to half of the sampling rate. Sigma-Delta, another major class of converters, handles narrow band input signals and employs oversampling and noise shaping to increase resolution.

In advanced technologies it is increasingly difficult to achieve the necessary ADC performance just by resorting on “analog solutions”. For example the small intrinsic transistor gain ($g_m.r_{ds}$) they present, reduces the achievable DC gain of opamps, thus limiting the performance of ADCs employing amplifiers. In this case the “analog solution” consists on using multi-stage opamps, but this solution is power hungry (and complex if more than two stages are needed).

On the other hand, advanced technologies provide faster devices and increased digital processing power at a lower cost [5]. So this has been motivating the adoption of ADC architectures that are less dependent on transistors' analog properties, and/or that use digital calibration to recover from the drop of performance caused by devices' impairments [5, 9].

Section 2 reviews the typical requirements of ADCs used in some of the most widespread mobile communication, wireless and wireline connectivity systems. This section also discusses how these requirements evolved during the last decade, and the main trends of published ADC performance evolution. Finally we compare the figures of merit (FOMs) of several Synopsys ADC IP products, and relate them with technology and architecture evolutions.

Analog-to-digital conversion solutions are then reviewed. We purposely discuss ADC architectural aspects (fundamental operations and limitations), and the circuit solutions that implement its sub-blocks (along with their design trade-offs) in two separate Sects. 3 and 4). This separation helps distinguishing *architecture shortcomings* from *limitations on the circuit solutions*. This distinction is important, since optimized solutions can only be found by carefully and ingeniously selecting the circuits that best address architecture limitations, while crafting the architecture to ease the trade-offs found in practical circuit solutions.

Section 4 also describes 12 bit digitally calibrated, high-speed, pipeline and SAR ADCs [10] that were implemented in 65 nm, 40 nm and 28 nm CMOS technologies. Finally, conclusions are presented in Sect. 5.

2 ADC Requirements and Evolution

2.1 Requirements for ADCs Used in Communication Systems

The two fundamental parameters of an ADC are the *sampling frequency*, (f_s—rate at which the input is examined, and the corresponding output is produced), and the *resolution* (N—number of output bits). Naturally the requirements for the ADC depend on the system where it will be used, as illustrated in Fig. 2.

Data converters embedded in communication applications are tailored according to needs of the systems implementing each standard. Though any two independently designed SoCs targeting the same standard will necessarily differ, the key performance specifications of data converters are clustered around certain

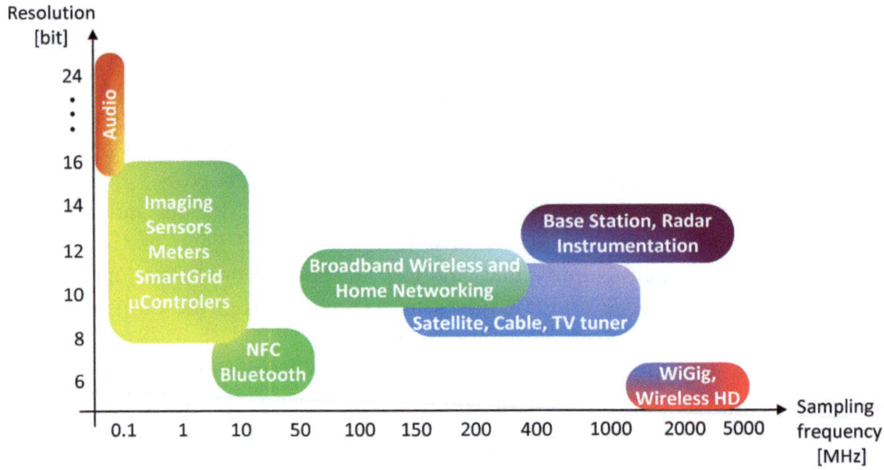

Fig. 2 ADC performance and speed requirements in different applications

values. Each standard defines the RF channel's width, which determines the bandwidth of the signal applied to the ADC. For example if quadrature demodulation is used, the bandwidth of the baseband signal applied to the ADC equals half of that of the RF channel [11]. Then, in order to ease anti-aliasing filtering and improve the signal-to-noise ratio (SNR) by digital filtering [13], the sampling frequency is typically 4 to 8X larger than the bandwidth of the signal applied to the ADC.[2] So the RF channel's bandwidth, defined in each standard, has a strong influence in the choice of ADC sampling frequency.

More complex modulations (e.g. using QAM-64 instead of QAM-16) achieve superior spectral efficiency, but require higher SNR to ensure the same bit error rate (BER). So, this is one of the main factors determining the necessary ADC resolution. As shown in Fig. 1, programmable gain amplifiers (PGAs) precede the ADC to ensure the input signal stays within its input range. However strong out-of-band interferers (the amount by which they are filtered is necessarily limited), fading, and the high peak-to-average ratio that is typical of certain modulation schemes (e.g. OFDM), cause the ADC's average input signal amplitude to be set at a level significantly below its full scale input range [13]. This limitation on signal power must also be considered when determining the necessary ADC resolution that allows achieving the intended BER.

[2] Orthogonal Frequency-Division Multiplexing (OFDM) is used in a variety of systems (802.11, WiMAX, LTE, G.hn, etc.), to cope with severe non-idealities of transmission mediums, such as high frequency attenuation in copper wires or frequency dependent fading due to multipath propagation [3, 12]. In OFDM each channel is composed of a number of closely spaced orthogonal sub-carriers, and the exact ADC sampling frequency is typically a multiple of the sub-carrier separation.

Though some of these factors depend on receiver architectural details, the necessary ADC sampling frequency and resolution typically do not change dramatically between different SoCs that implement the same standard. Table 1 shows the typical ADC specifications for some of the most widespread mobile communication (LTE/LTE-Advanced [14–16]) and wireless connectivity standards (WiFi [3, 17]).

WiMAX is a broadband wireless standard that provides up to 63 Mbps (802.16e) and 1 Gbps (802.16m) over a range of several km [18]. In 802.16e the RF channels have a width up to 20 MHz, which means the baseband signals applied to the I/Q ADCs have a bandwidth not higher than 10 MHz. Typically, ADCs with a resolution up to 10 bit and a sampling rate of 44.8 MHz are employed in the receive channel of these systems. The 802.16m release specifies RF channel bandwidths up to 100 MHz through the aggregation of 5 channels [19, 20], which may or may not be contiguous (this is also done in LTE-A). Depending on how the transceiver is implemented, carrier aggregation may or may not require the ADCs to process more than one channel simultaneously [21–23].

WiGig (802.11ad) uses OFDM to provide up to 6.76 Gbps [24] in indoor environments (the standard also considers a single-carrier, low power mode, that achieves 4.62 Gbps). Signals are transmitted in the 60 GHz frequency band, where there is much more spectrum available than in the 2.4 GHz and 5 GHz bands used in WiFi. As a result of that, and compared to 802.11ac which features similar (maximum) transmission rate, WiGig has much wider RF channels (2.16 GHz instead of 160 MHz), but a simpler modulation scheme (QAM-64 instead of QAM-256) and a single spatial stream instead of 8 [24]. Evolution of WiGig considering a more complex modulation and several spatial streams, should allow reaching much higher data rates in the next few years [25]. Typically 5–6 bit ADCs sampling at several GHz are required for these systems.

Though currently home networks and internet access are mostly supported on wireless technology, it provides limited performance when obstructions are present or when large range is needed. An alternative is transmitting over the existing power line, coaxial or phone line infrastructure. There are a number of standards—e.g. HomeGrid/G.hn, HomePlug, MoCA—that support transmission at different media, with data rates extending from hundreds of Mbps up to 1 Gbps [26–28]. The receive paths of these systems typically employ 12 bit ADCs with sampling frequencies in the 200–500 MHz range. The sampling frequency required in these systems is typically higher than in mobile applications, but the extra power dissipation is less of an issue since these equipment are plugged.

2.2 *Evolution of ADC Performance*

With the exception of WiGig, the requirements for ADCs embedded in SoCs implementing the above mentioned mobile communication, wireless and wireline connectivity standards are not very different. Furthermore since standards have

Table 1 ADC requirements for common mobile communication standards

Application	WiFi 802.11			LTE	LTE-A
	a/b/g	n	ac		
Date of release	1999–2003	2009	2012	2007	2011
Peak data rate [Mbps] (downlink)	54/11/54	600	6933	300	1000
Frequency bands [GHz]	5/2.4/2.4	2.4, 5	5	0.7–2.7	(country dependent)
RF channel width [MHz]	20	40	80[a]	20	20–100[b]
ADC input signal bandwidth [MHz]	10	20	40	10	10–50
ADC sampling rate [MHz]	40/80	80	160	61.44	61.44/122.88
Resolution [bit]	10	10–12	12	10–12	12

[a] It can optionally be 160 MHz, in which case the ADC sampling rate should be 2X higher
[b] Through carrier aggregation

long lifetimes, the necessary ADC performance does not change over short periods of time.

As a result of this, the teams developing these ADCs are typically focused on power and area optimization—which are paramount to increase battery life and reduce die costs—rather than in achieving higher dynamic range and sampling frequency, since that is not (immediately) needed. Note that this optimization process may not be straightforward, as it may imply significant changes to the original design. Furthermore, when the ADC needs to be implemented in a more advanced technology node, the reduced supply voltage or the degradation on the analog characteristics of the transistors may force a somewhat profound architecture change.

Naturally, from time to time new or revised standards are released that target higher data rates, typically by resorting on increased channel bandwidth and a more complex digital modulation (to improve spectral efficiency). This obviously requires faster and higher resolution ADCs, but the difference may not be dramatic—the discussion and approval of a new standard considers what is feasible with current technologies, given the acceptable power dissipation for the complete system. In fact, as illustrated in Table 1, the necessary ADC performance for mobile communication and wireless connectivity standards has not increased dramatically in the last decade. In any way, once a new standard is released, the more aggressive performance can typically be fulfilled with non-optimal solutions in terms of power consumption and occupied die area, which are later subject of optimization, as discussed above.

The larger pressure to increase energy efficiency rather than in improving performance (i.e. achieve high SNR or spurious-free dynamic range, SFDR) has been identified as one of the drivers for the observed trends of ADC performance. Several surveys and analysis of trends have been published in [29–31], and [4] presents a good overview of all of them, along with some additional considerations. Here, we will simply highlight some of the critical conclusions. ADC energy efficiency has been improving consistently and rapidly (about 2X in every 2 years), whereas the improvement of dynamic performance has been irregular and characterized by sporadic advances. A factor contributing to this situation is that only a few applications push for higher performance (e.g. software defined radio) [29], but the pressure to reduce costs (lower area) and optimize efficiency to extend battery life is enormous. Thus commercial motivations end up influencing overall ADC performance evolution and research efforts [4].

Another factor is that high-resolution ADCs are noise limited, whereas the architectures typically employed in high-speed converters rely on the speed capabilities of the transistors. This means technology scaling has a more profound (positive) impact on the performance of medium- or low-resolution/high-speed ADCs. In fact their energy efficiency doubles every 1.6 years, while it takes about 5.4 years to see the same energy efficiency improvement in high-resolution ADCs [4].

Figure 3 shows the energy efficiency evolution for some of the Nyquist rate Synopsys ADC IP products, all of them designed with core devices in 65, 40 and 28 nm CMOS technologies. The comparison is based on Walden's figure of merit, $FOM = P/(2^{ENOB} f_s)$, where P is the total power dissipation, f_s is the sampling frequency and ENOB is the effective number of bits featured by the ADC [32]. The figure shows the FOMs normalized to that of the 10 bit 80 MS/s pipeline ADC fabricated in 65 nm. That converter is used as reference because it relied on pure analog techniques to achieve the required performance.

A 12 bit 200MS/s pipeline ADC was later developed in the same 65 nm technology, employing a gain and capacitor mismatch digital calibration, to enable the use of very simple amplifiers. This implementation is described in Sect. 4.2. A power efficiency improvement of 2.5X was obtained which is quite significant, particularly considering that both the resolution and sampling frequency were increased. Most probably a 12 bit 200 MHz ADC that relied completely on analog techniques would feature a FOM considerably worse than that of our 10 bit reference ADC.

For reasons that will later become clear, it is easier to take advantage of technology scaling in SAR ADCs than in pipeline converters. Thus a new product line was conceived around a 12 bit 80 MS/s SAR ADC, which was time interleaved to yield 12 bit ADC products at 160 MS/s and 320 MS/s. The power efficiency improvement between 40 nm and 28 nm implementations is evident. Some details about this implementation are provided in Sect. 4.3.

2.3 Other Requirements for Embedded ADCs

In addition to achieving previously mentioned performance values, and featuring low power and area, embedded ADCs must also:

- *Maintain performance if temperature and supply voltage changes*: if digital calibration is used and is only done at startup, ADC performance may later drop when temperature or supply voltage changes [33]. So either the system

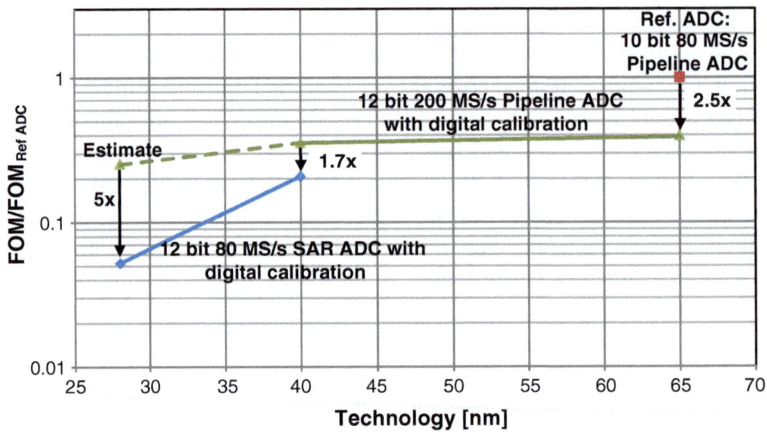

Fig. 3 FOM evolution in some of Synopsys' Nyquist rate ADCs

allows the ADC to be put offline from time to time to be re-calibrated, or there needs to be a background calibration mechanism (preferable). Another possibility is to choose conversion architectures whose analog limitations, addressed by calibration, are not supply voltage or temperature dependent [34]—the 12 bit SAR ADC discussed in Sect. 4.3 is an example.

- *Include a standby mode*: in some communication systems, ADCs only need to be on from time to time, after which they go into a low power mode. Frequently, this low power mode cannot be a complete shutdown due to startup time limitations (see next point).
- *Feature a fast "startup time" (from complete power-down) and even faster "standby-to-normal operation time"*: when the ADC is turned on from complete power-down, bias current and reference voltage generators, as well as switched capacitor circuits must startup and, possibly, digital calibration must be performed. All this can take a significant amount of time, and so typically there is a standby mode, whose transition to normal operation takes much less time. Still, it is desirable that the ADC does not take too much time during power on as this may, for example, impose serious limitations in production testing [5]. This may not be straightforward to achieve if correlation based digital calibration algorithms are used.
- *Scale power dissipation with sampling frequency*: possible solutions are to avoid using circuits having static consumption, or to bias internal circuits with a frequency dependent current.
- *Avoid strict clock duty cycle requirements*: for example in pipeline ADCs the outputs of some stages settle when clk $= 0$ while others settle when clk $= 1$. If clock duty cycle deviates from 50 %, the settling time of some of them decreases, affecting overall ADC performance; a possible solution is the use of a clock duty cycle corrector.

- In recently published SAR ADCs the clock is required to have a small duty cycle [35, 36]. This is not convenient, and it is preferable to conceive the architecture to avoid this limitation, as discussed in Sect. 4.3.
- *Use as few external components as possible*: typically only decoupling capacitors should be required.
- *Use as few pads/bondings/pins as possible*: in high-speed ADCs a significant number of these may be needed, since bondwire inductance on the supplies, inputs and references degrades performance. Internal decoupling strategies should be used to minimize the number of pads/bondings/pins, but cannot occupy a significant amount of die area.

3 ADC Architectures

Existing ADC architectures are typically presented as different and somewhat unrelated alternative solutions, each with its own pros and cons, each best suited to certain resolution and sampling frequency range (see Fig. 4). An objective of this section is to show how they relate to one another. This is done by describing ADC architectures in terms of the fundamental operations realized inside them. This focus on the *operations* allows highlighting the fundamental limitations and trade-offs of different conversion architectures, and provides a unified treatment that relates all of them.

On the other hand, architecture *shortcomings* are sometimes confused to *limitations on the circuit solutions* used in typical implementations. For example one of the drawbacks attributed to the flash architecture is the static current drawn by its resistive reference ladder. However, we will see that the resistive ladder can be regarded as DAC that provides all its 2^N outputs simultaneously. The above mentioned power inefficiency thus arises from the way this DAC is *implemented* rather than being a fundamental limitation of the flash architecture. Indeed, there are flash ADC implementations without a resistive ladder [37–39].

The discussion of non-idealities on different architectures will focus on those causing static deviations to the ADC transfer function. Discussing also noise, limited bandwidth/settling speed, memory effects, etc., would require a much longer text.

3.1 Single and Multi-Bit ADC Architectures

An ADC executes two fundamental operations, *sampling* and *quantization*, which happen in this order. The first is implemented by the sample-and-hold (S/H), and so the following figures will always show v_I connecting to it. (As discussed later, the S/Hs are also used to allow pipelining in multi-step ADCs.) *Quantization* is

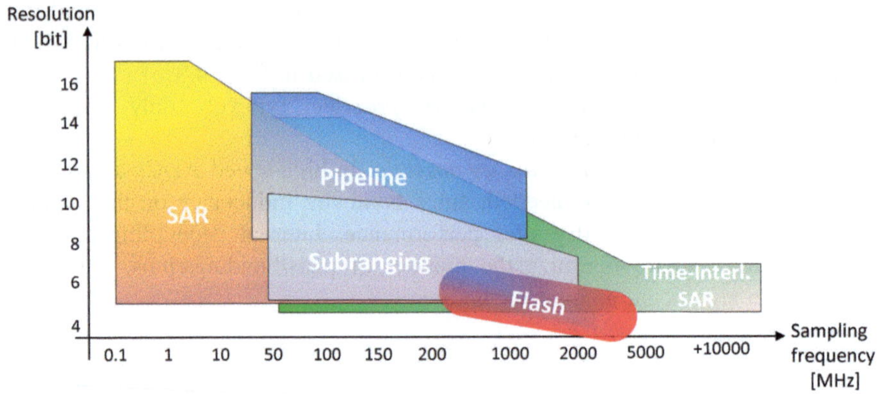

Fig. 4 ADC architectures versus resolution versus sampling frequency

implemented by the *comparator,* whose digital output indicates the polarity of its input signal (as discussed below, in multi-bit ADCs, a DAC and a subtractor are also part of the quantizer). The full digital output of any ADC is always derived from the outputs of its comparators.

Figure 5a illustrates the simplest ADC (1 bit): the S/H implements the sampling operation and the comparator performs quantization.[3] Ideally the output code of this ADC changes from 0 to 1 at $v_I = 0$. As shown in Fig. 5b offset voltages of the S/H and comparator cause that transition to occur at a different voltage (this is the only non-ideality that can defined for a 1 bit ADC[4]). Note that the gain error and non-linearity of the S/H are irrelevant, because the comparator is only sensitive to the *sign* of S/H's output voltage.

Figure 6 shows the core operations found inside a multi-bit ADC. Compared to the 1 bit example just discussed, there is now a subtractor between the S/H and the comparator, which subtracts the sampled input from an estimation of it provided by the DAC. The comparator determines if the output voltage of the subtractor, i.e.

[3] Note that actual 1 bit ADC implementations may not include a dedicated S/H circuit. However the sampling operation happens anyway, because when the comparator is triggered it decides based on the input signal value at *that* time. (There are publications, mainly in the 80's, where no dedicated S/H exists in front of flash ADCs [40, 41].) So the presence of the S/H in Fig. 5 is meant to imply that this operation is performed, and *not* that there is necessarily a dedicated circuit implementing it. Similar considerations apply to the discussion in the remainder of this section. For example [42] does not use dedicated comparator circuits, but the *comparison operations* still take place.

[4] We are referring to those affecting the ADC static transfer function. There is also noise, limited bandwidth, memory effects, etc., but as previously mentioned those are beyond the scope of this text.

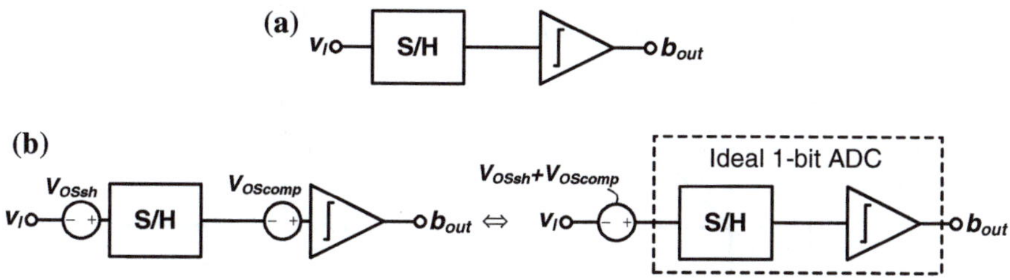

Fig. 5 1-bit ADC: **a** ideal, **b** effect of offsets

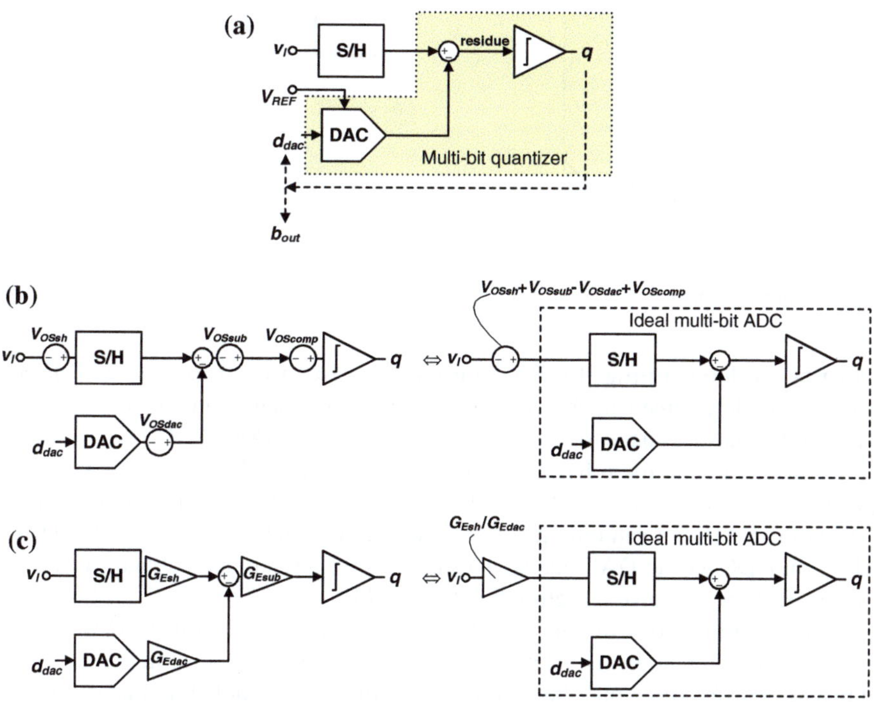

Fig. 6 Multi-bit ADC: **a** Ideal. **b** Effect of offsets. **c** Effect of gain and non-linear errors (G_{Esh}, G_{Esub}, G_{Edac} equal to 1 means "no error")

the *residue,*[5] is positive or negative (1 bit quantization). **The ADC conversion result, b_{out}, corresponds to the input code of the DAC that minimizes the residue. So, fundamentally, the conversion process on a multi-bit ADC**

---

5 The *residue* is the error corresponding to the difference between the input signal and the result from the quantization performed so far, which is provided by the DAC. Figure 6a shows the residue being applied to a single-bit quantizer (comparator), but it may be applied to multi-bit quantizers (see Sect. 3.4.).

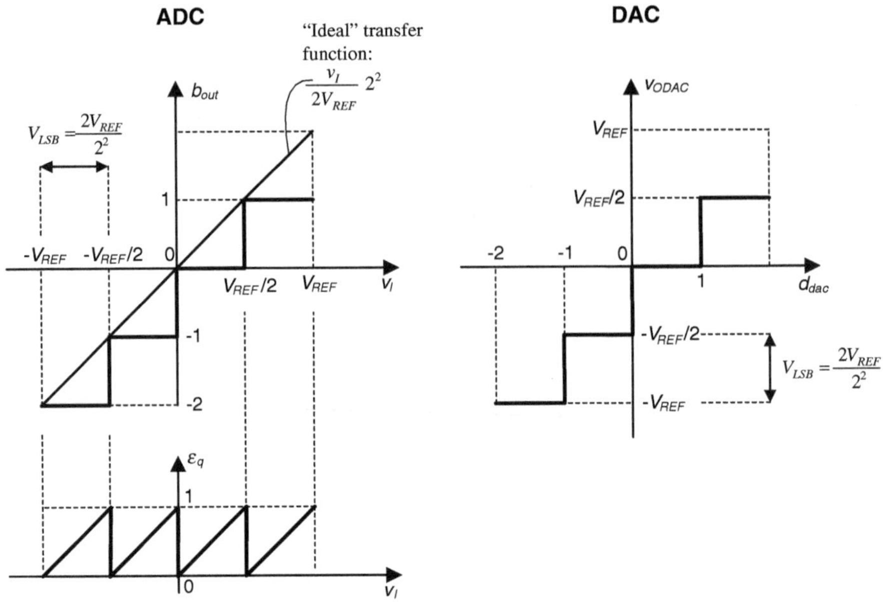

Fig. 7 Transfer function of a 2-bit ADC and DAC

consists on determining the DAC output nearest to v_I. (Later we will discuss *how* b_{out} is determined in the two most direct implementations of this structure: flash ADC and the successive-approximation-register, SAR, ADC.)

Finally, note that the reference voltage, V_{REF}, is applied to the DAC: it sets the maximum output voltage of the DAC, and therefore defines the full-scale input range of the ADC (it can only process signals in the range generated by the DAC). The 1 bit ADC has no DAC and so does not receive a reference voltage. As a result of this, it does not have a well-defined gain [43].

Static non-idealities of linear circuits are decomposed in offset, (linear) gain errors and non-linear errors. ADCs and DACs are inherently non-linear due to their staircase transfer functions, but can be approximately modeled as linear systems. Considering the input signal, v_I, varies between $-V_{REF}$ and $+V_{REF}$, the digital output of an N-bit ADC, which ranges from -2^{N-1} to $+2^{N-1}-1$, is given by

$$b_{out} = \frac{v_I}{2V_{REF}}2^N - \varepsilon_q \tag{1}$$

where ε_q is the quantization error, that ideally varies between 0 and 1 (see Fig. 7). The output voltage of a DAC is related to its input code, d_{dac}, by

$$v_{ODAC} = \frac{d_{dac}}{2^N}2V_{REF}. \tag{2}$$

The DAC does not perform quantization since its input is already quantized, and so (2) does not have a quantization error term. The width of the quantization steps is $V_{LSB} = \frac{2V_{REF}}{2^N}$.

We will now determine the effects caused by non-idealities of the ADC sub-blocks. Figure 6b shows that the offsets of the S/H, DAC, subtractor and comparator are equivalent to an offset of the input signal. (Throughout this manuscript we consider that the S/H and subtractor have a unitary nominal gain). Figure 6c shows that the gain errors of the DAC and S/H introduce a gain error on the overall transfer function of the ADC. In fact, the output of the comparator, q, is related to its input voltage, v_{Icomp}, by the Heaviside function, $q = u[v_{Icomp}]$, and so

$$\begin{aligned} q &= u[G_{Esub}(G_{Esh}v_I - G_{Edac}v_{ODAC})] = u[G_{Esh}v_I - G_{Edac}v_{ODAC}] \\ &= u\left[G_{Edac}\left(\frac{G_{Esh}}{G_{Edac}}v_I - v_{ODAC}\right)\right] = u\left[\frac{G_{Esh}}{G_{Edac}}v_I - v_{ODAC}\right] \end{aligned} \tag{3}$$

where G_{Esh}, G_{Esub}, and G_{Edac} are the gain errors of the S/H, subtractor and DAC, v_I is the input voltage of the ADC and v_{ODAC} is the output of the ideal DAC. Note the gain error of the subtractor is irrelevant, which is expected because the comparator is only sensitive to the sign of its output voltage.

In communication systems, offset and gain errors on the overall ADC transfer function are not critical because no harmonics of the input signal are generated [4]. The same is not true with the non-linearity of the S/H and DAC transfer functions, which can modeled as input dependent gain errors: $G_{Esh}(v_I)$, $G_{Edac}(d_{dac})$. The equivalence shown in Fig. 6c still holds, which means it is as if the input signal is pre-distorted by a non-linear block having a gain $G_{Esh}(v_I)/G_{Edac}(d_{dac})$. Sources of S/H non-linearity depend on implementation details, but are typically: signal dependent charge injection in MOS switches, tracking non-linearity due to signal dependent resistance of sampling switches and non-linearity of the active amplifier/buffer (in case there is one). The non-linearity of the DAC is mainly determined by matching of internal unit elements (capacitors and resistors, in the vast majority of the DACs implemented inside ADCs).

In Fig. 6c and in (3) the gain errors/non-linearity of the S/H and DAC were referred to the ADC input. Figure 8 shows it is also possible to refer them to the input of the DAC. As mentioned above, the conversion process occurring inside the ADC consists on finding the d_{dac} which corresponds to the DAC output that best approaches the sampled input value. However, now d_{dac} is not a good digital representation of the input signal due to the non-linear gain $G_{Edac}(d_{dac})/G_{Esh}(v_I)$, which means the ADC presents distortion if we use $b_{out} = d_{dac}$. Nonetheless, the fictitious signal d_{dac_ideal} is a faithful representation of the input signal. So, by obtaining estimates for the S/H and DAC non-linearity, $\tilde{G}_{Esh}$ and $\tilde{G}_{Edac}$, it is possible to correct the digital output, as shown in Fig. 8. This was done in [44] to correct S/H linearity, and Sect. 4.3 describes an implementation where this principle is applied to correct DAC non-linearity in SAR ADCs.

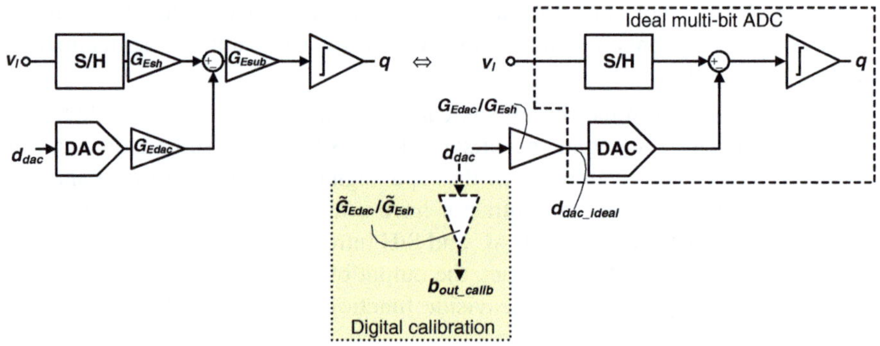

Fig. 8 Digital calibration of gain error and non-linearity of the S/H and DAC

We will now justify analytically this DAC calibration strategy. Let us consider the case depicted in Fig. 6c, with only the DAC nonlinearity present (i.e. $G_{Esh} = 1$, and G_{Esub} is irrelevant as previously discussed). Using (2) allows writing the output voltage of the DAC in this situation,

$$v_{ODAC} = \frac{G_{Edac}(d_{dac})d_{dac}}{2^N} 2V_{REF}. \tag{4}$$

On the other hand, at the end of the conversion v_{DAC} equals v_I, apart from a small error, ε_V, (due to the discrete nature of its output),

$$v_{ODAC} = v_I - \varepsilon_V. \tag{5}$$

Applying (4) in this case, suggests that a DAC output ε_V corresponds to an input $G_{Edac}(d_{dac})\frac{\varepsilon_q}{2^N}2V_{REF}$, where ε_q is the quantization step of the ADC. Substituting this in (5), and equating (4) to (5) leads to

$$d_{dac} = \frac{v_I/G_{Edac}(d_{dac})}{2V_{REF}} 2^N - \varepsilon_q. \tag{6}$$

Comparing (6) to (1) allows concluding it is as if the ADC is converting an input signal $v_I/G_{Edac}(d_{dac})$, which agrees with the equivalence shown on Fig. 6c. Performing the digital calibration as depicted in Fig. 8 leads to (assuming the DAC nonlinearity is correctly estimated, $\tilde{G}_{Edac}(d_{dac}) = G_{Edac}(d_{dac})$)

$$b_{out\_calib} = G_{Edac}(d_{dac})d_{dac} = \frac{v_I}{2V_{REF}} 2^N - G_{Edac}(d_{dac})\varepsilon_q, \tag{7}$$

which equals the ideal ADC output, (1), apart from the shaping of the quantization error by $G_{Edac}(d_{dac})$ function. In case of small errors $G_{Edac}(d_{dac})$ is near 1, so this shaping is not problematic.

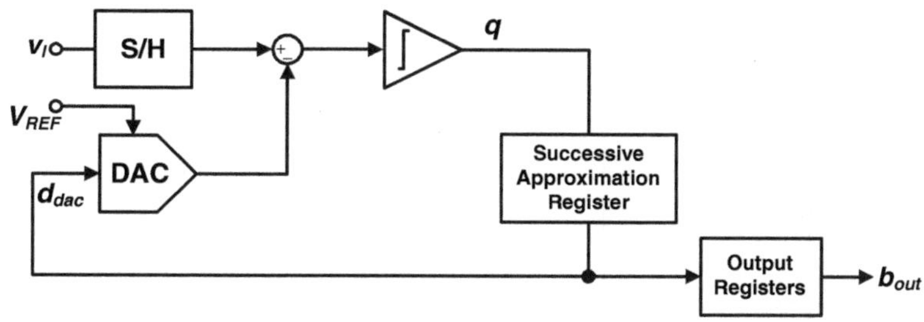

Fig. 9 Successive-approximation-register (SAR) ADC

3.2 Direct Implementations of the Multi-Bit Architecture: SAR and Flash ADCs

The SAR ADC, shown in Fig. 9, is the direct implementation of the multi-bit architecture represented in Fig. 6a. It uses a successive approximation algorithm to determine the N bits of d_{dac}, in N cycles. In each cycle, a bit of d_{dac} is set to 1 (starting from the MSB), and the comparator indicates if the corresponding output of the DAC is above or below the sampled input, thus determining if that bit should remain at 1 or be set to 0. The d_{dac} found at the end of this process, is then output as b_{out}. This architecture is very efficient as it reuses the same hardware in each cycle, and the necessary number of cycles grows linearly with resolution.

The analysis of the effect caused by offsets, gain errors and non-linearities just presented (Figs. 6, 8 and associated text) applies directly to the SAR ADC.

Figure 10a depicts the flash ADC in the way it is typically presented. It is also an implementation of the multi-bit architecture shown in Fig. 6a, but since this is less obvious let us start by considering Fig. 11a, which shows a resistive ladder based DAC. The resistive ladder generates all possible DAC output levels and the switches steer the desired voltage to the output, according to the input bits. *So, the resistive ladder alone can be regarded as a DAC that provides all outputs simultaneously* (see Fig. 11b). This allows representing the flash ADC as shown in Fig. 10b, which highlights the fact that it is an implementation of the multi-bit architecture shown in Fig. 6a at which the input signal is compared to all DAC outputs simultaneously. The output of the comparators (thermometer code) indicates which of the DAC output voltages is nearest to the input, v_I. In this way, no feedback is needed between the output of the comparator and the DAC.

The impact of offsets, gain error and non-linearity of the sub-blocks is illustrated in Fig. 12. Note that the offset of the DAC deviates all outputs by the same amount, V_{OSdac}, whereas its gain error causes all of them to be multiplied by the same factor, G_{Edac}. The DAC non-linearity is modeled with an output dependent gain factor,[6] similarly to what was done in Fig. 6c. The effects caused by the non-idealities of the

[6] Alternatively, an output dependent offset could be considered, which would add to the offsets of the subtractors/comparators.

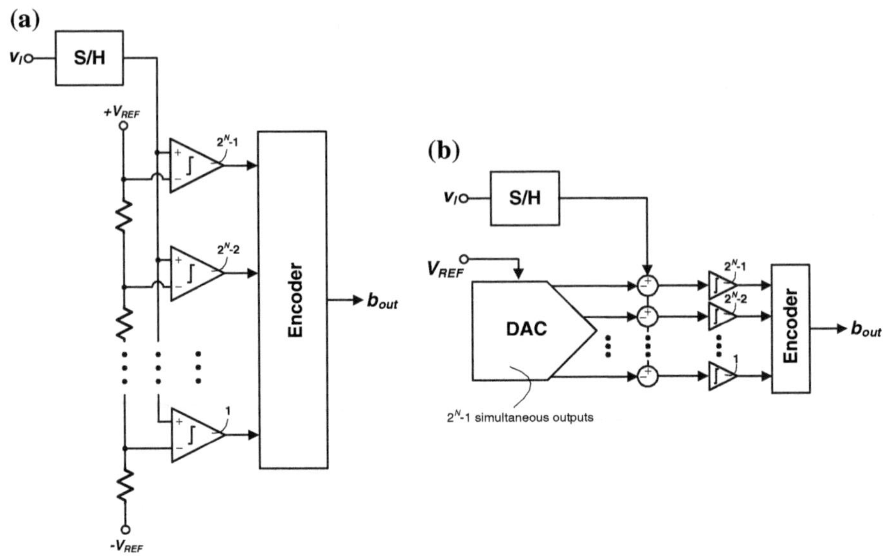

Fig. 10 Flash ADC: **a** Typical implementation. **b** Block diagram showing it is an implementation of the structure shown in Fig. 6a

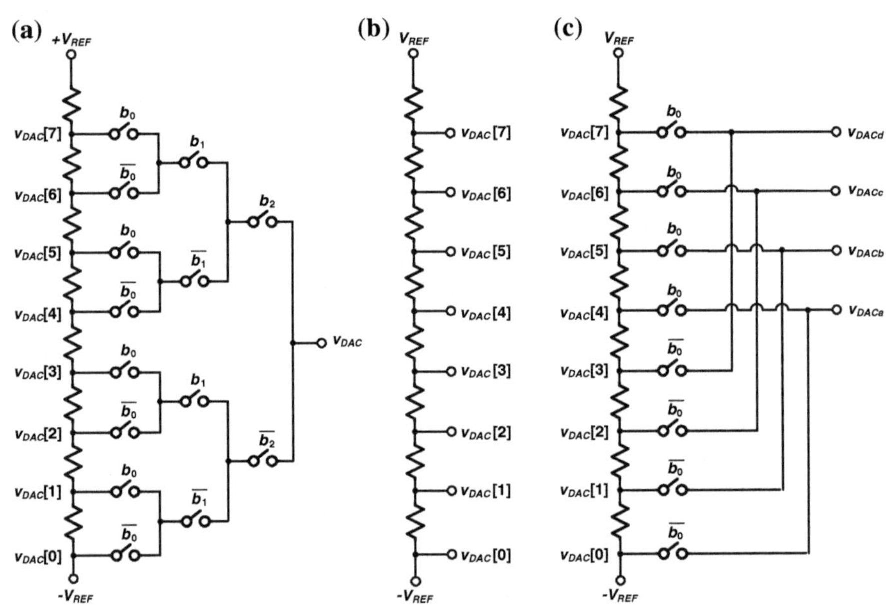

Fig. 11 **a** Resistive ladder based DAC. **b** Resistive ladder as a DAC that provides all outputs simultaneously. **c** Resistive ladder as a DAC that provides groups of outputs

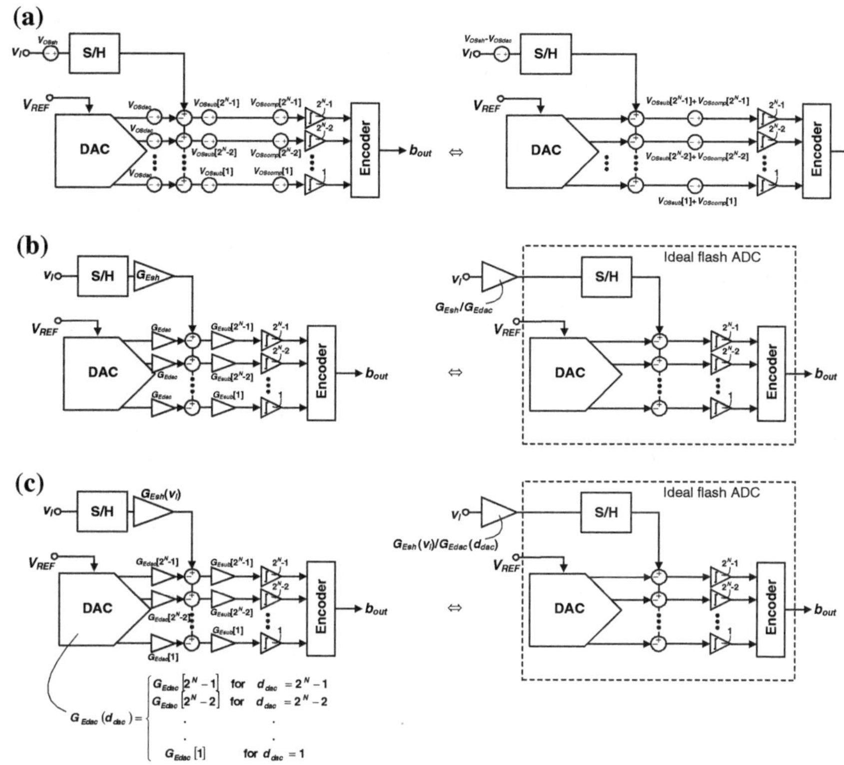

Fig. 12 Flash ADC: **a** Effect of offsets.**b** Effect of gain errors. **c** Effect of non-linearities

S/H and DAC in the ADC transfer function are those previously discussed when the general multi-bit ADC structure was introduced (Figs. 6, 8 and associated text).

However, since each output code is now determined by different subtractors/ comparators, their offsets generate different deviations on the ADC code transition levels, causing distortion (and not a harmless shift on the overall transfer function, as in the SAR ADC). So, to ensure the deviations on the code transition levels of the ADC are below 1 LSB, the offsets of the subtractors/comparators must all be lower than that value [32]. Since the number of internal elements increases exponentially with resolution, in practice flash ADCs are only used up to around 6 bit. It is not difficult to ensure enough linearity for S/Hs and DACs at this resolution level, so the offsets of subtractors/comparators pose the main linearity limitation on this ADC architecture.

The flash resorts to parallel processing to obtain the output code, while the SAR uses a serial process, whereby bits are determined one by one. Naturally, for a given resolution the SAR ADC is more power and area efficient, since the same hardware is used to determine all bits, whereas in the flash a different subtractor and comparator determines each output code transition ($2^N - 1$ are needed for an N bit ADC). However, the flash is faster since it completes the conversion in a single cycle—in fact it is frequently considered the fastest analog-to-digital conversion architecture. The SAR needs several cycles (and in each of them d_{dac} needs to be set, the output signal of the DAC must settle, and finally the comparator takes a decision), and so it has historically been relegated to low speed applications. Even so, the increased speed provided by advanced technologies, along with a number of techniques developed in the last few years, allowed implementing high speed SAR converters: e.g. the implementations described in [36, 45–47] featured $\sim$6 bit ENOB (effective number of bits) at 1.2 GS/s, >10 bit ENOB at near 100 MS/s, and >11.5 bit ENOB at 40MS/s ([47] uses 2X time interleaving to achieve 80MS/s, so each SAR operates at 40MS/s).

Furthermore, using SAR ADCs in time-interleaving allows overcoming their inherent speed limitations, without an excessive increase of complexity. For example, an 8-bit ADC may be constituted by 9 time-interleaved SAR ADCs, where, at a certain clock cycle, one is sampling, another is ending a conversion, and the others are converting different samples. In this way, in each cycle the input signal is sampled and a conversion result is provided, just like in a flash ADC (there now is latency, though). However, this solution is more power efficient than a full flash ADC, which would require 255 comparators and subtractors, instead of 9.

The recent publication [48] of a 64X time-interleaved SAR ADC sampling at 90 GHz and featuring up to 36 dB of signal-to-noise and distortion ratio (SNDR)—it is the highest sampling rate published so far with this level of performance—demonstrates that it is a viable and power efficient high-speed analog-to-digital conversion solution. A few other very high speed, time interleaved SAR ADCs were described, e.g. [49–51]. This certainly defies the argument that a flash ADC is the fastest analog-to-digital conversion architecture.

In fact, we have shown that the SAR ADC is the direct implementation of the basic multi-bit architecture shown in Fig. 6a, but takes several cycles to complete the decision. **This limitation is overcome by using parallelization, which is what both the time-interleaved SAR (time parallelization—*several samples are processed in parallel*) and the flash ADCs (parallelization in the code-searching process—*all codes are searched in parallel*) do**. However, parallelization always comes with a price: as previously mentioned, in the flash ADC the offsets of subtractors/comparators introduce non-linearity in the ADC transfer function; in time interleaved ADCs differences of offset, gain, and sampling time between the converters introduce spurs at the output spectrum [52].

Note that time-interleaving can be used with any ADC architecture, but it is advantageous to choose one that is both power and area efficient. The SAR ADC is a natural candidate. The Sect. 3.3 discusses an architecture (cyclic subranging) that bridges between the flash and the SAR, and allows trading between the two kinds of parallelization mentioned above: *time-parallelization* (maximum in the time-interleaved SAR) and *parallelization in the code-searching process* (maximum in the flash ADC).

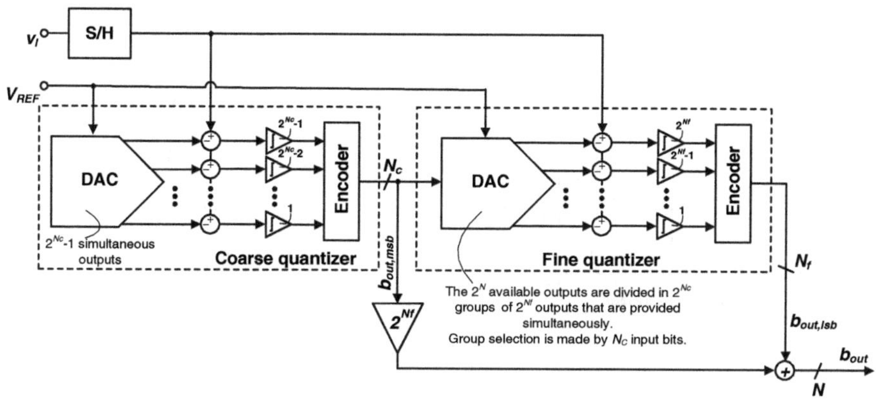

Fig. 13 Two-step subranging ADC using flash ADCs as quantizers

3.3 Multi-step Subranging ADC Architectures

A subranging ADC is constituted by several quantizers (aggregation of DAC + subtractors + comparators) that successively refine the conversion, i.e. find a DAC output that is nearer the input signal. Figure 13 shows the most common implementation (two-step subranging), that has two quantizers implemented as flash ADCs. The first one performs a coarse quantization (N_c bits), indicating a (somewhat broad) range where the input signal lies; then the DAC of the fine flash ADC (N_f bits) provides output voltages in that range, which are compared to the sampled input, thereby refining the range indicated by the coarse ADC. The DAC in the coarse quantizer only provides $2^{N_c} - 1$ outputs, but the DAC in the fine quantizer needs to provide all the 2^N outputs, that are arranged in 2^{N_c} groups of 2^{N_f}outputs. Figure 11c shows an example of such a DAC, for $N = 3$, $N_f = 2$ and $N_c = 1$.

The LSB of the coarse quantizer corresponds to 2^{N_f} LSBs of the fine quantizer, and so to obtain b_{out} the digital output from the former must be multiplied by 2^{N_f}.

The subranging ADC shown in Fig. 13 needs two cycles to complete the conversion, but uses much less subtractors/comparators than the full flash ADC ($2^{N_c} + 2^{N_f} - 1$ instead of $2^N - 1$): here it is, again, the trade-off between sampling speed and parallelism. As in the SAR ADC case discussed in the previous subsection, another kind of parallelism (time-interleaving) can be used to address this limitation. Figure 14 shows an implementation with a coarse ADC and two time-interleaved fine ADCs. In each clock cycle one of the fine ADCs samples the input along with the coarse ADC, while the other fine ADC is finishing the quantization of the sample taken in the previous clock cycle [32, 53]. This solution uses $2^{N_c} + 2^{N_f+1} - 1$ subtractors/comparators which is still significantly less than in a flash ADC.

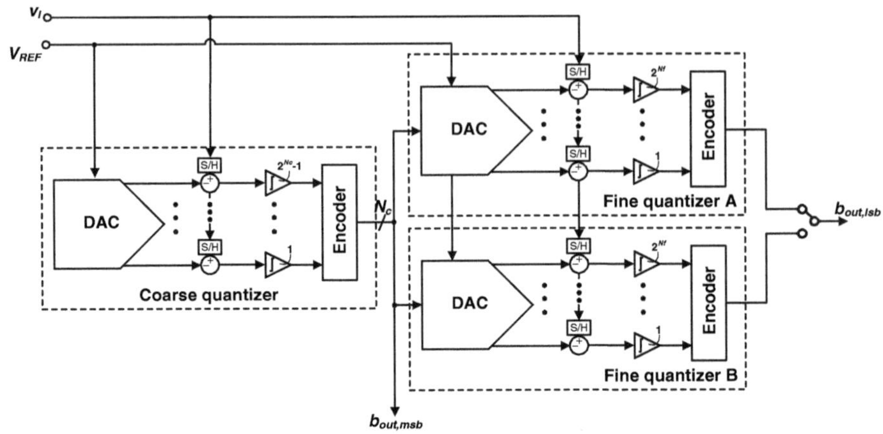

Fig. 14 Two-step subranging ADC with time-interleaved fine ADCs and distributed S/H

In order to drive all subtractors/comparators, frontend S/Hs typically use an amplifier or a source follower, which may introduce distortion and consume a significant amount of power. This can be avoided by employing passive switched capacitor circuits[7] before the comparators, which sample the input and subtract if from the references [32, 53]. This distributed S/H solution is also illustrated in Fig. 14. Lastly, note that though Fig. 14 shows 3 DACs and Fig. 13 shows 2 DACs, a single resistive ladder along with decoding switches (see Fig. 11a) may implement them all [32].

In the examples just described quantizers are implemented by flash ADCs, and so the effects of sub-blocks' non-idealities in the overall ADC transfer function are similar to those discussed in the previous sub-section for the flash ADC. There is, though, one additional issue: if the code transition voltages of the coarse quantizer are deviated by more than 1 LSB (due to errors at the output voltage of its DAC and offsets of its subtractors/comparators), for certain input signal ranges it provides wrong range indications to the fine quantizer. When this happens, the input signal lies outside the range of voltages provided by the DAC of the fine quantizer, and there will be missing codes (see Fig. 15). A power efficient solution to this problem is the use of *redundancy*, whereby wrong most-significant-bit decisions are corrected when determining the least-significant bits (this technique is used in all multi-step architectures). In the subranging ADC, redundancy enables the fine quantizer to convert input signals that lie in (half of) the regions adjacent to the one indicated by the coarse ADC (see Fig. 15). This is accomplished by increasing 1 bit in the resolution of either the coarse or the fine quantizers. In this case the

[7] These have the downside of introducing attenuation, which increases the input referred values of the comparator's offsets [32].

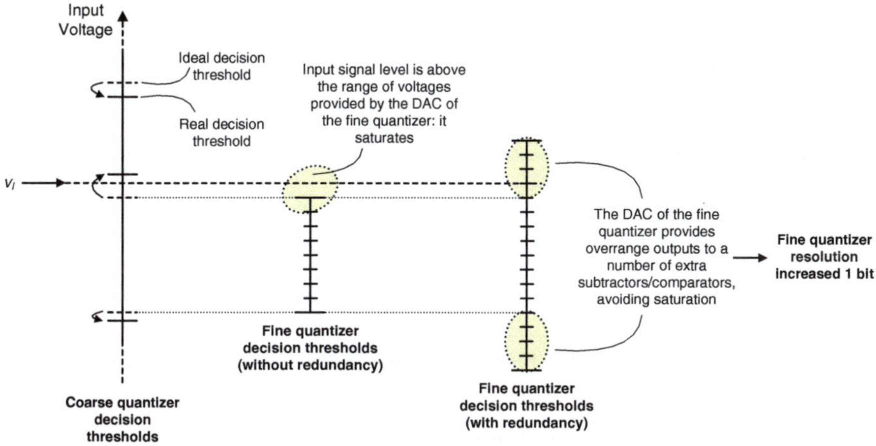

Fig. 15 Use of redundancy to prevent saturation in the fine quantizer

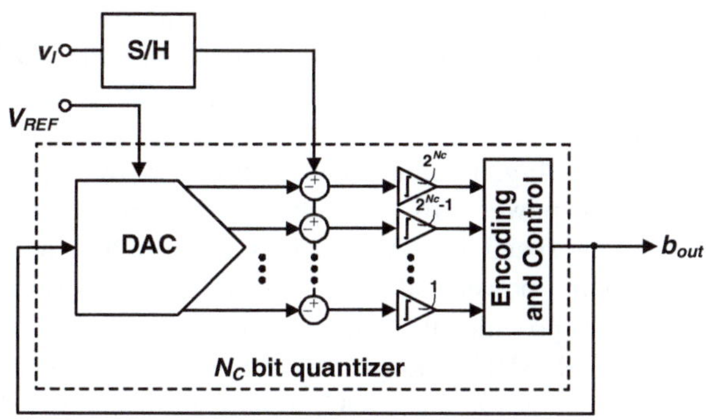

Fig. 16 Cyclic multi-step subranging ADC with a N_C bit quantizer, which takes N/N_C cycles to complete a conversion

output code of the coarse quantizer needs to be multiplied by 2^{N_f-1} to calculate b_{out} (instead of 2^{N_f} as shown in Fig. 13).

Redundancy relaxes significantly the specifications for the internal blocks of the coarse quantizer, since it may have deviations on its code transition levels up to $1/2{\cdot}2V_{REF}/2^{N_c}$, instead of $V_{LSB} = 2V_{REF}/2^N$. Note that the offsets of the subtractors/comparators belonging to the fine quantizer still need to be lower than the LSB—this is the primary limitation on this architecture. The linearity of the S/H and the fine quantizer's DAC also become a concern for higher resolutions. Finally, note that redundancy can also accommodate differences between the values sampled by the S/Hs in the coarse and fine ADCs [32].

Though two step architectures are more common, three step architectures have been proposed [54]. Figure 16 illustrates another variation of the subranging architecture, where the same quantizer is re-used cyclically [55, 56]. Note that in case of $N_C = 1$ bit this reduces to the SAR ADC (conversion takes N cycles), and

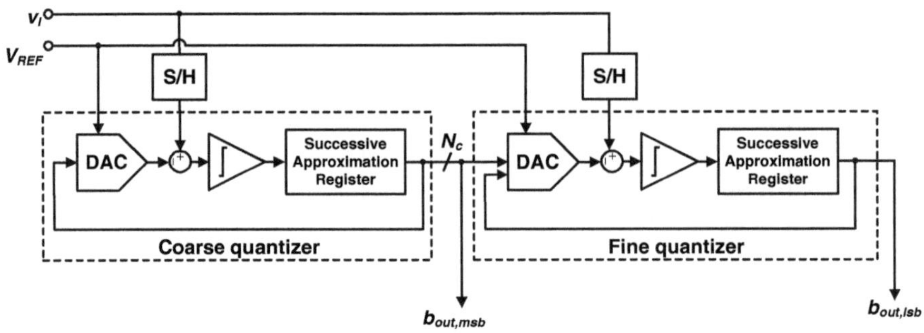

Fig. 17 Two-step subranging ADC using SAR ADCs as quantizers

for $N_C = N$ bit it becomes the full-flash ADC (conversion takes 1 cycle). So the cyclic sub-ranging converter is the bridge between those two architectures. Using N/N_C ADCs in time-interleaving, allows sampling the input and providing the result of a conversion in each clock cycle. Adjusting N_C allows trading between time-parallelization (maximum in the time-interleaved SAR) and parallelization in the process of finding the code (maximum in the flash ADC). Recent publications of "2 bit/cycle SAR ADCs" (i.e. $N_C = 2$) are a good example of this—see for example [57].

Finally, Fig. 17 illustrates the use of SAR ADCs as quantizers in a slow, but very low power two-step subranging converter [58]. Naturally, the achievable operating speed is lower than if flash ADCs are used, but the objective was minimizing power and area. As discussed earlier, the DAC in the fine ADC must be able to provide all 2^N output voltages, and so the fine quantizer is actually a full N bit SAR ADC. Though this seems an unnecessary complication (compared to simply using the fine ADC, alone), the coarse SAR quantizer has lower noise requirements due to redundancy, thus determining the most significant bits faster while consuming less power.

Figure 17 also illustrates the usage of two S/Hs, which is a possibility different from those shown in Figs. 13 and 14. It is a natural choice in this case, given that in practical SAR ADC implementations the S/H, DAC and subtractor operations are performed by a single switched capacitor circuit. A variation of this architecture is the combination of a flash coarse quantizer with a SAR fine quantizer [59],[8] [60]. The objective is to increase operating speed, with respect to that obtained with just the SAR, since the first few MSBs are determined in a single cycle.

[8] This is used in the quantizer of the last stage, on the pipeline ADC described in that reference.

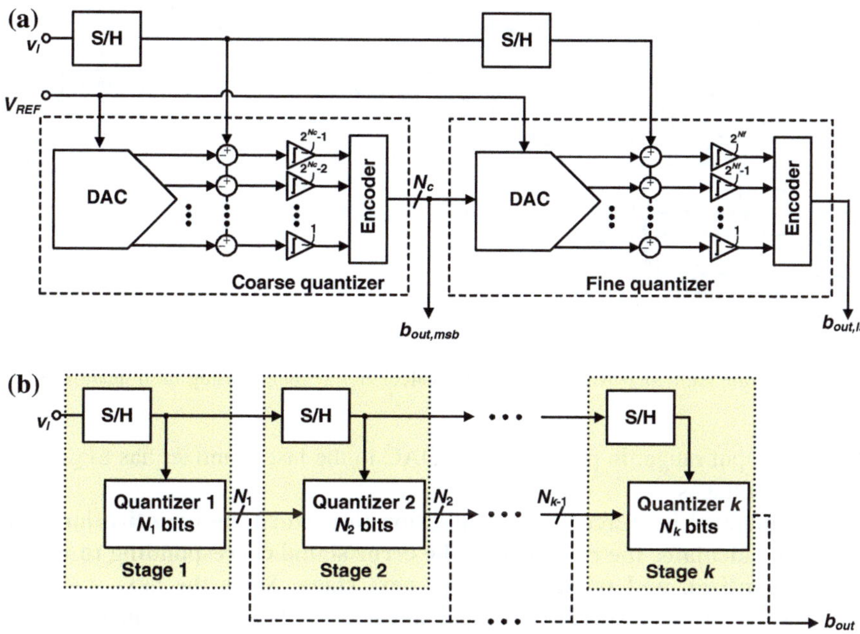

Fig. 18 **a** Two-step subranging ADC using pipelining. **b** Multi-step subranging ADC using pipelining

3.4 Multi-step ADC Architectures with Pipelining and Residue Amplification

In order to sample the input signal and provide a conversion result in every clock cycle, previous sub-sections shown that the flash, as well as the time-interleaved SAR and subranging converters resort to "time parallelization" and/or to "parallelization in the code-searching process". In non-cyclic subranging converters there is actually another possibility—*pipelining*—that is exemplified in Fig. 18a. Compared to Fig. 13, there is now a second S/H that provides the input for the fine quantizer, while the coarse quantizer is already taking the next sample. The two quantizers work concurrently, so it is not necessary to wait for the completion of the fine quantization to take the next sample. ADCs using this principle are described in [61, 62].

Since performing the conversion in multiple steps using several quantizers reduces the resolution of each of them, and that pipelining allows all the stages to work concurrently, it seems logic to use many pipelined stages each with a low resolution quantizer (see Fig. 18b). This decreases the number of subtractors/comparators in each quantizer, but not the complexity of their DACs. In fact, since all stages need to quantize the full input range, their DACs must generate voltages

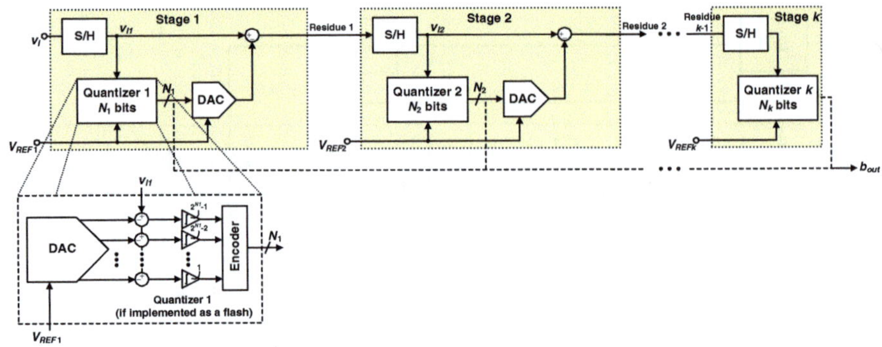

Fig. 19 Pipeline multi-step ADC with residue calculation (example using flash quantizers)

in the full input range; in particular, the DAC in the last quantizer has to generate all the 2^N voltages.

This limitation is overcome if, **in addition to performing quantization, each stage also calculates the residue (i.e. the error signal corresponding to what is left to quantize), and passes it to the next stage**. Since the next stage now quantizes the residue instead of the input signal, the DAC inside its quantizer does not need to generate voltages in the full input range.

The residue is obtained as shown in Fig. 6a: a subtractor calculates the difference between the input signal and the result of the quantization performed so far, provided by the DAC; the difference to Fig. 6a is that now the residue is applied to the next stage instead of a single comparator. This is illustrated in Fig. 19, where the quantizers are flash ADCs (as it is typically the case).

Comparing Fig. 18b to 19 suggests that, in the later, stages are more complex. Each one now has two DACs—one inside the quantizer and another to calculate the residue—and a subtractor. However, in Fig. 19 the DACs inside the quantizers are much simpler because they only need to provide 2^{N_j} output voltages (N_j: quantizer resolution of stage j), and not several *groups* of 2^{N_j} voltages. Furthermore, in practice the functions of S/H, DAC, subtractor (and residue amplification that will be discussed shortly) are implemented by a single switched capacitor circuit. So this complexity is more apparent than real.

Just as in subranging converters, the use of redundancy allows accommodating larger deviations in the code transition voltages of all quantizers except the last one: any error on that quantizer impacts the overall ADC transfer function and, since there is no gain along the pipeline, it must be lower than the ADC's LSB.

In the architecture of Fig. 19 the residue calculated by a stage has a maximum amplitude 2^{N_j} times smaller than the one it received—see Fig. 20a. So the reference voltage provided to each stage—which defines its input range—must scale by that amount. Therefore it is necessary to provide different reference voltages to each stage, which need to scale precisely. This is not convenient, and is circumvented by amplifying the residue 2^{N_j} times, before passing it to the next stage, as

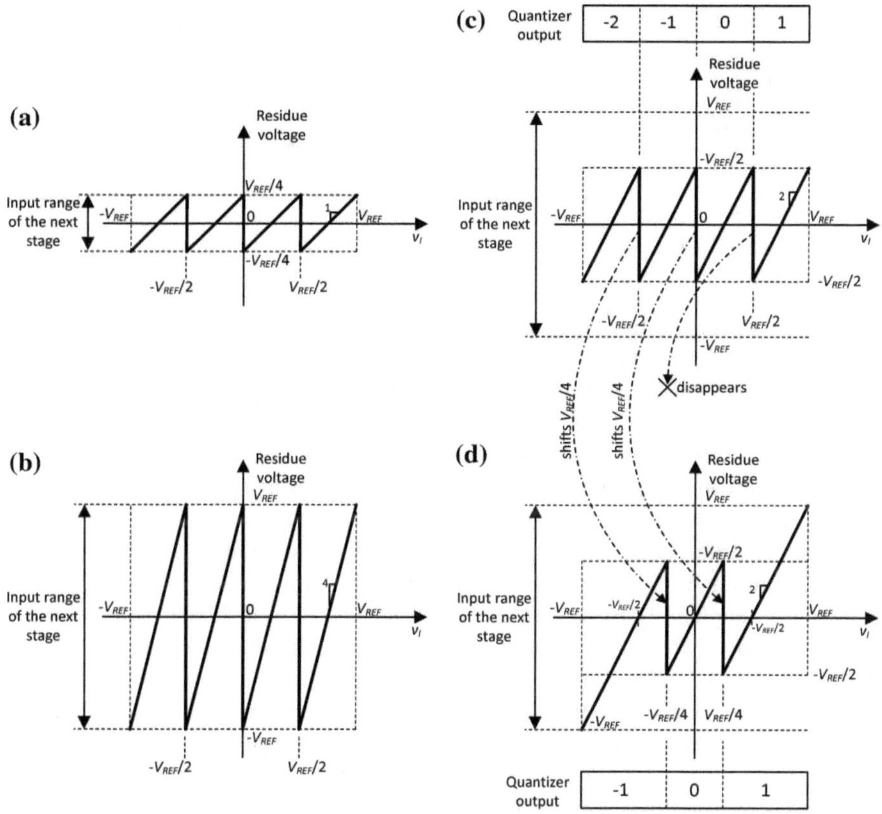

Fig. 20 Residue signal for a 2 bit stage ($N_j = 2$). The input range of the stage is from $-V_{REF}$ to $+V_{REF}$ in all cases: **a** No residue amplification. **b** Residue amplification by 2^{Nj}. **c** Residue amplification by 2^{Nj-1}. **d** Residue of a 1.5 bit stage

illustrated in Fig. 20b. Now all stages receive (the same) V_{REF} and have the same input range.[9] This is an essential principle in pipeline ADCs.

In the case illustrated in Fig. 20c the residue is actually only amplified 2^{N_j-1} times to implement redundancy: this ensures the next stage has an input range twice as large as the residue generated by the present stage, when its quantizer is ideal. The code transition voltages of the quantizers (in the example of Fig. 20c, $\pm V_{REF}/2$ and 0), can be deviated by as much as $1/2\frac{2V_{REF}}{2^{N_j}} = V_{REF}/2^{N_j}$, without causing the residue to go out of the input range of the next stage. Typically, pipeline ADCs use low resolution stages, and so the design the DAC, subtractors and comparators inside the quantizers is straightforward. Furthermore, there is

[9] This is the most frequent case on pipeline ADC. However, as discussed at the end of this sub-section there are, for example, (two-stage) implementations with low gain residue amplification, where the stages receive different reference voltages.

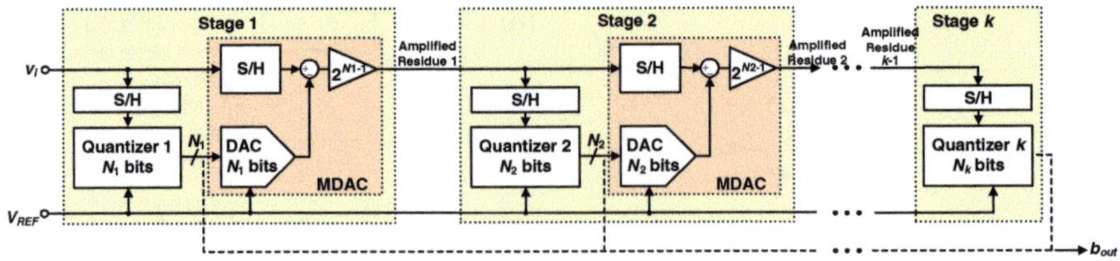

Fig. 21 Pipeline ADC

typically so much gain along the pipeline, that the contributions from the non-idealities of the last stage's quantizer can be neglected.[10] **This is a major difference to previously presented architectures, whose linearity was limited by quantizer errors.**

Reference [65] shows that if the thresholds of the quantizer are all shifted by $V_{REF}/2^{N_j}$, and the uppermost is removed—see Fig. 20d—the allowed deviations on the code transition levels of the quantizer remain the same, but the calculations of the ADC output code are simplified. The residue shown in Fig. 20d is produced by a 1.5 bit stage,[11] whose quantizer provides 3 output codes instead of 4. This principle is applicable to higher resolution stages—e.g. a 2.5 bit stage[12] can be obtained from a 3 bit stage in the same way. This is the solution typically used on the pipeline ADC, depicted in Fig. 21. This figure already reflects the fact that in practical implementations, the operations yielding the residue are implemented by a single switched capacitor circuit called the *Multiplying DAC*[13]—MDAC; moreover there is a separate S/H operation in front of the quantizer, eventually distributed by its subtractors as illustrated in Fig. 14.

The pipeline ADC is the most modular of all ADCs. For example a 12 bit ADC can be made by simply adding a 2.5 bit stage in front of an existing 10 bit ADC.

---

[10] There are pipeline ADC implementations employing a small number of stages, each with a higher resolution. In those cases the quantizers are more complex, must comply with tighter specifications, and consume more power. References [33, 63] address these limitations by using SAR ADC quantizers instead of the traditional flash ADCs, thereby improving power efficiency. Furthermore there is only one comparator whose offset must be made low enough.

Reference [59] implements the 7 bit quantizer in its last stage as a two-step subranging ADC composed of a coarse flash ADC and a fine SAR ADC. The 4.5bit quantizers (the meaning of this will be explained shortly) used on the other stages are also two-step subranging converters, but with coarse and fine flash ADCs. There is redundancy *inside* all these two-step quantizers to relax the specifications of their coarse ADCs. Naturally, there is also redundancy between the pipelined stages, as just explained.

[11] The "1.5 bit stage" designation may seem odd but this simply means its flash ADC has 3 output codes, which is halfway between the number of codes encountered on a "1 bit" and on a "2 bit" ADC.

[12] Some authors [64] call this a 2.8bit stage, because the quantizer provides 7 output codes, and $\log_2(7) \approx 2.8$.

[13] Naturally the non-idealities of the MDAC can be traced to those of the basic operations it implements.

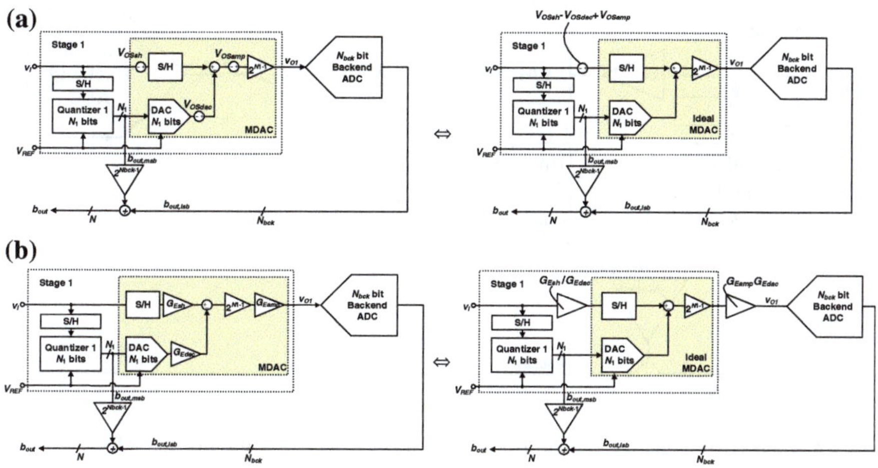

Fig. 22 Pipeline ADC: **a** Effect of offsets. **b** Effect of gain errors and non-linearities

Furthermore, note that the effects caused by the non-idealities inside a stage (e.g. gain errors, non-linearity, settling errors and noise in S/Hs, DACs and amplifiers, etc.) are reduced by an amount equal to the gain existing between the ADC input and that stage. This relaxes the specifications of later stages, which results in important power savings.

The impacts of offsets, gain error and non-linearity of the sub-blocks of the MDAC are illustrated in Fig. 22, for the first stage (the remaining stages of the pipeline act as an N_{bck} bit "backend" ADC.) Offset/gain errors/non-linearity of the residue amplifier is indistinguishable from those of the subtractor, so only the former are shown explicitly. Quantizer errors are neglected for the reasons mentioned above.

Figure 22a shows that the offsets voltages of the sub-blocks cause an offset on the ADC overall transfer function,[14] which is not critical in communication systems. Figure 22b shows that the effects caused by the gain errors of the S/H, DAC and residue amplifier (in the ideal case $G_{Esh} = G_{Edac} = G_{Eamp} = 1$) are modeled by an ideal MDAC preceded and succeeded, respectively, by gain factors G_{Esh}/G_{Edac} and $G_{Eamp}G_{Edac}$. (Non-linearity can also be modeled in this way, by considering the dependence of the gain error on the input of the respective sub-block: $G_{Esh}(v_I)$, $G_{Edac}(d_{dac})$, etc.) We will now discuss the issues originated by gain errors, and indicate how they can be overcome using digital calibration. The Sect. 4 describes an implementation where these principles are applied.

[14] The quantizer is not affected by this offset error. From the perspective of the MDAC it is as if the code transition voltages of the quantizer are shifted, but this is not a problem due to redundancy.

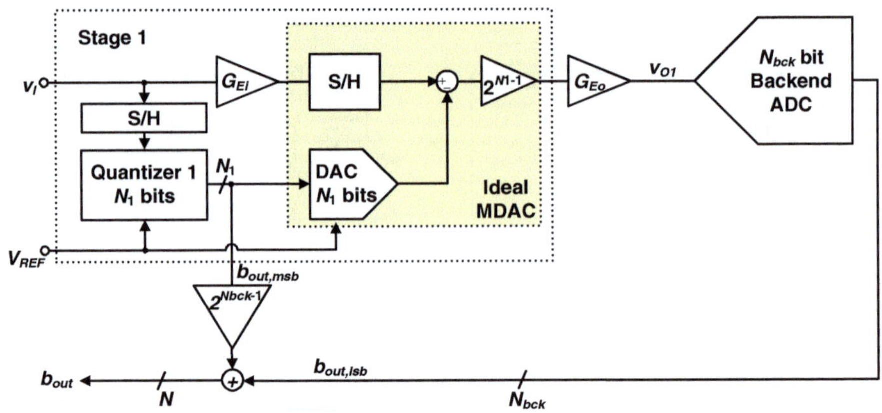

Fig. 23 Gain error modeling in the first stage of a $N_I + N_{bck} - 1$ bit pipeline ADC

The input gain factor of the first stage, G_{Esh}/G_{Edac}, causes a gain error on the overall transfer function of the ADC.[15] This is not critical in communication systems. Similarly, the input gain factor of the second stage, (i.e. the first stage of the backend ADC), introduces a gain error on the transfer function of the backend ADC, which multiplies by the output gain factor of the first stage, $G_{Eamp}G_{Edac}$—this is aggregated as G_{Eo} in Fig. 23, which we will consider to perform our analysis. Naturally G_{Ei} shown on that figure equals G_{Esh}/G_{Edac} of the first stage.

Equation (1) is still applicable to calculate the output of quantizer 1, even though its thresholds have been shifted in the way illustrated in Fig. 20(d).[16] We have, therefore,

$$b_{out,msb} = \frac{G_{Ei} v_I}{2V_{REF}} 2^{N_1} - \varepsilon_{q1}, \tag{8}$$

where ε_{q1} is the quantization error that now varies between -1 and +1. It can be shown that $b_{out,msb}$ ranges between $\pm(2^{N_1-1} - 1)$. Using (2) to calculate the output of the DAC, and then considering the subtraction and residue amplification, leads to the input voltage of the backend ADC

$$v_{O1} = G_{Eo}\varepsilon_{q1} V_{REF}, \tag{9}$$

[15] The quantizer is not affected by this gain error. From the perspective of the MDAC it is as if the code transition voltages of the quantizer are shifted, but this is not a problem due to redundancy.

[16] Figure 20 illustrates how the 1.5 bit stage is derived from a 2 bit stage. The 1.5 bit stage is addressed by considering $N_1 = 2$ in the model shown in Fig. 23. Likewise, for a 2.5 bit stage one should consider $N_1 = 3$ and so on.

which, as expected, is proportional to the quantization error and varies between $\pm G_{Eo} V_{REF}$ (see Fig. 20d, at which $G_{Eo} = 1$ was being considered). The output of the backend ADC is obtained by applying (1)

$$b_{out,lsb} = \frac{v_{O1}}{2V_{REF}} 2^{N_{bck}} - \varepsilon_{qbck} = G_{Eo}\varepsilon_{q1} 2^{N_{bck}-1} - \varepsilon_{qbck}. \tag{10}$$

The output code is, finally

$$b_{out} = 2^{N_{bck}-1} b_{out,msb} + b_{out,lsb} = \underbrace{\frac{G_{Ei} v_I}{2V_{REF}} 2^{N_1+N_{bck}-1} - \varepsilon_{qbck}}_{\substack{\text{Ideal } N_1+N_{bck}-1 \text{ bit ADC that} \\ \text{quantizes } G_{Ei}v_I}} - \underbrace{\varepsilon_{q1} 2^{N_{bck}-1}(1 - G_{Eo})}_{\text{Non-ideal term}}. \tag{11}$$

If $G_{Eo} = 1$ the resulting transfer function corresponds, according to (1), to that of a $N_1 + N_{bck} - 1$ bit ADC that quantizes an input $G_{Ei} v_I$ (as mentioned earlier G_{Ei} causes a non-problematic gain error on the overall transfer function). If $G_{Eo} \neq 1$ the quantization error of the first stage introduces an error on the overall ADC transfer function, originating distortion. So, although incorporating gain in the pipeline chain brought great advantages (architecture modularity, alleviating the specifications that must be met by later stages, simplification of the quantizers to the point where they no longer limit performance contrarily to what happens in all other architectures), the gain must be precisely set to avoid degrading performance. The long established solution to this problem consists on using negative feedback around a high-gain opamp but, as will be discussed in the Sect. 4, this is not power efficient. One way of addressing this limitation is through the use of digital calibration: the digital output of the backend ADC is multiplied by $1/\tilde{G}_{Eo}$, where $\tilde{G}_{Eo}$ is an estimation of G_{Eo} (see Fig. 24). Considering $1/\tilde{G}_{Eo} = 1/G_{Eo} + \Delta g$, where Δg is the error on the digital calibration coefficient, yields

$$\begin{aligned} b_{out} &= 2^{N_{bck}-1} b_{out,msb} + \tfrac{1}{\tilde{G}_{Eo}} b_{out,lsb} = \\ &= \underbrace{\frac{G_{Ei} v_I}{2V_{REF}} 2^{N_1+N_{bck}-1} - \frac{\varepsilon_{qbck}}{\tilde{G}_{Eo}}}_{\substack{N_1+N_{bck}-1 \text{ bit ADC that} \\ \text{quantizes } G_{Ei}v_I}} + \underbrace{\varepsilon_{q1} 2^{N_{bck}-1} \Delta g G_{Eo}}_{=0 \text{ if } \Delta g = 0 \Leftrightarrow \tilde{G}_{Eo} = G_{Eo}}. \end{aligned} \tag{12}$$

If $\Delta g = 0$ ($\tilde{G}_{Eo} = G_{Eo}$), the unwanted term disappears, but the quantization noise changed from ε_{qbck} to $\frac{\varepsilon_{qbck}}{G_{Eo}}$. Typically $G_{Eo} < 1$ which means digital calibration increases quantization noise. This may imply raising the resolution of the backend ADC to lower its quantization noise. For example the 12 bit implementation described in the Sect. 4 is actually a 16 bit ADC, to deal with the accumulated gain error of its 10 calibrated stages (low gain amplifiers are used in each of them, causing G_{Eo} to be in the range of 0.85–0.9).

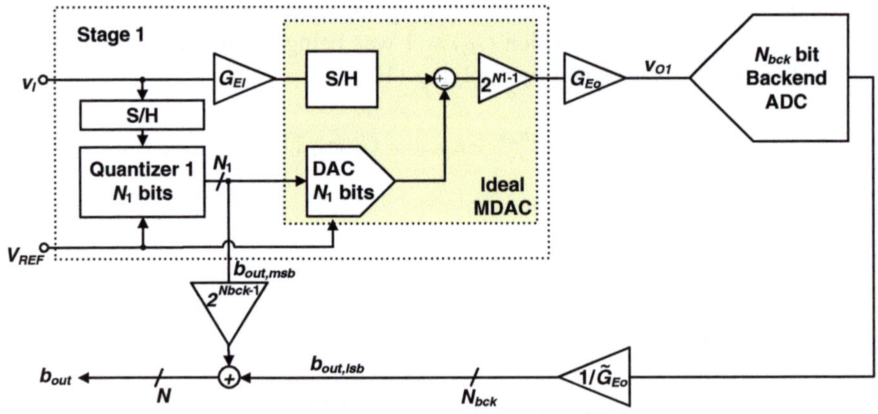

Fig. 24 Digital gain error calibration

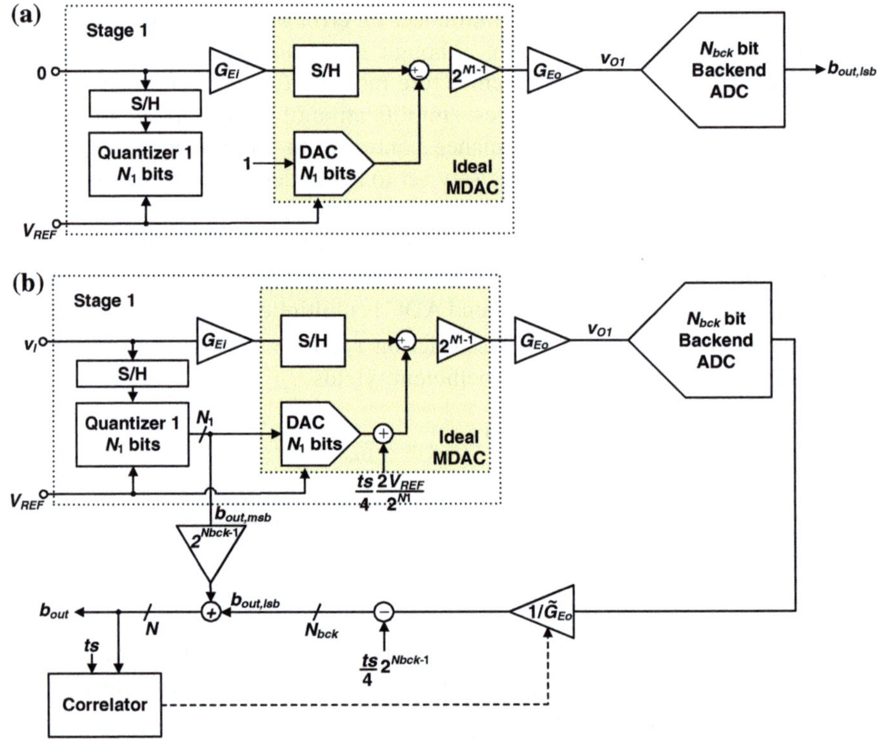

Fig. 25 Methods of estimating the gain error: **a** Foreground calibration. **b** Background calibration

We will now describe the two methods of determining $\tilde{G}_{Eo}$ shown in Fig. 25. The first one consists on sampling $v_I = 0$, ignoring the quantizer output, and applying digital code 1 to the DAC, whose output voltage then becomes $v_{ODAC} = \frac{2V_{REF}}{2^N}$. The signal applied to the backend ADC is

$$v_{O1} = G_{Eo} V_{REF}. \tag{13}$$

For $G_{Eo} = 1$ this is the positive full-scale of the backend ADC, corresponding to code $b_{out,lsb} = 2^{N_{bck}-1}$. So G_{Eo} may be estimated by comparing the digital output of the backend ADC when it receives (13) as input, against the full-scale value,

$$\tilde{G}_{Eo} = \frac{b_{out,lsb}}{2^{N_{bck}-1}}. \tag{14}$$

The Sect. 4 presents a digitally calibrated pipeline ADC that, at the startup, uses this principle to determine the gain correction coefficients. Let us now describe the background calibration process that adapts the digital coefficients in response to temperature or supply voltage variations while the converter is operating, illustrated in Fig. 25b. A pseudo-random binary sequence (PRBS), ts, controls the injection of an analog signal that shifts the output of the DAC by ±1/4 of its LSB. The input voltage of the backend ADC, changes from (9) to

$$v_{O1} = G_{Eo}\varepsilon_{q1} V_{REF}\left(\varepsilon_{q1} + \frac{ts}{4}\right). \tag{15}$$

So the PRBS causes the output code of the backend ADC to change by $\frac{ts}{4} G_{Eo} 2^{N_{bck}-1}$. The PRBS is digitally removed from the backend output, after the gain calibration (multiplication by $1/\tilde{G}_{Eo}$). If $\tilde{G}_{Eo} = G_{Eo}$ this removal is totally effective. In case $\tilde{G}_{Eo} \neq G_{Eo}$, the output code becomes

$$b_{out} = \underbrace{\frac{G_{Ei} v_I}{2V_{REF}} 2^{N_1+N_{bck}-1} - \frac{\varepsilon_{qbck}}{\tilde{G}_{Eo}}}_{\substack{N_1+N_{bck}-1 \text{ bit ADC that} \\ \text{quantizes } G_{Ei} v_I}} + \underbrace{\left(\varepsilon_{q1} + \frac{t_s}{4}\right) 2^{N_{bck}-1} \Delta g G_{Eo}}_{=0 \text{ if } \Delta g = 0 \Leftrightarrow \tilde{G}_{Eo} = G_{Eo}}, \tag{16}$$

which now contains the leakage from the PRBS, in addition to the quantization error of stage 1. A correlator detects this leakage, and the calibration coefficient is adjusted to minimize it, thereby forcing $\Delta g \to 0$ and thus removing also the quantization error of stage 1. To calibrate several stages of the pipeline ADC using this approach, statistically independent PRBSs must be applied to each of them.

We stated in the beginning of this sub-section, that pipelining allows the different stages to perform quantization (and residue amplification) concurrently. This is an alternative to "time" or "code search" parallelizations employed in flash, time-interleaved SAR and subranging converters, to sample the input signal and provide an output sample in every clock cycle. And, in fact, high-speed pipeline ADCs have been described [66].

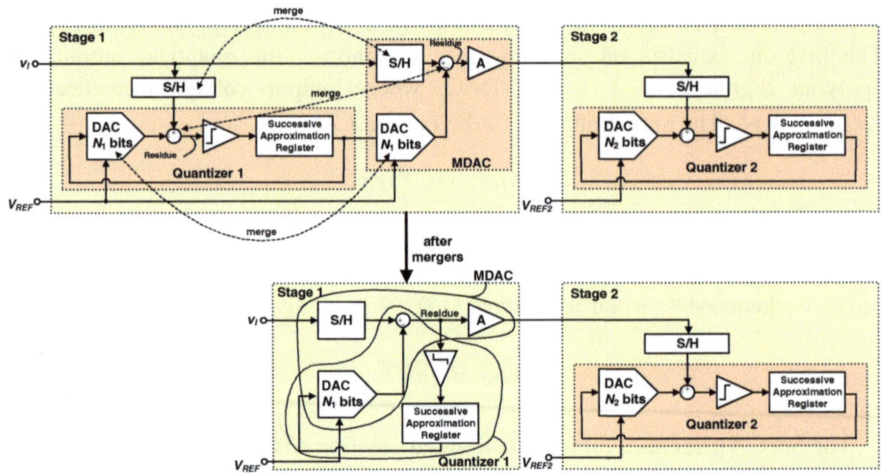

Fig. 26 Obtaining the "SAR-assisted pipeline ADC" from a two stage pipeline ADC that employs SAR quantizers in both stages

We also mentioned that most pipeline ADCs are composed of a number of low resolution stages (1.5–3.5 bit), whose quantizers have relaxed specifications, due to redundancy and the gain introduced in each stage. Note that simplifying the quantizers was the objective of introducing residue calculation in multi-step ADCs. However, as explained in Sect. 3.2, it is now possible to implement high-speed, medium resolution (5–7 bit) SAR ADCs, which are very power efficient. These can be employed as quantizers in pipeline ADCs, enabling the use of a small number of medium resolution stages. In particular, implementations composed of only two stages featured very good power efficiency [33, 63]. Moreover in the first stage the S/H, DAC and subtractor are shared by the quantizer and the MDAC—see Fig. 26. These have been named *SAR-assisted pipeline ADCs* [67].

Given the low power featured by SAR quantizers, the challenge is on implementing a power efficient residue amplifier: dynamic open-loop amplifiers have been used, whose gain error is corrected through digital calibration and whose non-linearity is mitigated by passing a small amplitude residue to the second stage: it is amplified by a factor, A, lower than 2^{N-1}, meaning that a lower reference voltage (V_{REF2}) needs to be generated since the second stage must feature a reduced input range.

4 ADC Implementations in Advanced CMOS Technologies

4.1 Consequences of Technology Scaling on Analog Circuits

The down scaling of CMOS transistor feature sizes, along with the supply voltage from the mid 90s onwards [68], has consistently enabled digital circuits to operate faster with lower power, while occupying less silicon area. However analog circuits do not necessarily benefit from scaling. For example the available signal headroom is smaller with lower V_{DD}, which means more power has to be spent to decrease noise, so as to maintain SNR. In addition, stacking (cascode) transistors was an efficient way of increasing the DC gain of amplifiers, because it required no extra current; but with reduced V_{DD} one must resort to folded or multi-stage architectures which have extra branches, and possibly compensation capacitors that need to be charged, thereby consuming more power.

In addition, the intrinsic gain of the MOS transistors ($g_m.r_{ds}$) is degraded by *drain induced barrier lowering*, which becomes more prominent as transistor length decreases and its gate-to-source voltage is reduced[17] [69, 70]. Furthermore *impact ionization*, which can be modeled with a conductance between the drain and the substrate [70], decreases drain output resistance limiting even the effectiveness of cascode transistors. Finally the effective gate resistances, caused by tunneling currents across the gate oxide, set the ultimate limit to amplifier gain.

The threshold voltage of the MOS transistors, V_{th}, has also been scaling down, but not as fast as V_{DD}, due to the necessity of limiting leakage currents [4]. Since V_{th} corresponds to an increasing percentage of V_{DD}, the on resistance of CMOS switches is impacted, especially for signals near $V_{DD}/2$ [71]. This motivates the use of architecture/circuit solutions where most of the switches connect to signals near 0 or V_{DD}. When this is not possible, this limitation can be circumvented at the expense of additional power and complexity, by using clock bootstrap circuits [8, 71, 72].

Although a lower V_{th} is desirable to reduce the on resistance of MOS switches, the transistor's drain-source leakage current increases exponentially [70]. It prevents MOS switches from turning off completely, which discharges capacitors in switched capacitor circuits, thereby limiting their low frequency performance.

Two components fabricated near each other and having the same dimensions do not present the exact same electrical characteristics. These *mismatches* are of random nature and occur due to unavoidable variations on the fabrication process. The widely used Pelgrom's first order model [73] suggests the variance of transistor mismatches is inversely proportional to their *channel area* (similar considerations apply to resistors and capacitors). However devices with either short W or short L experience larger mismatch due to higher order terms [74]. So, the use of minimum size devices is avoided in mismatch limited circuits such as the comparators, meaning that one cannot take full advantage of technology scaling. It has been verified theoretically and experimentally that V_{th} matching is proportional to the oxide thickness of the transistors [75], thus reducing as technology shrinks. However a number of effects seriously limit matching improvement with scaling in sub-100 nm devices [76], thereby preventing a significant reduction of transistor

[17] With the supply voltage reduction at more advanced technology nodes, one must bias the MOS transistors with lower overdrive voltages, $v_{OVD} = v_{GS} - V_{th}$. In the past it was necessary to bias transistors far into strong inversion ($v_{OVD} > 0.2$ V), to achieve the necessary operating speed. However transistors in modern technologies present large transit frequencies even for low or negative v_{OVD} [30], which enables designers to use reduced overdrive voltages.

sizes.[18] Also, device matching is becoming more dependent on its surroundings [7, 76]. As transistors further scale, the quantum effects will eventually dominate and degrade matching considerably [6]. Also, it has been pointed out that gate leakage mismatches may become relevant for devices with very thin oxides [77].

Mismatches are responsible for offset voltages (e.g. in subtractors/comparators), and non-linearity in the DACs. Higher accuracy requires larger devices which impacts either the power consumption or the speed of the circuits [78]. This speed-power-accuracy trade-off may be eased through the use of calibration, as we will see.

In spite of these difficulties, technology scaling does bring advantages. The **availability of faster transistors** allows designing ADCs with higher sampling rates: the core of any comparator is composed by two cross-coupled CMOS inverters [79], which become faster (as any digital circuit does); furthermore, passive switched capacitor circuits—i.e. those composed only of capacitors and switches, without amplifiers—also get faster (as long as the switches connect to voltages near $0/V_{DD}$). So architectures that mainly rely on these elements, such as the SAR ADC, benefit directly from scaling, which explains the massive adoption of this architecture in the last few years [80].

Even architectures resorting to amplification (pipeline ADCs) benefit from having faster transistors, because the secondary poles of the amplifiers are pushed to higher frequency. However this improvement is invalidated if a more complex structure needs to be used, to achieve the necessary gain (see discussion above). A way of dealing with this problem is resorting to digital calibration, whereby simpler and faster amplifiers may be used. This is, actually, the second important advantage offered by technology scaling: **digital processing, that is increasingly powerful, cheap and low power, is available to overcome limitations in the analog sub-blocks of ADCs** [5, 9]. This trend of *digitally aided analog design* transfers the complexity from analog circuits to digital processing blocks that fully benefit from technology scaling.

4.2 ADCs with Residue Amplification

As described in the Sect. 3.4, each stage of the pipeline ADC has an MDAC that implements sampling, DAC, subtraction and amplification operations, to calculate the residue passed to the next stage for further quantization. Since in these ADCs the specifications of the quantizers are relaxed, most of the power is dissipated on the MDACs, which therefore set the overall power efficiency.

[18] Since MOS gate capacitance is inversely proportional to oxide thickness, it increases for more advanced technologies. If transistor sizes cannot be significantly reduced, the capacitances will become larger, which may eventually increase the power consumption.

Figure 27 depicts the most common implementation of the 1.5 bit MDAC, which produces the residue show in Fig. 20d. It uses a high-gain amplifier enclosed in a negative feedback loop to set the gain very precisely. Throughout this section we will review alternative solutions that are more power efficient, typically at the expense of increased gain errors or non-linearity, which are addressed by digital calibration.

The load of the amplifier is composed of the sampling capacitors of the next stage, as well as the parasitic capacitance at the output, C_{Lp}, and the capacitors of the feedback path. The input signal is sampled on both C_R and C_F during ϕ_1 (ideally $C_R = C_F = C$). Assuming the time constant associated to the sampling network is low, and does not depend significantly on the input signal level, the gain error and non-linearity associated to the sampling operation is negligible, i.e. $G_{Esh} = 1$ (see Fig. 22b). Bottom plate sampling [32, 81] is typically used to avoid signal dependent charge injection when the switches open at the end of the sampling phase.

During the amplification phase, ϕ_2, C_F closes the negative feedback loop around the amplifier, and C_R connects to $\pm V_{REF}$ or 0, depending on the output of the quantizer, $b = \pm 1$ or 0. If capacitors are equal and the amplifier has infinite gain, the output of the MDAC is

$$v_O = \underbrace{2}_{\substack{\text{Residue amplifier}\\ \text{gain}}} \underbrace{(v_I - \underbrace{b\frac{V_{REF}}{2}}_{\text{DAC output}})}_{\substack{\text{Subtraction between input}\\ \text{signal and DAC output}}} . \tag{17}$$

If, due to mismatches, capacitors are not exactly equal and the DC open loop gain of the amplifier is finite, A_0, the output becomes

$$v_O = G_{Eamp} 2\left(v_I - G_{Edac} b\frac{V_{REF}}{2}\right) = G_{Eamp} G_{Edac} 2\left(\frac{1}{G_{Edac}} v_I - b\frac{V_{REF}}{2}\right), \tag{18}$$

where

$$G_{Eamp} = \frac{1}{2}\frac{1}{\frac{1}{A_0} + \frac{C_F}{C_F + C_R}}, \tag{19}$$

$$G_{Edac} = \frac{2C_R}{C_F + C_R}. \tag{20}$$

The last expression in (18) is in agreement to the equivalence shown in Fig. 22b, ($G_{Esh} = 1$ is being considered). G_{Eamp} depends both on capacitor mismatches and the finite gain of the amplifier. G_{Edac} is only caused by capacitor mismatches, and in the case of the 1.5 bit stage does not depend on b (the same does not happen for higher resolution stages).

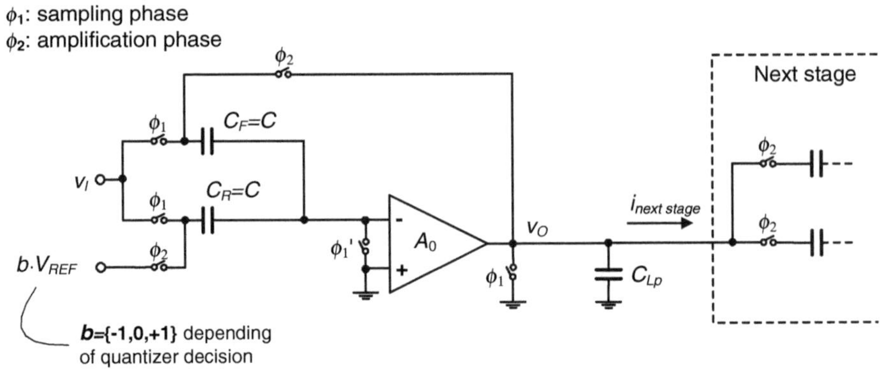

Fig. 27 1.5 bit MDAC switched capacitor implementation (single-ended example)

Amplifiers with one or more gain stages, operating in class-A and employing cascode transistors and eventually gain boosting (e.g. see [66]) are frequently used in pipeline ADCs. But as explained in the previous sub-section, technology scaling is making it increasingly difficult to attain the high gain that is necessary to ensure proper performance of these converters. Furthermore, though the MDAC circuit being discussed resorts to the well-known and proven concept of using negative feedback to set, precisely, the amplifier's gain, this solution is not very power efficient. We will now discuss this point and review some of the solutions proposed in the literature.

In classical amplifier implementations, the currents in each branch of the amplifier are set by current sources, and stay practically unchanged in the different operating phases (class A operation). However, it is clear from Fig. 27 that the amplifier is only used during ϕ_2 and so it is simply wasting energy during the sampling phase. A possible solution is to turn the amplifier off during ϕ_1, by either forcing v_{GS} of internal current source transistors to 0 [82], or adding switches in series with the current sources [64, 83]. The first solution powers-down the amplifier very quickly, but re-starting is too slow because the gate capacitances of the current sources must the re-charged. Adding switches in series with the current sources provides a faster restart, but adds parasitic capacitances and voltage drops that degrade amplifier performance. In the pipeline ADC implementation described at the end of this section, we disclose a solution [84] that does not suffer from speed and/or signal swing limitations.

Another solution meant to avoid wasting energy during the sampling phase, consists on sharing the amplifier by two consecutive pipeline stages [85, 86], because when one is sampling the other is on the amplification phase. This is typically done by adding multiplexing switches in series with the inputs of the amplifier (this must be done very carefully in order not to disturb, during switching, the charge sampled on the capacitors of the two stages). In the case considered in Fig. 27, the amplifier must be able to settle from 0 (reset) to $\pm V_{REF}$ (full-scale output) during ϕ_2; but since its output is not reset when it is shared

Fig. 28 Supply current versus current in the next stage's sampling capacitors

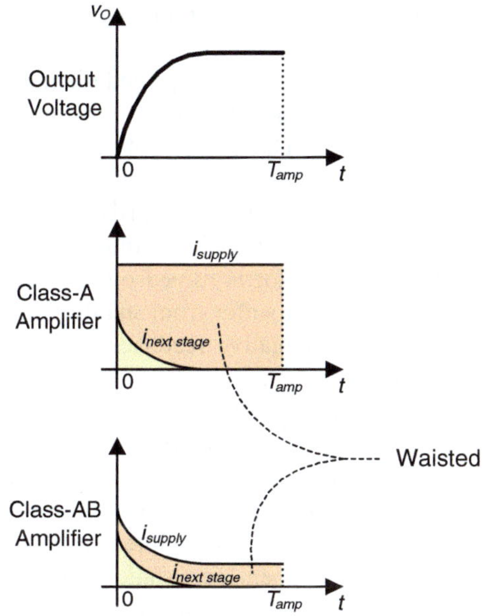

between consecutive stages, the amplifier may need to settle from $\pm V_{REF}$ to $\mp V_{REF}$, thereby requiring extra power.

As mentioned in Sect. 3.4, the gain of a stage allows relaxing the specifications of the ones following it: capacitors may be smaller and amplifiers dissipate less power. However, the specifications of an amplifier that is shared between consecutive pipelined stages are set by the first one, thus dissipating more than needed when performing the amplifications of the second stage. This is another point limiting the amount of power saved with this solution. An alternative, which does not suffer from this last limitation, consists on sharing the amplifiers between the corresponding stages of two time-interleaved ADCs [87, 88].

However, the efficiency obtained using class-A amplifiers is rather poor even if we just consider the amplification phase. As illustrated in Fig. 28, the supply current does not decrease, even when the output voltage is already near enough the ideal value. Furthermore, the peak current provided to the sampling capacitors of the next stage may be significantly smaller than the total supply current because of the need to charge the capacitors in the feedback path, the parasitic capacitance at the output, the compensation capacitors when they exist and bias all internal amplifier branches. Class-AB amplifiers, whose output current increases when higher input voltages are applied to the amplifier, have been used at the expense of extra complexity [89]. However, since the input voltage of the amplifier is an attenuated version (by more than 50 % in the implementation shown in Fig. 27) of the output voltage error, the amount of output current increase with respect to the quiescent condition is limited. Another solution [84] consists of increasing amplifier current during the initial part of the amplification phase, boosting slew-rate and amplifier bandwidth when it is most needed, and requires almost no extra complexity.

Figure 29 depicts a simplified diagram of the comparator based 1.5 bit MDAC [59, 90], where the amplifier is substituted by a continuous-time comparator and a current source. The output voltage is set to a value higher than V_{REF} before the amplification phase starts, during which it is discharged by a current source. The comparator detects when v_O arrives to the correct value because its input voltage crosses zero—at that point in time the sampling switch of the next stage is opened and the comparator is turned off, no longer consuming power. Furthermore, this solution does not suffer from stability and gain-bandwidth product limitations that are typical of negative feedback amplifiers [59]. All this makes it a very power efficient technique.

An alternative that allows reducing power while still using amplifiers enclosed in a negative feedback loop is to employ low gain amplifiers that are as simple as possible, and resort to digital gain calibration in the way illustrated in Fig. 24. The implementation described at the end of this section follows this trend.

A step forward was explored in [91], where the residue amplifier was implemented by an open-loop static amplifier (differential pair with load resistors). Power consumption is reduced in comparison to the closed loop solution shown in Fig. 27, due to the lack of stability constraints, much simpler amplifier structure, and because there are no feedback or compensation capacitors to charge. Since there is no feedback, the gain of the MDAC is not well controlled, being for example sensitive to parasitic capacitances at the input nodes of the amplifier (where the subtraction between the sampled input and DAC output occurs). Moreover, a non-linear digital calibration was required, since there is no amplifier non-linearity reduction mechanism (such as feedback).

Power consumption can also be reduced if the amplifier bandwidth is made lower, and the errors arising from insufficient settling are digitally calibrated. The critical point is ensuring the settling is linear, because the situation is then equivalent to a gain error. This has been done in [92] to reduce power in the above mentioned open loop amplifier, and also in a very high frequency pipeline ADC using closed loop MDACs [66].

References [33, 63] employed an open loop integrator as a residue amplifier. It is basically a differential pair that charges the next stage sampling capacitors, and whose input voltage (i.e. the residue before amplification) is set by a passive switched capacitor circuit that implements the S/H, DAC and subtraction operations. It combines the advantages mentioned above for the open-loop static amplifier (i.e. no stability constraints, simple circuit solution, no need to drive feedback capacitors), with the main advantage of the comparator based solution depicted in Fig. 29: power is only dissipated until the output voltage reaches the desired value, after that the amplifier is turned off. These advantages have a price: the gain is highly sensitive to parasitic capacitances, as well as to supply voltage and temperature variations [33]. This has been addressed by tracking gain variations through the injection of a PRBS, in ways similar to what was shown in Fig. 25b [63, 93]. Furthermore, the lack of feedback implies that these amplifiers

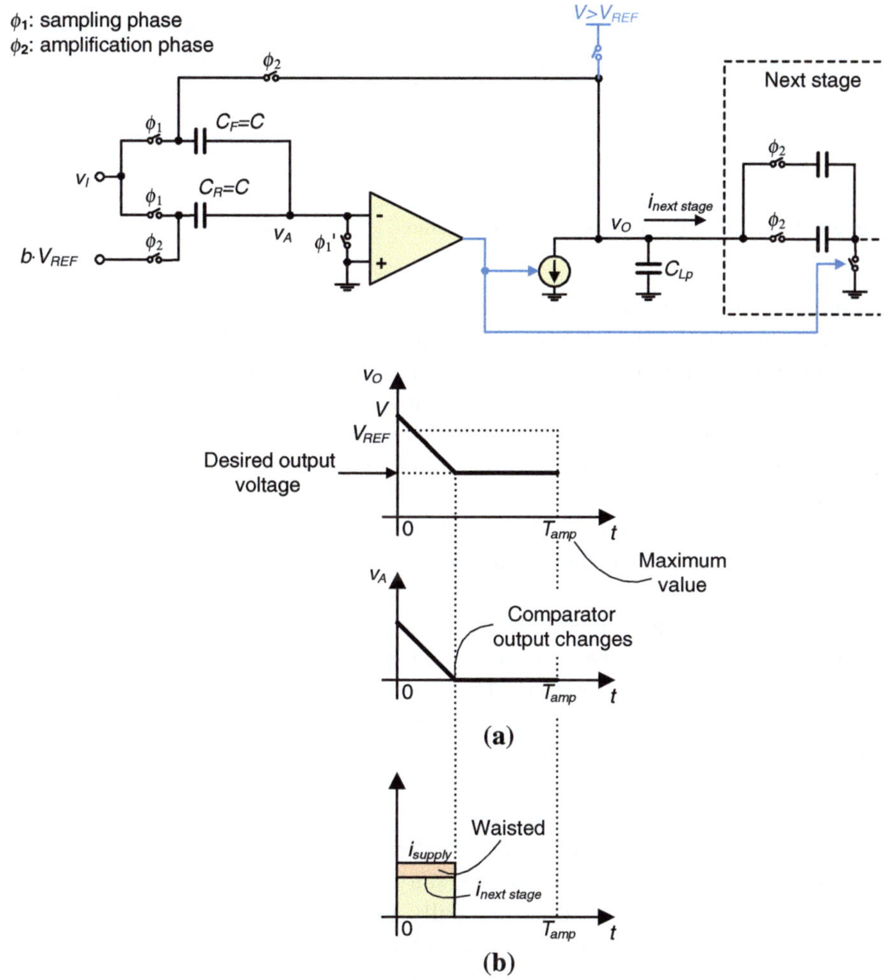

Fig. 29 Comparator based 1.5 bit MDAC: **a** Simplified circuit and some of the signals. **b** Supply current vs current in the next stage's sampling capacitors

will present a significant non-linearity. This has been mitigated by passing only a small amplitude residue to the next stage which, of course, needs to have a reduced input range [33].

Apart from the solution shown in Fig. 29, the alternatives discussed so far all consisted of amplifying structures using the MOS device as a transconductor. But other possibilities have been described. Reference [94] discloses an implementation based on parametric amplifiers, which consists of manipulating the inversion layer of MOS transistors to achieve voltage amplification [95, 96]. This amplification mechanism is noise free, but highly dependent on parasitic capacitances, and not easy to implement with low supply voltages. This ADC does not employ

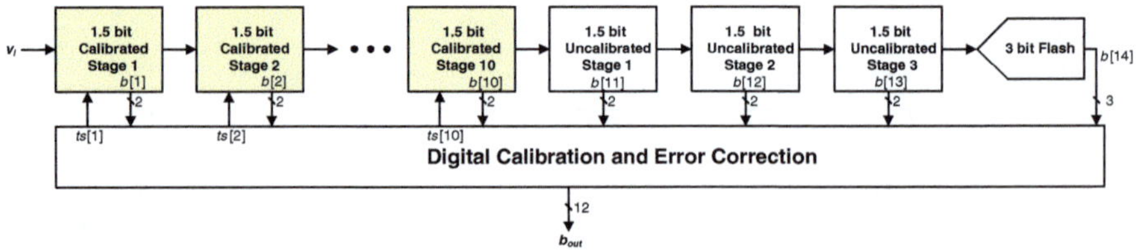

Fig. 30 12 bit 200 MS/s pipeline ADC architecture

calibration and features around 6 bit performance. Reference [97] described a different solution, but which uses similar principles.

Another possibility is the use of *bucket brigade circuits* that realize voltage gain by moving the charge stored on a large capacitor to a small capacitor [98, 99]. This solution is also highly sensitive to parasitic capacitances. Both these implementations resort to calibration to overcome limitations in this amplification mechanism.

We will now describe a 12 bit 200 MS/s pipeline ADC, whose main enabling techniques are the use of digital calibration—which compensates finite amplifier gain and all capacitor mismatches, while ensuring a reduced output swing at the stages—and an opamp switching technique that does not introduce speed or signal swing limitations. This ADC uses a single supply voltage, only regular V_{th} (core) transistors and no analog options. The digital calibration features a fast startup time, ensures robustness against variations of supply voltage or temperature without periodic offline re-calibration, and is independent of input signal statistics or amplitude.

The ADC architecture is illustrated in Fig. 30. The digital calibration principle depicted in Fig. 24 is extended to all 10 calibrated stages, in a nested fashion. The digital coefficients are obtained at the startup (foreground calibration), starting from the last stage and going sequentially up the first, by following a method similar to that shown on Fig. 25a. Thereafter, the digital coefficients are updated using a continuous background process, based on the injection of PRBS, *ts* [1] …*ts* [10], as illustrated in Fig. 25b. This combination of foreground and background calibration processes ensures a fast startup time, and the continuous adjustment of the coefficients as temperature or supply voltage changes, thus maintaining performance.

The analysis of this digital gain calibration process, as well as of the foreground and background methods of obtaining the coefficients, was already presented in Sect. 3.4. Here we will disclose some of the implementation details.

The MDAC schematic is presented in Fig. 31a, being equal to that already shown in Fig. 27 except for the $C_D = C/4$ capacitor that is used for the injection of the PRBS.

Figure 31b compares the standard transfer function of a 1.5 bit stage with the one employed [100]. Contrarily to previously published work [101–103], the output voltage of the MDAC is affected by its PRBS only at the middle segment,

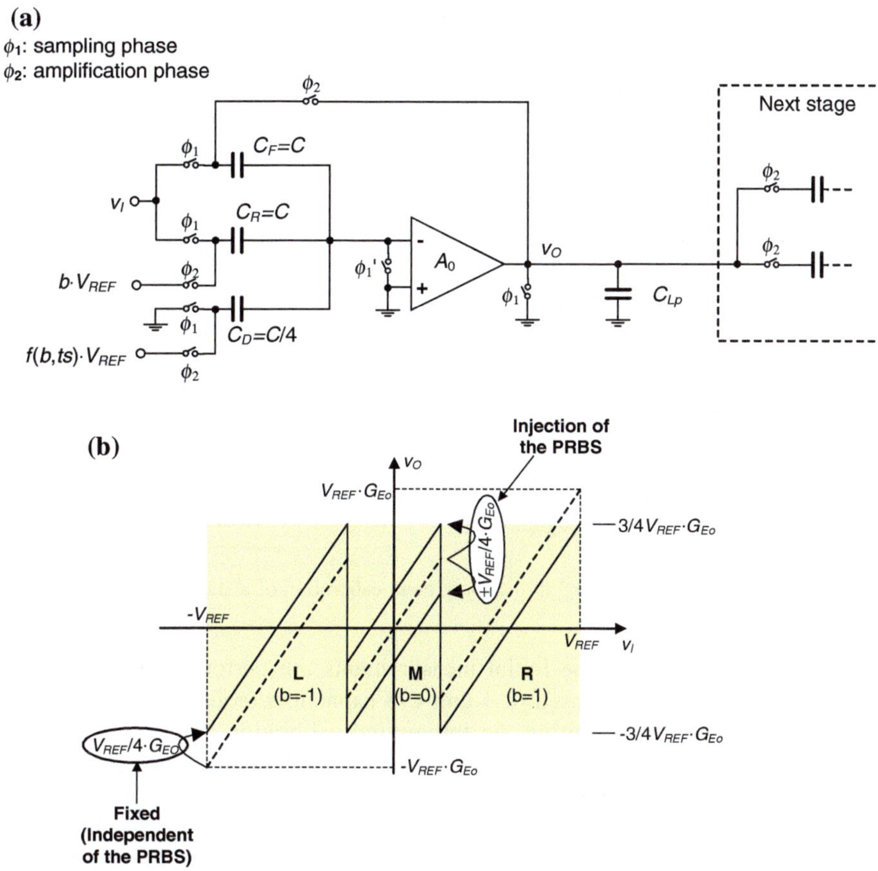

Fig. 31 **a** 1.5 bit MDAC circuit. **b** Comparison between the standard 1.5 bit stage transfer function (non-continuous lines) and the one used in this ADC (continuous lines). The PRBS only shifts the middle segment, M, and the signal swing in segments L/R is reduced

M, dislocating it randomly by $\pm V_{REF}/4 \cdot G_{Eo}$. Segments L/R remain fixed, but are shifted with respect to the standard transfer function in order to minimize the output signal excursion ($\pm 3/4 V_{REF} \cdot G_{Eo}$ instead of $\pm V_{REF} \cdot G_{Eo}$). This allows reducing power and mitigates amplifier non-linearity, which is not calibrated. The fixed shift in segments L/R is implemented using C_D.

The only limitation to the background process is that C_D must be exactly $C_R/4$, otherwise even if the digital gain correction coefficients are initially correct, the removal of the PRBSs in the digital domain is not perfect, and they will leak to the output. In that case the background calibration deviates the coefficients from the right values, causing a deficient correction of the MDACs' gain errors. We avoid this by measuring the C_D/C_R ratio during the foreground calibration, and using that information in the (digital domain) removal of the PRBSs.

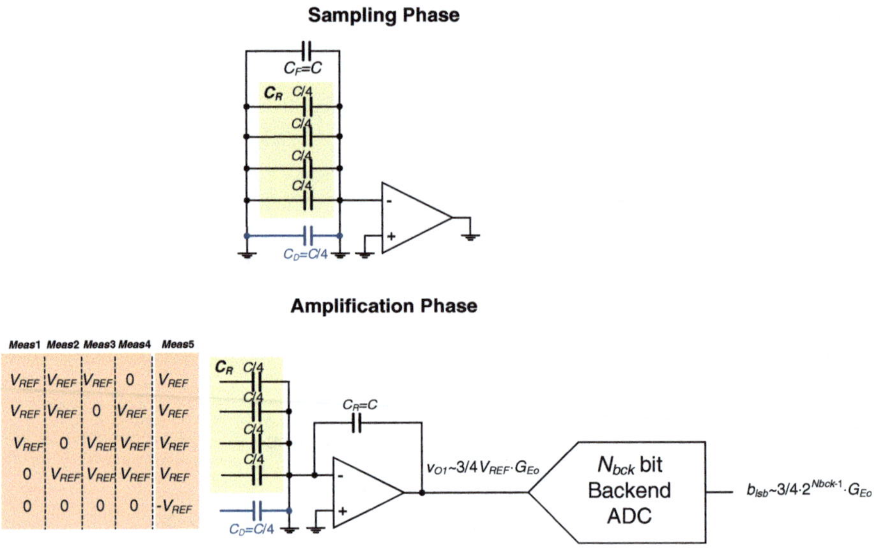

Fig. 32 Measurements performed in the foreground calibration of a stage

Figure 32 shows how the initial measurements are performed for each stage [104]. C_R is actually composed of 4 unit $C/4$ capacitors (equal to C_D), which are controlled independently during the foreground calibration. 5 measurements are taken by connecting the C_D and the 4 sub-units of C_R to either 0 or V_{REF}: the gain error, G_{Eo}, is obtained directly from the average of *Meas*1..4. Then, dividing *Meas*5 by the average of *Meas*1..4 allows determining the C_D/C_R ratio, that is used by the background calibration to avoid the digital coefficients from converging to wrong values. The output voltage of the amplifier is similar in all these measurements ($\approx 3/4 V_{REF} G_{Eo}$), which ensures calibration accuracy is not affected by its non-linearity.

Figure 33 shows the amplifier used in the calibrated stages, whose low DC gain leads to a loop gain on the MDAC, $A_0\beta$, of only ≈ 16 dB. Note that $A_0\beta$ around 75 dB is needed in the first stage of a 12 bit ADC without calibration, hence requiring a much more complex and power hungry amplifier.

Since the opamps are not used during the sampling phases of the MDACs, a switched-opamp technique reduces the supply current during those phases [84]: during the amplification phase of the MDAC, ϕ_2, S_1 is closed and C_B connects to $v_B = v_{diode}$ and $\phi_B = V_{DD}$; during the sampling phase S_1 opens and ϕ_B goes to 0, effectively reducing v_B and, consequently, the supply current of the amplifier. Then, just before the next amplification starts, ϕ_B is pushed to V_{DD}—restituting v_B back to the initial value, v_{diode}—and then S_1 closes. This solution adds no signal swing limitation to the amplifier, and is very fast because it depends only on the transition times of the inverter that generates ϕ_B, and on the charge redistribution time at node v_B.

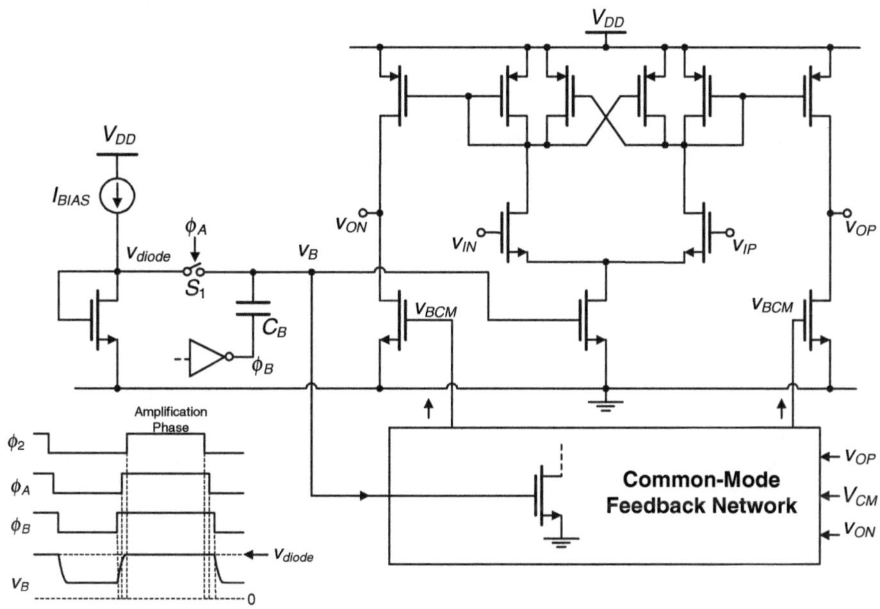

Fig. 33 Single-stage amplifier used in the calibrated stages, which employs a capacitive switching technique to reduce supply current during the sampling phase ($\phi_2 = 0$)

Figure 34 shows measured INL and DNL with calibration enabled, demonstrating that 12 bit resolution is effectively achieved. Figure 35 shows the situation when the digital calibration is disabled: there are many missing codes (those where DNL $= -1$), and the INL is nearly 100X worse, mainly due to the low open loop gain of the amplifiers. These results were obtained from the 65 nm implementation (see Fig. 3).

4.3 ADCs Without Residue Amplification

As discussed in Sects. 3.2 and 3.3, the performance of ADCs without residue amplification is limited by the non-idealities of quantizers. Since they do not require highly linear[19] and gain accurate blocks, they lend themselves better to be used in advanced technologies. This is one of the reasons why SAR ADCs became widely adopted in the last few years [80].

We have seen in Sects. 3.2 and 3.3 that in flash and subranging architectures the DACs need to provide a number of output voltages simultaneously, and so they are typically implemented with resistive ladders. In SAR ADCs the DAC only

[19] The S/H must be linear, but it can frequently be implemented without active elements.

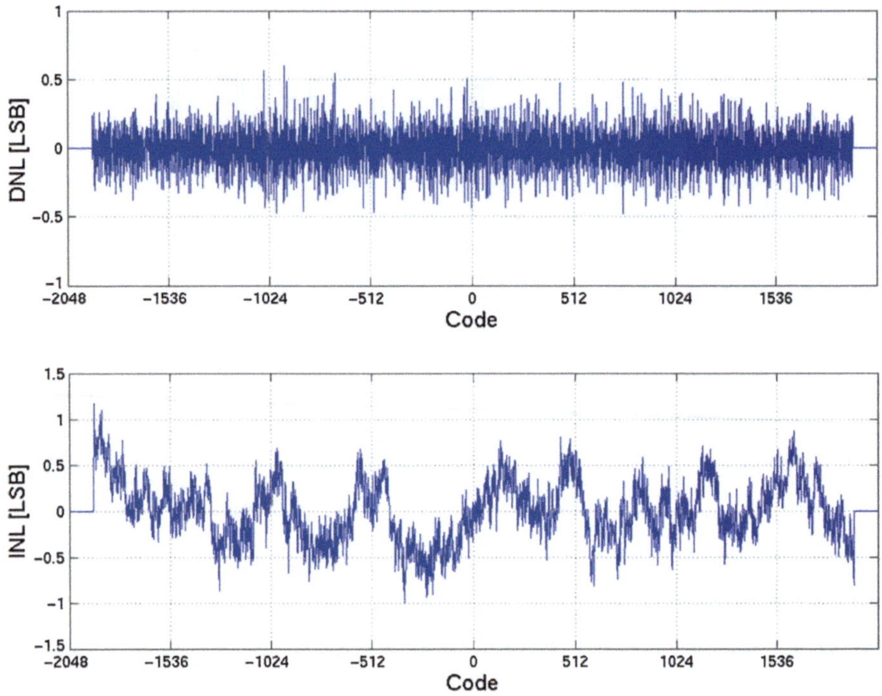

Fig. 34 DNL and INL with the calibration enabled ($f_s = 200$ MHz and $f_{in} = 10$ MHz)

provides one output, and is frequently implemented by a switched capacitor circuit that also performs sampling and subtraction operations. Switched capacitor implementations possess the advantage of having no static dissipation.

Random deviations on the constitutive elements of the DAC cause deviations on its output voltages (i.e. cause DAC non-linearity). Mismatches between DAC elements can be made smaller by using larger area devices [73]. Increasing the area of the resistors in a resistive ladder, increments the parasitic capacitances at its outputs, and may imply reducing the resistances to maintain proper settling. However, in switched capacitor DACs this minimum area constraint sets, directly, a minimum limit for the value of the capacitors. Since the sampling noise power is given by kT/C_S (where k is the Boltzmann constant, T is the temperature and C_S is the total sampling capacitance), it sets another minimum limit for capacitor sizes, and the largest of the two must be used.

As discussed in Sects. 3.1 and 3.2, the DAC non-linearity determines the overall non-linearity of SAR ADCs. In the 12 bit SAR ADC implementation described at the end of this section, DAC non-linearity is corrected through digital calibration, leaving only the noise constraint to be fulfilled. Reference [105] combines oversampling, chopping and deterministic dithering to, among other things, minimize the contribution of DAC non-linearity to the overall ADC noise and distortion that occurs in the signal band of interest.

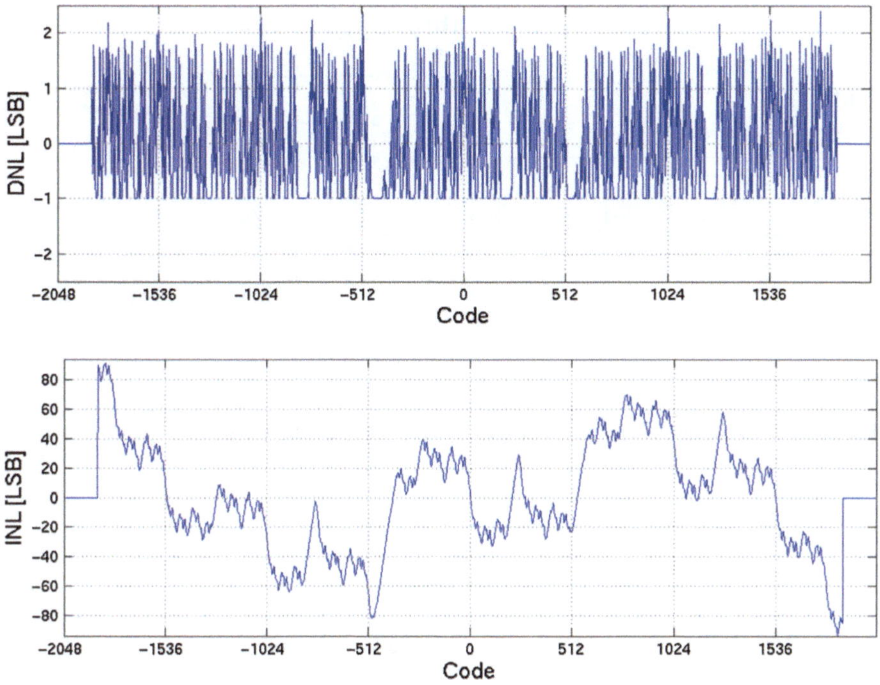

Fig. 35 DNL and INL with the calibration disabled ($f_s = 200$ MHz and $f_{in} = 10$ MHz)

We have seen in Sects. 3.2 and 3.3, that the linearity of flash and sub-ranging architectures, as well as of time-interleaved SAR ADCs, also depends on the offsets of subtractors and comparators. Note that in the flash with single-ended inputs shown in Fig. 10a, the subtraction between the input and the DAC outputs is automatically performed by the comparators.

However this is not the case with differential input signals. The subtraction may still be embedded in the comparator, if its input differential pair is replaced by a double differential pair arrangement [32], illustrated in Fig. 36a; or there may be a dedicated switched capacitor circuit before the comparator, that either also samples the input voltage (S/H operation) or the references voltages (i.e. the DAC outputs) [32, 53], as depicted in Fig. 36b.

Latched comparators used to be preceded by static pre-amplifying stages [32, 62] to reduce kickback noise, offset voltage, noise and metastability. However now that techniques have been proposed to decrease the kickback noise [32, 106, 107], and to calibrate directly the offset of the comparators [32, 37, 53, 108], and that noise and metastability have been conveniently characterized [79, 109, 110], it became possible to employ dynamic comparators without any static pre-amplification. We will now describe the main comparator offset reduction alternatives.

Figure 37 illustrates two alternative offset reduction techniques, which are based on the introduction of a pre-amplifier before the latched comparator. The offset of

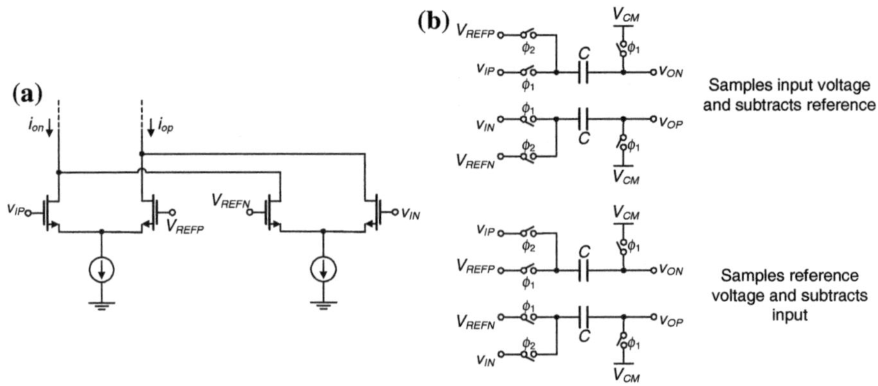

Fig. 36 Subtractor implementations: **a** Double differential pairs. **b** Switched capacitor circuits

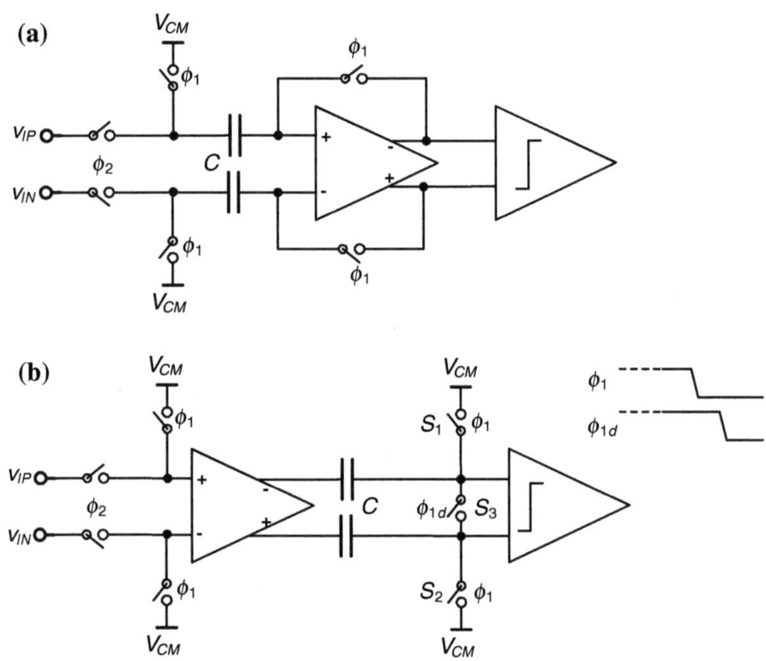

Fig. 37 Offset reduction using static pre-amplifiers: **a** Input offset storage. **b** Output offset storage

the pre-amplifier is stored on a capacitor, and removed during the comparison phase, while the offset of the comparator is attenuated by the gain of the pre-amplifier. Examples of ADCs resorting to these techniques are, for example [38, 62].

As detailed in [32] these solutions have residual offsets, which can be partially mitigated by employing several pre-amplifying stages. However, such approach does

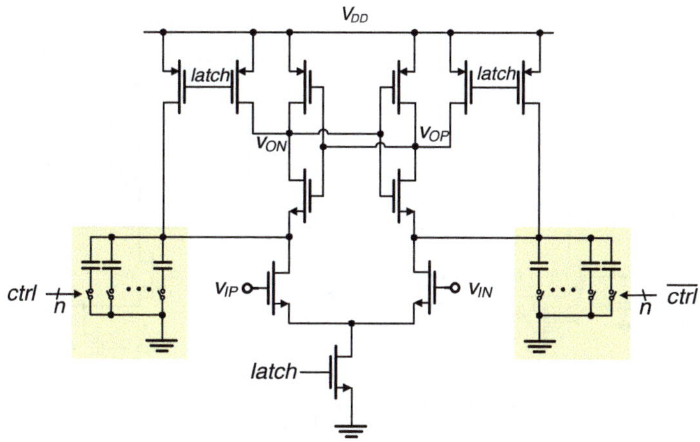

Fig. 38 Example of offset calibration through the addition of programmable capacitor arrays

not reduce the contributions from the mismatch of the charges injected by the switches that open at the end of the offset sampling phases [32]. Reference [111] proposed a sequential clocking scheme which attenuates the effect of charge injection mismatches, but it cannot be used in high speed applications. Luckily, in case of *output offset storage* there is an easy solution [32]: add a switch between the p and n sides (S_3), and open first the switches that set the common mode at the input of the comparator (S_1/S_2). S_3 removes the charge injection mismatch of S_1/S_2 and, when it opens, no mismatch is created as there is a single element. This technique was also used in [53] to remove the offset contribution of the switches on the distributed input sampling/subtractor circuits (whose architecture was depicted in Fig. 14).

In the attempt to maximize power efficiency, recent works calibrate the offset of the comparator directly. The authors of [37] added arrays of small capacitors connected to internal nodes of the comparator. The right capacitor imbalancement between the p/n sides of the comparator cancels its offset. This is illustrated in Fig. 38.

The arrays must contain a significant number of capacitors, in order to have a small offset correction step and be able to correct large offsets. This increases power consumption and slows down the comparator. An alternative is using current sources instead of capacitors [112], but the trade-offs are similar.

A better alternative is illustrated in Fig. 39a, where a scaled down version of the main differential pair is added in parallel with it, and whose input voltage is adjusted to eliminate the offset voltage. A model is shown in Fig. 39b: in the offset sensing phase the inputs of the main differential pair are shorted and the comparator is triggered, taking a decision based solely on its offset voltage. The small capacitance C_P is pre-charged to either V_{MAX} or V_{MIN} depending on that decision, and is later connected to the C_{CAL} ($C_{CAL} \gg C_P$), adjusting the calibration voltage, v_{CAL}, so as to place the comparator at its threshold point for zero input voltage.

After several offset sensing phases, v_{CAL} reaches the value that eliminates the offset. The comparator will thereafter decide alternatively 0 and 1, thus

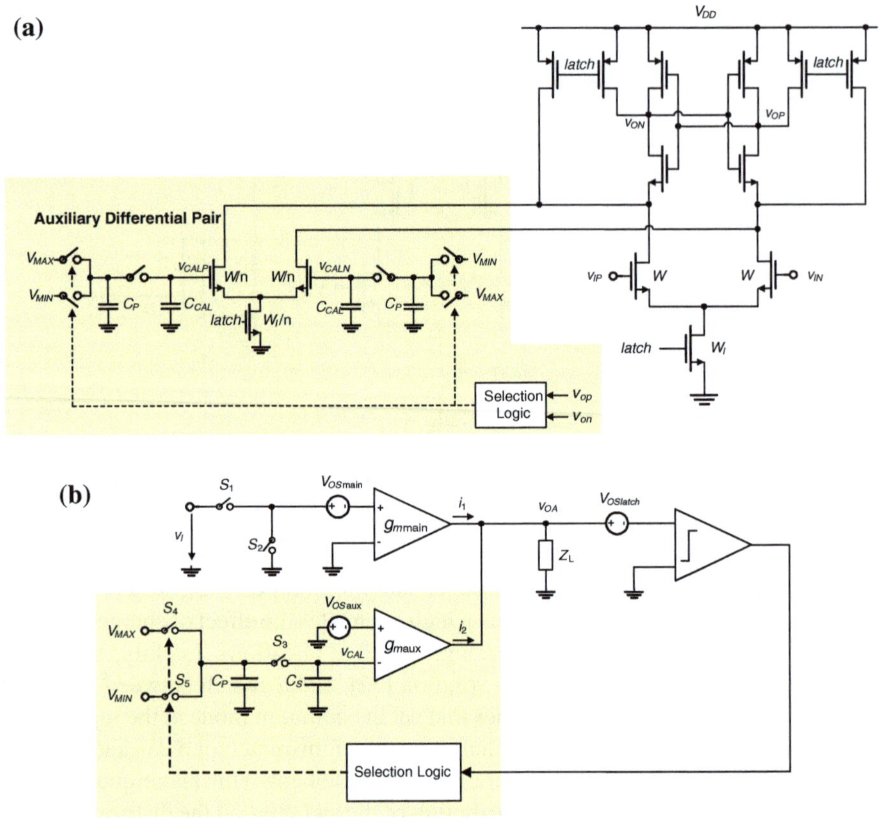

Fig. 39 Offset calibration technique with additional differential pair and switched capacitor integrator: **a** Circuit implementation. **b** Single-ended model

maintaining v_{CAL} around the correct value. The offset calibration range is defined by $V_{MAX} - V_{MIN}$ and g_{maux}/g_{mmain} and can easily be made high enough. The offset calibration step is, roughly, the calibration range divided by C_{CAL}/C_P, which can be made quite high since C_P is the parasitic capacitance at the node between the switches, and C_{CAL} is implemented with a MOS capacitor. This was proposed in [53], discussed in detail in [32], and used for example in [36, 48, 113, 114], and in the 12 bit SAR ADC that will be discussed at the end of this section.

Figure 40 illustrates a technique that is frequently employed to reduce the offsets in flash ADCs[20]—*averaging* [115]. Section 3.2 mentioned that these ADCs have $2^N - 1$ parallel paths, which simultaneously compare the input voltage against all threshold levels (outputs of the DAC). We have also seen that the offset in *each* of those paths, constituted by a subtractor and a comparator, must be lower

[20] And also in folding and interpolation ADCs, which were not discussed in this manuscript because currently they are not frequently used.

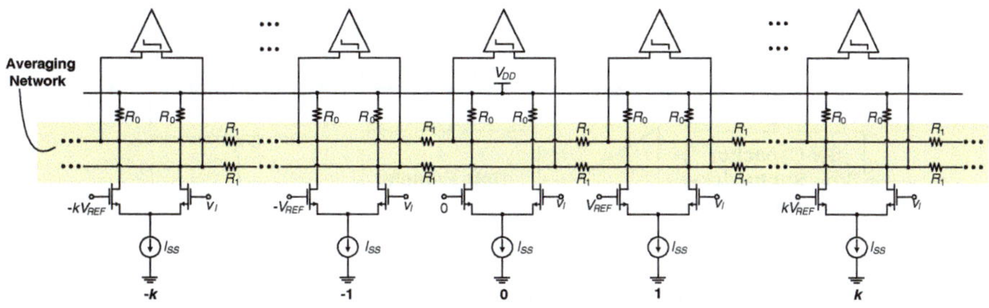

Fig. 40 Averaging technique

than the LSB. Offsets are caused by component mismatches, and are made smaller by increasing their area [32]. So, all subtractors/comparators may need to use large area devices, leading to a speed-power-accuracy trade-off [78].

Averaging is typically implemented by connecting resistors between the outputs of pre-amplifiers preceding the latched comparators. In this way, the input voltage of each comparator is now determined by the devices of a few pre-amplifiers instead of just one, and this allows reducing the area of individual components. Examples of recent ADCs resorting to these techniques are [62, 116–118]. Reference [119] provided an exhaustive study on the application of this technique to flash ADCs. Reference [32] expanded that study, covering also folding and interpolation ADCs, the transient response, termination strategies, etc.

Reference [120] described a flash ADC implementation, at which comparator matching is decoupled from overall ADC performance. It consists on having several minimum size comparators connected to each reference voltage, and then selecting those that yield the ADC transfer function best approaching the ideal one. The comparator selection is made on the ADC power-up, after which the unused comparators are turned off. This is another example where relaxed analog complexity is traded favorably by extra digital complexity (in the encoder and control of the comparators).

The *stochastic flash ADC* [121] also uses many minimum size comparators ($\approx 2{\cdot}4^N$ for N bit resolution), with a large offset spread. But the basic idea is different from that of [120]: the comparators do not receive different reference voltages—their trip points are uniquely determined by the individual offsets. By counting the number of '1' output by the comparators, one obtains an analog-to-digital conversion following a nonlinear transfer function, described by a Gaussian cumulative distribution. Different methods to circumvent this non-linearity are presented in [120, 122]. This last reference described a stochastic flash ADC that is fully synthesized from Verilog code and a standard cell library. It is the true all-digital ADC, with the input being the only analog signal present.

We will now briefly describe the Synopsys 12 bit 80 MS/s SAR ADC IP product. Two other Synopsys ADC IP products—a 12 bit 160 MS/s ADC and another sampling at 320 MS/s—are based on this one, by time-interleaving two and four of them. We will focus our attention on the 80 MS/s version, whose block diagram is shown in Fig. 41.

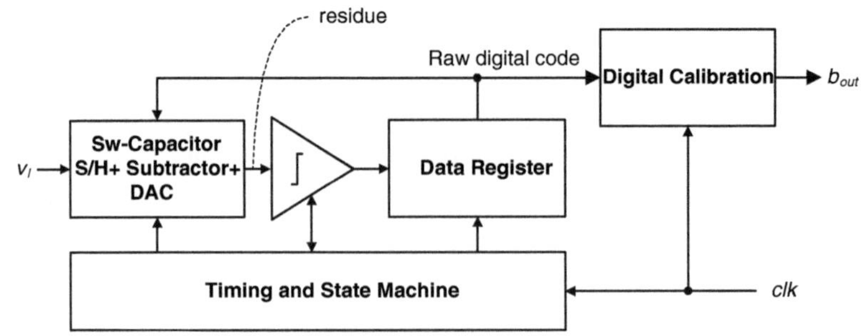

Fig. 41 Block diagram of the 12 bit 80 MS/s SAR ADC

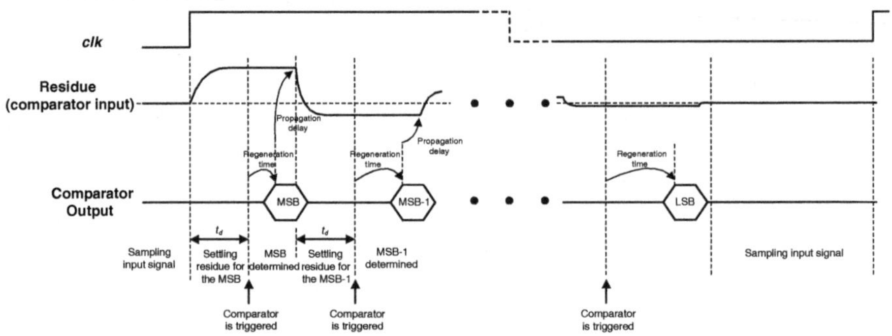

Fig. 42 Simplified timing diagram of the 12 bit SAR ADC

The sampling switches open shortly after the rising edge of the input clock, and the successive approximation search process begins (see Fig. 42). The conversion process is self-timed [35]: after it has been detected that the comparator completed a decision, the DAC input is updated according to that decision and must settle during a certain amount of time, defined by a delay line; the comparator is then triggered to determine the next bit, and the process just described is repeated. The ADC starts sampling the next input immediately after all bits have been found. This differs from most published asynchronous SAR ADC implementations [36, 35], that require an input clock with a small duty cycle to define the sampling window, which is not very convenient.

The comparator uses a fully dynamic structure, properly designed to achieve the necessary low thermal and kickback noise. The offset calibration method shown in Fig. 39 is employed.

A switched capacitor circuit implements the S/H, DAC and subtraction operations. A straightforward realization of a 12-bit DAC would require 4096 unit capacitors, which would have to be very small ($\sim$0.15 fF) in order to ensure the total capacitance was not higher than the value dictated by the sampling noise constraint. In order to avoid using such small capacitor units, this implementation uses capacitive dividers to avoid the exponential increase of the number of unit

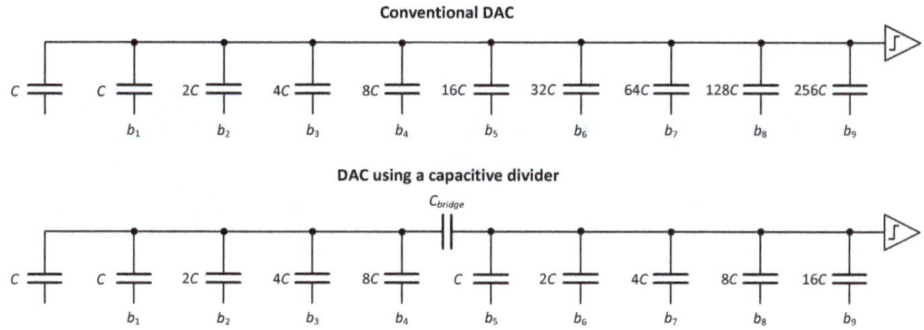

Fig. 43 Comparison between the conventional DAC and the solution employing a capacitive divider (example for a 9-bit ADC)

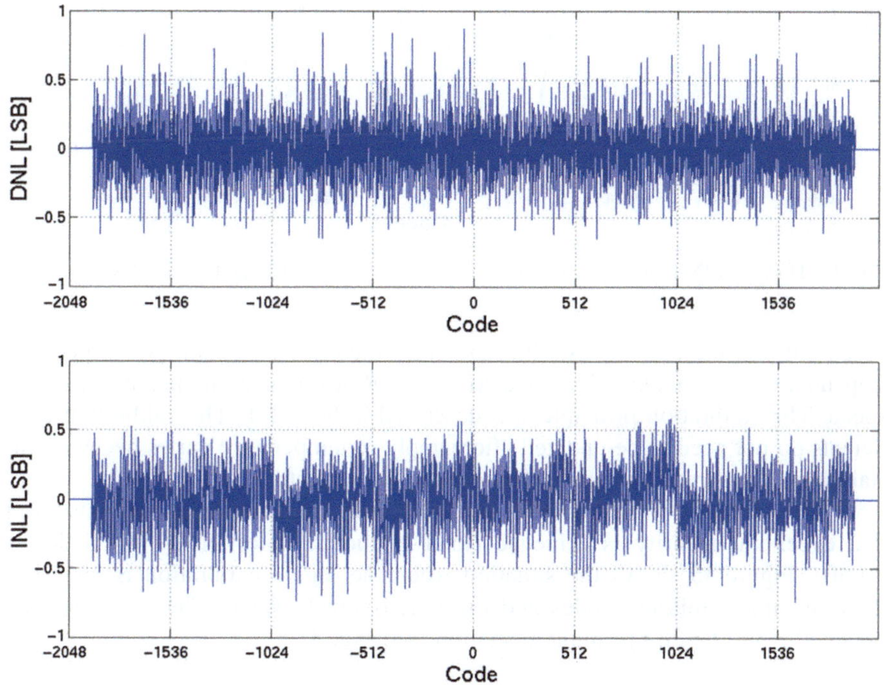

Fig. 44 DNL and INL with the calibration enabled (f_s = 80 MHz and f_{in} = 20 MHz)

capacitors [34], as illustrated in Fig. 43. Also, proper sizing of C_{bridge} allows implementing redundancy.

Digital calibration is employed to address random capacitor mismatches and the sensitivity exhibited by this solution to parasitics in the capacitive divider nodes. It measures actual capacitor weights at ADC startup, and during conversion corrects the raw digital code provided by the SAR ADC. This allows sizing the capacitors

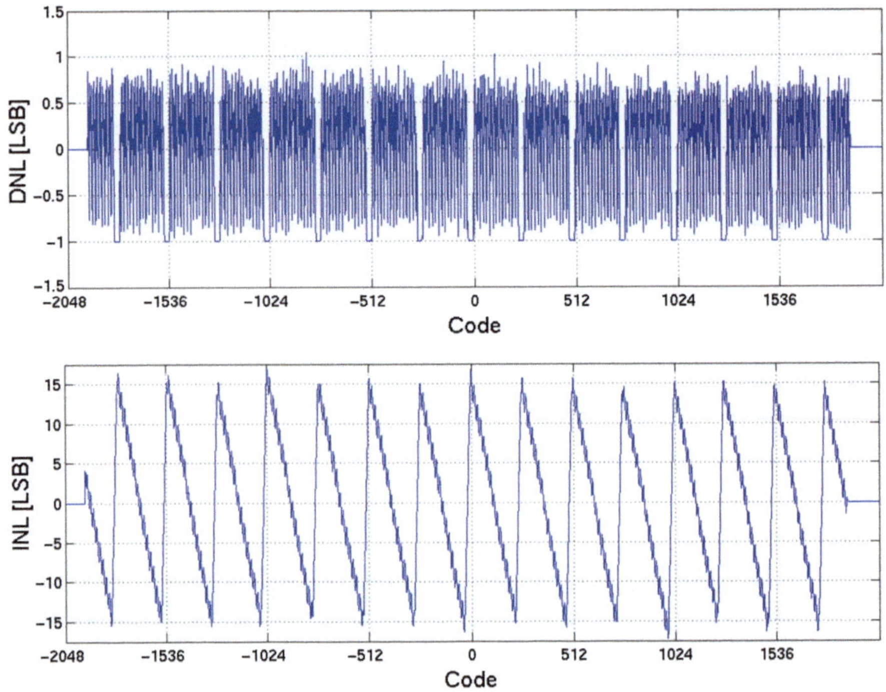

Fig. 45 DNL and INL with the calibration bypassed ($f_s = 80$ MHz and $f_{in} = 20$ MHz)

based only on the noise constraint, and since the calibration is a post-processing step no extra complexity is added to the SAR itself, maximizing its operating speed. The calibration principle was described in Sect. 3.1. The calibration coefficients do not need to be updated after supply or temperature varies because only stable and drift-free capacitor ratios need to be corrected [34].

Figure 44 shows measured INL and DNL in normal operation, demonstrating that excellent linearity is achieved (total harmonic distortion is well below −80 dB). Figure 45 shows the situation when the digital calibration is bypassed: there are many missing codes and the INL is much worse. These results were obtained from the 28 nm implementation (see Fig. 3).

5 Conclusions

This manuscript discussed the requirements and architectures for ADCs currently being embedded in communication SoCs. As new standards targeting higher data rates are released, ADC speed and resolution specifications have been increasing, but not dramatically. However the pressure to reduce costs (use lowest possible silicon area), and optimize energy efficiency to increase battery life, is enormous.

Furthermore, these converters need to be fabricated in advanced CMOS technologies, which are often at an immature state of development and feature lower supply voltages, less process options, higher variability, more layout dependent effects, higher leakage currents, and worse analog device properties. This reality fuels the search for more efficient ADC solutions, found by ingeniously selecting the circuit solutions that best address architecture limitations, while crafting the architecture to ease the trade-offs found in practical circuit solutions.

This chapter discussed ADC architectures in terms of the fundamental operations realized inside them. This constitutes a unified treatment that relates all architectures, thus providing a deeper understanding of their fundamental limitations and trade-offs. It follows from this analysis that Nyquist rate ADC architectures can broadly be separated in two groups:

1. The ones based **only on quantizers** (whose non-idealities limit performance), and that to increase operating speed resort to "time" and/or "code search" parallelizations—*flash, SAR, subranging ADCs*.
2. The *pipeline* ADC, which in addition to the quantizers (that have relaxed specifications), **includes residue calculator/amplifier blocks (MDACs)**, whose non-idealities limit performance. Furthermore, pipelining is used instead of the above mentioned parallelization techniques, to sample the input signal and provide an output sample in every clock cycle.

Of course this distinction constitutes no hard boundary, and published implementations joining features from both groups exist. For example SAR assisted pipeline ADCs [33, 63], used both time-interleaving and pipelining to increase operating speed, while the residue gain is just high enough to ease quantizer requirements while still keeping residue amplifier non-linearity low enough.

After discussing ADC architectures, this chapter overviewed some of the most relevant circuit solutions to implement their sub-blocks. There is a clear trend towards *digitally aided analog design*, where analog circuits' complexity is relaxed and traded favorably by extra digital complexity. This is enabled by technology scaling, where increasingly powerful, cheap and low power digital processing circuits are available.

Finally, we disclosed implementation details from two 12 bit digitally calibrated, high-speed ADCs, and discussed the evolution of energy efficiency observed when implementing them in different CMOS technology nodes.

References

1. G. Moore, "No exponential is forever: but 'forever' can be delayed!" *in Proc. ISSCC Dig. Tech. Papers*, pp. 20-23, Feb. 2003.
2. IC Insights (2013, Oct.), Wireless networking remains a strong IC market driver. [Online]. Available: http://www.icinsights.com/news/bulletins/Wireless-Networking-Remains-A-Strong-IC-Market-Driver/
3. A. Behzad, *Wireless LAN Radios*. IEEE Press/John Wiley & Sons, 2007.

4. G. Manganaro, *Advanced Data Converters*. Cambridge University Press, 2012.
5. K. Bult, "Embedded analog-to-digital converters," *in Proc. ESSCIRC*, pp. 14-18, Sep. 2009.
6. A. Asenov, "Simulation of statistical variability in nano MOSFETs," *in Proc. IEEE Symp. on VLSI Techn.*, pp. 86–87, Jun. 2007.
7. J. Faricelli, "Layout-dependent proximity effects in deep nanoscale CMOS," *in Proc. CICC*, pp. 1–8, Sep. 2010.
8. Y. Chiu, B. Nikolic, and P. Gray, "Scaling of analog-to-digital converters in ultra-deep-submicron CMOS," *in Proc. CICC*, pp. 375–385, Sep. 2005.
9. B. Murmann, "Digitally assisted data converter design," *in Proc. ESSCIRC*, pp. 24–31, Sep. 2013.
10. http://www.synopsys.com/dw/analogip.php?pg=analog-to-digital_converters
11. B. Razavi, *RF Microelectronics*. 2nd edition, Pearson, 2011.
12. M. Sternard *et al.*, "Towards systems beyond 3G based on adaptive OFDMA transmission," *Proc. of the IEEE*, pp. 2432–2455, Dec. 2007.
13. M. Mota (2009, Nov.), How system-level trade-offs drive data converter decisions. [Online] Available: https://www.synopsys.com/dw/doc.php/wp/dcs_decisions_wp.pdf
14. T. Bhandare, (Dec. 2008), LTE and WiMAX comparison. [Online]. Available: http://www.halcyonwireless.com/LTE%20and%20WiMAX%20Comparison-TejasBhandare.pdf
15. M. Sawahashi *et al.*, "Broadband radio access: LTE and LTE-advanced," *in Proc. of the ISPACS*, pp. 224–227, Jan. 2009.
16. D. Bai *et al.*, "LTE-Advanced modem design: challenges and perspectives," *IEEE Com. Magazine*, pp. 176–186, Feb. 2012.
17. Qualcomm (2012), IEEE802.11ac: The next evolution of WiFi standards. [Online]. Available: http://www.qualcomm.com/media/documents/files/ieee802-11ac-the-next-evolution-of-wi-fi.pdf
18. I. Papapanagiotou *et al.*, "A survey on next generation mobile WiMAX networks: objectives, features and technical challenges," *IEEE Com. Surveys & Tutorials*, pp. 3–18, 4th quarter 2009.
19. S. Ahmadi, "An overview of next-generation mobile WiMAX technology," *IEEE Com. Magazine*, pp. 84–98, Jun. 2009.
20. M. Muller (Jul. 2010), IEEE 802.16m Technology introduction, white chapter. [Online]. Available: http://cdn.rohde-schwarz.com/dl_downloads/dl_application/application_notes/1ma167/1MA167_3e_IEEE_80216m_technology.pdf
21. A. Cattoni *et al.*, "Multi-user MIMO and carrier aggregation in 4G systems: the SAMURAI approach," *in Proc. WCNCW*, pp. 288–293, Apr. 2012.
22. C.-M. Lai *et al.*, "Compact router transceiver architecture for carrier aggregation systems," *in Proc. of the 41th EuMC*, pp. 693–696, Oct. 2011.
23. C.-M. Lai *et al.*, "CMOS RF T/R router switch ICs for LTE carrier aggregation transceivers," *in Proc. of the 42nd EuMC*, pp. 904–907, Nov. 2012.
24. E. Perahia and M. Gong, "Gigabit wireless LANs: and overview of IEEE 802.11ac and 802.11ad," *Mobile Computing and Communications Review*, pp. 23–33, Jul. 2011.
25. M. Grodzinsky (Dec. 2013), Understanding where 802.11ad WiGig fits into the gigabit Wi-Fi picture. [Online]. Available: http://www.networkworld.com/news/tech/2013/120413–80211ad-wigig-276569.html
26. V. Oksman and S. Galli, "G.hn: The new ITU-T home networking standard," *IEEE Com. Magazine*, pp. 138–145, Oct. 2009.
27. HomePlug Powerline Alliance (2013), HomePlug^{TM} AV2 Technology. [Online]. Available: http://www.homeplug.org/tech/whitechapters/HomePlug_AV2_whitechapter_130909.pdf
28. A. Monk, R. Lee, and Y. Hebron, "The multimedia over coax alliance," *Proceedings of the IEEE*, pp. 2322–2338, Nov. 2013.
29. R. Walden, "Analog-to-digital survey and analysis," *IEEE J. on Selected Areas in Communications*, pp. 539–550, Apr. 1999.

30. B. Murmann, "A/D converter trends: power dissipation, scaling and digitally assisted architectures," *in Proc. of the CICC*, pp. 105–112, Sep. 2008.
31. B. Jonsson, "A survey of A/D-converter performance evolution", *in Proc. IEEE ICECS*, pp. 766-769, Dec. 2010.
32. P. Figueiredo and J. Vital, *Offset Reduction Techniques in High-Speed Analog-to-Digital Converters*. Springer, 2009.
33. B. Verbruggen, M. Iriguchi, and J. Craninckx, "A 1.7 mW 11b 250 MS/s 2-times interleaved fully dynamic pipelined SAR ADC in 40 nm digital CMOS," *IEEE J. Solid-State Circuits*, pp. 2880–2887, Dec. 2012.
34. M. Yoshioka, "A 10-b 50-MS/s 820mW SAR ADC with on-chip digital calibration," *IEEE Trans. on Biomedical Circuits and Systems*, pp. 410–416, Dec. 2010.
35. S.-W. Chen and R. Brodersen, "A 6-bit 600-MS/s 5.3-mW asynchronous ADC in 0.13-μm CMOS," *IEEE J. of Solid-State Circuits*, pp. 2669–2680, Dec. 2006.
36. L. Kull *et al.*, "A 3.1 mW 8b 1.2 GS/s single-channel asynchronous SAR ADC with alternate comparators for enhanced speed in 32 nm digital SOI CMOS," *IEEE J. Solid-State Circuits*, pp. 3049–3058, Dec. 2013.
37. G. Plas. S. Decoutere, and S. Donnay, "A 0.16pJ/conversion-step 2.5mW 1.25GS/s 4b ADC in 90 nm digital CMOS process," *in Proc. ISSCC Dig. Tech. Papers*, pp. 566–567, Feb. 2006.
38. C. Sandner *et al.*, "A 6-bit 1.2-GS/s low-power flash-ADC in 0.13-μm digital CMOS," *IEEE J. Solid-State Circuits*, pp. 1499–1505, Jul. 2005.
39. J. Pernillo and M. Flynn, "A 1.5-GS/s flash ADC with 57.7-dB SFDR and 6.4-bit ENOB in 90 nm digital CMOS," *IEEE J. Solid-State Circuits*, pp. 837–841, Dec. 2011.
40. B. Peetz, B. Hamilton, and J. Kang, "An 8-bit 250 megasample per second analog-to-digital converter: operation without a sample and hold," *IEEE J. Solid-State Circuits*, pp. 997–1002, Dec. 1986.
41. R. Plassche, *CMOS Integrated Analog-to-Digital and Digital-to-Analog Converters*. 2nd edition, Kluwer, 2003.
42. Y. Lim and M. Flynn, "A 100MS/s 10.5b 2.46mW comparator-less pipeline ADC using self-biased ring amplifiers," *in Proc. ISSCC Dig. Tech. Papers*, pp. 202–203, Feb. 2014.
43. L. Risbo, "Stability predictions for high-order Σ-Δ modulators based on quasilinear modeling," *in Proc. IEEE ISCAS*, pp. 361–364. Jun. 1994.
44. P. Nikaeen and B. Murmann, "Digital correction of dynamic track-and-hold errors providing SFDR $\gg$ 83 dB up to fin = 470 MHz," *in Proc. CICC*, pp. 21–24, Sep. 2008.
45. Y. Zhun *et al.*, "A 10.4-ENOB 120MS/s SAR ADC with DAC linearity calibration in 90 nm CMOS," *in Proc. A-SSCC*, pp. 69–72, Nov. 2013.
46. J.-W. Nam *et al.*, "A 95-MS/s 11-bit 1.36-mW asynchronous SAR ADC with embedded passive gain in 65 nm CMOS," *in Proc. CICC*, pp. 1–4, Sep. 2013.
47. R. Kapusta *et al.*, "A 14b 80MS/s SAR ADC with 73.6 dB SNDR in 65 nm CMOS," *IEEE J. Solid-State Circuits*, pp. 3059–3066, Dec. 2013.
48. L. Kull *et al.*, "A 90GS/s 8b 667mW 64 × interleaved SAR ADC in 32 nm digital SOI CMOS," *in Proc. ISSCC Dig. Tech. Papers*, pp. 378–379, Feb. 2014.
49. P. Schvan *et al.*, "A 24GS/s 6b ADC in 90 nm CMOS," *in Proc. ISSCC Dig. Tech. Papers*, pp. 544–545, Feb. 2008.
50. E. Janssen *et al.*, "An 11b 3.6GS/s time-interleaved SAR ADC in 65 nm CMOS," *in Proc. ISSCC Dig. Tech. Papers*, pp. 464–465, Feb. 2013.
51. S. Tual *et al.*, "A 20 GHz-BW 6b 10GS/s 32mW time-interleaved SAR ADC with master T&H in 28 nm UTBB FDSOI technology," *in Proc. ISSCC Dig. Tech. Papers*, pp. 382–383, Feb. 2014.
52. C. Vogel, "The impact of combined channel mismatch effects in time-interleaved ADCs," *IEEE Trans. on Instrumentation and Measurement*, pp. 415–427, Feb. 2005.
53. P. Figueiredo *et al.*, "A 90 nm CMOS 1.2 V 1GS/s two-step subranging ADC," *in Proc. ISSCC Dig. Tech. Papers*, pp. 568–569, Feb. 2006.

54. H. Ploeg *et al.*, "A 15-bit 30-MS/s 145-mW three-step ADC for imaging applications," *IEEE J. Solid-State Circuits*, pp. 1572–1577, Jul. 2006.
55. S. Hosotani *et al.*, "An 8-bit 20-MS/s CMOS A/D converter with 50-mW power consumption," *IEEE J. Solid-State Circuits*, pp. 167–172, Feb. 1990.
56. R. Taft and M. Tursi, "A 100-MS/s 8-b CMOS subranging ADC with sustained parametric performance from 3.8 V down to 2.2 V," *IEEE J. Solid-State Circuits*, pp. 331–338, Mar. 2001.
57. H. Hong *et al.*, "A 8.6 ENOB 900MS/s time-interleaved 2b/cycle SAR ADC with a 1b/ cycle reconfiguration for resolution enhancement," *in Proc. ISSCC Dig. Tech. Papers*, pp. 470–471, Feb. 2013.
58. H. Tai *et al.*, "A 0.85fJ/conversion-step 10b 200kS/s subranging SAR ADC in 40 nm CMOS," *in Proc. ISSCC Dig. Tech. Papers*, pp. 196–197, Feb. 2014.
59. D.-Y. Chang *et al.*, "A 21mW 15b 48MS/s zero-crossing pipeline ADC in 0.13 μm CMOS with 74 dB SNDR," *in Proc. ISSCC Dig. Tech. Papers*, pp. 204–205, Feb. 2014.
60. S. Lee *et al.*, "A 1GS/s 10b 18.9mW time-interleaved SAR ADC with background timing-skew calibration," *in Proc. ISSCC Dig. Tech. Papers*, pp. 384–385, Feb. 2014.
61. B. Brandt and J. Lutsky, "A 75-mW, 10-b, 20-MSPS CMOS Subranging ADC with 9.5 effective bits at Nyquist," *IEEE J. Solid-State Circuits*, pp. 1788–1795, Jul. 2005.
62. J. Mulder *et al.*, "A 21-mW 8-b 125-MSample/s ADC in 0.09-mm^2 0.13-μm CMOS," *IEEE J. Solid-State Circuits*, pp. 2116–2125, Dec. 2004.
63. F. Goes *et al.*, "A 1.5mW 68 dB SNDR 80MS/s 2 × interleaved SAR-assisted pipeline ADC in 28 nm CMOS," *in Proc. ISSCC Dig. Tech. Papers*, pp. 200–201, Feb. 2014.
64. H. Kim, D. Jeong, and W. Kim, "A 30mW 8b 200MS/s pipelined CMOS ADC using a switched-opamp technique," *in Proc. ISSCC Dig. Tech. Papers*, pp. 284–285, Feb. 2005.
65. S. Lewis *et al.*, "A 10-b 20-Msample/s analog-to-digital converter," *IEEE J. Solid-State Circuits*, pp. 351–358, Mar. 1992.
66. A. Ali *et al.*, "A 14b 1GS/s RF sampling pipelined ADC with background calibration," *in Proc. ISSCC Dig. Tech. Papers*, pp. 482–483, Feb. 2014.
67. C. Lee and M. Flynn, "A SAR-assisted two-stage pipeline ADC," *IEEE J. Solid-State Circuits*, pp. 859–869, Apr. 2011.
68. M. Horowitz *et al.*, "Scaling, power and the future of CMOS," *in IEEE IEDM Techn. Digest*, pp. 7–15, Dec. 2005.
69. B. Murmann *et al.*, "Impact of scaling on analog performance and associated modeling needs," *IEEE Trans. on Electron Devices*, pp. 2160–2167, Sep. 2006.
70. Y. Tsividis and C. McAndrew, *Operation and Modeling of the MOS Transistor*. 3rd edition, Oxford University Press, 2011.
71. A. Abo and P. Gray, "A 1.5-V, 10-bit, 14.3-MS/s CMOS pipeline analog-to-digital converter," *IEEE J. Solid-State Circuits*, pp. 599–606, May 1999.
72. K. Bult, "Analog design in deep sub-micron CMOS," *in Proc. ESSCIRC*, pp. 126–132, Sep. 2014.
73. M. Pelgrom, A. Duinmaijer, and A. Welbers, "Matching properties of MOS transistors," *IEEE J. Solid-State Circuits*, pp. 1433–1440, Oct. 1989.
74. M. Steyaert *et al.*, "Threshold voltage mismatch in short-channel MOS transistors," *Electronic Letters*, pp. 1546–1548, Sep. 1994.
75. P. Stolk and D. Klaassen, "The effect of statistical dopant fluctuations on MOS device performance," *in Proc. of the IEDM*, pp. 627–630, Dec. 1996.
76. L. Lewyn *et al.*, "Analog circuit design in nanoscale CMOS technologies," *Proc. of the IEEE*, pp. 1687–1714, Oct. 2009.
77. A. Annema *et al.*, "Analog circuits in ultra-deep-submicron CMOS," *IEEE J. Solid-State Circuits*, pp. 133–143, Jan. 2005.
78. K. Uyttenhove and M. Steyaert, "Speed–Power–Accuracy tradeoff in high-speed CMOS ADCs," *IEEE Trans. Circuits Syst. II*, pp. 280–287, Apr. 2002.
79. P. Figueiredo, "Comparator metastability in the presence of noise," *IEEE Trans. Circuits Syst. I*, pp. 1286–1299, May. 2013.

80. S. Cho *et al.*, "A 550-mW 10-b 40-MS/s SAR ADC with multistep addition-only digital error correction," *IEEE J. of Solid-State Circuits*, pp. 1881–1892, Aug. 2011.
81. D. Haigh and B. Singh, "A switching scheme for switched capacitor filters which reduces the effect of parasitic capacitances associated with switch control terminals," *in Proc. of the IEEE ISCAS*, vol. II, pp. 586–589, 1983.
82. J. Crols and M. Steyaert, "Switched-Opamp: an approach to realize full CMOS switched-capacitor circuits at very low power supply voltages," *IEEE J. of Solid-State Circuits*, pp. 936–942, Aug. 1994.
83. H. Choi *et al.*, "A 15mW 0.2 mm^2 10b 50MS/s ADC with wide input range," *in Proc. ISSCC Dig. Tech. Papers*, pp. 842–843, Feb. 2006.
84. P. Figueiredo and P. Cardoso, "Dynamic biasing of an amplifier using capacitive driving of internal bias voltages," U.S. Patent 8 610 422, Jan. 24, 2012.
85. K. Nagaraj *et al.*, "A 250-mW, 8-b, 52-Msamples/s parallel-pipelined A/D converter with reduced number of amplifiers," *IEEE J. of Solid-State Circuits*, pp. 312–319, Mar. 1997.
86. B.-M. Min *et al.*, "A 69-mW 10-bit 80-MSample/s pipelined CMOS ADC," *IEEE J. of Solid-State Circuits*, pp. 2031–2039, Dec. 2003.
87. L. Sumanen, M. Waltari, and K. Halonen, "A 10-bit 200-MS/s CMOS parallel pipeline A/D converter," *IEEE J. of Solid-State Circuits*, pp. 1048–1055, Jul. 2001.
88. Y. Yao, D. Ma and F. Dai, "A 12-bit interleaved opamp-sharing pipeline ADC for extreme environment applications," *in Proc. IEEE ICSICT*, pp. 394–396, Nov. 2010.
89. J. Kim and B. Murmann, "A 12-bit, 30-MS/s, 2.95-mW pipelined ADC using single-stage class-AB amplifiers and deterministic background calibration," *in Proc. ESSCIRC*, pp. 378–381, Sep. 2010.
90. L. Brooks and H.-S. Lee, "A 12b 50MS/s fully differential zero-crossing-based ADC without CMFB," *in Proc. ISSCC Dig. Tech. Papers*, pp. 166–167, Feb. 2006.
91. B. Murmann and B. Boser, "A 12-bit 75 MS/s pipelined ADC using open-loop residue amplifier, " *IEEE J. of Solid-State Circuits*, pp. 2040–2050, Dec. 2003.
92. E. Iroaga and B. Murmann, "A 12-Bit 75-MS/s pipelined ADC using incomplete settling," *IEEE J. of Solid-State Circuits*, pp. 748–756, Apr. 2007.
93. B. Verbruggen *et al.*, "A 2.1 mW 11b 410 MS/s dynamic pipelined SAR ADC with background calibration in 28 nm digital CMOS," *in Proc. of the IEEE Symp. on VLSI Circuits*, pp. 268–269, Jun. 2013.
94. J. Oliveira *et al.*, "An 8-bit 120-MS/s interleaved CMOS pipeline ADC based on MOS parametric amplification," *IEEE Trans. Circuits Syst. II*, pp. 105–109, Feb. 2010.
95. S. Ranganathan and T. Tsividis, "Discrete-time parametric amplification based on a three-terminal MOS varactor: analysis and experimental results," *IEEE J. of Solid-State Circuits*, pp. 2087–2093, Dec. 2003.
96. P. Figueiredo and J. Vital, "The MOS capacitor amplifier," *IEEE Trans. Circuits Syst. II*, pp. 111–115, Mar. 2004.
97. J. Hu *et al.*, "A 9.4-bit, 50-MS/s, 1.44-mW pipelined ADC using dynamic residue amplification," *in Proc. of the IEEE Symp. on VLSI Circuits*, pp. 216–217, Jun. 2008.
98. M. Anthony *et al.*, "A process-scalable low-power charge-domain 13-bit pipeline ADC," *in Proc. of the IEEE Symp. on VLSI Circuits*, pp. 222–223, Jun. 2008.
99. N. Dolev, M. Kramer, and B. Murmann, "A 12-bit, 200-MS/s, 11.5-mW pipeline ADC using a pulsed bucket brigade front-end", *in Proc. of the IEEE Symp. on VLSI Circuits*, pp. 98–99, Jun. 2013.
100. P. Figueiredo, "Pipeline analog-to-digital converter stages with improved transfer function," U.S. Patent Application 2013/0187802, Jan. 23, 2013.
101. B. Hernes *et al.*, "A 92.5mW 205MS/s 10b pipeline IF ADC implemented in 1.2 V/3.3 V 0.13 μm CMOS," *in Proc. ISSCC Dig. Tech. Papers*, pp. 462–463, Feb. 2007.
102. P. Bogner *et al.*, "A 14b 100MS/s digitally self-calibrated pipelined ADC in 0.13 μm CMOS," *in Proc. ISSCC Dig. Tech. Papers*, pp. 832–833, Feb. 2006.
103. K. Hsueh *et al.*, "A 1 V 11b 200MS/s pipelined ADC with digital background calibration in 65 nm CMOS," *in Proc. ISSCC Dig. Tech. Papers*, pp. 546–547, Feb. 2008.

104. P. Figueiredo, G. Minderico and C. Fachada, "Gain and dither capacitor calibration in pipeline analog-to–digital converter stages," U.S. Patent 8 742 961, Jun. 3, 2014.
105. P. Harpe, E. Cantatore and A. Roermund, "An oversampled 12/14b SAR ADC with noise reduction and linearity enhancements achieving up to 79.1 dB SNDR," *in Proc. ISSCC Dig. Tech. Papers*, pp. 194–195, Feb. 2014.
106. P. Figueiredo, "Kickback noise reduction techniques for CMOS latched comparators," *IEEE Trans. Circuits Syst. II*, pp. 541–545, Jul. 2006.
107. G. Zhan *et al.,* "A low-kickback-noise and low-voltage comparator for folding and interpolation ADC," *IEICE Electron. Exp.*, pp. 943–948, Nov. 2008.
108. N. Verna and A. Chandrakasan, "A 25 μW 100kS/s 12b ADC for wireless micro-sensor applications," *in Proc. ISSCC Dig. Tech. Papers*, pp. 222–223, Feb. 2006.
109. P. Nuzzo *et al.,* "Noise analysis of regenerative comparators for reconfigurable ADC architectures," *IEEE Trans. Circuits Syst. I*, pp. 1441–1454, Jul. 2008.
110. J. Kim *et al.,* "Simulation and analysis of random decision errors in clocked comparators," *IEEE Trans. Circuits Syst. I*, pp. 1844–1857, Aug. 2009.
111. D. Allstot, "A precision variable-supply CMOS comparator," *IEEE J. of Solid-State Circuits*, pp. 1080–1087, Dec. 1982.
112. T. Danjo *et al.,* "A 6b, 1GS/s, 9.9mW interpolated subranging ADC in 65 nm CMOS," *in Proc. of Int. Symp. on VLSI Design, Automation and Test*, pp. 1–4, Apr. 2012.
113. M. Miyahara *et al.,* "A low-noise self-calibrating dynamic comparator for high-speed ADCs," *in Proc. of A-SSCC*, pp. 269–272, Nov. 2008.
114. Z. Cao, S. Yan, and Y. Li, "A 32mW 1.25GS/s 6b 2b/step SAR ADC in 0.13 μm CMOS," *in Proc. ISSCC Dig. Tech. Papers*, pp. 542–543, Feb. 2008.
115. K. Kattmann and J. Barrow, "A technique for reducing differential non-linearity errors in flash A/D converters," *in Proc. ISSCC Dig. Tech. Papers*, pp. 170–171, Feb. 1991.
116. Y. Zhang *et al.,* "A single channel 2GS/s 6-bit ADC with cascade resistive averaging," *in Proc. of the Int. Conf. on ASIC*, pp. 195–198, Oct. 2009.
117. C.-C. Lee, C.-M. Yang and T.-H. Kuo, "A compact low-power flash ADC using auto-zeroing with capacitor averaging," *in Proc. of the IEEE Int. Conf. of EDSSC*, pp. 1–2, Jun 2013.
118. M. Miyahara *et al.,* "A 2.2GS/s 7b 27.4mW time-based folding-flash ADC with resistively averaged voltage-to-time amplifiers," *in Proc. ISSCC Dig. Tech. Papers*, pp. 388–389, Feb. 2014.
119. P. Figueiredo and J. Vital, "Averaging technique in flash analog-to-digital converters," *IEEE Trans. Circuits Syst. I*, pp. 233–253, Feb. 2004.
120. C. Donovan and M. Flynn, "A 'digital' 6-bit ADC in 0.25-μm CMOS," *IEEE J. of Solid-State Circuits*, pp. 432–437, Mar. 2002.
121. S. Weaver *et al.,* "Stochastic flash analog-to-digital conversion," *IEEE Trans. Circuits Syst. I*, pp. 2825–2833, Nov. 2010.
122. S. Weaver *et al.,* "Digitally synthesized stochastic flash ADC using only standard digital cells," *IEEE Trans. Circuits Syst. I*, pp. 84–91, Jan. 2014.

High-Performance DACs: Unifying 16-Bit Dynamic Range with GS/s Data-Rates

Joost Briaire, Pieter van Beek, Govert Geelen, Hans Van de Vel, Harrie Gunnink, Yongjie Jin, Mustafa Kaba, Kerong Luo, Corné Bastiaansen, Bang Pham, William Relyveld, Peter Zijlstra and Edward Paulus

Abstract Analysis of recent publications reveals that high performance DAC design can be sub-divided into two types of design approaches. In essence these approaches differ as far as the control of mismatch related effects is concerned. On the one hand one can design such that these effects are intrinsically sufficiently under control but then additional advanced design techniques are required to limit the side-effects of this intrinsic approach. On the other hand one can also rely on mismatch calibration to simplify the design itself. For this approach we will also focus on a more recent specific calibration concept for high performance DACs to unify 16-bit dynamic range with GS/s data-rates. Both approaches have their strengths and weaknesses and depending on the application either one could be more favorable.

1 Introduction

The domain of high-performance digital-to-analog converters (DACs) is typically driven by a push from the demands in the wired- and especially wireless infrastructure to maximize sample-rate and output frequency while maintaining a high output linearity and dynamic range. In this domain one finds 12-bit, 14-bit and especially 16-bit DACs which are capable to deliver a high linearity, e.g. an IMD3 <-80 dBc, with sample-rates of multiple GS/s at output frequencies of 100s of MHz.

Especially the push in sample-rate of 16-bit DACs has been remarkable over the past decade: from 400 MS/s in 2003 as shown by Schofield [1] to 3.2 GS/s in

J. Briaire (✉) · P. van Beek · G. Geelen · H. Van de Vel · H. Gunnink · Y. Jin · M. Kaba · K. Luo · C. Bastiaansen · B. Pham · W. Relyveld · P. Zijlstra · E. Paulus
Integrated Device Technology (IDT), Eindhoven, The Netherlands
e-mail: Joost.Briaire@idt.com

P. Harpe et al. (eds.), *High-Performance AD and DA Converters, IC Design in Scaled Technologies, and Time-Domain Signal Processing*,
DOI: 10.1007/978-3-319-07938-7_6,

2014 as demonstrated by Van de Vel [2]. This has taken place while the high frequency linearity has remained or even improved as well.

In this chapter we would like to compare the different approaches which have led to this performance increase. To do so, we will continue with a literature comparison of different high performance DACs and show that there basically have been two types of design approaches in recent years. Due to the fact that the solution provided by Van de Vel is relatively new [2] and therefore less discussed, we will then provide more insight into this approach and its robustness. Finally we will arrive at a strengths /weaknesses comparison of the two types of design approaches and end with conclusions.

2 Comparison of Key High Performance DACs

Although the progression of the high performance DACs throughout the years has taken place across multiple CMOS process nodes, the used CMOS process node is not a dominant factor in the explanation of the sample-rate improvement. It primarily influences the required power consumption of the digital pre-processing of the on-chip data and the analog latches within the DAC, but is less influential as far as the performance of the analog output stage of a DAC is concerned. To demonstrate this we compared different designs in Table 1 and also plotted the IMD3 performance vs. output frequency in Fig. 1. Now for comparison purposes we also added a previously unpublished design in 140 nm which otherwise has an identical architecture- and design-approach as the 65 nm result of Van de Vel [2].

Based on Table 1 it is hard to extract a structural relation between process node and sample-rate and it even becomes more complex when the performance, e.g. the max. frequency at which the IMD3 remains below −80 dBc, is also included.

When the threshold is placed at the aforementioned −80 dBc it is clear that the designs of Engel [3] and Van de Vel [2] are superior and equal from both a performance and sample-rate perspective but their technology nodes differ. But when the threshold would be placed at −85 dBc (see Fig. 1) then Schofield and the 140 nm design also share a similar maximum frequency of about 200 MHz, comparable to Engel and Van de Vel, however both sample-rate and process node of these designs differ significantly (see Table 1). This shows that all these designs, including the design of Lin [4], e.g. if the threshold would be −70 dBc, share similar high frequency performance results regardless of the technology node which is used.

The power consumption of the DAC core, or better the power consumption per sample rate, does depend on technology: from 200 mW/GS for the 1.8 V technologies down to approximately 70 mW/GS for the 65 nm CMOS. Note that this obviously is a very coarse comparison, for instance because the presented DAC itself may potentially be able to run at higher sample-rates but the digital input logic and/or I/Os does not allow this. Such kind of external circumstances could

Table 1 An overview of different high performance DAC designs

	Schofield [1]	Engel [3]	140 nm 3D-SC DAC	Lin [4]	Van de Vel [2]
Output resolution	16-bit	14-bit	16-bit	12-bit	16-bit
Year	2004	2012	2014	2009	2014
CMOS technology node	250 nm	180 nm	140 nm	65 nm	65 nm
Sample rate	400 MS/s	3.0 GS/s	1.6 GS/s	2.9 GS/s	3.2 GS/s
Max output freq with IMD3 < -80dBc	300 MHz	640 MHz	350 MHz	–	600 MHz
Power consumption	400 mW	600 mW	308 mW	188 mW	240 mW
Power consumption per sample rate	1000 mW/GS	200 mW/GS	193 mW/GS	65 mW/GS	75 mW/GS
Key enabling features	DC calibration, shadow-latches & back-gate tracking	quad-switch output stage	3D-SC calibration & local output cascodes	Local output cascodes	3D-SC calibration & local output cascodes

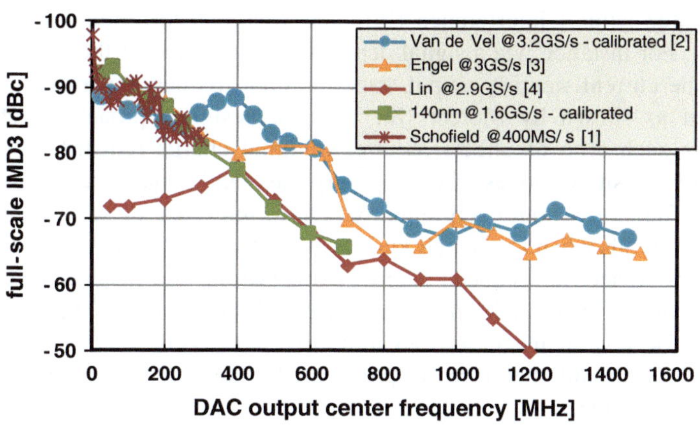

Fig. 1 IMD3 versus output frequency of different high performance DACs

for instance explain the resulting power consumption per sample rate of the design of Schofield.

The sample-rate increase of high performance DACs is also not caused by a drastic change in DAC architecture. During the past decades the current-steering architecture has been by far the most successful and is for instance used in all discussed designs.

Key enabler of the sample-rate increase is a set of advanced design techniques which are able to control a sub-set of the total range of limitations within a high performance DAC such that, in combination with careful design, the overall solution has been able to advance in recent years. The wide range of root-causes which lead to performance limitations has been discussed extensively in literature. For instance more recently two excellent reviews have been published by Manganaro [5] and Balasubramanian [6] which discuss the primary trade-offs in high performance digital-to-analog converters. In the next section we show that some of these root-causes have contradicting consequences and therefore advanced techniques have to be applied to limit the impact of the overall trade-offs.

3 Two High Performance DAC Design Approaches

Root-causes with opposing consequences limit the high frequency performance of DACs and without additional efforts such as for instance shadow-latches [1], calibration [2] and/or a quad-switching output stage which was introduced by Park [7] it does not seem possible at present to minimize all performance limitations simultaneously.

The reason for this is that, on the one hand one has performance limiting factors such as inter-symbol interference (ISI) and charge feed-through which lead to distortion. For instance, via a signal dependent modulation of the common-source node of the current-steering output stage the high frequency performance can be influenced as shown by Mercer [8]. Using a design approach which directly minimizes such type of effects requires minimal charge redistribution and fast settling of sensitive nodes. Consequently fast and small transistors for key switching functions should substantially help to minimize these types of issues.

But on the other hand it is well known from Pelgrom [9] that device mismatch increases as devices are reduced in area. Doris [10] has for instance calculated the impact of device mismatch on timing mismatch and consequentially DAC performance for several use-cases. To minimize device-mismatch, well controlled large devices should therefore be preferred for key switching functions in an analog output stage.

Clearly these straight forward solutions for the two types of error sources are opposing each other and therefore designers typically seem to have chosen to start with one of the two possible solution-approaches in recent designs. If one chooses large devices for key functions such as the output switches then there will be less high frequency degradation due to device mismatch. Examples of such an approach are given by Engel [3] who uses the 3 V thick-oxide switches and Schofield [1] who uses 250 nm CMOS technology with intrinsically large devices. However, in such a case one does require alternatives to minimize the associated common-source node effects, e.g. via the utilization of back-gate biasing [1] and/or a quad-switching output stage [7].

On the other hand, if the approach is taken to use fast and small devices for critical functions, for example switches with low swing switching signals at the gate such as Van de Vel [2], then the common-source node effects can be sufficiently controlled by design. Typically thick-oxide output cascodes [2, 4] which shield the switches from the large signal-swing output itself are then also used. Van de Vel has demonstrated that for such type of high-performance DACs the random timing mismatch can limit the high frequency performance of the DAC and that this can be overcome via calibration [2].

Based on the design-details provided by Lin [4] it is not possible to conclude which design approach has been taken for this design. However, the performance results of Fig. 1 show that this design's maximum performance is limited compared to the others and therefore a compromise between the two design approaches may have been applied here.

Given that the specific design techniques such as for instance shadow-latches, calibration in general and quad-switching are already frequently discussed in literature [1, 3, 5–8], there is no need to further detail their pro's and con's here. But since the specific calibration approach which was demonstrated by Van de Vel [2] is new, we will discuss its impact in the next section.

4 DAC Calibration

Calibration of DACs to improve the amplitude precision of the output signal is well established [1, 5, 6, 8]. However, for high performance DACs not only the constructed amplitude levels which form the output signal are critical, but also the switching precision. For current steering DACs the key output signal is not the current itself, but more generally the current integrated over the sample-period which is the generated charge in the output. This is a result of the fact that DACs generate a sampled data signal which is continuous in time and therefore the information is represented by the total signal during a sample.

As sample rate and output frequencies of high performance DACs increases, the contribution of the transients in between samples also increases and therefore calibration of the amplitude or static signal alone is no longer sufficient. It is required that the total switching transient is proportional to the data-change to ensure a linear operation. For modeling purposes this switching transient can be simplified to an ideal step response with a variable effective switching moment.

Random timing mismatch is a result of device mismatch, for instance a random threshold voltage, which leads to an uncertainty in the exact switching moment when this device is used as a switch. Typically such uncertainties are a result of imperfections during manufacturing of the device and one could therefore consider random timing mismatch as time-invariant or fixed per device after manufacturing, similar to random amplitude mismatch per device. If the timing- and amplitude mismatch of each DAC output stage or slice would be known, it could therefore be possible to reduce the impact.

In case of random timing mismatch, both the switching moments of the rising edge and the falling edge can vary compared to each other and therefore the mismatch of an output stage can be generalized to random amplitude-, delay- and duty-cycle mismatch. Tang [11] has demonstrated that it is indeed possible to measure this three dimensional mismatch vector per output slice by periodically switching each output unit against a reference source to obtain an error-signal. For instance by mixing this signal back to DC one can then easily measure small variations *in between* different output units even when the errors of the reference itself and the measurement circuitry will lead to a common offset error for all measured units. Obviously it is not sufficient to only know the difference between the errors if one wants to eliminate these errors via a classical calibration which reduces the errors to zero. However, this is not the only approach because the errors do not need to become zero, but only equal compared to each other to achieve a linear function.

One could consider the set of measured mismatch vectors as a large random set in which for each mismatch vector there also is a counter-part with similar amplitude but opposite sign. When such a pair of mismatch vectors is combined, the summed result will have much less mismatch [12]. Van de Vel has proven that multi-dimensional mismatch reduction or calibration is indeed possible via this so-called three-dimensional sort-and-combine method (3D-SC) which does not require the correction of any signal in the analog domain [2].

Doris has shown that such random timing mismatch will typically lead to a performance decrease which is inversely proportional to the output frequency [10]. In Fig. 2 we repeated the results of Fig. 1, but now on a semi-log scale and we also added uncalibrated results of the 140 nm DAC and the uncalibrated results of Van de Vel [2]. Clearly the uncalibrated performance results are inversely proportional to the output frequency as predicted, indicating that the uncalibrated results are indeed performance limited due to random timing mismatch.

Note that it only requires a single measurement of each three dimensional mismatch vector at a single frequency to reduce the effects of mismatch over the whole frequency range. In practice, it is therefore not required to measure the three dimensional mismatch vectors at multiple frequencies as Manganaro [5] suggested when discussing the results of Tang [11].

Finally, for practical usage it is key that the results of the three-dimensional sort-and-combine method remain correct even when the temperature of the DAC varies. To test this we calibrated the 140 nm DAC at three different ambient temperatures: −10, 30 and 70 °C. For each of the three calibration results we made a performance sweep vs temperature as shown in Fig. 3. Clearly the calibration results are robust against such temperature variations, especially when taking into account that the uncalibrated IMD3 results at 250 MHz are around −70 dBc (see Fig. 2).

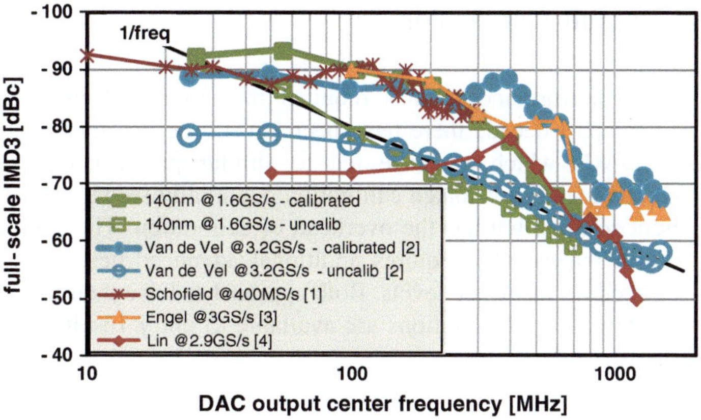

Fig. 2 Results of Fig. 1 on a semi-log scale, with uncalibrated results and the roll-off trend

Fig. 3 Same 140 nm DAC calibrated at three different temperatures to demonstrate calibration robustness against temperature variations

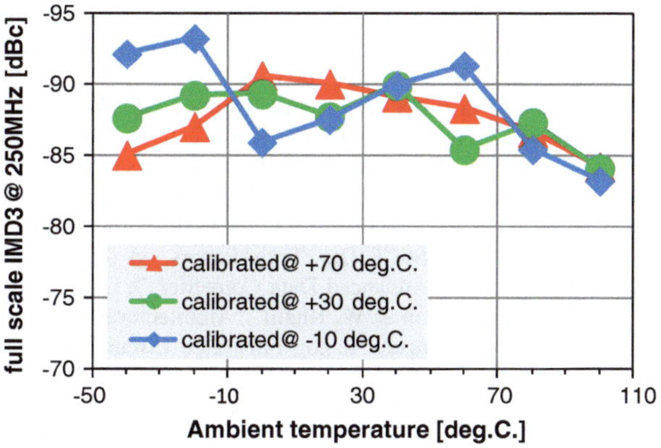

5 Comparison of Design Approaches

The two design approaches can be simplified into the choice: will timing calibration be used or not? On the one hand calibration requires calibration circuitry [11] and/or requires additional steps during production. This is not the case for the design approach in which the timing accuracy is sufficient by design. However, this approach also seems to lead to a result in which the power-benefit, e.g. when designing high performance DACs in advanced CMOS technologies, will be limited since the design will require relatively large devices for key functions. Clearly it will depend on the specific target use-case if this penalty in power-consumption is preferred compared to the overhead of calibration. The proof-of-concept of these two complimentary approaches can therefore only strengthen future high performance DAC design.

6 Summary and Conclusions

Designing high performance DACs requires the control of mismatch related effects. Over the past decade there have been several design approaches for high performance DACs which all have led to similar performance results. The approach which includes mismatch calibration potentially seems to lead to a more power-efficient solution but has the overhead of calibration. The approach which limits mismatch intrinsically requires additional advanced design techniques to reach the required performance levels. Both approaches have their pro's and con's but it is clear that multiple solutions are available to unify 16-bit dynamic range with GS/s data-rates.

References

1. W. Schofield, D. Mercer & L. St. Onge, "A 16 b 400 MS/s DAC with <-80 dBc IMD to 300 MHz and <-160dBm/Hz Noise Power Spectral Density", IEEE, ISSCC Dig. Tech. Papers, pp. 126–127, Feb. 2003.
2. H. Van de Vel et al., "A 240 mW 16 b 3.2 GS/s DAC in 65 nm CMOS with <-80 dBc IM3 up to 600 MHz", IEEE, ISSCC Dig. Tech. Papers, pp. 206–207, Feb. 2014.
3. G. Engel, S. Kuo & S. Rose, "A 14 b 3/6 GS/s Current-Steering RF DAC in 0.18 um CMOS with 66 dB ACLR at 2.9 GHz", IEEE, ISSCC Dig. Tech. Papers, pp. 458–459, Feb. 2012.
4. C-H. Lin et al., "A 12 b 2.9 GS/s DAC with IM3 <-60 dBc beyond 1 GHz in 65 nm CMOS", IEEE, ISSCC Dig. Tech. Papers, pp. 74–75, Feb. 2009.
5. G. Manganaro, "Advanced Data Converters", Cambridge University Press, 2012.
6. S. Balasubramanian & W. Khalil, "Architectural trends in current-steering digital-to-analog converters", Springer, Analog Integrated Circuits & Signal Processing, vol. 77, pp. 55–67, May 2013.
7. S. Park et al., "A Digital-to-Analog Converter Based on Differential-Quad Switching", IEEE, J. of solid-state circuits, vol. 37, no. 10, pp. 1335–1338, Oct. 2002.
8. D. Mercer, "A Study Of Error Sources In Current Steering Digital-to-Analog Converters", IEEE, CICC Dig. Tech. Papers, pp. 185–190, Oct. 2004.
9. M. Pelgrom, A. Duinmaijer & A. Welbers, "Matching Properties of MOS transistors", IEEE, J. of solid-state circuits, vol. 24, no. 5, pp. 1433–1440, Oct. 1989.
10. K. Doris, A. van Roermund & D. Leenaerts, "Wide-Bandwidth High Dynamic Range D/A Converters", Springer, 2006.
11. Y. Tang, et al., "A 14 bit 200 MS/s DAC with SFDR $>$ 78dBc, IM3 <-83 dBc and NSD <-163 dBm/Hz across the whole Nyquist band enabled by dynamic mismatch mapping", IEEE, J. of solid-state circuits, vol. 46, no. 6, pp. 1371–1381, June 2011.
12. J. Briaire, "Error reduction in a digital-to-analog (DAC) converter", US patent 7394414, April 2005.

Part II
IC Design in Scaled Technologies

Andrea Baschirotto

The topic of this second session regards the exploitation of scaled technologies in analog circuit design. The six chapters address the topic from six different point-of-views, trying to compile a wide-band evaluation report.

Mixed-Signal IP Design Challenges in 20 nm, FinFET and Beyond from Brent Beacham (Synopsys) deals with basic design challenges led by the deep sub-micron (including 20nm and FinFET) processes. The author discusses about process spread, I/O voltage limitations, transistor reliability, limitations in transistor W and L, restrictive physical design rules, device matching, simulation verification including parasitics, and electromigration. Some possible solutions to these problems are introduced, like cascoded transistors to prevent overvoltage stress on devices, running LPE simulation verification, which includes metal resistances, and using compound transistors, which have DC characteristics similar to long-channel transistors.

After this chapter dealing with basic aspects, all the other chapters demonstrate that the design of analog blocks in scaled technologies requires innovative solutions and not a straightforward design porting.

Continuous Time Analog Filters Design in Nanometerscale CMOS Technologies from Marcello De Matteis (University of Milan-Bicocca) deals with the design of Continuous Time Analog Filters in Nanometerscale CMOS Technologies. Analog filters are widely used in several kinds of integrated mixed-signal systems, since they are intrinsically needed for inband signal selection, out-of-band noise rejection, and anti-aliasing for the A/D and D/A conversion. The performances of these blocks are affected by scaled MOS device performance degradations, like lower V_{DD}/V_{TH} ratio, and transistor intrinsic gain decreasing. The chapter shows that these limitations reduce the robustness of standard Active-RC and g_m-C topologies, and they can be overcome by designing new filter topologies able to guarantee robust performance even in scaled technologies.

Silicon Innovation Exploiting Moore Scaling and "More than Moore" Technology by Patrick J. Quinn (Xilinx) discusses FPGAs, which have been leading the industry in always being among the first to the node with commercial products and adopting new technologies such as 3D-IC. The chapter covers such items as process selection, 3D-IC technology, SERDES and high-performance AMS features, as well as power reduction strategies and the migration steps to 20 nm and 16 nm CMOS.

The Impact of CMOS Scaling on the Design of Circuits for mm-Wave Frequency Synthesizers by Francesco Svelto (University of Pavia) reports the design of mm-wave frequency synthesizers in scaled technologies. The question addressed in the chapter is whether scaling is just providing advantages at mm-waves or otherwise. Switches used in VCOs for tank components tuning, MOM and AMOS capacitors, inductors. f_T and f_{MAX} increase though slower than in the past, $r_{on}{\cdot}C_{off}$, a Figure-of-Merit for switches, improves correspondingly. As a consequence, wide-band circuits benefit from technology scaling. On the contrary, passive components do not improve and eventually degrade their performances. As a consequence, a conventional LC VCO, relying on tank quality factor, is not expected to improve. New topologies can then be discussed, and, in the chapter, a new Voltage Controlled Oscillators, based on inductor splitting, showing low noise and wide tuning range in ultra-scaled nodes is proposed.

Digital Enhanced Transmitter Concepts for Nanometer-CMOS Technologies by Michael Fulde (Intel) shows that also if telecom DAC design the technology scaling-related drawbacks for analog and RF design can be mitigated by novel digital enhanced transmitter concepts. Moving analog requirements from voltage to time domain allows to fully benefit from high switching speed in scaled CMOS. Innovative digital RFDAC concepts are presented based on current mode and capacitive operation. Pre-distortion, calibration, distributed mixers, and novel decoding schemes are employed to fulfill tough cellular and coexistence specifications and to allow multi-mode operation in future digital TX architectures.

Finally Design of a DC/DC Controller IP in 28 nm by Roberto Pelliconi (Silicon and Software Systems) discusses the design of power management IPs in deep sub-micron technologies to optimize the power efficiency of today's complex chips. The development of a 28 nm analog buck DC/DC controller is presented here, showing the main design challenges and architectural choices. The main idea is to take advantage of technology scaling, so the controlling engine is purely digital, generating the discrete PWM driving signals to the off-chip power FETs.

Mixed-Signal IP Design Challenges in 20 nm, FinFET and Beyond

Brent Beacham

Abstract Deep sub-micron (including 20 nm and FinFET) processes have led to significant design challenges. These include process spread, I/O voltage limitations, transistor reliability, limitations in transistor W and L, restrictive physical design rules, device matching, simulation verification including parasitics, and electromigration. This chapter discusses these challenges and ways to address them, including using cascoded transistors to prevent over-voltage stress on devices, running LPE simulation verification which includes metal resistances, and using compound transistors which have DC characteristics similar to long-channel transistors.

1 Introduction

As processes continue to scale to smaller feature size, it is more challenging to develop high-quality, high-speed mixed-signal IP. 20 nm and FinFET processes pose additional challenges than were not found in 40–28 nm technologies. For 20 nm and FinFET processes, this chapter discusses trends in the process performance and spread, and limitations in the voltages of I/O devices. Transistor reliability and degradation concerns and verification are addressed. Design restrictions resulting from limits in the maximum width and length of transistors are covered, and the use of compound transistors (composed of multiple individual transistors) is proposed as a work-around to those limits. The DC, random variation, and large-signal characteristics of these compound transistors are also discussed. Furthermore, challenges from physical design rules and layout effects which cause matching errors are addressed. Finally, the design verification required in FinFET processes is reviewed, including design simulation with important layout parasitics, and electromigration limit and IR drop checks.

B. Beacham (✉)
Synopsys Inc., 1 City Centre Drive, Suite 900, Mississauga, ON L5B 1M2, Canada
e-mail: beacham@synopsys.com

P. Harpe et al. (eds.), *High-Performance AD and DA Converters, IC Design in Scaled Technologies, and Time-Domain Signal Processing*,
DOI: 10.1007/978-3-319-07938-7_7,

2 Process Performance and Spread

All processes have performance characteristics that change over process manufacturing corners, voltage, and temperature (PVT). For robust performance, a circuit must be designed to work over all these conditions. Corner spread is the difference in performance between the fastest corner and the slowest corner. As the corner spread increases it becomes more challenging to design circuitry with a given performance across PVT, as different circuit failures occur in either the fast or slow corner; for instance, custom digital blocks can have setup issues in the slow corner and hold time issues in the fast corner. In analog blocks, there might be low headroom in the slow corner, but not enough overdrive voltage ($V_{gs}-V_{th}$) in the fast corner for the transistors to be properly biased.

A free-running ring oscillator can give a measure of the speed of a process, because the oscillation frequency is a function of the delay through each inverter in the oscillator. The oscillation frequencies in the typical, slow, and fast corners for several processes are shown in Fig. 1. Each inverter in the oscillator uses minimum length (L) transistors, and the number of stages was adjusted such that the oscillation frequency in the slowest corner in all processes was about the same. From 40 to 20 nm, the trend is for corner spread to increase as feature size decreases. In the sub-20 nm FinFET technologies, the spread is observed to reduce to levels below 28 nm. With less corner spread there can be circuit optimization, because circuit characteristics do not vary as much. Therefore, designs in FinFET technologies should be able to have better power and area optimization than in 28 or 20 nm.

Another trend that is observed is that as feature size decreases both the supply voltage and transistor threshold voltage (V_{th}) decrease (Fig. 2). However, the V_{th} is decreasing at a slower rate than the supply, and as a result the difference between the two is decreasing. This leaves less headroom for circuits and reduces the maximum overdrive voltage ($V_{gs}-V_{th}$) possible.

The effect of corner spread can be reduced by setting tighter tolerance on off-chip power supplies, by using bandgap-referenced on-chip regulated supplies (which have tighter tolerances than off-chip supplies), or by selective use of low-V_{th} devices. Low-V_{th} devices can have better circuit performance at low supply voltages (due to decreased V_{th}). For instance, when used in a comparator, low-V_{th} devices can improve sensitivity over PVT [1]. A drawback of low-V_{th} devices is increased leakage, which is particularly noticeable in low-power circuit states.

The trend for reduced core supply voltages can also have an effect on circuit architecture. For example, to save power, serializer/deserializer (SERDES) drivers may use a voltage-mode driver instead of a current-mode driver. Many high-speed SERDES specifications require a launch amplitude of 0.8 V or greater. When the supply is a nominal 0.8 V, with 5–10 % tolerance, and a 1–2 dB package loss, there is insufficient voltage to meet a 0.8 V launch amplitude specification. Sometimes there is an option for overdriven devices (that can operate at a higher voltage), but these are not generally available in all process nodes. Even if they are

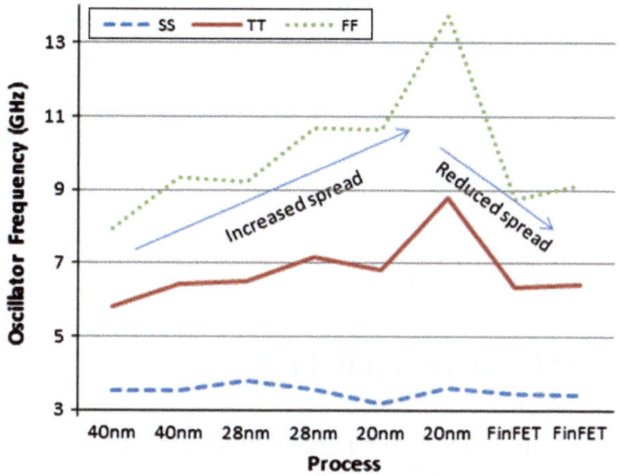

Fig. 1 Process spread

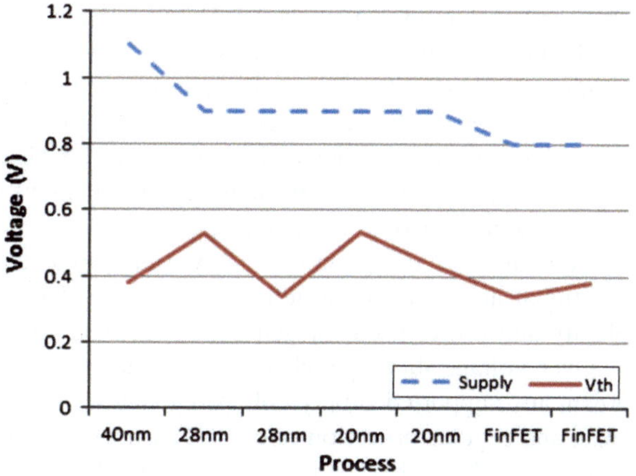

Fig. 2 Supply and threshold voltage versus process

available, the extra supply voltage likely will not be enough to meet the specification, once supply tolerance and package loss are subtracted.

One solution is to drive the line using the 1.8 V I/O supply and use I/O devices in the driver. However, a driver using all I/O devices might not be fast enough for high speed (e.g., 10 Gbps) SERDES applications. There is also a significant power penalty with using a 1.8 V supply instead of a 0.8 V supply when the same current is used by the driver. Another solution to increase the I/O differential swing above the core supply is to use a hybrid mode transmitter (Fig. 3). With this circuit, core devices can still be used in a voltage mode driver supplied by 0.8 V, but there is a current-mode boost applied in low-supply corners, so that the transmitted differential swing meets specification.

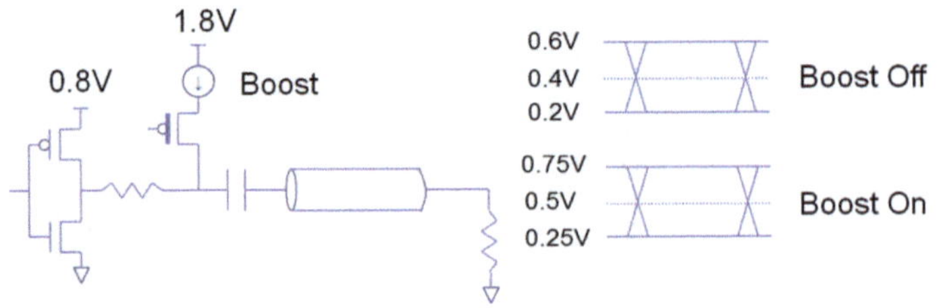

Fig. 3 Hybrid driver

3 I/O Device Availability

A circuit design is fundamentally limited by the transistor types that are available, particularly with I/O. As the feature size has reduced, the maximum voltage of I/O devices has also reduced. In 40 nm technologies and above, 2.5 V I/O devices are generally available. In 28 nm and below, often only 1.8 V I/O devices are available, and the trend is toward 1.5 V I/O only being available. This limitation is challenging when I/O levels higher than 1.8 V are required, such as the 3.3 V required for USB 2.0 full-speed and low-speed signaling. I/O levels of 3.3 V can still be achieved without device stress using 1.8 V transistors by cascoding transistors so that no single device has more than 1.8 V across any of the drain-source, gate-source, or gate-drain terminals. Figure 4 shows an example where cascoding is used to prevent over-voltage stress.

Compared to using native 3.3 V transistors, cascoding 1.8 V devices uses additional area due to the added cascode devices. There is also additional complexity and power used for generating the bias voltages on the cascode devices.

Over-voltage should be checked for states that the circuit is in: consider power-down, any supported supply collapsing states, power-up, overshoot/undershoot on I/Os, and regular operation over PVT. Checks can be done during transient simulations to check that all junction voltages (V_{gd}, V_{gs}, V_{ds}, V_{gb}) do not exceed a specified limit, such as with the biaschk command in HSPICE®. If over-voltage is reported, the design can be modified to reduce stress on the flagged devices.

3.1 *Transistor Reliability*

Reporting over-voltage is one reliability check, but there are other degradation effects which affect transistor reliability which should be considered and checked. These include negative bias temperature instability (NBTI), positive bias temperature instability (PBTI), and hot carrier injection (HCI).

NBTI predominately affects PMOS devices under negative gate bias conditions. It is usually modeled as a transistor V_{th} shift. PBTI primarily affects NMOS devices under positive gate bias. This degradation can be modeled with Eq. (1) and

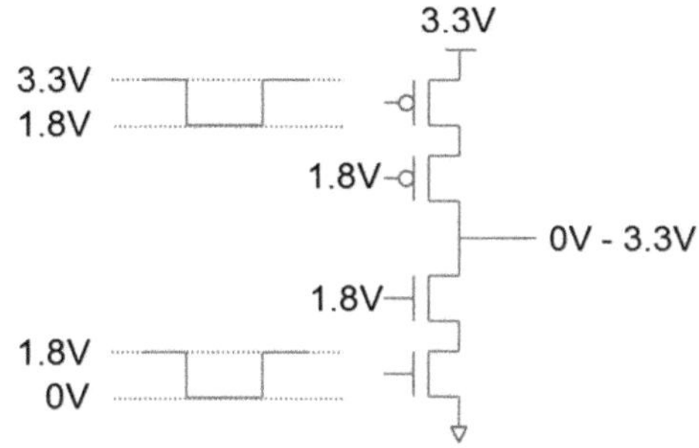

Fig. 4 Over-voltage protection

is a strong function of the voltage stress (V_{gs}) and time (t). (Parameters E_a, K, γ, a, and n are constants for the process.)

$$\Delta V_{th} = A \cdot f(W, L) \cdot e^{-\frac{E_a}{kT}} \cdot \frac{V_{gs}{}^{\gamma}}{a} \cdot t^n \tag{1}$$

HCI degradation occurs when there is a large electric field across the drain and source of a transistor while the transistor is conducting current. It is usually modelled as a shift in I_{dsat}. HCI degradation can be modeled with Eq. (2) and is a strong function of the voltage stress (V_{ds}) and the time (t) that the stress is applied. (The parameters E_a, k, b, A, n are constants for the process.)

$$\Delta I_d = A \cdot f(W, L) \cdot t^n \cdot e^{-\frac{b}{V_{ds}}} \cdot e^{-\frac{E_a}{kT}} \tag{2}$$

Even with reasonable bias levels, some device stress and degradation still occurs and verification is required to ensure that the design will continue to work through its lifetime. Circuit performance degradation over time can be predicted with reliability simulations, such as the MOS reliability analysis (MOSRA®) feature in HSPICE®. These reliability simulations are two-pass, with the first simulation used to measure the beginning-of-life (BOL) performance and to calculate the device stress that all transistors are subjected to at each simulation time-step. The degradation is accumulated and extrapolated to a calculated end-of-life (EOL) V_{th} and I_{dsat} degradation. In the second simulation, the calculated degradation is applied to each transistor, and the EOL performance simulated. It is important to check all simulation modes, because the worst-case degradation could occur in the active state, supply collapsing states, in power-down or during a transition between states. The results from an aging simulation of a ring oscillator are shown in Fig. 5, where the frequency of a free-running ring oscillator drops over time due to device degradation. Much of the degradation occurs early in the lifetime of the part.

For matched differential structures to remain matched over time, it is important that degradation is minimized and kept symmetrical. This should be considered in the design of the circuit. As an example, the worst case stress and degradation in a receiver may occur in sleep mode if the receiver inputs are parked at different

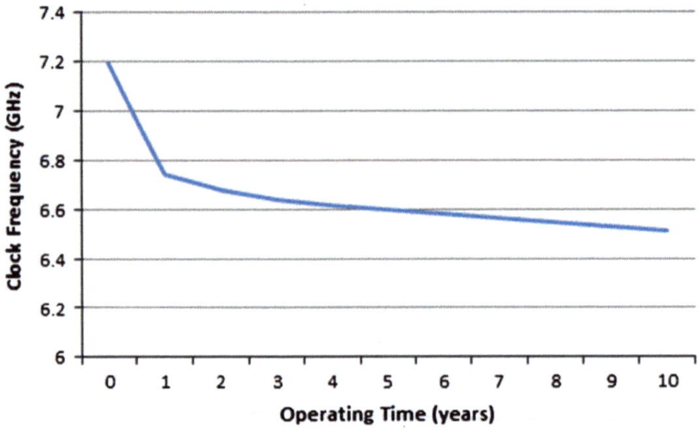

Fig. 5 Clock frequency versus operating time

levels, due to NBTI degradation. The design can be modified in this case to disconnect the receiver inputs in sleep from the input, and connect them to the same voltage (Fig. 6).

3.2 Device Geometry Restrictions

The available device sizes have become more restrictive as the feature size has decreased. In 40 nm processes and above, there are generally broad limits for W and L, and transistors could be made almost any size within those limits. In 28 nm and below, the available transistor sizes generally become more restrictive, with limits to maximum transistor area, reduced maximum transistor W or L, and in some cases only several transistor L sizes available. In 20 nm processes, these restrictions continue, with fewer transistor L sizes available. In FinFET processes, instead of a relatively continuous drawn transistor W, the effective width is increased by adding an integer number of fins (nfin), which limits the effective W to fewer discrete steps. There are also fewer L sizes available, and a significant reduction to the maximum L, in some cases to 100 nm or less.

4 Compound Transistors

A transistor can be designed with a particular W/L ratio or total gate area for various reasons including matching (from random variation), gain, or output impedance. With rules limiting L and/or W, those specific characteristics might not be able to be achieved with a single transistor. A transistor with a similar effective W/L can be created using a compound transistor that has several

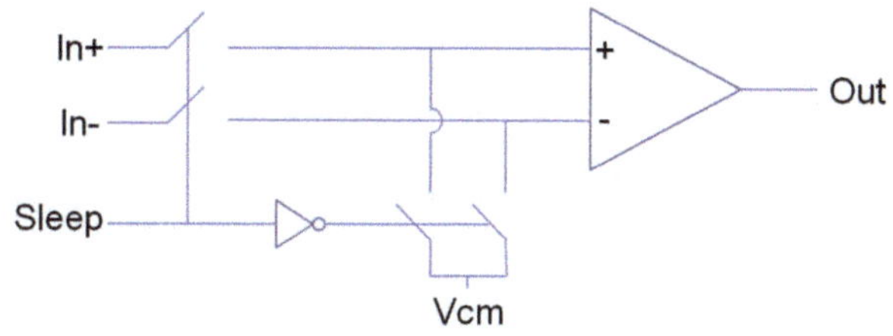

Fig. 6 Receiver with NBTI protection

transistors combined in series and in parallel. An example in Fig. 7 shows how a transistor with an effective W/L = 1.5 μm/300 nm can be created using nine W/L = 0.5 μm/100 nm transistors.

4.1 Compound Transistor DC Characteristics

The DC characteristic of these compound transistors is similar to those of a long channel device. The plot in Fig. 8 shows the I_{ds} versus V_{ds} when n transistors are stacked in series to form an n-stack compound device. In each case, the effective W/L was adjusted to keep the drain current constant. Increasing the n-stack increases the output impedance in a similar way that increasing the L increases the output impedance of a single transistor. Monte Carlo simulations on these compound transistors also report less random variation as more transistors are added in series and parallel. This effect is similar to the reduction in random variation that is observed in a single transistor as the gate area is increased.

When a silicon-proven design is retargeted to a FinFET process from a 28-nm (or above) technology, a major design challenge is how to convert existing analog blocks to the target process while keeping the same performance characteristics as in the source technology. The same performance is desired to reduce risk and design porting effort, because the behavior of the converted circuit in the FinFET process is expected to match that of the silicon-proven design. Figure 9 shows the I_d versus V_{ds} characteristics of an L = 1 μm transistor in a 28 nm technology. If the transistor in the target FinFET process has a matching curve, the DC characteristics are similar.

Of the n-stack options from Fig. 8, the one that best matches the curve in Fig. 9 is the 100 nm 5-stack. The 5-stack and 1 μm transistor I_{ds} versus V_{ds} curves are overlaid in Fig. 10. This 5-stack compound FinFET matches the DC characteristics of the source design using the fewest transistors and therefore the smallest area.

Creating compound stacked transistors has an area overhead compared to single transistors due to the extra area used for the source/drain regions between the stacked transistors. In this case, the additional area is offset by the smaller total L of the 5-stack transistor (500 nm for the 5-stack versus 1 μm transistor) and the higher drive strength of the FinFET transistor for a drawn W. Even with the

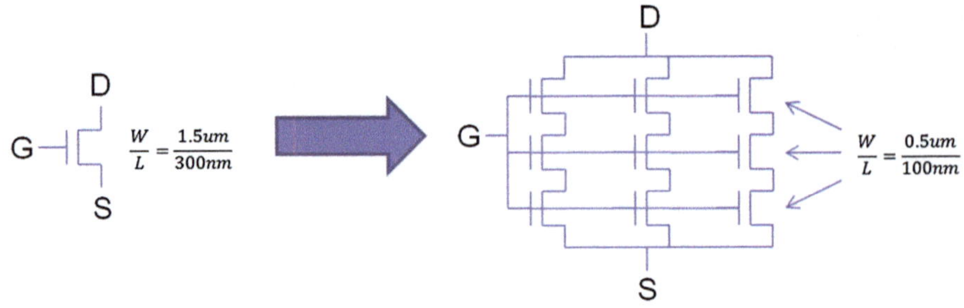

Fig. 7 Compound transistor

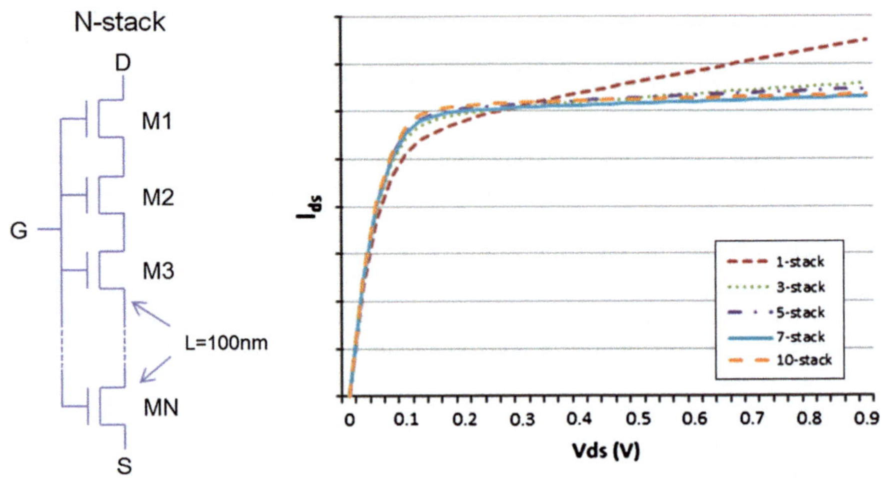

Fig. 8 N-stack I_{ds} versus V_{ds}

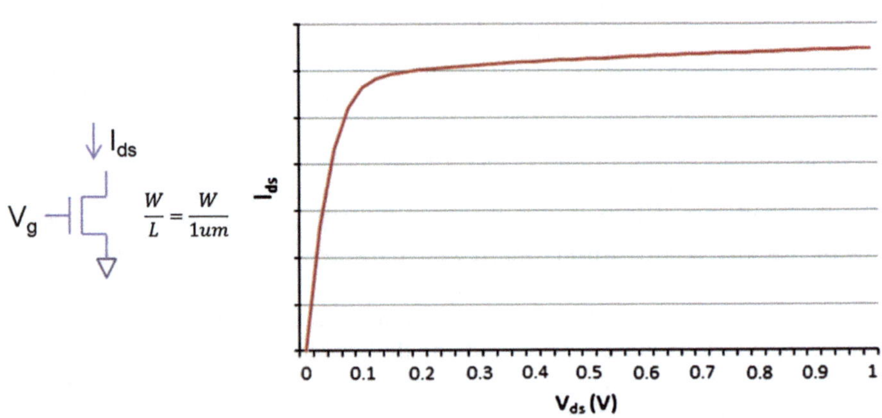

Fig. 9 I_{ds} versus V_{gs} of a 28 nm L = 1 μm transistor

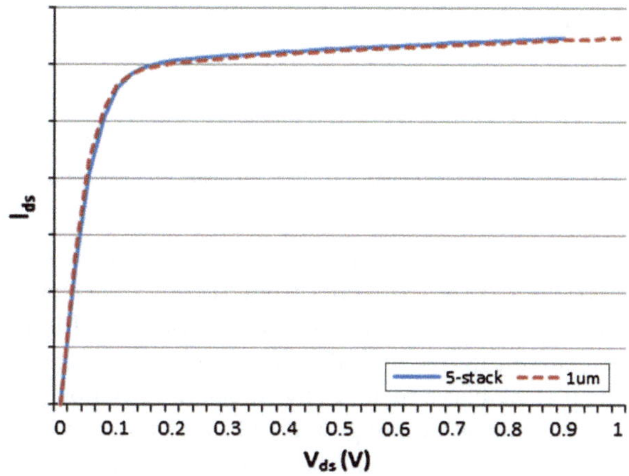

Fig. 10 FinFET 100 nm 5-stack versus 28 nm L = 1 μm

additional overhead, the total area of a compound FinFET transistor can be 25 % or less of a 28 nm transistor, for equivalent DC characteristics.

4.2 Compound Transistor Random Mismatch

Another important effect to consider is the random variation or mismatch that can occur from part-to-part. This variation cannot be measured by regular PVT corner simulations, but must instead be measured with Monte Carlo simulations. The effect that transistor geometry and biasing has on random variation varies depending on the circuit. For the specific case of current mismatch in a current mirror (Fig. 11), it can be described with Eqs. (3), (4), and (5) [2].

$$\sigma^2\left(\frac{\Delta I_d}{I_{ds}}\right) = \frac{4\sigma^2(\Delta V_{th})}{\left(V_{gs} - V_{th}\right)^2} + \sigma^2\left(\frac{\Delta\beta}{\beta}\right) \tag{3}$$

$$\sigma^2(\Delta V_{th}) = \left(\frac{A_v}{\sqrt{WL}}\right)^2 \tag{4}$$

$$\sigma^2(\Delta\beta) = \left(\frac{A_\beta}{\sqrt{WL}}\right)^2 \tag{5}$$

The mismatch can be minimized by increasing ($V_{gs}-V_{th}$) and increasing $\sqrt{WL}$. Generally, increasing the effective L of devices first will give the best results because it improves both parameters, but ($V_{gs}-V_{th}$) is limited by the headroom and the supply voltage available.

Fig. 11 Current mirror

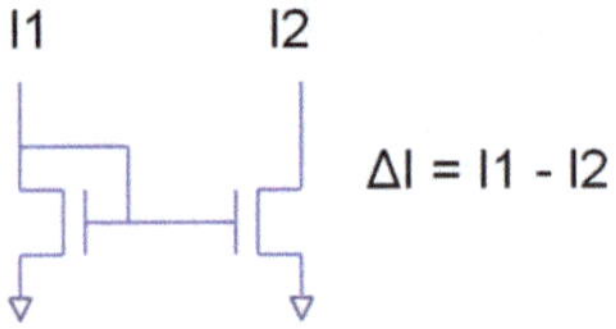

The standard deviation of the current mismatch of current mirrors using the 100 nm n-stack transistors from Sect. 4.1 is shown in Fig. 12. To keep the same bias point, when n series transistors were used in the stack, the number of fins in the current mirror was also scaled by n. Also shown in Fig. 12 is the point where the n-stack curve matches the standard deviation in the reference L = 1μm 28 nm current mirror. Although the DC characteristics match with a 5-stack, an 11-stack current mirror is needed here for the random variation to be the same. When the overdrive of the compound transistors in the current source is kept constant, the matching depends on the number of fins used in the compound device. Therefore, an alternative to increasing the n-stack is to instead increase both the bias current and number of fins proportionally, until the desired matching result is obtained. Including the overhead for the additional source/drain regions, the total area of a compound FinFET transistor can be 85 % or less of a 28 nm transistor for equivalent DC and matching characteristics.

Random variation can be reduced by increasing area and/or power, but this comes at a cost. The best design is one that meets the performance requirements with the minimum amount of area and power consumption. This is only achieved by identifying the transistors that have the greatest effect on the variation in the circuit and then optimizing total size and biasing of those transistors to reduce their contribution to the total variation. For any measurement made in simulation, HSPICE has additional AC, DC, and transient Monte Carlo simulation features that list the amount that specific transistors in the circuit contribute to the random variation of that measurement. These features can be used to identify the transistors which contribute the most to the total variation, then those transistors can be targeted for optimization.

4.3 Large-Signal Behavior of Compound Transistors

Although the DC characteristics of a stacked compound device are similar to a long-channel device, the large-signal transient behavior is different. For example, with the circuit in Fig. 13, there is a difference in the transient behavior of the output current (I) after the enable switch is turned on.

The plot in Fig. 14 shows the output current for several n-stack cases when the enable switch is turned on at t = 0. As the n-stack increases, there is an increased delay before the steady-state current is reached at the output. The reason for the delay can be determined by examining the circuit and the current and voltages in

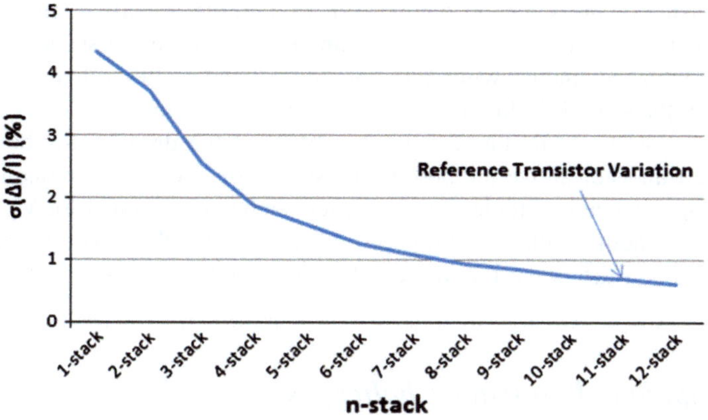

Fig. 12 N-stack current mismatch

Fig. 13 Compound transistor circuit

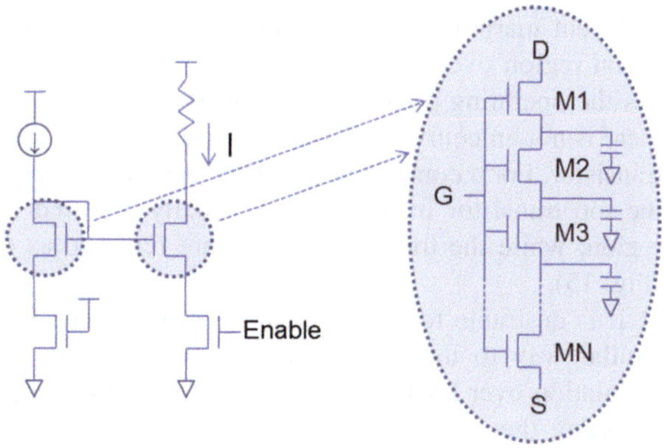

Fig. 14 Large-signal switching of compound transistors

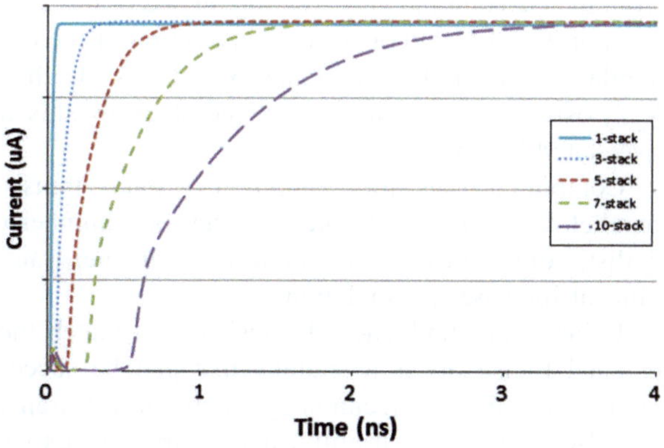

each transistor of the stack. In comparison to a single transistor, the n-stack has extra capacitance from the source/drain junctions between the stacked transistors (Fig. 13). After the enable switch turns on, the voltage on the source of the bottom transistor in the stack is pulled to ground and the bottom transistor starts to conduct current. This starts to discharge the capacitance at the drain of the bottom transistor. The transistor above the bottom transistor does not start to conduct current until the capacitance is sufficiently discharged such that the transistor V_{gs} is greater than V_{th}. This adds a delay, and this process repeats for the next transistor up the stack until all of the transistors are conducting current in a DC steady state.

4.4 Compound Transistor Challenges

It is more difficult to measure the operating point of a stacked device than a single long-channel transistor. With a single transistor, the V_{dsat} of the device can be reported by the simulator, and the V_{ds} can be checked to verify that there is sufficient margin (e.g., 100 mV) to ensure the transistor is operating in the saturation region over PVT. A compound transistor with multiple stacked transistors has the operating point of each transistor reported separately by the simulator, so there is not an equivalent effective measurement reported for the whole compound transistor. For a compound transistor operating effectively in the saturation region, the top transistor in the stack is usually reported as operating in the saturation region, while the transistors below are reported as operating in the linear region (Fig. 15).

It is desirable to measure the operating point of the compound transistor in a similar way to the single transistor, so that the measurement can be made in simulation over PVT. An approximation of the V_{dsat} of the stack can be made by reporting the V_{dsat} of the top device in the stack. This approximation becomes more inaccurate as more transistors are stacked in series. From the example in Fig. 8, the single-L transistor has a reported v_{dsat} of 190 nm, which is reasonable from the curve in the plot. The top device of the 10-stack transistor has a reported V_{dsat} of 275 mV, although from inspection of the curve the actual V_{dsat} should be similar to the single-stack value of 190 mV. In this example, approximating the V_{dsat} using the top transistor in the stack by this method gives a result that is overstated by 85 mV.

Compound transistors composed of many transistors connected in series and parallel can significantly increase the total number of transistors in a simulation netlist, compared to using single long-channel transistors. This can cause a significant increase in simulation time.

Using compound stacked transistors to match the DC characteristics of long-channel transistors is a strategy that can be successfully used when designing circuits in a FinFET technology where large L transistors are not available. They are also particularly useful when retargeting existing designs which use long-channel devices to a FinFET process. Compared to single transistors, stacked

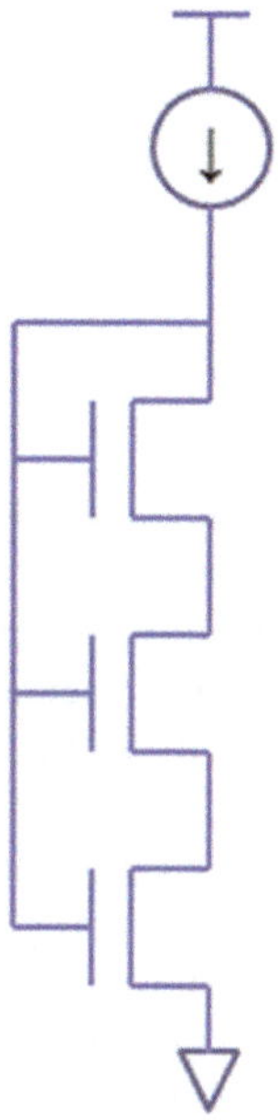

Fig. 15 Compound transistor in saturation

transistors have different large-signal transient behavior and should only be used where appropriate in a design.

5 Physical Design Rules

In 40 nm technologies and above, transistors can be placed with a vertical or horizontal orientation. In 28 nm and below, all transistors must be oriented in the same direction. FinFET technologies have further restrictions, with resistor orientation fixed and some metal spacing rules direction dependent.

With IP, it is preferable to allow placement on any side of the chip, as this allows the maximum usage flexibility. This is not possible with the same layout when devices are restricted to either vertical or horizontal orientation. Mixed-signal IP typically provides an interface between on-chip and off-chip and needs to be on the edge of the chip, to minimize the route length of I/Os. Dual-edge support is possible with single device orientation by complete re-layout, translation, or by block layout that allows block re-use. Re-layout requires the most effort, as the entire layout is created again for the new orientation.

Translation involves creating the layout such that it can be placed against either edge of the chip (Fig. 16), with changes only to the top-layer route distribution layer (RDL) of the chip. It requires the least amount of layout effort, but does not work well for IP with many I/Os, such as a multi-lane SERDES transceiver. (If there were more TX and RX blocks extending along the chip edge in the left side

Fig. 16 PHY translation

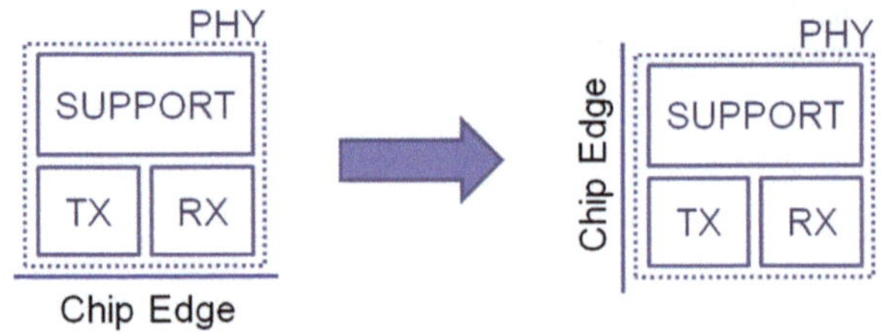

of Fig. 16, they would be far away from the chip edge, after the translation shown in the right side of Fig. 16.)

Block layout allowing reuse is shown in Fig. 17, where individual blocks can be reused in the same orientation, but with different arrangement. The individual blocks are re-used but the interconnections between them are redone.

Density rules for poly, diffusion, and metal are also more restrictive as the feature size decreases. The maximum and minimum density limits are closer together, and the checking window is smaller.

The lowest metal layers in 20 nm and FinFET technologies use double pattern lithography. The design rules associated with this cause additional complications for hand-drawn layout because the required spacing between metal shapes on the same metal layer changes depending on if they are on the same mask layer or not. For metal, there are also spacing rules which vary depending on the voltage difference between nets.

5.1 Device Matching

Systematic mismatch is caused by effects such as OD spacing effects (OSE), poly-spacing effects (PSE), well proximity effects (WPE), and shallow trench isolation (STI) and stress effects.

OSE can be reduced by having the same OD distance to surrounding devices and PSE can be reduced by having the same poly spacing between matched devices. In some cases, very tight control of the poly density gradient is also required for optimal matching between transistors.

STI causes systematic variation in the transistor threshold and I_{dsat}. The variation is caused by mechanical stress and is a function of the distance between the gate and the edge of the diffusion. The transistors near the edge have a shift characteristics compared to those in the center. The effect of STI can be reduced by adding dummy devices (Fig. 18) on the ends of the rows of transistors, at the expense of additional area.

WPE affects the threshold voltage of devices and is another source of deterministic variation. It is caused by scattering of dopant (N-type and P-type), and devices near the edge of wells have a different dopant density (and a different

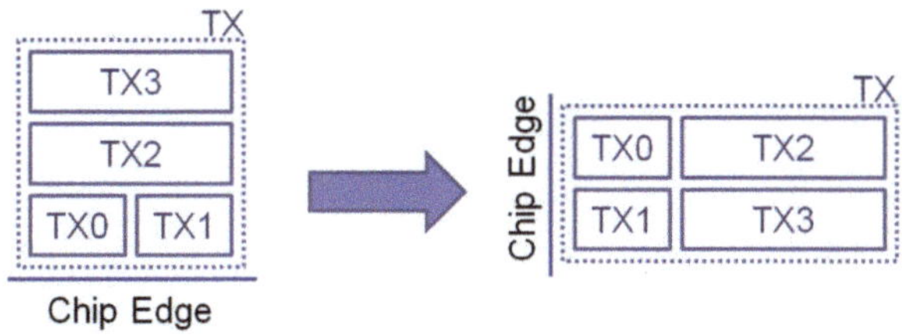

Fig. 17 Block re-use

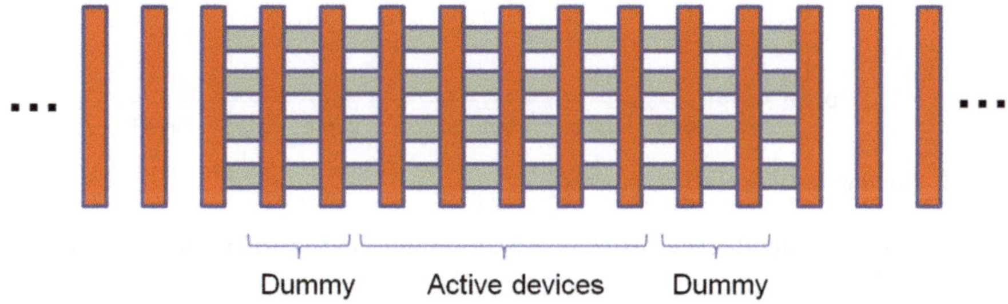

Fig. 18 Reducing STI with dummy transistors

threshold voltage). For matched devices, care should be taken in the layout to avoid unequal distances to the edge of wells.

To avoid issues with deterministic mismatch, the layout should be reviewed and modified to prevent these effects. It is still critical to verify the final design with a netlist containing full parasitics that include these effects.

6 Design for Yield

IP in 20 nm and FinFET processes needs to consider performance from both a design and a manufacturing perspective. This section discusses the verification needed and issues to take into account to ensure working and manufacturable IP.

6.1 Layout Extracted Parasitic (LPE) Simulations

In processes 40 nm and above, the metal resistance of wires for signal nets can be omitted in most simulations. The simulated voltage and current waveforms are usually only affected by the capacitive load and transistor drive strength. In 28 nm, the effect of the metal resistance starts to be noticeable, and in 20 nm and FinFET

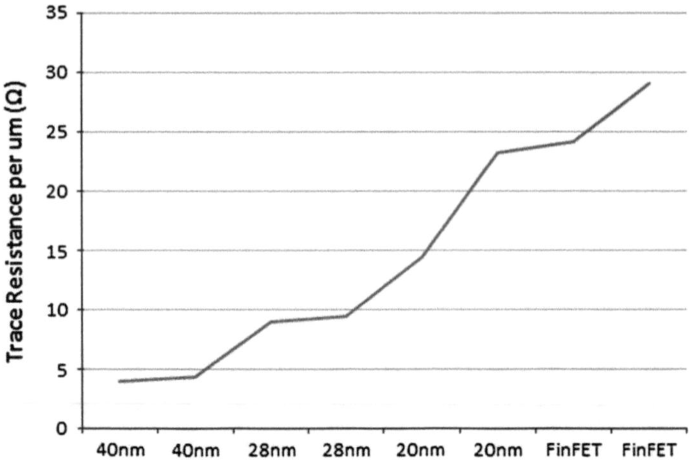

Fig. 19 Routing resistance versus process

the effect of the metal trace resistance is dominant and must be included in LPE simulations.

The trend of resistance in a minimum-width M1 route per μm of route length is shown in Fig. 19. Due to the resistance, even relatively short traces of 20 or 30 μm can have significant RC filtering, which will affect high-speed clock and data routes. Care must also be taken with routes carrying DC current, such as bias currents or locally regulated supplies, because the voltage drop along these routes can cause headroom problems for the circuits following them.

Block level simulation verification in 20 nm and FinFET processes must be done with an RCC extracted simulation netlist, which includes parasitic R and ground and coupling capacitors. Problems can also occur at the PHY level when high-speed clocks or signals are routed over longer distances between blocks, or due to resistance in the supplies or I/O routing. For these reasons, PHY level simulation verification should also include metal resistance.

Including route resistance in LPE simulations significantly increases the circuit simulation time. For large blocks or PHY level simulation, a fast-SPICE simulator such as Synopsys® CustomSim™ or FineSim® is needed. Simulations can be accelerated by selectively adjusting simulation accuracy per sub-block, omitting non-critical nets from parasitic back-annotation, and setting parasitic extraction and back-annotation options to filter out small parasitics that will not affect the simulation result.

6.2 Electromigration and Supply IR Drop

As design geometries shrink, metal thickness also decreases. The metal stack used is generally optimized for using the lowest layers for local signal routes and for power routing at the upper levels. The total number of metal layers is often

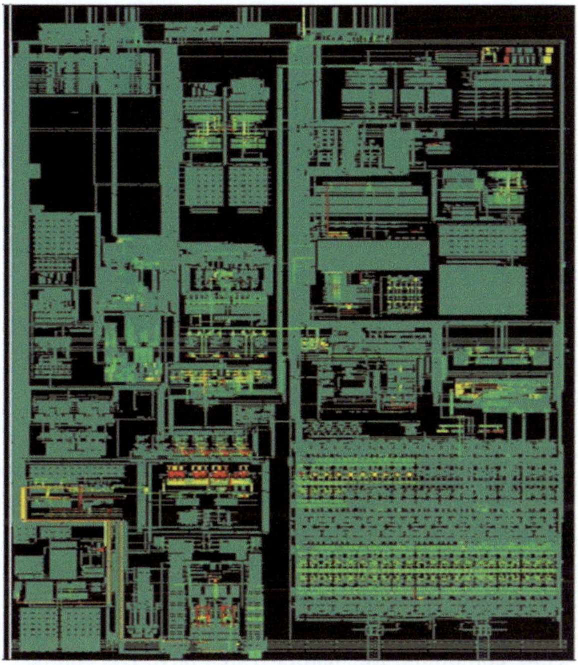

Fig. 20 EM violations

minimized due to manufacturing cost, and this often limits the number of thick top metal layers available.

Electromigration (EM) is the gradual movement of metal atoms due to the momentum transfer between conducting electrons and diffusing metal atoms. It reduces reliability because over time short- or open-circuits can form in the metal interconnect. With thinner metal in 20 nm and FinFET technologies, the maximum current permitted is also lower and meeting electromigration limits is more difficult.

A simulator such as Synopsys CustomSim can be used to check for EM rule violations in the design. A separate analysis is done for peak, RMS, and average currents, because these all have different limits. The layout is highlighted with colors indicating the maximum current density. Areas that are over the current density limit can be quickly identified and the layout updated to correct the issues. (See Fig. 20 for an example which shows electromigration violations, which are highlighted in red.)

The limited number of thick metal layers also makes it more difficult to route power to each circuit. Power routing becomes more complicated when there are many power domains on-chip and only one or two thick metal layers available to route them. The IR drop in these cases must be carefully verified to ensure that there are no excessive voltage drops. A simulator such as Synopsys CustomSim can also be used to measure IR drop in power domains during a mixed-signal simulation. In a manner similar to EM simulations, the IR drop can be displayed graphically and areas which exceed a specified value (e.g. 25 mV) highlighted.

With simulation verification, it is important to account for the IR drop on each supply (including ground.) The budgeted amount of IR drop on supply and ground should be subtracted from the minimum supply voltage and that value used in low-supply simulations.

It is preferable to follow a correct-by-construction approach and create the initial circuit layout considering electromigration and IR drop. This reduces the amount of iteration required to fix errors.

7 Conclusions

This chapter highlighted some significant challenges when designing high-speed, mixed-signal IP in 20 nm and FinFET processes. These challenges include process spread, I/O device availability, transistor reliability, device geometry restrictions, physical design rules and matching, metal trace resistance, and electromigration and supply IR drop. Techniques were described which address these challenges, including using cascoded transistors to prevent over-voltage stress on devices, using compound transistors which have DC characteristics similar to long-channel transistors, and running LPE simulation verification which includes metal resistance. The techniques discussed in this chapter can be used to create successful mixed-signal IP in 20 nm and FinFET processes and they are used throughout the Synopsys DesignWare® mixed-signal IP portfolio.

References

1. B. Beacham, P. Hua, C. Lacy, M. Lynch, and D. Toffolon, "Mixed-Signal IP Design Challenges in 28 nm and Beyond", Design Reuse Conference, 2012
2. M. J. M. Pelgrom, A. C. J. Duinmaijer and A. P. G. Welbers "Matching properties of MOS transistors", IEEE J. Solid-State Circuits, vol. 24, no. 5, pp. 1433–1440 1989

Continuous Time Analog Filters Design in Nanometerscale CMOS Technologies

Marcello De Matteis and Andrea Baschirotto

Abstract Analog filters are widely used in several kinds of integrated mixed-signal systems, since they are intrinsically needed for in-band signal selection, out-of-band noise rejection and anti-aliasing for the A-to-D (D-to-A) conversion. CMOS technological scaling down has already entered in nm-range scenario, leading to severe degradation of some of the most important MOS device performance for analog design. Among them, lower V_{DD}/V_{TH} ratio (supply voltage and threshold voltage, respectively), higher power consumption, and transistor intrinsic gain decreasing. On the other side the increasing MOS transistor transconductance (g_m) and transition frequency (f_T) enables tens of GHz applications, which take advantage of cheaper and smart CMOS processes. Specific continuous-time analog filters will be presented in this chapter, with the aim to introduce novel circuital topologies or optimizations and techniques suitable to mitigate the severe issues present in sub-90 nm CMOS technological nodes, and at the same time to exploit the opportunity of higher technological scaling-down.

1 Introduction

Nanometer range CMOS technologies represent a key opportunity for smart and efficient mixed-signal systems. Integration of analog and digital circuits in the same die area is then sustained by the technological scaling-down, since lower power consumption can be achieved in mostly-digital systems due to the lower supply voltage [1].

M. De Matteis (✉) · A. Baschirotto
Physics Department, University of Milano Bicocca, Milan, Italy
e-mail: marcello.dematteis@unimib.it

A. Baschirotto
e-mail: andrea.baschirotto@unimib.it

P. Harpe et al. (eds.), *High-Performance AD and DA Converters, IC Design in Scaled Technologies, and Time-Domain Signal Processing*,
DOI: 10.1007/978-3-319-07938-7_8,

Even though digital signal processing is replacing several analog functions, the first interface of the mixed-signal systems with the external world is intrinsically analog. In addition mixed-signal systems require a conversion step from analog (digital) to digital (analog) domain. For this reason analog filters play a key role for signal conditioning, in terms of bandwidth selection, out-of-band noise/interferes rejection, or anti-aliasing. As a consequence, in order to achieve efficient and reliable circuits in nm-range CMOS technologies, analog designers have to be able to manage the poor analog performance of the MOS transistors and at the same time to take advantage of the opportunities given by the increasing technological scaling-down [2, 3].

The scaling-down of the physical (length, oxide thickness, etc.) and electrical (supply voltage) parameters leads to a significant improvement for digital circuits in terms of speed and power. This trend is not so automatic for analog circuits, since they experience severe design drawbacks, like V_{DD}/V_{TH} decreasing (supply voltage and threshold voltage, respectively), higher sensitivity to Process-Voltage-Temperature (PVT) variations, transistor intrinsic gain decreasing and in general higher power consumption.

Nevertheless, increasing transition frequency in CMOS nm-range technologies represents an evident opportunity for high-data rate wireless transceivers implementation [4], and as a consequence analog designers must be able to operate in such contrasting scenario, where wide bandwidth analog signals have to be processed with acceptable frequency mask accuracy, and maintaining the Signal-to-Noise ratio.

1.1 V_{DD} and V_{TH} Trend in Nanometer Scale IC Technologies

Technological scaling-down sustains the development of low cost and low power mixed-signal systems, suitable to operate at higher clock frequencies. Low power consumption and higher clock frequency can be easily obtained in digital circuits thanks to the lower supply voltage for CMOS standard logic devices and the smaller parasitic capacitances.

Figure 1 shows that standard supply voltage (V_{DD}) of the analog devices decreases with the minimum transistor channel length. Low voltage supply is a necessity in scaled technologies. In fact electromigration process, leakage currents (I_{OFF}) and the breakdown events are related with the intensity of the inside-silicon electric fields. Thus lower V_{DD} bounds these physical 2nd-order effects, which affect the reliability and the robustness of the microelectronics circuits.

Despite that, the intensity of the I_{OFF} currents increases with the technological scaling-down. Large I_{OFF} can be detrimental for portable and not telecommunications devices, which are in power off for the most of time. One of the possible

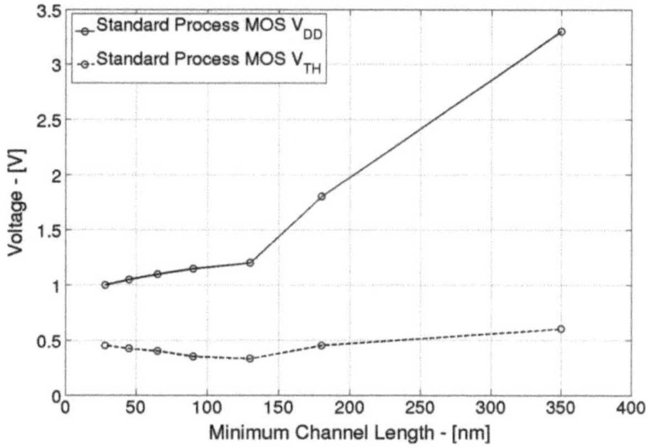

Fig. 1 V_{DD} and V_{TH} trend versus CMOS minimum channel length

approaches in order to break the I_{OFF} currents increasing is to slightly increase the CMOS transistors threshold voltage V_{TH}.

This situation is illustrated in Fig. 1 where V_{DD} and V_{TH} are plotted versus CMOS technological nodes. Starting from 0.35 μm down to 0.13 μm the V_{TH} scaling down has been a feasible option. Entering in sub-90 nm CMOS technologies, V_{TH} increases. At the same time V_{DD} has not be so scaled, like for 0.35 μm down to 0.18 μm.

These considerations are intentionally generic, because several nm-CMOS foundries consider the possibility to have different technological options, like High-Performance devices where standard supply voltage can be about 0.8/0.9 V and threshold voltage is typically lower, maintaining in any cases the V_{DD}/V_{TH} approximately equal to 2. As a conclusion the digital design requirements force to operate analog circuits with a strong reduction of the V_{DD}/V_{TH} ratio.

In 0.35 μm CMOS technology this ratio was about 6, while in 28 nm is about 2. The main consequences for analog design of the V_{DD}/V_{TH} trend versus nanometerscale technologies are:

- Operating point issues. MOS transistors strong/moderate region biasing is problematic, so that weak inversion region may be used, improving transistor efficiency (the g_m/I ratio is maximized at low overdrive voltage). At the same time current mismatch increases, and the exponential law between drain-source current and gate-source voltage degrades linearity.
- No cascode topologies can be used.

For the sake of clarity, take into account the simple differential stage in Fig. 2 (a key building block for continuous time filters design). It can be use as input stage for Operation Amplifiers (for Active-RC topologies) or integrators (in g_m-C filters). In order to maximize the signal swing, the differential stage input nodes are typically biased at $V_{DD}/2$. As a consequence the minimum supply voltage needed for a stable biasing is:

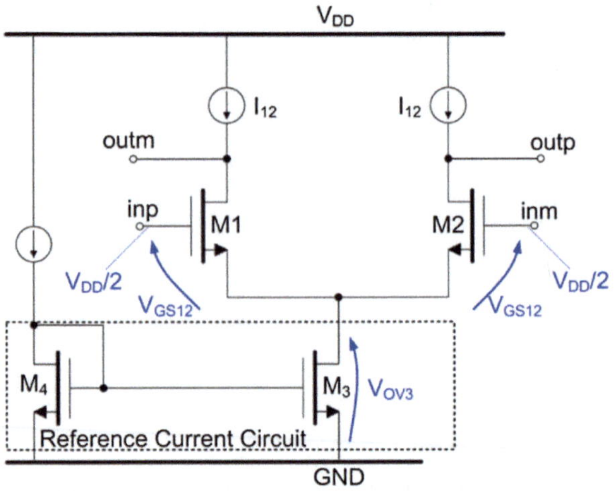

Fig. 2 CMOS differential stage

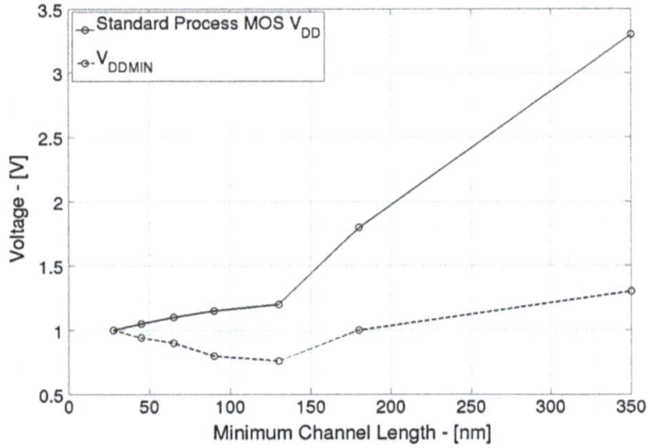

Fig. 3 V_{DD} and $V_{DD,MIN}$ versus minimum channel length

$$V_{DD,MIN} > 2 \cdot (2 \cdot V_{OV} + V_{TH}) \tag{1}$$

Assuming the overdrive voltage equal to 0.1 V, the $V_{DD,MIN}$ trend versus CMOS minimum channel length approaches the standard supply voltage trend, as illustrated in Fig. 3. The main consequence is that there is no any margin in nm-range CMOS technologies for cascode topologies, or larger overdrive.

In addition, a further voltage supply technological scaling down becomes detrimental for analog circuits design. These considerations introduce to two interesting topics from Microelectronics research point of view.

- To develop circuital solutions able to guarantee a stable operating point at standard supply voltage in sub-90 nm CMOS technologies, where the V_{DD}/V_{TH} ratio is about 2.

Fig. 4 Transistor intrinsic gain versus CMOS minimum channel length [3]

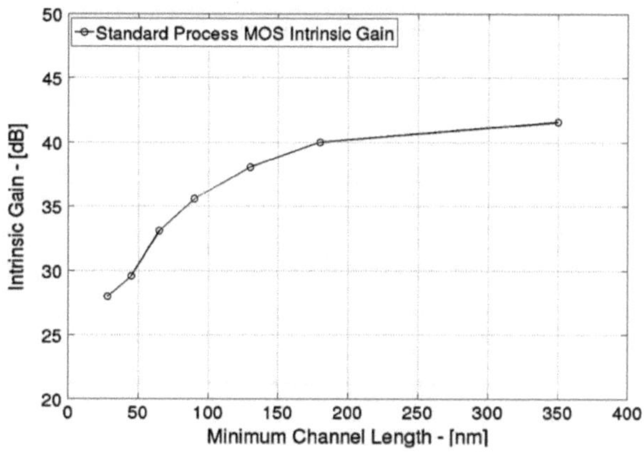

- To explore the possibility to enhance the operation of such circuits, optimizing them for sub-1 V supply voltage operation. In particular, sub-1 V biasing represents a key opportunity for power consumption reduction at system level in mostly-digital mixed-signal systems.

1.2 Analog Power Consumption in Sub-90 nm CMOS Technologies

One of the most important aspects of the nm-range technologies is the significant power consumption reduction at system-level in mixed-signal circuits. This power performance improvement is due to the digital circuits, because their power consumption is (almost) totally dynamic power. So that it is proportional to the supply voltage square (V_{DD}^2) and to the overall parasitic capacitance connected at the output of a specific logic port. The situation is significantly different in analog circuits, where technological scaling-down does not lead automatically to lower power consumption. Among the main reasons, transistor intrinsic gain ($g_m \cdot r_{ds}$) and Signal-to-Noise ratio reduction.

1.2.1 Transistor Intrinsic Gain

In sub-90 nm technologies, the MOS transistors experience a significant gain reduction, as illustrated in Fig. 4. Advanced CMOS processes are firstly developed and optimized for digital circuits, where switching capability is one of the most important parameter.

In order to minimize Short Channel Effects (SCE) and DIBL (Drain Induced Lowering Barrier) the doping density is increased close to the Drain and Source

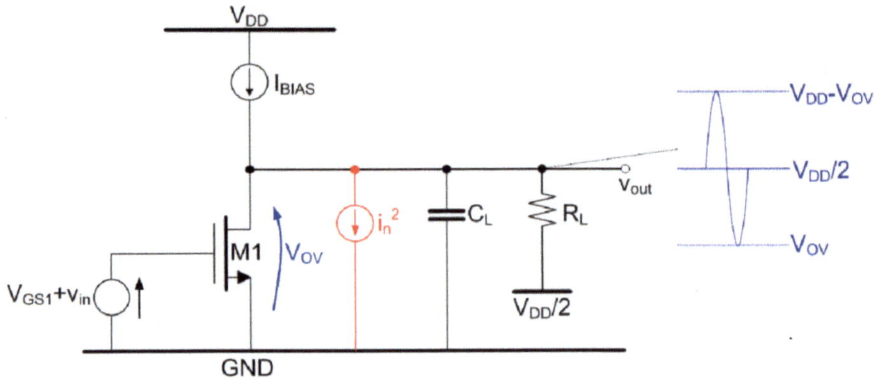

Fig. 5 Common source stage

wells. This is for sure a reliable approach for leakage current minimization during the switching, but it becomes detrimental for analog circuits because the transistor output resistance strongly reduces.

Even though MOS transistor transconductance (g_m) tends to be larger when the scaling down increases (the transistors have larger transition frequency), the output resistance drop is so high that the transistor intrinsic gain decreases of about 20 dB, as illustrated in Fig. 4.

In order to mitigate the output resistance drop in sub-90 nm technologies, MOS channel length in analog design should be increased of about 3 times the minimum channel length [3].

This increases the physical distance between Source and Drain wells, reducing the DIBL and SCE unwanted effects. Moreover, analog designers are forced to distribute horizontally the gain, increasing circuital complexity and the number of poles (critical also for stability in closed-loop circuits).

1.2.2 Signal-to-Noise Ratio

Lower supply voltage results in lower Signal-to-Noise ratio (SNR) at constant noise power, because the maximum output signal swing processed by a specific analog circuit is smaller. In order to understand the strict relation between Signal-to-Noise ratio and power consumption in CMOS analog circuits, take into consideration one of the most simple analog amplifiers: the common source circuit in Fig. 5.

The SNR is defined as:

$$SNR = \frac{V^2_{OUT,RMS}}{V^2_{NOISE,RMS}} \tag{2}$$

where $V^2_{OUT,RMS}$ is the power of the signal and $V^2_{NOISE,RMS}$ is the overall in-band integrated noise power. Both powers are taken at the output node of the common source stage. For the sake of simplicity, assume M1 in moderate inversion region ($V_{OV} = 0.1$ V) and a very accurate I_{BIAS} current given by:

$$I_{BIAS} = I_1 = k_n \cdot (V_{GS1} - V_{TH})^2 \tag{3}$$

where I_1 is the M_1 drain-source current, V_{GS1} is the M1 gate-source voltage, and V_{TH} is the M1 threshold voltage. As a consequence at dc no current is flowing trough the R_L resistive load, and the output common mode voltage is fixed at $V_{DD}/2$. In these conditions the signal power at the output of the stage is:

$$V^2_{OUT,RMS} = \frac{\left(\frac{V_{DD}}{2} - V_{OV}\right)^2}{2} \tag{4}$$

The noise due to the MOS transistor is assumed dominated by the thermal noise contribution, and it is modeled as an equivalent current noise source connected between drain and source (neglect the load R_L thermal noise contribution). The value of the output in-band integrated noise of the common source is approximately given by:

$$V^2_{NOISE,RMS} \cong \frac{2}{3} \cdot 4 \cdot k \cdot T \cdot g_m \cdot R_L^2 \cdot BW = \frac{\frac{2}{3} \cdot 4 \cdot k \cdot T \cdot A_V^2 \cdot BW}{g_m} \tag{5}$$

k and T are Boltzmann constant and temperature, respectively. g_m is the M1 transconductance, A_V and BW are the amplifier dc-gain and −3 dB bandwidth, respectively. Resuming, the SNR can be written as:

$$SNR = \frac{V^2_{OUT,RMS}}{V^2_{NOISE,RMS}} = \frac{\frac{\left(\frac{V_{DD}}{2} - V_{OV}\right)^2}{2}}{\frac{\frac{2}{3} \cdot 4 \cdot k \cdot T \cdot A_V^2 \cdot BW}{g_m}} = \frac{\left(\frac{V_{DD}}{2} - V_{OV}\right)^2}{\alpha_{kTABW} \cdot V_{OV}} \cdot I_1 \tag{6}$$

where the α_{kTABW} includes the constant terms. Notice that A_V and BW are also assumed to be constant, since they typically are system level specifications. Concluding at constant gain and bandwidth (and overdrive voltage), the common source amplifier SNR is proportional to I_1 by V^2_{DD}. Lower is V^2_{DD}, lower is SNR. Suppose to estimate the most important design parameters of the common source amplifier (g_m, I_1, noise, output swing) for each technological node (from 0.35 μm and down) starting from the specifications in Table 1 (for sake of simplicity given only for 0.35 μm and 28 nm CMOS nodes).

The design parameters are then reported in Table 2. Starting from the same IRN requirement for each technological node, the common source can be easily designed, with one main difference: the lower output swing, forced by the lower supply voltage in 28 nm with respect to the 0.35 μm technological node. As a consequence lower SNR is achieved in 28 nm implementation.

Table 1 Common source specifications

Technology	CMOS 0.35 μm	CMOS 28 nm
Supply voltage—V_{DD} (V)	3.3	1.1
Threshold voltage—V_{TH} (V)	0.6	0.45
Dc-gain—A_V (dB)	20	20
−3 dB bandwidth (MHz)	10	10
IRN PSD (evaluated as spot noise at 1 MHz) (nV/$\sqrt{\text{Hz}}$)	5	5

Table 2 Common source design parameters at constant noise

Design parameters	CMOS 0.35 μm	CMOS 28 nm
g_{m1} (mA/V)	0.45	0.45
$I_1(I_{BIAS})$ (μA)	11	11
M1 overdrive—V_{OV} (V)	0.1	0.1
R_L (kΩ)	22.6	22.6
C_L (pF)	0.78	0.78
Output swing	$3.1V_{PEAK\text{-}TO\text{-}PEAK}$	$0.8V_{PEAK\text{-}TO\text{-}PEAK}$
SNR (dB)	77	66

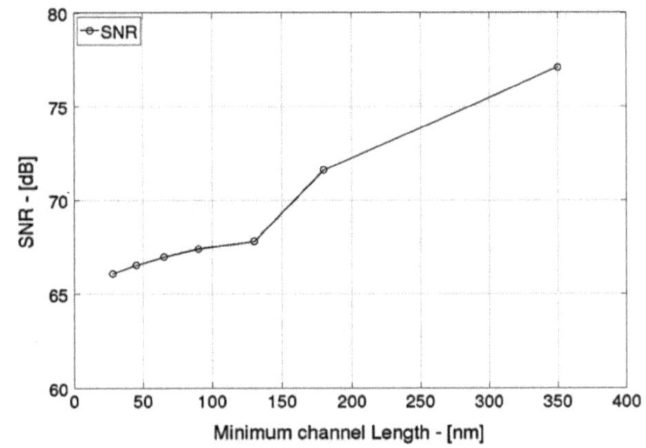

Fig. 6 Common source stage SNR at constant g_m

This is illustrated in Fig. 6 where the SNR versus the CMOS technological nodes has been plotted. For every CMOS process, signal swing is limited by the supply voltage. Since in Table 2 noise is constant (the same g_m is used for M1), then SNR decreases because the output swing is lower. A possible approach in order to maintain the same SNR achieved in CMOS 0.35 μm technology (where supply voltage is higher) is to reduce noise power spectral density over the amplifier bandwidth. That means to increase g_m.

At constant overdrive voltage, larger g_m implies larger current, and as a consequence larger power.

These last considerations are validated by Fig. 7, where the M1 g_m and power are plotted versus CMOS technological nodes.

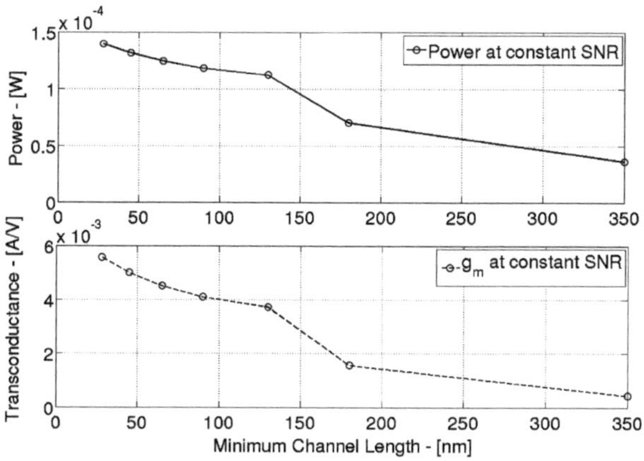

Fig. 7 Common source stage power and M1 g_m at constant SNR

Table 3 Common source design parameters at constant SNR

Design parameters	CMOS 0.35 μm	CMOS 28 nm
g_{m1} (mA/V)	0.45	5.5
$I_1(I_{BIAS})$ (μA)	11	140
R_L (kΩ)	22.6	1.79
C_L (pF)	0.78	8.8
Output swing	$3.1V_{PEAK\text{-}TO\text{-}PEAK}$	$0.8V_{PEAK\text{-}TO\text{-}PEAK}$
SNR (dB)	77	77

In Table 3 the main design parameters are then reported for 0.35 μm and 28 nm CMOS nodes in case of constant SNR requirement. The output swing remains lower in 28 nm implementation, but at the same time g_m is larger, so that the in-band integrated noise power decreases and as a consequence SNR is 77 dB. Notice also that R_L and C_L are adapted in order to maintain the same gain and bandwidth in both implementations.

1.3 Transition Frequency in Nanometer Scale IC Technologies

The increasing growth of wireless communications is pushing towards broad band, low power and portable transceivers. In the last years a significant effort has been concentrated on mixed-signal systems development for 10 GHz spectrum (in particular voice and data fro mobile phones and portable computers) regulated by the Ultra-Wide-Band standard. Unfortunately the adjacent spectrum is relatively populated by many unwanted signals due to the Wi-Fi bands [5]. The Federal Communications Commission introduced in 2001 the possibility to use the 57 and 66 GHz band.

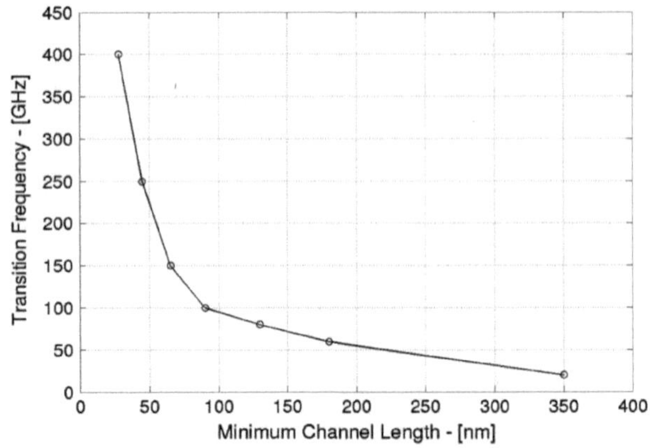

Fig. 8 MOS transition frequency versus minimum channel length

At the same time sub-90 nm CMOS technologies achieve MOS transistor transition frequency significantly higher than in the past, entering the hundreds of GHz domain (see Fig. 8, where MOS transition frequency trend is plotted versus minimum channel length). As a result, CMOS technologies have become even more attractive, due to the cheaper process cost with respect to the InP and GaAs technologies (typically used in the past for mm-wave applications).

From analog circuits design point of view, the increasing transition frequency enables the opportunity to implement broad band analog filters and amplifiers, embedded in the base band section of the 60 GHz transceivers. Obviously at the cost of severe design challenges.

- Entering in GHz domain leads to higher power consumption and noise, since the power spectral density is integrated in a larger bandwidth.
- Active-RC filters design—widely used in the past in telecommunications transceivers due to their capability to have large linearity and frequency response accuracy—becomes critical, because they require Operation Amplifiers with very much larger unity gain bandwidth than the filter poles.

The scenario for analog design is challenging and stimulating at the same time, because the research of the low power and low noise circuital solutions becomes more and more important.

1.4 Conclusions

The topic of this Chapter is to provide possible circuital topologies suitable to mitigate the severe issues for analog design in nanometerscale CMOS technologies, and exploiting the definitive opportunities given by the most advanced CMOS processes. This chapter is organized as follows.

Section 2 introduces two analog filters operating in telecommunications transceivers (for DVB-T and WLAN). The first filter is a low-pass (LP) filter with

8.2 MHz −3 dB frequency in CMOS 65 nm node [6]. A proper common mode circuit and specific design choices have been used to achieve a stable operating point for the Opamps, overcoming the V_{DD}/V_{TH} issues highlighted in Sect. 1.1. An improved version of such common mode circuit has been used in the second filter presented in this Section, making the operating point insensitive to the PVT variations. A 4th-order filter has been designed to operate at 0.55 V supply voltage in CMOS 0.13 μm technology, with $V_{TH} = 0.35$ V [7].

Since power is one of the main concerns in scaled technologies, Sect. 3 proposes two low-pass filters designs, based on source follower [8] and diode-C architectures [9], and suitable to strongly minimize power, while meeting the specifications for telecommunications base band section.

In Sect. 4, two analog filters circuits implemented in CMOS 45 nm and 28 nm process, currently under development, will be presented for operating in next generation 60 GHz transceivers. In particular the first one is a complete design of a closed-loop 4th order Active-RC low-pass filter able to operate a 1 GHz −3 dB frequency [10, 11]. The second filter is based on a super-buffer architecture [12] adequately improved for synthesizing biquadratic cell and amplifiers.

2 Analog Filters for Low V_{DD}/V_{TH} Operation

2.1 A 65 nm-CMOS 1.3 mW 52 dB-DR Filter-and-Amplifier for DVB-T Receivers

A 4th-order 8.2 MHz −3 dB bandwidth Filter-and-Amplifier for Digital Video Broadcasting-Terrestrial receivers is hereby presented. A silicon prototype of the filter has been integrated in 65 nm CMOS technological node, managing a V_{DD}/V_{TH} (supply/threshold voltage) ratio as low as 2.

The filter uses as reference biquadratic cell architecture the Rauch multipath cell. A proper noise/linearity/transfer-function optimization is developed at filter architecture level, with the aim of minimizing power consumption. In this way, the filter design meets the minimum power operation for a given transfer function and noise/linearity specifications set. For this reason a proper low-voltage bias circuit has been used in the filter, optimizing the operational amplifiers operating point. The device consumes 1.3 mW from a single 1.2 V supply voltage, features −10 dBm-Input-IP3 at 32 dB pass-band gain, and 1.6 mV_{rms} output integrated noise over the filter bandwidth (300 kHz ÷ 8 MHz).

2.1.1 DVB-T Filter Circuital Topology

A complete schematic view of the 4th-order filter is presented in Fig. 9 in fully-differential topology. The most important performance of the filter are then

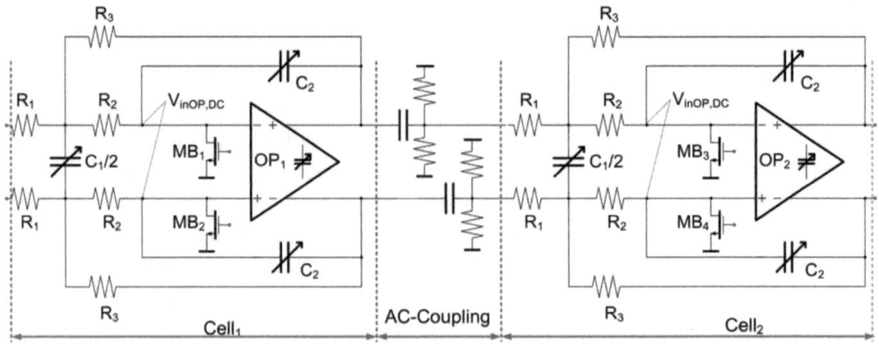

Fig. 9 MOS transition frequency versus minimum channel length

Table 4 DVB-T filter performance

Parameter	Value
Supply voltage—V_{DD}	1.2 V
CMOS technology	65 nm
Filter dc-gain—G	31 dB
Cell1/Cell2 poles frequency—$f_{@-3dB}$	8.2 MHz
Cell1 poles quality factor	0.5412
Cell2 poles quality factor	1.3066
Power consumption	1.3 mW
Output integrated noise (100 kHz ÷ 10 MHz)	1.63 mV_{rms}
IRN spectral density@7 MHz	18 $nV/\sqrt{Hz}$
1 dB-compression-Point	$0.9V_{zero\text{-}peak}$
THD ($v_{out} = 1.05V_{zero\text{-}peak}$@3 MHz)	40dBc
DR@THD = 40dBc	52 dB
Input IP3—$v_{in} = v_{in1} + v_{in2}$ (v_{in1}@2 MHz, v_{in2}@3 MHz)	−10 dBm
Output IP3—$v_{in} = v_{in1} + v_{in2}$ (v_{in1}@2 MHz, v_{in2}@3 MHz)	21.3 dBm

summarized in Table 4 [13, 14]. For power reduction, a single-opamp biquadratic cell has been selected. Rauch cell is preferred to Sallen-Key for better noise/substrate decoupling and simpler fully-differential implementation. Moreover R-C input net performs some filtering of the out-of-band interferers, resulting in an improvement of the out-of-band linearity that is critical due to the presence of large blockers.

One of the most important aspects of this design is the presence of two current sources connected at the Opamps input nodes (MB_1–MB_2, MB_3–MB_4). By sinking an additional small current to ground, it is possible to lower the opamp input node common mode voltage to 0.3 V, guaranteeing larger overdrive for Opamp differential stage [6].

The biasing current (I_{B12}) flowing by MB_1–MB_2 (MB_3–MB_4) is related with the Opamp input common mode voltage ($V_{inOP,DC}$) by Eq. 7:

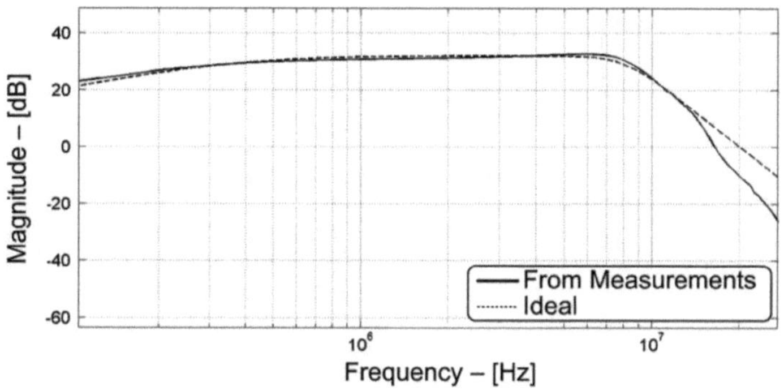

Fig. 10 Filter frequency response

$$V_{inOP,DC} = \frac{V_{DD}}{2} - I_{B12} \cdot \left(R_2 + \frac{R_1 \cdot R_3}{R_1 + R_3} \right) \quad (7)$$

This technique based on a simple level shifter has been widely used in the past for low voltage operational amplifier design [15], but never applied to a second order filter where quality factor accuracy and noise requirements have to be maintained. Actually, MB_1–MB_2 (MB_3–MB_4) feature finite output resistance and generate an additional noise contribution.

Since MB_1–MB_2 are connected at the Opamp input node, the virtual ground principle limits the signal swing across the current sources. About noise, a careful evaluation of the MB_1–MB_2 noise is here presented. For each Rauch biquadratic cell in the cascade, the input referred noise expression, including the dependence on MB_1–MB_2 noise (modeled as noise current source i_n), is given by:

$$IRN^2 = 8 \cdot k \cdot T \cdot R_1 + 8 \cdot k \cdot T \cdot R_3 \cdot \left(\frac{1}{G}\right)^2 + \left(8 \cdot k \cdot T \cdot R_2 + IRN^2_{OPAMP}\right) \cdot \left(1 + \frac{1}{G}\right)^2 + \frac{i_n^2}{\Delta f} \cdot (R_1 + R_2 \cdot (1 + 1/G))^2 \quad (8)$$

2.1.2 DVB-T Filter Experimental Results

Frequency Response

The DVB-T filter prototype has been tested in terms of frequency response, available in Fig. 10, where 31.3 dB is the passband gain, and 8.2 MHz is the nominal cut-off frequency, obtained after variable capacitors tuning. A slight slope increasing is observed at high frequency (about 20 MHz), due to the equivalent capacitor load present in the PCB used for test (about 5 pF).

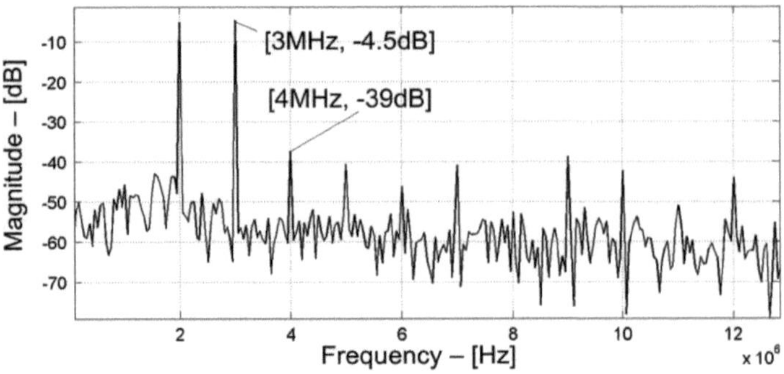

Fig. 11 IM3—$v_{in} = v_{in1@2\ MHz} + v_{in2@3\ MHz}$

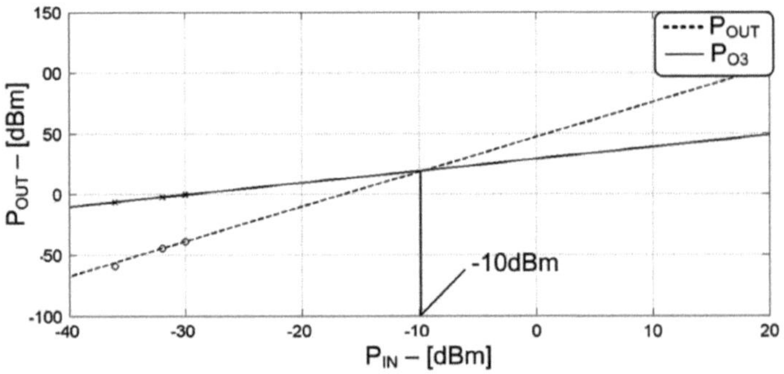

Fig. 12 IIP3

Linearity Performance

Linearity performance have been evaluated in terms of total harmonic distortion and IP3. In Fig. 11 two in-band tones (2 and 3 MHz) have been used as input signal, resulting in a total voltage swing at the filter output of about 1 V. The resulting input IP3 is plotted in Fig. 12 and it is −10 dBm (Fig. 13).

2.2 A 0.55 V 60 dB-DR 4th-Order Analog Baseband Filter

A 0.55 V supply voltage 4th-order low-pass continuous-time filter is presented. The low-voltage operating point is achieved by an improved bias circuit that uses different opamp input and output common-mode voltages. The 4th–order filter architecture is composed by two Active-G_m-RC biquadratic cells, which use a single opamp per-cell with a unity-gain-bandwidth comparable to the filter cut-off frequency. The −3 dB filter frequency is 12 MHz. The −3 dB frequency can be adjusted by means of a digitally-controlled capacitance array. In a standard

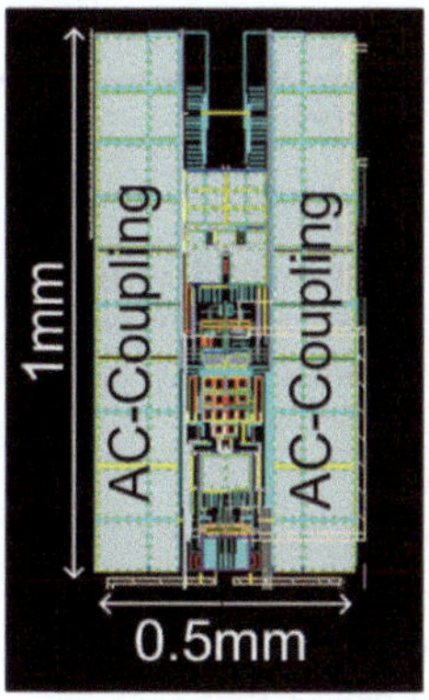

Fig. 13 Layout photo

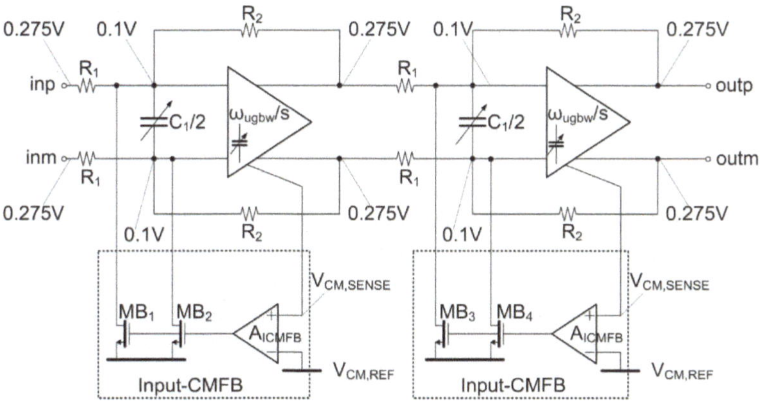

Fig. 14 0.55 V filter top view schematic

0.13 μm CMOS technology with $V_{THN} \approx 0.3$ V and $V_{THP} \approx 0.35$ V, the filter operates with a supply voltage as low as 0.55 V. The filter (total area = 0.47 mm^2) consumes 3.4 mW. A 8 dBm-in-band IIP3 and a 13.3 dBm-out-of-band IIP3 demonstrate the validity of the proposal.

2.2.1 0.55 V Filter Circuital Topology

The complete schematic of a 4th order filter meeting the WLAN receivers specifications is illustrated in Fig. 14. The filter is composed by the cascade of two biquadratic cells, whose main performance are summarized in Table 5. Each biquadratic cell is implemented using the Active-G_m-RC topology [16], where the Opamp unity gain bandwidth is used as filter transfer function parameter. With reference to the generic biquadratic cell, the low-pass transfer function is given by:

Table 5 0.55 V filter performance

Parameter	Value
Filter order	4th
G	0 dB
Cell1/Cell2 poles frequency—$f_{@-3dB}$	11.3 MHz
Cell1 poles quality factor	0.5412
Cell2 poles quality factor	1.3066
V_{DD}	0.55 V
CMOS	0.13 μm ($V_{THNMOS} = 0.3$ V)
Current cons.	5.8 mA
Power cons.	3.5 mW
In-band IIP3	10 dBm
Out-band IIP3	13 dBm
1dBcP	0.5 dBm
Output noise	110 μV_{rms}
SNR—THD@40dBc	60 DB

$$T(s) = G \cdot \frac{1}{s^2 \cdot C_1 \cdot \frac{R_2}{\omega_{ugbw}} + s \cdot \frac{1}{\omega_{ugbw}} \cdot \left(1 + \frac{R_2}{R_1}\right) + 1} \tag{9}$$

where G is the biquadratic cell dc-gain, R_1-R_2-C_1 are the passive components highlighted in Fig. 14, and ω_{ugbw} is the Opamp unity gain bandwidth.

As in DVB-T filter, a current source pair composed by the MB_1–MB_2 (MB_3–MB_4) transistors allows to lower the Opamp input common mode voltage to 0.1 V. As a result, the current (I_{B12}) flowing by the MB_1–MB_2 fixes the Opamp input common mode voltage ($V_{inOP,DC}$) as indicated in Eq. 10:

$$V_{inOP,DC} = \frac{V_{DD}}{2} - I_{B12} \cdot \left(\frac{R_1 \cdot R_2}{R_1 + R_2}\right) \tag{10}$$

The novelty of this filter with respect to the DVB-T implementation, is the presence of the Input-Common-Mode-Feedback-circuit. The main task of the I-CMFB is to reduce the sensitivity of the biquadratic cell operating point with respect to PVT variations. Actually, since in CMOS process physical and electrical characteristics of the MOS transistors change due to the process and temperature, the Opamp input common mode voltage (in theory fixed by MB_1–MB_2 current) could divert from nominal 0.1 V biasing value. The simple resistor variation modifies the $V_{inOP,DC}$ value, as reported in Eq. 10.

The operating point margin for Opamp input differential pair at 0.55 V is very limited, so that every unwanted variation can be detrimental for the circuit operating point. The main task of the I-CMFB is to sense eventual common mode variations and to reregulate the MB_1–MB_2 current in order to maintain the Opamp dc input voltage approximately equal to 0.1 V (Fig. 15).

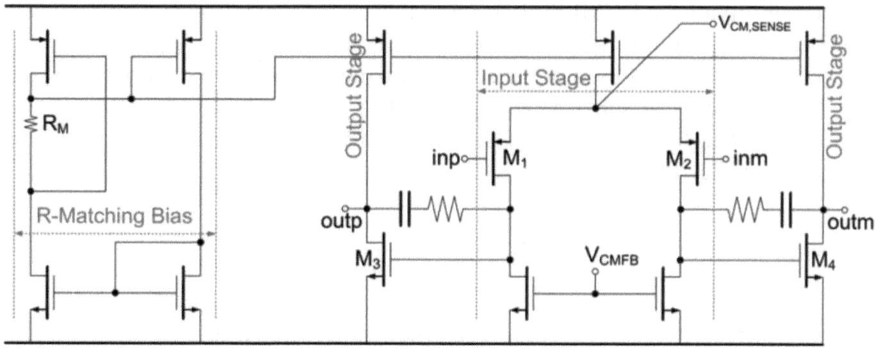

Fig. 15 0.55 V Opamp top view schematic

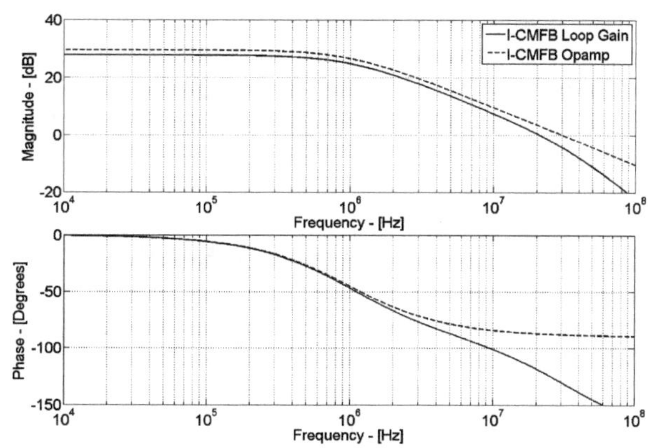

Fig. 16 Single-ended Opamp and I-CMFB loop gain frequency response

The Input-CMFB circuit is a closed-loop circuit, whose main tasks are:

- To sense the variation of the Opamp input common mode voltage (fixed at 0.1 V in nominal conditions).
- To increase/decrease the MB_1–MB_2 biasing current as a function of the sensed Opamp input common mode variation.

The sensing task is performed connecting the source of the fully-differential Opamp input MOS with the plus node of the single-ended Opamp, as illustrated in Fig. 16 (see V_{CM_SENSE} node). In this way no additional components connected at the Opamp input nodes are needed for implementing the I-CMFB sensing.

This is a very important point for this design, since in Active-G_m-RC circuits Opamp unity gain bandwidth is for definition close to filter poles frequency. This implies that for input signal frequencies close to the pole, the signal swing at the Opamp input nodes increases, because the virtual ground principle becomes weak. Evaluating the maximum swing (versus frequency) at the Opamp input is crucial for correctly biasing the MB_1–MB_2 pair [7].

The stability of the I-CMFB circuit depends on the single ended opamp transfer function $A_{ICMFB}(s)$, on the current source transconductance $g_{m,MB12}$, on the passive components values used in the cell and on the current source output resistance $r_{ds,cs}$ (unfortunately at sub-1 V voltage supply $r_{ds,cs}$ is comparable with the filter resistances). The loop gain—$G_{loop}(s)$—of the Input-CMFB is in first approximation given by Eq. 11:

$$G_{loop}(s) = g_{m,MB12} \cdot \left(R_1//R_2//r_{ds,cs}\right) \cdot \frac{A_{ICMFB}(s)}{1 + s \cdot C_1 \cdot \left(R_1//R_2//r_{ds,cs}\right)} \tag{11}$$

The dominant pole in the $G_{loop}(s)$ depends on the $A_{ICMFB}(s)$ first pole. But the second pole is determined by the time constant $C_1 \cdot (R_1//R_2//r_{ds,cs})$, and it is comparable with the filter cut-off frequency. If the single ended opamp features a very low unity gain frequency—$\omega_{UGBW,ICMFB} << 1/(C_1 \cdot (R_1//R_2//r_{ds,cs}))$—and a phase margin of about 90°, then the stability of the I-CMFB circuit is then guaranteed. Obviously, this approach allows to strongly reduce the I-CMFB additional power consumption, so that the dominant I-CMFB power contribution is practically given by MB_1–MB_2 currents. Figure 16 shows the I-CMFB loop gain and single-ended Opamp frequency response (phase margin is about 60°).

2.2.2 0.55 V Filter Experimental Results

A prototype of the proposed filter has been realized in a 0.13 μm CMOS technology. The chip photo is then shown in Fig. 17. The core area is 0.45 mm^2 and is widely dominated by the capacitances.

As illustrated in Fig. 14, capacitor arrays are used in order to align the filter cut-off frequency, thus compensating the CMOS process deviation. The generic schematic of the Opamps used in both biquadratic cell is shown in Fig. 15, where on the left side of the figure, the input stage bias circuit is also shown. It correlates the transconductance variations of the opamp input pair transistors (due to temperature, aging, and technological process spread) to the variation of the resistances of each cell. It forces the opamp input stage g_m to be proportional to R_M. In this way the unity gain frequency of the opamp is inversely proportional to the $R_M \cdot C_c$ time constant (where C_c is the opamp Miller capacitance and R_M is the reference resistance used in bias circuit), and it can be adjusted by properly setting the capacitors array. This aspect is very important in Active G_m-RC cells, where the poles frequency and quality factor depend on the opamp unity gain frequency, as reported in Eq. 9.

Figure 18 shows the filter gain frequency response for $V_{DD} = 0.55$ V, $V_{DD} = 0.525$ V, and $V_{DD} = 0.5$ V. A 11.3 MHz cut-off frequency is obtained, performing only 5 % of deviation with respect to the nominal −3 dB frequency, as reported in Table 1. The dc gain is −1.4 dB. In fact each biquadratic cell introduces some drop at the dc, dependent on the finite dc gain of low voltage opamps. In addition the frequency responses are plotted also for values below $V_{DD,min}$ (525 and 500 mV). The selectivity of the filter is approximately maintained also at

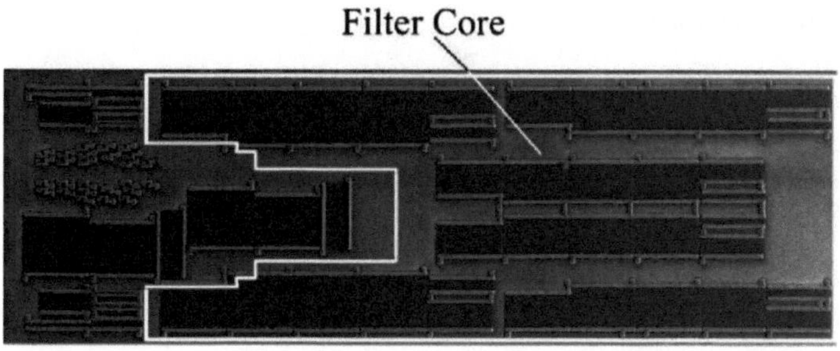

Fig. 17 Prototype filter photo

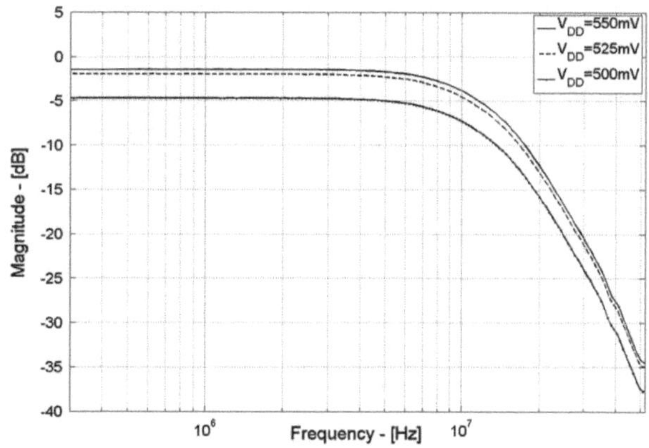

Fig. 18 Filter transfer function versus V_{DD}

Table 6 Filter design parameters

Parameter	Value
R_{in}	4 kΩ
R_1	3.2 kΩ
R_3	2 kΩ
R_4	356 Ω
R_5	32 kΩ
R_6	4 kΩ
R_7	500 Ω
C_1	0.135 pF
C_2	0.2 pF
C_3	0.315 pF
ω_{u2}	$2\pi \cdot 3.8$ Grad/s
ω_{u3}	$2\pi \cdot 1.81$ Grad/s

0.5 V supply voltage. Unfortunately, a dc-gain drop appears for 0.5 V supply voltage, due to the reduced Opamps gain at lower V_{DD} (Table 6).

A 0.5 dBm 1 dB-Compression Point is achieved, as reported in Fig. 19. In band IIP3 has been also evaluated setting two input tones at 3 and 4 MHz.

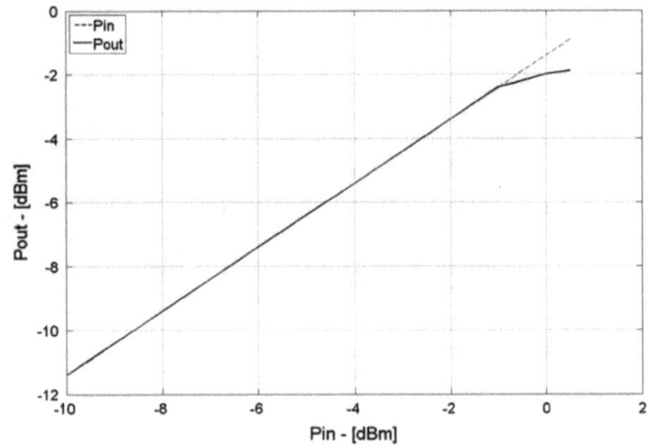

Fig. 19 1 dB compression point

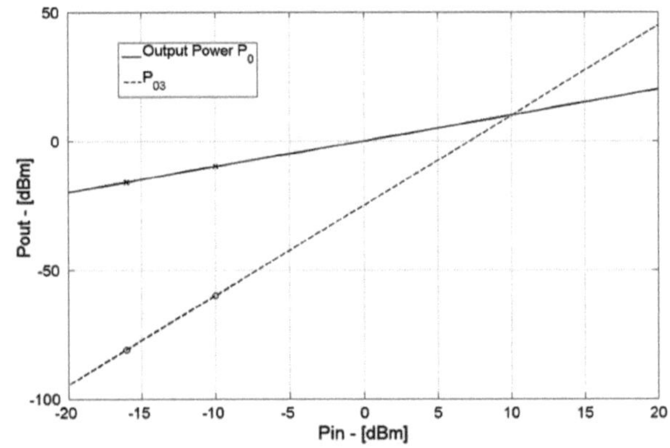

Fig. 20 In-band IIP3

Measured points are plotted in Fig. 20, resulting in 10 dBm IIP3. As a conclusion, the filter meet the specifications imposed in the base band section of WLAN receivers, operating at 0.55 V supply voltage.

These experimental results represent a fully characterization of the filter in terms of frequency response, linearity and power, validating the operation at 0.55 V in CMOS 0.13 μm technology node.

3 Low Power Circuital Topologies: Source Follower and Diode-C Filters

To get higher operation frequency at low power consumption, open loop filters, like G_m-C filters, are generally employed, as they do not require opamps with a bandwidth quite larger than the cut-off frequency. However, they suffer for low linearity and weak robustness against parasitic. In the last years different solutions of open loop filters have been proposed with improved linearity and power

Fig. 21 Source-follower

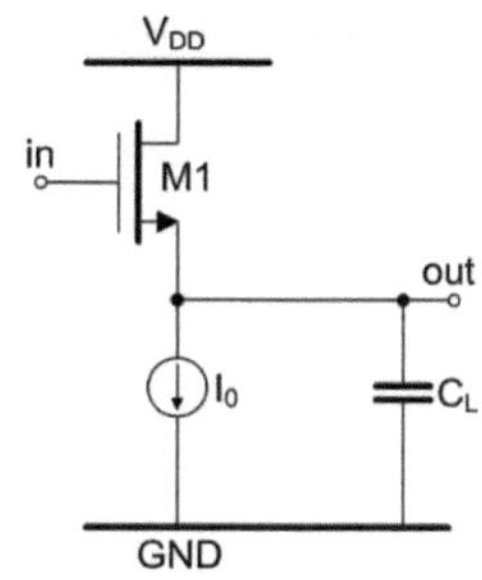

efficiency. In [8] an alternative architecture of open loop analog filter based on a complex source-follower topology is proposed. Figure 21 shows the source-follower circuit, largely used in analog signal processing. It has some interesting features that make it very attractive to design analog filters. First, it has a good in-band linearity, due to its inherent local feedback. As any feedback structure, its linearity improves with a large closed-loop gain (G_{loop}) which for the source-follower at low frequency can be approximated to:

$$G_{loop} = \frac{g_{m1}}{g_{m1} + g_{ds1} + g_{ds0}} \tag{12}$$

where g_{m1} and g_{ds1} are transconductance and output resistance of the M1 transistor, respectively, while g_{ds0} is the output resistor of the current source I_0.

This means that a larger transconductance value (achieved for lower overdrive V_{OV}, since for long channel length devices $g_m = 2 \cdot I_0/V_{ov}$) results in a larger loop gain and then in a better linearity, as given in Eq. 12.

This basic conclusion completely differs from other active filters (like G_m-C filters, for instance), where the linearity is improved at the cost of larger overdrives, and then larger currents (for a given g_m) and power consumption. Breaking the dependence V_{ov}-versus-linearity immediately has a large impact on the power performance. Minimizing V_{OV} corresponds to reduce the current level to achieve the same value. This is reflected in a substantial power saving for the same linearity level. The above concept corresponds to the fact that the source-follower processes the signal directly in the voltage domain, avoiding conversion into a current and then back to the voltage. In this way, it is possible to increase the linearity range, as the main source of distortion in G_m-C filters comes from the conversion of the voltage signal into a current signal. This operation is performed by the transconductors in a G_m-C filter. Each filter transconductor processes large signals in an open-loop configuration. Therefore, they introduce distortion (Fig. 22).

Moreover, a time constant is obtained depending only on the transistors transconductance and on the capacitive load. This means that the circuit does not need to drive any resistive load, avoiding additional current consumption under signal regime. The absence of a resistive load to be driven gives higher power efficiency to the source-follower. Furthermore, the source-follower has more advantages.

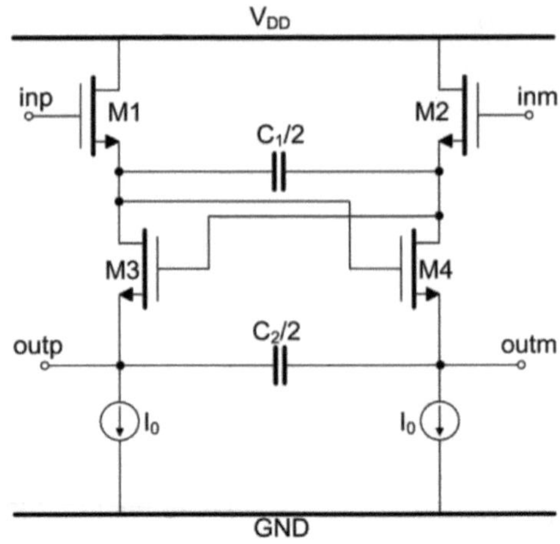

Fig. 22 Source-follower-based biquadratic cell [8]

- No circuital parasitic poles are present. In fact, each circuit node corresponds to a pole included in the source follower transfer function. Therefore, there is no need of extra current to push parasitic poles to higher frequencies.
- The output common-mode voltage is self-biased by the transistor, without adding any additional circuit, as the common-mode feedback circuit.
- In addition, the source-follower can drive a resistive load without substantially modifying its linearity performance and its pole frequency.

3.1 Source-Follower-Based WLAN Filter in CMOS 0.18 μm

The advantages of the source-follower above described are reported in the second-order low-pass cell shown schematically in Fig. 12.

This cell presents an optimized single-branch fully differential structure and operates like a "composite" source-follower. It exploits a positive feedback obtained thanks to the cross coupling connection, to get complex poles. The transfer function of the biquadratic cell is given by:

$$T(s) = \frac{1}{s^2 \cdot \frac{C_1 \cdot C_2}{g_m^2} + s \cdot \frac{C_1}{g_m} + 1} \qquad (13)$$

The proposed source-follower-based biquadratic cell has been used to design a 4th-order Bessel low-pass filter, satisfying typical WLAN 802.11.a/b/g receiver specifications in 0.18 μm CMOS technology. The overall 4th-order schematic is shown in Fig. 23.

It consists of a cascade of two biquadcells. The first cell is made up of p-MOS transistors and the second one is made up of n-MOS transistors. The use of

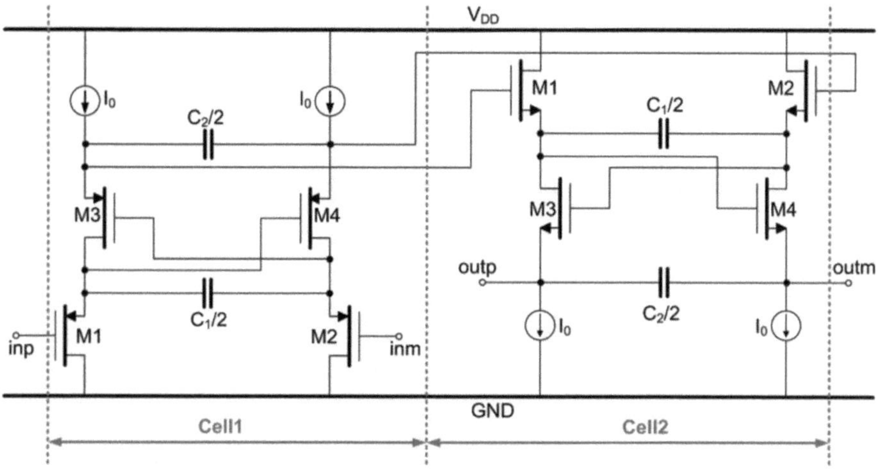

Fig. 23 Source-follower-based 4th-order filter for WLAN [8]

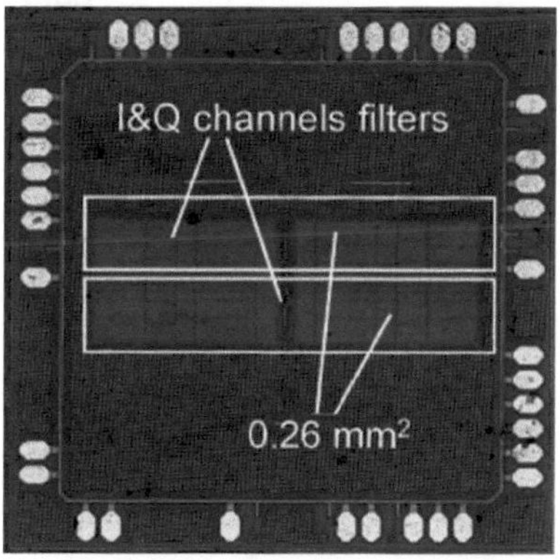

Fig. 24 Source-follower-based 4th-order filter for WLAN—chip photo

cascading pMOS and nMOS structures allows compensating for the input-to-output common-mode voltage difference, generally typical of the source-follower and more specifically of the proposed source-follower-based 2nd-order cell.

The filter performs at 17.5-dBm IIP3 and 40-dB HD3 for a 600-mV input signal amplitude. A 24 μV_{rms} noise gives a 79 dB dynamic range, with 2.25-mA current consumption. A chip photo is shown in Fig. 24.

The minimum supply voltage required by source-follower-based filter presented in Sect. 3.1 is quite higher than the standard supply voltage used in sub-90 nm CMOS technology. Actually such $V_{DD,MIN}$ is given by:

$$V_{DD,MIN} = 3 \cdot V_{OV} + 2 \cdot V_{TH} \tag{14}$$

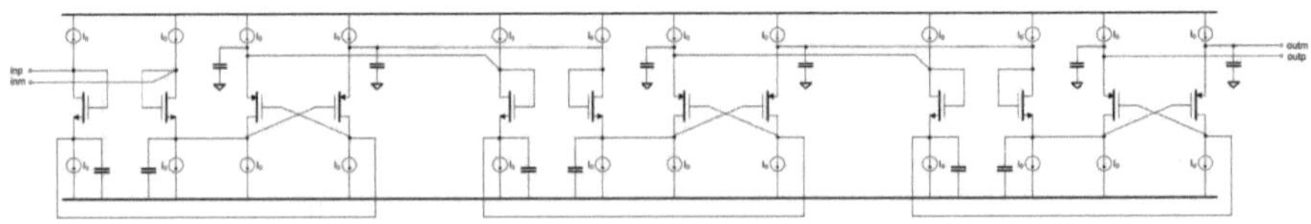

Fig. 25 6th-order low-pass filter based on a diode-C architecture [9]

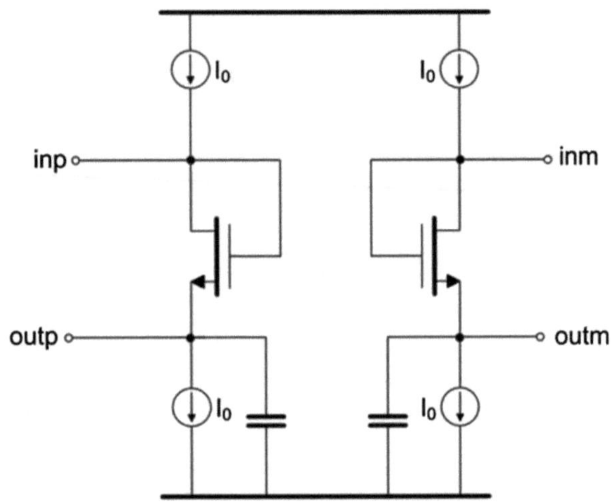

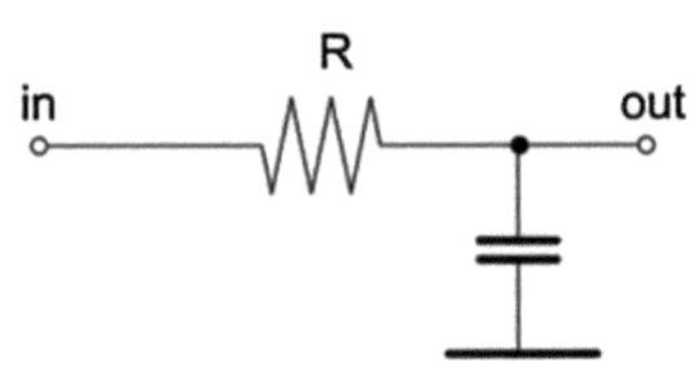

Fig. 26 Diode-C for R [9]

3.2 *Diode-C-Based Ultra-Wide-Band Filter in CMOS 0.13 μm*

In [9] a very efficient high frequency analog filters is reported. The filter features a 6th-order low-pass transfer function with 280 MHz cut-off frequency. It consumes 120 μW at 53.6 dB SNR. The overall filter architecture is reported in Fig. 25. It exploits a diode-C architecture. The basic structure of this filter is the diode–C stage reported in Fig. 26. The diode-C stage has an (almost) unitary DC-gain. The dominant pole is due to the time constant at the output node. Ideally, the correspondent transfer function is:

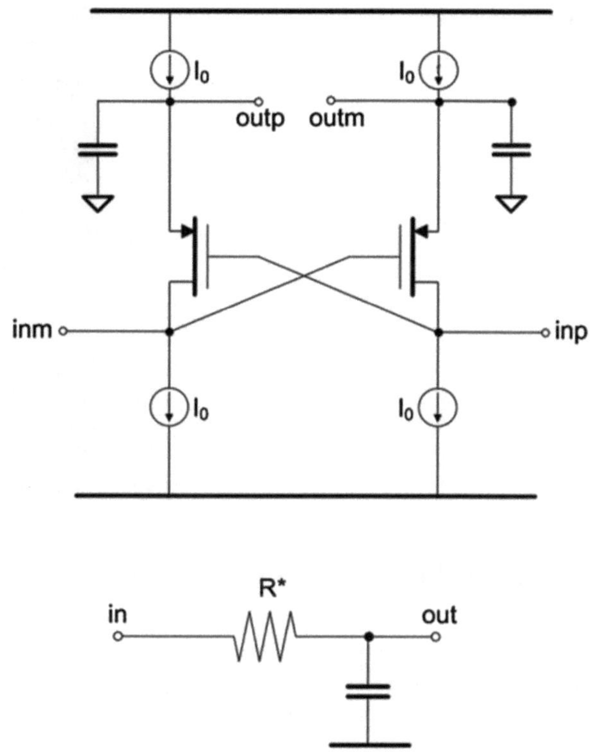

Fig. 27 Diode-C for $R^* = -R$ [9]

$$T_1(s) = \frac{1}{1 + s \cdot C_1/g_{m1}} \tag{15}$$

where C_1 is the capacitance connected at the output node and g_{m1} is the M1 transistor transconductance. The cascade of several basic diode-C single-pole cell does not allow to implement higher order complex-poles filter, since it definitively is a cascade of R-C nets. Using the cross-coupled topology in Fig. 27, allows to invert the resistor from $R = 1/g_{m1}$ to $R^* = -1/g_{m1}$. As a consequence it is possible to synthesize complex poles by cascading p-MOS/n-MOS diode-C cells as in Fig. 25.

As for the source follower, the diode-C filters exhibit high linearity. In fact a 11 dBm IIP3 is measured at low frequency. In the analog filter reported in [9], the output poles of each diode-C stages are exploited to synthesize all the filter poles. However, this kind of architecture does not enable DC-gain higher than 0 dB. Figure 28 shows the chip micrograph. The chip area is as small as $200 \times 90\ \mu m^2$. In this filter the total amount of capacitance is 500 fF, which occupies 0.01 mm^2. On the other hand, the use of limited V_{OV} allows the circuit to achieve large gm with low current. In this prototype, the total current is about 100 μA for a total power dissipation of 120 μW.

Fig. 28 Chip photo [9]

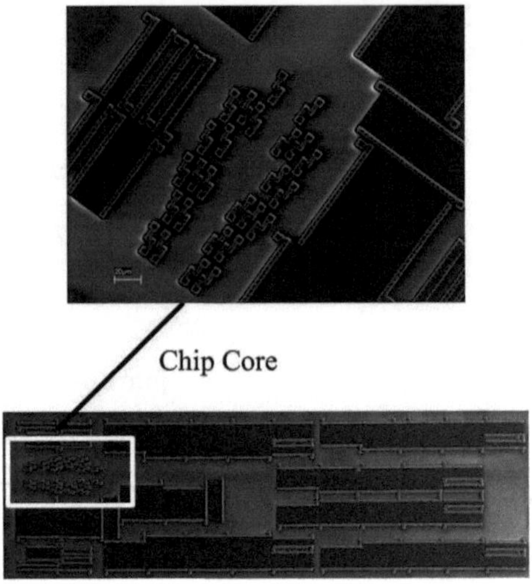

4 Analog Filter for 60 GHz Transceivers in CMOS 28 nm

The 57–66 GHz bandwidth availability represents an important opportunity for high-data rate wireless transceivers and for wide bandwidth CMOS integrated circuits. One of the most important aspect for analog design is the possibility to exploit the larger transition frequency in sub-90 nm CMOS technologies, enabling the design of broad band amplifiers and filters in a cheaper technological node. The transceivers typically use zero-IF architectures [4], with the RF-Section responsible for the acquisition of the signal coming from the antenna and low-noise amplification. After the down-conversion, the input signal of the base band section has about 0.88/1.76 GHz bandwidth.

From base band section analog design the main concerns are power and noise, for several reasons.

- The filter is usually the first base band section processing block, so that it has more severe noise constraints than the following Programmable-Gain-Amplifier.
- The filter pass-band is in the GHz order, so that low thermal noise is typically required, comparing with the other (at lower bandwidth) telecommunications standards. Actually, the target is to avoid significant output in-band integrated noise increasing due to the larger bandwidth, and as a consequence the specification in terms of pass-band Input Referred Noise (IRN) power spectral density is significantly stringent.
- In 60 GHz base band section transceivers, the power consumption tends to be higher than other telecommunications standards due to the higher bandwidth and to the lower required IRN.

In this scenario the main efforts for analog designers are focused on in-band noise and power consumption reduction.

4.1 A 54 dB-DR 1-GHz-Bandwidth Continuous-Time Low-Pass Filter with In-band Noise Reduction

The Active-RC topologies are often selected for continuous-time filter design, due to the capability to guarantee large linearity and high frequency response accuracy, comparing with g_m-C filters where power consumption and noise are typically lower, while linearity and frequency response can be critical. Actually, a typical design issue in broad band Active-RC filters is thermal noise power spectral density, since when it is integrated over the entire bandwidth it can affect critically the Signal-To-Noise-Ratio. This issue is further stressed in 57–66 GHz transceivers, where base-band chain bandwidth requirements are in the GHz order.

The bandwidth request imposes deep integration scale, so that typically CMOS 65 nm and down nodes are required, since they exhibit larger transition frequency. Moreover, technological scaling down, forced by the GHz bandwidth, leads to lower supply voltage, meaning that Signal-To-Noise-Ratio degradation (due to the larger in-band integrated noise power) cannot be recovered by processing a larger output voltage signal.

In this scenario the analog filter here presented, proposes an interesting circuital solution, suitable to reduce the in-band integrated noise power [11], and to perform 55 dB SNR.

The key concept of this filter is to synthesize a 5th-order 1 GHz-bandwidth low-pass filter using a single compact Active-RC closed-loop cell. Most of the thermal noise sources are high-pass filtered, while maintaining the low-pass filtering of the in-band signal.

A prototype of the filter has been designed in CMOS 45 nm technology, with 1 GHz –3 dB-bandwidth. The simulation results shows as the noise power spectral density reduction leads to Signal-To-Noise-Ratio (versus power) improvement comparing with the state-of-the-art [9, 17].

4.1.1 5th-Order 1 GHz-Bandwidth Filter Schematic and Transfer Function

Figure 29 shows the functional scheme of the filter. There are three main blocks: an integrator (whose transfer function is $-\omega_{u1}/s$) and two biquadratic cells ($T_2(s)$ and $T_3(s)$). The integrator is fully defined by its unity gain frequency. The two biquadratic cells must be adequately sized in order to comply with the transfer function requirements in Fig. 29. Two feedbacks are then needed to synthesize a 5th order low-pass transfer function. The complete schematic implementation of the fully-differential 5th order filter is then illustrated in Fig. 30.

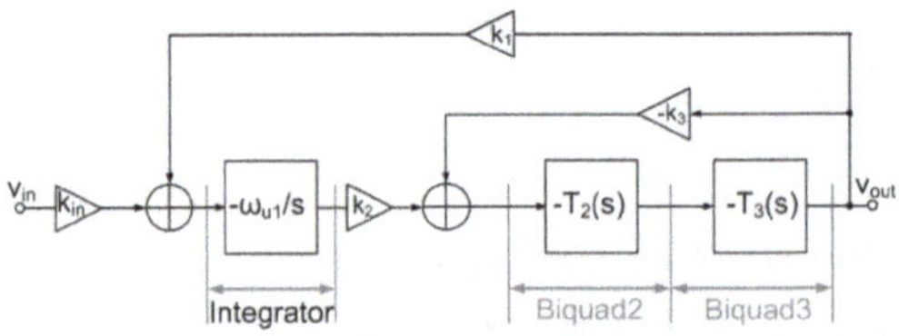

Parameter	Value	
Transfer Function (Low-Pass)	5th Order Butterworth	
Pass-Band Gain - G	0dB	
f@-3dB	1GHz	
IRN_{Filter}	15nV/√Hz	
SNR@(THD≥40dBc)	>50dB	
Parameter	**Poles Pair 1**	**Poles Pair 2**
Poles Frequency – ω_0	1GHz	1GHz
Quality Factor – Q	Q_1=1.3066	Q_2=0.7
Real pole Frequency	1GHz	

Tab. I – Analog Filter Requirements

Fig. 29 5th order filter functional scheme and requirements

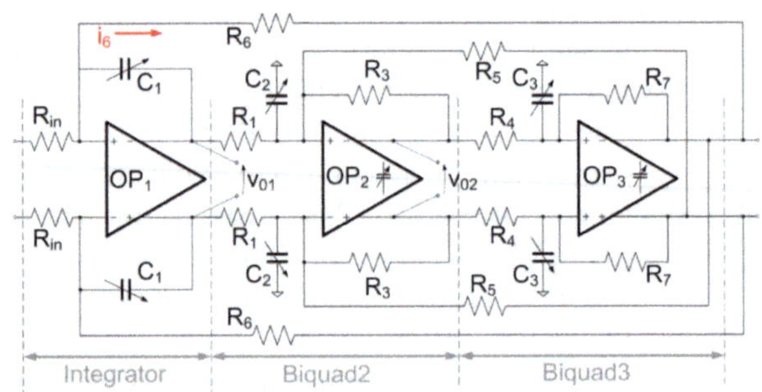

Fig. 30 5th order filter top-view schematic

The analog filter has to comply with the requirements shown in Fig. 29. Fifty dB-SNR is required, while the output signal swing must perform 40 dBc of Total-Harmonic-Distortion. At least a 5th order filter transfer function is needed for channel selection. The transfer function is composed by two complex poles pairs (with the same frequency ω_0, and quality factor Q_1–Q_2) and one negative real pole (having the same frequency ω_0).

4.1.2 Filter Transfer Function

The most common implementation approach for a 5th order filter is the cascade of three Active-RC cells (one for the real pole and one cell for each complex poles pair). At 0 dB dc-gain every cell contributes to the overall thermal noise power. The maximum allowable noise is about 15 nV/$\sqrt{\text{Hz}}$, resulting in very low resistors (<1 kΩ), due to the 1 GHz bandwidth. In these conditions, large power is expected for (low) resistive load driving, and Opamp input stage g_m (for noise power reduction). The presented design approach proposes to use a single compact cell for an overall 5th order transfer function, where in-band noise reduction is implemented, and power consumption is 9 mW.

The integrator is designed by using a single-Opamp topology, and the two biquadratic cells are implemented in Active-G_m-RC configuration [16, 18]. That allows to synthesize a complex poles pair using only one additional Opamp, comparing with the most popular 2nd-order Tow-Thomas biquad. The filter ideal transfer function is:

$$T(s) = \frac{G}{1+\frac{s}{\omega_0}} \cdot \frac{1}{1+\frac{s}{\omega_0 \cdot Q_1} + \left(\frac{s}{\omega_0}\right)^2} \cdot \frac{1}{1+\frac{s}{\omega_0 \cdot Q_2} + \left(\frac{s}{\omega_0}\right)^2} \quad (16)$$

The filter transfer function can be expressed as a function of the design parameters: R-C values, and OP_2–OP_3 unity gain bandwidth (ω_{u2} and ω_{u3}, since in Active-G_m-RC cell the Opamp is used as integrator). The resulting transfer function based on the circuit in Fig. 30 is:

$$T(s) = \frac{R_6}{R_{in}} \cdot \frac{1}{a_{rc5} \cdot s^5 + a_{rc4} \cdot s^4 + a_{rc3} \cdot s^3 + a_{rc2} \cdot s^2 + a_{rc1} \cdot s + 1} \quad (17)$$

where a_i parameter are function of R-C values, and OP_2–OP_3 unity gain bandwidth (ω_{u2} and ω_{u3}) [19].

Noise

The only thermal noise sources affecting the overall filter noise power spectral density are dependent on R_i, R_6 and OP_1. In fact, due to the presence of C_1 capacitance, no feedback is present between the OP_1 output and its inverting input (at dc). So that if the signal is not coming from v_{in}, i_6 current is always zero. As a result, every internal loop noise contribution is totally high-pass filtered out by the same circuital topology. The final expression of the in-band IRN for the filter (IRN_{FILTER}) in Fig. 30 is given by Eq. 18:

$$IRN^2_{FILTER} = 8 \cdot k \cdot T \cdot \left(R_{in} + R_6 \cdot \left(\frac{1}{G}\right)^2\right) + IRN^2_{OP1} \cdot \left(1 + \frac{1}{G}\right)^2 \quad (18)$$

Obviously, increasing frequency, the C_1 impedance decreases and the inside-loop noise contributions become critical. However, the in-band noise is strongly attenuated.

4.1.3 5th-Order 1 GHz-Bandwidth Filter Simulations Results

A prototype of the filter has been designed in CMOS 45 nm technology. The nominal frequency response is plotted in Fig. 31, with v_{01}/v_{in} and v_{02}/v_{in} frequency response. Note that at dc the output swing of the first and second operational amplifier (v_{01}/v_{in} and v_{02}/v_{in}, respectively) are intentionally maintained lower than the v_{out}/v_{in} (of about −3 dB). The frequency response features 1 GHz −3 dB-bandwidth, and it is externally tuned using variable capacitors.

Note also the compensation capacitances in OP_2–OP_3 are also tuned (see Opamp in Fig. 32). In this way a very precise −3 dB-frequency (±5 % deviation from PVT simulations) has been obtained. The Opamps are implemented by a simple class-A Miller-Compensation scheme. The main performance are then available in Table 7.

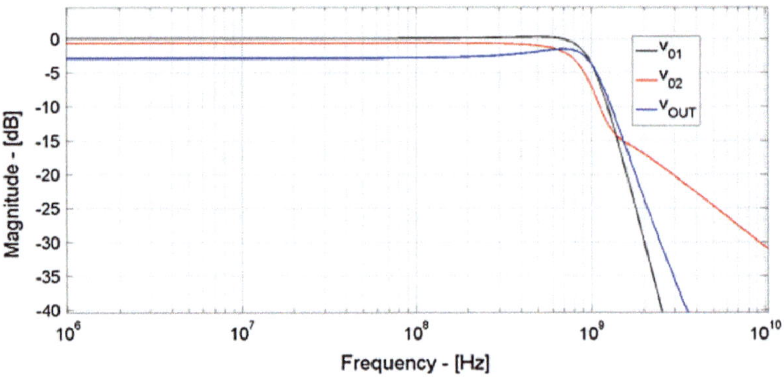

Fig. 31 Frequency response

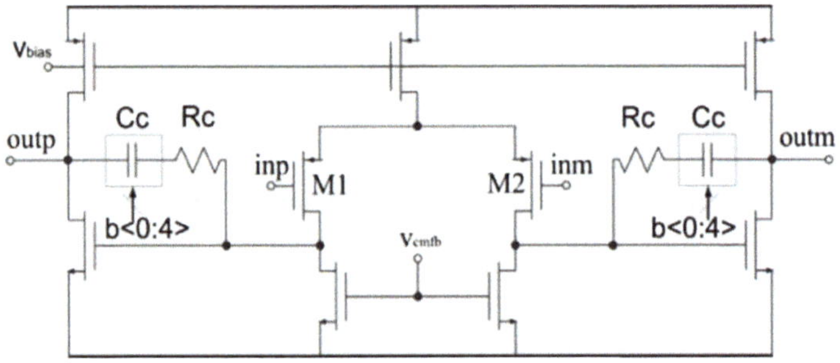

Fig. 32 Opamps schematic

Table 7 Opamps design parameters

Parameter	OP1	OP2	OP3
DC-gain (dB)	55	37	42
Unity gain BW (GHz)	2	3.8	1.81
IRN_{OPi} (nV/$\sqrt{}$Hz)	2	2	3.5
Current consumption (mA)	2	4	3

Noise Power Spectral Density

Figures 33 and 34 show the noise transfer function of the resistors used in the filter, while in Fig. 35 the noise transfer function for OP_1–OP_2–OP_3 is plotted. Note that the only thermal noise contributions constant over the entire bandwidth are dependent on R_{in}, R_6 and IRN_{OP1} (as in Eq. 18). Due to the presence of C_1 capacitance, the remaining noise sources are high-pass filtered. Notice that some peaking in noise transfer function is present around the poles frequency (see overall noise PSD at the filter output, in Fig. 36), due to the other resistors and OP_2–OP_3. This peaking leads to a small noise increasing with respect to the Eq. 18.

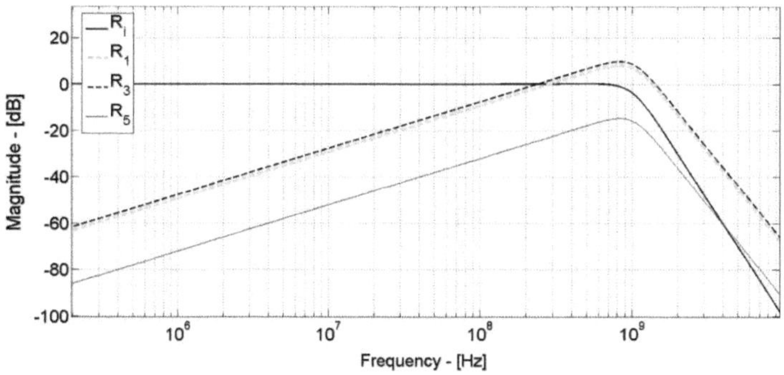

Fig. 33 R_i-R_1-R_3-R_5 noise TF

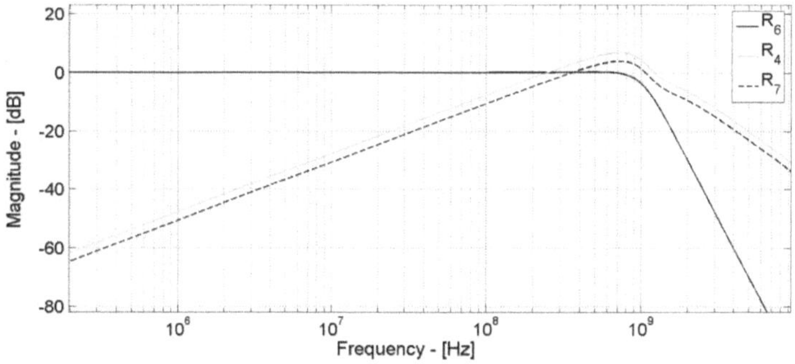

Fig. 34 R_4-R_6-R_7 noise TF

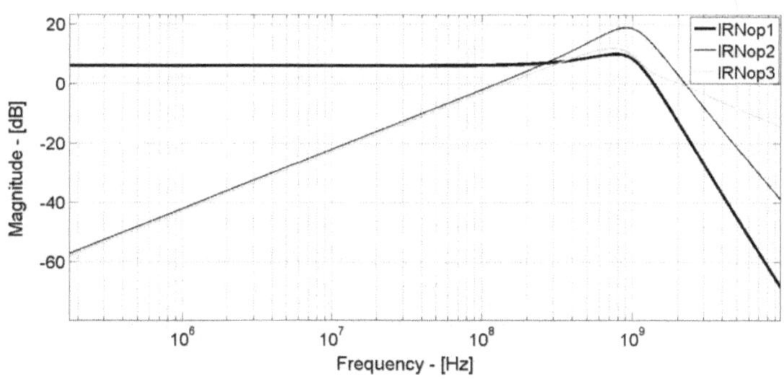

Fig. 35 OP_1–OP_3 noise TF

The overall in-band output integrated noise is 500 μV_{rms}. The effective IRN calculating considered also the peaking effect (14 nV/$\sqrt{}$Hz) due to the noise shaping is a little bit higher than the IRN at low frequency (12.9 nV/$\sqrt{}$Hz). All that demonstrates the in-band noise reduction operates rejecting all thermal noise

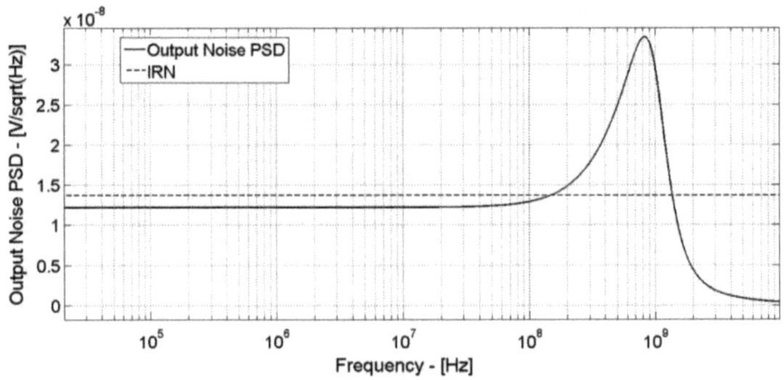

Fig. 36 Output noise PSD

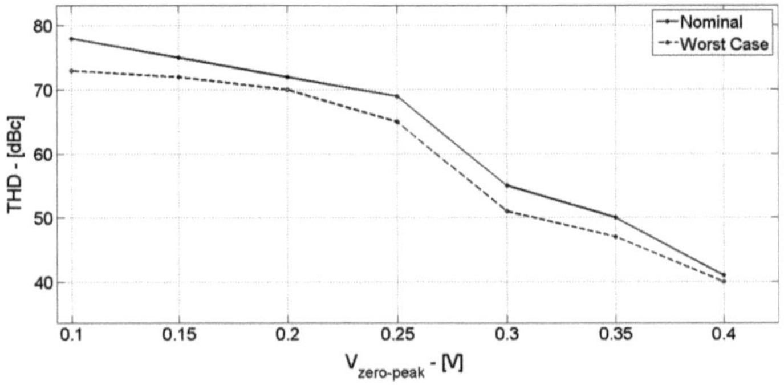

Fig. 37 Total harmonic distortion versus output signal

Table 8 Filter performance resume

Parameter	Value
Filter dc-gain—G	0 dB
−3 dB-Bandwidth—$f_{@\text{-3dB}}$	1.01 GHz
Supply voltage—V_{DD}	1.2 V
CMOS technology	45 nm
Power consumption	9 mW
Output integrated noise (100 kHz ÷ 20 MHz)	500 μV_{rms}
IRN spectral density@10 MHz	12 nV/$\sqrt{}$Hz
THD—$v_{out} = 0.45V_{zero\text{-}peak}$@100 MHz	40 dBc
SNR@THD = 40 dBc	55 dB

contributions present inside the R_{in}-R_6 loop, at the cost of a slight noise increasing of about 10 % (from 12.9 to 14 nV$\sqrt{}$Hz).

Linearity THD has been evaluated for different output voltage swings. Simulation results are available in Fig. 37. In the worst case scenario (obtained spreading PVT), a 0.45 Vzero-peak output in-band swing is processed by the filter with 40 dBc of THD (Table 8).

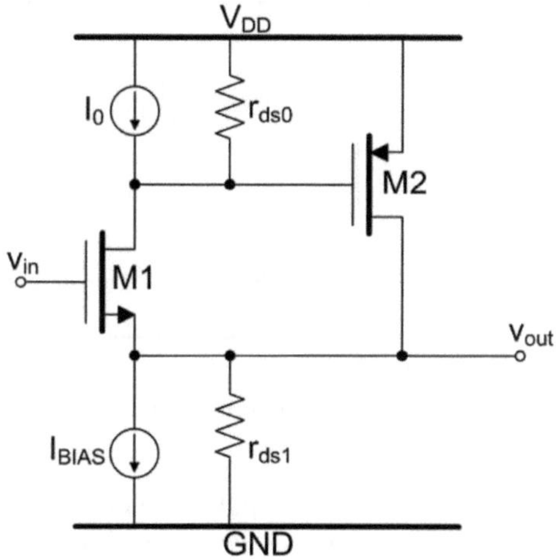

Fig. 38 Super-buffer

4.2 Super-Buffer-Based Baseband Section for 60 GHz Transceivers

The object of this section is a compact reconfigurable base band section for 60 GHz transceivers. The base band is composed by the cascade of two macro-blocks. The first one is a reconfigurable gain filter (5 and 25 dB), and the second one is a 4th-order Butterworth low-pass filter. Both blocks features variable bandwidth, 880 MHz and 1.76 GHz as in 60 GHz telecommunication standard.

The base band section is designed in 28 nm CMOS technology and it is currently under development, so that the main performance will be summarized as simulation results.

One of the main ideas of this base band section is to exploit a single compact cell as key building block for filtering and amplification. Such cell is the super-buffer circuital topology, whose basic schematic is shown in Fig. 38 [12].

The M2 transistor implements a local loop, where at dc the loop gain is given by Eq. 19 (the M1 drain-source resistance has been neglected, r_{ds0} and r_{ds1} are the overall resistance connected at M1 drain and source respectively).

$$G_{loop} \cong -\frac{g_{m1} \cdot r_{ds1} \cdot g_{m2} \cdot r_{ds0}}{1 + g_{m1} \cdot r_{ds1}} \cong -g_{m2} \cdot r_{ds0} \tag{19}$$

The dc-gain of the super-buffer topology results improved with respect to the simple follower, as indicated in Eq. 20:

$$G_{dc} \cong \frac{g_{m1} \cdot r_{ds1} \cdot g_{m2} \cdot r_{ds0}}{1 + g_{m1} \cdot r_{ds1} \cdot g_{m2} \cdot r_{ds0}} \cong 1 \tag{20}$$

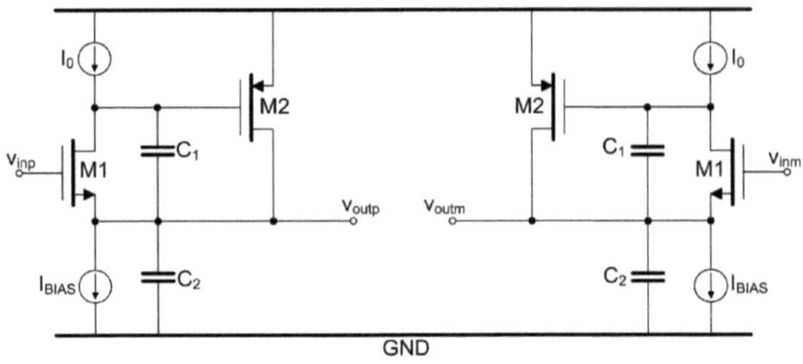

Fig. 39 Super-buffer-based biquadratic cell

In addition, the output resistance ($1/g_m$ in simple follower implementation) is here a factor $g_m \bullet r_{ds}$ lower, as in Eq. 21:

$$R_{OUT} \cong \frac{r_{ds1}}{1 + g_{m1} \cdot r_{ds1} + g_{m1} \cdot r_{ds1} \cdot g_{m2} \cdot r_{ds0}} \cong \frac{1}{g_{m1}} \cdot \frac{1}{g_{m2} \cdot r_{ds0}} \tag{21}$$

4.2.1 Super-Buffer-Based Baseband Section for 60 GHz Transceivers: Filter and Amplifier

Starting from the circuital topology in Fig. 38, a 2nd-order biquadratic cell can be synthesized as shown in Fig. 39. C_2 is connected between drain and source of M1, and C_1 is connected between M1 source and ground. The biquadratic cell in Fig. 39 is in (pseudo) differential topology. The poles frequency and quality factor of the single complex poles pair are given by:

$$\omega_0^2 = \frac{g_{m2} \cdot g_{m1}}{C_1 \cdot C_2} \tag{22}$$

$$Q = \sqrt{\frac{C_2}{C_1} \cdot \frac{g_{m1}}{g_{m2}}} \tag{23}$$

Resuming the v_{out}/v_{in} transfer function of the filter is:

$$T(s) = \frac{1}{1 + s \cdot \frac{C_1}{g_{m1}} + s^2 \cdot \frac{C_1 \cdot C_2}{g_{m2} \cdot g_{m1}}} \tag{24}$$

The same super-follower cell can be used for amplifier implementation, as shown in Fig. 40, where the transfer function at low frequency is:

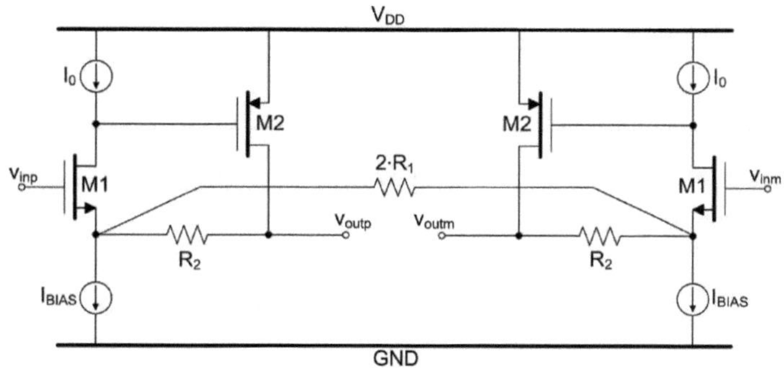

Fig. 40 Super-buffer-based 1st-order filter and amplifier

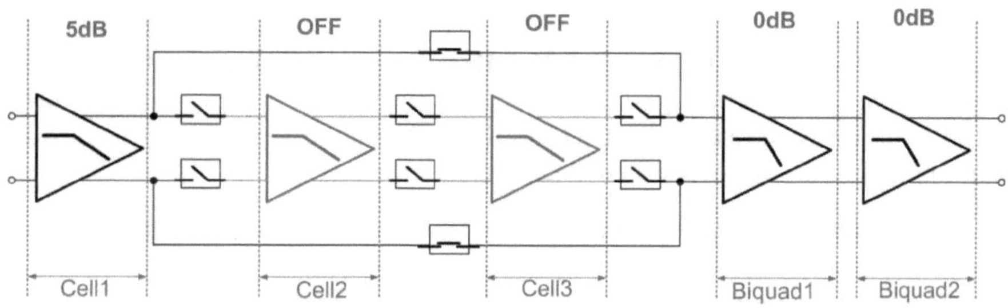

Fig. 41 Super-buffer-based base band at 5 dB-gain

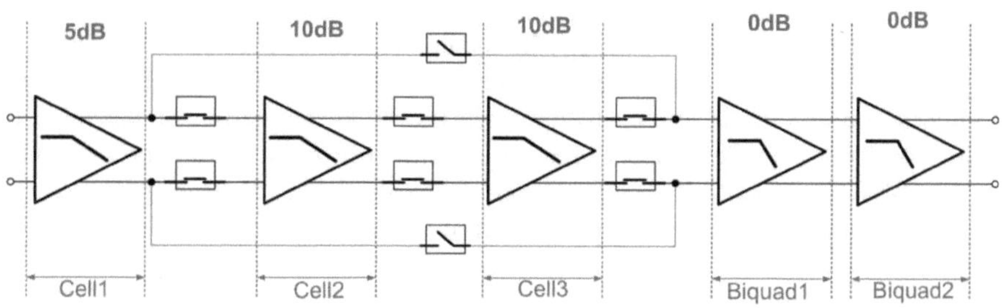

Fig. 42 Super-buffer-based base band at 25 dB-gain

$$T(s) = 1 + \frac{R_2}{R_1} \tag{25}$$

These two cells have been used to design a 7th-order low pass filter with programmable gain and bandwidth in CMOS 28 nm technology at 1.1 V supply.

The architecture of the entire base band is shown in Figs. 41 and 42, for 5 and 25 dB gain, respectively. The base band is reconfigurable in terms of gain (5 and

Table 9 Base band section performance

Bandwidth	1.76 GHz		870 MHz	
DC-gain	5 dB	25 dB	5 dB	25 dB
Supply voltage (V)	1.1	1.1	1.1	1.1
−3 dB bandwidth	1.87 GHz	1.81 GHz	889 MHz	868 MHz
IRN ($nV/\sqrt{Hz}$)	5.01	5.07	5	4.7
In-Band integrated output Noise (100 kHz ÷ 2 GHz)	285 μVrms	2.99 mVrms	283 μVrms	2.3 mVrms
$V_{OUT,SWING}$ [$V_{0\text{-peak,diff}}$] (THD ≥ 30dBc) (V)	0.3	0.28	0.3	0.2
SNR (dB)	57.13	36.5	57.4	35.77
Power (mW)	13	20	5	7
Output IP3 (dBm)	7	2	8	3

25 dB) and bandwidth (0.88 and 1.76 GHz). There are two main macro blocks. The first one is composed by the cascade of three 1st-order filters, with 5, 10, and 10 dB gain for each one. Cell1-Cell2-Cell3 have about 3 GHz −3 dB-bandwidth, and the main purpose of the first three cells are:

- To perform a low-noise amplifications of the weak input signal at the base band chain input nodes.
- To filter-out an amount of out-of-band interferers.

The second macro is a 4th-order low-pass filter with programmable cut-off frequency (0.88 MHz and 1.76 GHz), where each Biquad is designed using the circuital topology in Fig. 39. Frequency programmability is implemented at constant quality factor by acting on C_1–C_2 capacitors.

An overall resume of the base band performance is reported in Table 9. Proper switches configurations are set to optimize power as a function of the required gain. For 5 dB gain Cell2 and Cell3 are switched off, reducing power and noise.

The SNR varies accordingly to dc-gain from 36 dB, at 25 dB dc-gain case, up to 57 dB at 0 dB dc-gain. The input referred noise remains stable at 5 $nV/\sqrt{Hz}$.

5 Conclusions

A complete resume of the analog filters state-of-the-art is reported in Table 10. The main parameters for analog filters comparison are power, linearity, noise and bandwidth. Telecommunication standards typically impose severe requirements to the base band analog filters, and for this reason the intermodulation product performance versus noise are taken into account for linearity.

One of the most usual approach, in order to correctly compare different filters for different bandwidth and standards, is to use the following Figure-of-Merit [1]:

Table 10 Analog filters performance comparison

Work	Architecture	Tech.	Order	f_{-3dB}	Gain	Linearity	Noise	Supply (V)	Power (mW)
[16]	Active G_m-RC	0.13 μm	4th	2.11–11 MHz	4 dB	IIP3 = 11 dBm	IRN = 36 μV	1.2	3.4–14.1
[7]	Low-Voltage Active G_m-RC	0.13 μm	4th	11.3 MHz	0 dB	IIP3 = 0.5 dBm	IRN = 110 μV	0.55	3.5
[20]	Shunt-shunt feedback	90 nm	6th	255 MHz	−2.57 ÷ 39.02 dB	2/−33.8 dBm 1 dBcp	IRN(@40 dB) = 12.5 nV/$\sqrt{Hz}$	1.2	1.56 ÷ 2.3
[8]	Source follower	0.18 μm	4th	10 MHz	−3.5 dB	IIP3 = 13.5 dBm	7.5 nV/$\sqrt{Hz}$	1.8	4.1
[9]	Diode-C	0.13 μm	6th	280 MHz	0 dB	IIP3 = 11 dBm	IRN = 22 nV/$\sqrt{Hz}$	1.2	0.12
[21]	G_m-C	0.18 μm	5th	2 GHz	−16 dB ÷ 34 dB	N/A	N/A	1.8	40
[22]	G_m-C	0.18 μm	4th	900 MHz	−39 dB ÷ 55 dB	1dBcp = −10.8/−59.1 dBm	NF = 6.8 dB ÷ 67 dB	1.8	20.52
[23]	G_m-C	45 nm	4th	800 MHz	0 dB	THD = −47 dB@f_{-3dB}/3 and 200 mV_{pp}	SNDR = 33 dB	1.1	47.96
[24]	Active-RC	0.18 μm	5th	44–300 MHz	0 dB	N/A	ORN = 860 μVrms	1.8	54
[25]	G_m-C	0.35 μm	5th	70–500 MHz	0 dB	18 dBm@20 MHz (@70 MHz BW)	ORN = 366 μVrms	3.3	100
[26]	Active-RC	0.18 μm	5th	500 MHz–1 GHz	0 dB	IIP3 = 28.7dBm	IRN = 18 nV/$\sqrt{Hz}$	1.8	90
[27]	G_m-C	0.35 μm	4th	550 MHz	N/A	IIP3 = 10dBm	−151 dBm/Hz	1.65	140.25
[28]	Active-RC	0.13 μm	1st, 3rd, 5th	1–20 MHz	0 dB	IIP3 = 31.3dBm	IRN = 85-52 nV/$\sqrt{Hz}$	1	3–7.5
[29]	G_m-C with withΣ-Δ modulator	0.18 μm	5th	2 MHz	0 dB	IM3 > 82 dB (−13 dBm tones)	N/A	1.8	11
[30]	G_m-C	0.13 μm	2nd	200 MHz	0 dB	IIP3 = 14 dBm	IRN = 35.4 nV/$\sqrt{Hz}$	1.2 V	20.8
[31]	G_m-C	0.13 μm	6th	250 MHz	−9 ÷ 73 dB	IIP3 = −71 dBV @73 dB gain	IRN = 1.42 nV/$\sqrt{Hz}$	1.2 V	56.4
[32]	Active-RC	0.18 μm	6th	240 ÷ 500 MHz	1.2 dB	IIP3 = 15.9 dBm	IRN = 13.1 nV/$\sqrt{Hz}$	1.8 V	4.1
[33]	Ring oscillatore integrator based	90 nm	4th	7 MHz	0 dB	THD = 60 dB @FS(≈−8dBm)	SNR = 60 dB @FS(≈ −8 dBm)	0.55 V	2.9
[34]	Active-RC	0.18 μm	6th	7.1 ÷ 20.3 MHz	1.2 dB	OIP3 = 13.6 dBm	IRN = 66.2 nV/$\sqrt{Hz}$	1.8 V	0.47
[35]	G_m-C	90 nm	6th	8.1 ÷ 13.5 MHz	−3 dB	IIP3 = 21.7 ÷ 22.1 dBm	IRN = 75 nV/$\sqrt{Hz}$	1 V	4.35

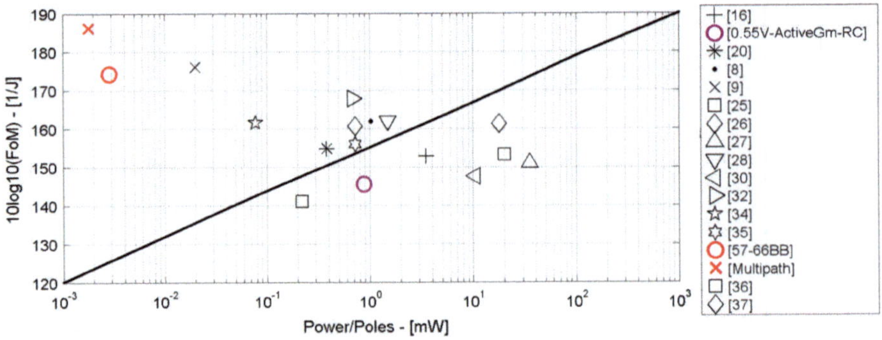

Fig. 43 Figure-of-merit versus power-per-poles

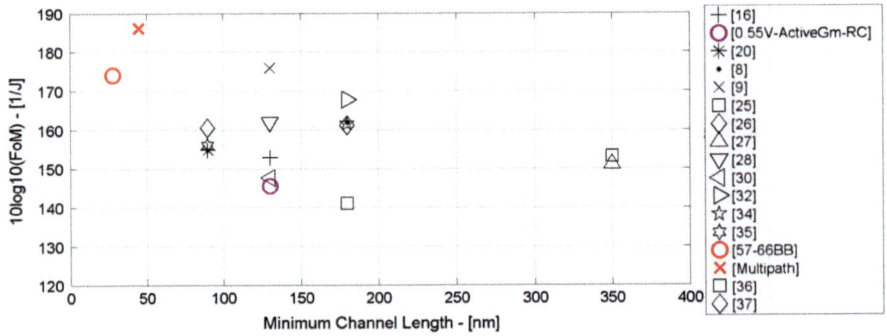

Fig. 44 Figure-of-merit versus supply voltage

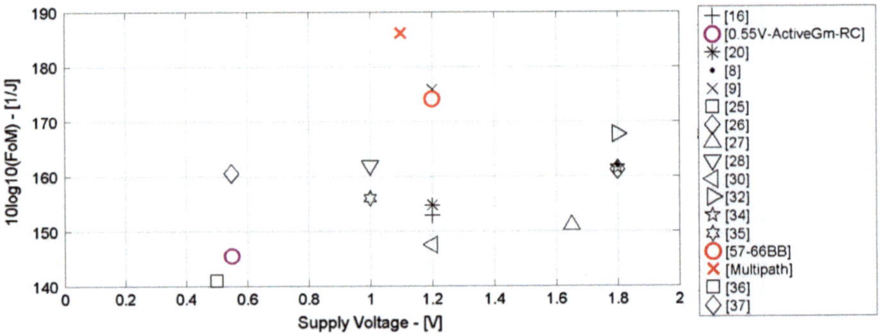

Fig. 45 Figure-of-merit versus technological node

$$FoM = 10 \cdot log_{10} \frac{IMFDR_3 \cdot f_{-3dB} \cdot N}{P_W} \tag{26}$$

where PW is the total power consumption, f_{-3dB} is the cut-off frequency, N is the number of poles, and $IMFD_3$ is the spurious-free IM3 and it is calculated as follows:

$$IMFDR_3 = \left(\frac{IIP3}{V_{N,in}}\right)^{4/3} \tag{27}$$

IIP_3 is the third order input intercept point, while $V_{N,in}$ is the in-band integrated input referred noise. Both IIP_3 and $V_{N,in}$ are expressed in V_{rms}.

The Figure-of-Merit in Eq. 26 is plotted versus Power-per-Poles in Fig. 43. In Figs. 44 and 45 the same Figure-of-Merit is plotted versus supply voltage and CMOS technological node. The analog filters presented in this chapter are favourably compared with the state-of-the-art.

References

1. Sansen, W. "Analog design challenges in nanometer CMOS technologies". Solid-State Circuits Conference, 2007. ASSCC '07. IEEE Asian. Publication Year: 2007, Page(s): 5–9.
2. Nauta, B.; Annema, A.-J. "Analog/RF circuit design techniques for nanometer scale IC technologies". Solid-State Device Research Conference, 2005. ESSDERC 2005. Proceedings of 35th European. Publication Year: 2005, Page(s): 45–53.
3. Dautriche, P. "Analog design trends and challenges in 28 and 20 nm CMOS technology". ESSCIRC (ESSCIRC), 2011 Proceedings of the. Publication Year: 2011. Page(s): 1–4.
4. Borremans, J.; Raczkowski, K.; Wambacq, P. "A digitally controlled compact 57-to-66 GHz front-end in 45 nm digital CMOS". Solid-State Circuits Conference - Digest of Technical Papers, 2009. ISSCC 2009. IEEE International. Publication Year: 2009, Page(s): 492–493.
5. Niknejad, A. M. and Hashemi, H. (2008). "mm-Wave Silicon Technology 60 GHz and Beyond". Springer, ISBN 978-0-387-76558-7, USA. Pages(s): 1–302.
6. De Matteis, M.; Cocciolo, G.; De Blasi, M.; Baschirotto, A. "A 1.3mW CMOS 65 nm 4th order 52 dB-DR continuous-time analog filter for DVB-T receivers". Electronics, Circuits and Systems (ICECS), 2011 18th IEEE International Conference on. Publication Year: 2011. Page(s): 21–24.
7. De Matteis, M.; D'Amico, S.; Baschirotto, A. "A 0.55 V 60 dB-DR Fourth-Order Analog Baseband Filter". Solid-State Circuits, IEEE Journal of. Volume 44, Issue: 9. Publication Year: 2009. Page(s): 2525–2534.
8. S. D'Amico, M.Conta, A. Baschirotto "A 4.1-mW 10-MHz Fourth-Order Source-Follower-Based Continuous-Time Filter With 79-dB DR" IEEE Journal of Solid-State Circuits. Volume 41, no. 12, December 2006, pages 2713–2719.
9. S. D'Amico, M. De Matteis, A. Baschirotto "A 6th-Order 100μA 280MHz Source-Follower-Based Single-loop Continuous-Time Filter" International Solid-State Circuits Conference 2008. Digest of Technical Papers. Page(s): 72–73.
10. De Matteis, M.; D'Amico, S.; Cocciolo, G.; De Blasi, M.; Baschirotto, A. "A 54dB-DR 1-GHz-bandwidth continuous-time low-pass filter with in-band noise reduction". Circuits and Systems (ISCAS), 2013 IEEE International Symposium on . Publication Year: 2013. Page(s): 1280–1283.

11. De Matteis, M.; Pezzotta, A.; Baschirotto, A. "4th-Order 84dB-DR CMOS-90nm low-pass filter for WLAN receivers". Circuits and Systems (ISCAS), 2011 IEEE International Symposium on. Publication Year: 2011. Page(s): 1644–1647.
12. Willingham, S.D., Martin, K.W., and Ganesan, A.: 'A BiCMOS low-distortion 8-MHz low-pass filter', IEEE J. Solid-State Circuits, 1993. Volume 28, no. 12. Page(s): 1234–1245.
13. Antoine, P.; Bauser, P.; Beaulaton, H.; Buchholz, M.; Carey, D.; Cassagnes, T.; Chan, T.K.; Colomines, S.; Hurley, F.; Jobling, D.; Kearney, N.; Murphy, A.; Rock, J.; Salle, D.; Tu, C.-T. "A Direct-Conversion Receiver for DVB-H". Journal of Solid-State Circuits, IEEE. Volume 40. Issue 12. Dec. 2005, pp.: 2536–2546.
14. Liang Zou; Kefeng Han; Youchun Liao; Hao Min; Zhangwen Tang. "A 12th order active-RC filter with automatic frequency tuning for DVB Tuner applications". Solid-State Circuits Conference, 2008. A-SSCC '08. IEEE Asian. Publication Year: 2008, Page(s): 281–284.
15. S. Karthikeyan et al., "Low-voltage analog circuit design based on biased inverting opamp configuration." IEEE Trans. Circuits and Systems II. Volume. 47, no. 3. 2000. Page(s): 176–184.
16. S. D'Amico, V. Giannini, A. Baschirotto, "A 4th-order active Gm-RC reconfigurable (UMTS/WLAN) filter". Solid-State Circuits, IEEE Journal of Volume 41, Issue 7, July 2006 Page(s): 1630–1637.
17. Wambacq, P.; Giannini, V.; Scheir, K.; Van Thillo, W.; Rolain, Y. "A fifth-order 880MHz/1.76GHz active low pass filter for 60GHz communications in 40 nm digital CMOS". ESSCIRC, 2010 Proceedings of the. Publication Year: 2010. Page(s): 350–353.
18. Baschirotto, et al. "Advances on analog filters for telecommunications". Advanced Signal Processing, Circuits, and System Design Techniques for Communications. 2006 IEEE International Symposium on Circuits and Systems, art. no. 4016440, pp. 131–138.
19. De Matteis, M.; D'Amico, S.; Cocciolo, G.; De Blasi, M.; Baschirotto, A. "A 54 dB-DR 1-GHz-bandwidth continuous-time low-pass filter with in-band noise reduction. Circuits and Systems (ISCAS), 2013 IEEE International Symposium on. Publication Year: 2013, Page(s): 1280–1283.
20. S. D'Amico, M. De Blasi, M. De Matteis, A. Baschirotto "A 255 MHz Programmable Gain Amplifier and Low-Pass Filter for Ultra Low Power Impulse-Radio UWB Receivers" Transactions on Circuits and Systems –I: Regular Paper, vol. 59, no.2, February 2012, pages 337–345.
21. Chia-Hsin Wu, Chang-Shun Liu, and Shen-Iuan Liu "A 2 GHz CMOS Variable-Gain Amplifier with 50dB Linear-in-Magnitude Controlled Gain Range for 10GBase-LX4 Ethernet" Proc. of ISSCC 2004.
22. H. D. Lee, K. A. Lee, S. Hong"A Wideband CMOS Variable Gain Amplifier With an Exponential Gain Control" IEEE Transactions on Microwave Theory and Techniques, vol. 55, no. 6, June 2007, pages 1363–1373.
23. E. O'hAnnaidh, E. Rouat, S. Verhaeren, S. Le Tual, C. Garnier"A 3.2 GHz-Sample-Rate 800 MHz Bandwidth Highly Reconfigurable Analog FIR Filter in 45 nm CMOS" Proc. of ISSCC 2010.
24. T. Laxminidhi, V.Prasadu, S.Pavan "Widely Programmable High-Frequency Active RCFilters in CMOS Technology"IEEE Transactions on Circuits and Systems-I: regular papers, vol. 56, no. 2, February 2009 pages 327–335.
25. S. Pavan, T. Laxminidhi "A 70-500 MHz Programmable CMOS Filter Compensated for MOS Nonquasistatic Effects" Proc. of ESSCIRC 2006.
26. J. Harrison, N. Weste "A 500 MHz CMOS Anti-Alias Filter using Feed-Forward Op-amps with Local Common-Mode Feedback" Proc. of ISSCC 2003.
27. Tien-Yu Lo, Chung-Chih Hung "A 1 GHz OTA-Based Low-Pass Filter with A High-Speed Automatic Tuning Scheme"Proc. of ASSCC 2007.
28. H. Amir-Aslanzadeh, E. J. Pankratz, E. Sánchez-Sinencio "A 1-V +31 dBm IIP3, Reconfigurable, Continuously Tunable, Power-Adjustable Active-RC LPF" IEEE Journal of Solid-State Circuits vol. 44, no. 2, February 2009, pages 495–508.

29. Y. Aiba, K. Tomioka, Y. Nakashima, K. Hamashita, B.-S. Song, "A Fifth-Order Gm–C Continuous-Time Σ-Δ Modulator With Process-Insensitive Input Linear Range" IEEE Journal of Solid-State Circuits vol. 44, no. 9, September 2009, pages 2381–2391.
30. M. Mobarak, M. Onabajo, J. Silva-Martinez, E. Sánchez-Sinencio "Attenuation-Predistortion Linearization of CMOS OTAs With Digital Correction of Process Variations in OTA-C Filter Applications" IEEE Journal of Solid-State Circuits vol. 45, no. 2, February 2010, pages 351–367.
31. H.-Y. Shih, C.-N. Kuo, W.-H. Chen, T.-Y. Yang, K.-C.Juang "A 250MHz 14dB-NF 73dB-Gain 82 dB-DR Analog Baseband Chain With Digital-Assisted DC-Offset Calibration for Ultra-Wideband" IEEE Journal of Solid-State Circuits, vol. 45, no. 2, February 2010.
32. Le Ye, Huailing Liao, Congyin Shi, Junhua Liu, and Ru Huang "A 2.3 mA 240-to-500 MHz 6th-order Active-RC Low- Pass Filter for Ultra-Wideband Transceiver", Proc. of ASSCC 2010.
33. B. Drost, M. Talegaonkar, P. K. Hanumolu "Analog Filter Design Using Ring Oscillator Integrators" IEEE Journal of Solid-State Circuits, vol. 47, no. 12, December 2012, pages 3120–3129.
34. L. Ye, C. Shi, H. Liao, R. Huang, Y. Wang "Highly Power-Efficient Active-RC Filters With Wide Bandwidth-Range Using Low-Gain Push-Pull Opamps" IEEE Transactions on Circuits and Systems—I: regular papers, vol. 60, no. 1, January 2013 pages 95–107.
35. M. S. Oskooei, N. Masoumi, M. Kamarei, H. Sjöland "A CMOS 4.35-mW +22-dBm IIP3 Continuously Tunable Channel Select Filter for WLAN/WiMAX Receivers" IEEE Journal of Solid-State Circuits, vol. 46, no. 6, June 2011, pages 1382–1391.
36. S. Chatterjee, Y. Tsividis and P. Kinget. "A 0.5 V filter with PLL-based tuning in 0.18 μm CMOS," ISSCC Dig. Tech. Papers, pp. 506–507, Feb. 2005.
37. B. Drost, M. Talegaonkar, P. K. Hanumolu. "A 0.55 V 61 dB-SNR 67 dB-SFDR 7 MHz 4th-Order Butterworth Filter Using Ring-Oscillator-Based Integrators in 90 nm CMOS" ISSCC Dig. Tech. Papers, pp. 360–361, Feb. 2012.

Silicon Innovation Exploiting Moore Scaling and "More than Moore" Technology

Patrick J. Quinn

Abstract The way the electronics industry has traditionally ensured the continued reduction in cost per function year-in, year-out is to aggressively follow Moore's Law process scaling, where the number of transistors per unit area doubles about every 2 years, coupled with a 25 % performance increase and 20 % power reduction per function with scaled V_{CC}. Increasingly complex process technologies are needed each chip generation to ensure continued scaling with the consequence that chip costs are going up enormously. It is also becoming more difficult to eke out performance gains each process node and to this end architectural innovation needs to go hand in hand with direct circuit porting to fully exploit the benefits of the technology. Unprecedented levels of system integration have become possible using advanced technology options, where highly complex logic, CPUs and analog processing can exist side by side in one package. FPGAs have been leading the industry in always being among the first to the node with commercial products and adopting new technologies such as 3D-IC. This chapter presents out the experience gained in following Moore's Law and the application of "More than Moore" technology. It covers such items as process selection, 3D-IC technology, high-performance AMS features, as well as power reduction strategies and the migration steps to 20 and 16 nm CMOS.

1 Moore's Law: A History of Opposing Forces

The IC Industry has, for the past 50 years, been in the grips of one of the most remarkable technological exponential improvements ever, a trend first observed by Gordon Moore of Intel. The drivers for the industry are the insatiable user

P. J. Quinn (✉)
Xilinx Ireland, Citywest, Dublin, Ireland
e-mail: patrick.quinn@ieee.org

P. Harpe et al. (eds.), *High-Performance AD and DA Converters, IC Design in Scaled Technologies, and Time-Domain Signal Processing*,
DOI: 10.1007/978-3-319-07938-7_9, © Springer International Publishing Switzerland 2015

demands for more bandwidth and increased performance densities at lower costs. This translates into an increasing number and variety of ever smaller and more powerful devices in our daily lives.

The increasingly rapid expansion in global internet traffic is putting increasing demands on equipment manufacturers to keep pace through continual upgrading of technology [1]. Enterprise bandwidth is required to grow at a rate of almost 3x every 2 years while power levels can only increase moderately. Innovative new applications, especially those based on high quality video streaming, are driving insatiable demand for more intelligent network bandwidth. Moreover, the network infrastructure is becoming more fragmented, despite efforts at industry convergence, and has to deal with a variety of media technologies (copper, optical fibre, air)—Fig. 1. Investment in internet infrastructure (at only 8 % annually) is not keeping pace with the rapid increase in IP traffic [2]. The consequence is that the cost per unit bandwidth must come down sharply so that future bandwidth needs can be met with the expected investment forecasts.

Cisco VNI usage (Fig. 2) provides quantitative insights into current activity on service provider networks [3]. In this cooperative program, more than 20 global service providers share anonymous, aggregate data with Cisco to analyse current network usage trends and gauge future infrastructure requirements. In March 2014, Cisco released the VNI Global Mobile Forecast spanning 2013–2018. Mobile data traffic is predicted to increase about 11-fold over the next 5 years and reach an annual run rate of over 200 exabytes (billions of gigabytes). By 2018, nearly 5 billion mobile users and more than 10 billion mostly smart mobile devices will drive mobile video to occupy almost 70 % of all mobile data traffic.

The IC industry has been on a 2 year rejuvenation path since the mid 1960s after the observation by Gordon Moore [4] that "the complexity for minimum component costs has increased at a rate of roughly a factor of 2 per year". This was revised in 1975 by Moore to a factor two increase in component count every 2 years and indeed the IC industry has maintained this trend ever since, governed by what's called "Moore's Law" [5]. Although not a physical law, the IC industry, dominated by Intel over most of that period, has been set to achieve this target each new process generation.

Figure 3 captures the main economic forces at work in the IC industry which is a combination of increasing performance and reductions in cost. What's clear from these graphs is that although the cost to design and manufacture IC's continues to grow at an enormous pace, the performance and functionality of these IC's grows even faster, so allowing the IC industry to continue to deliver more value for money each new generation.

The key costs in following Moore's Law arise in the facilities to manufacture the new processes. These are primarily the construction costs of a new fabrication facility (fab), the kitting out with the latest processing equipment and the infrastructure needed to support chip design. There remains only a small cohort of semiconductor manufacturers who can afford to build new fabrication facilities in order to get on to the next leg of scaled CMOS processes, namely Intel, TSMC, Samsung, GlobalFoundries and IBM (with partners such as UMC). The capital

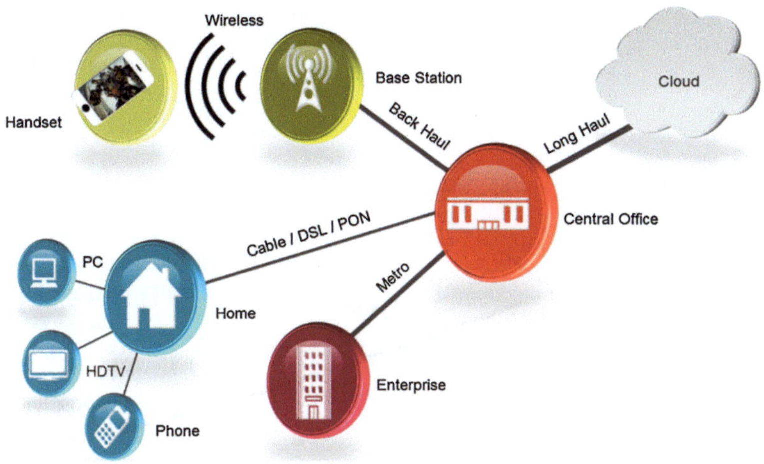

Fig. 1 Exponential growth in network bandwidth from home to core

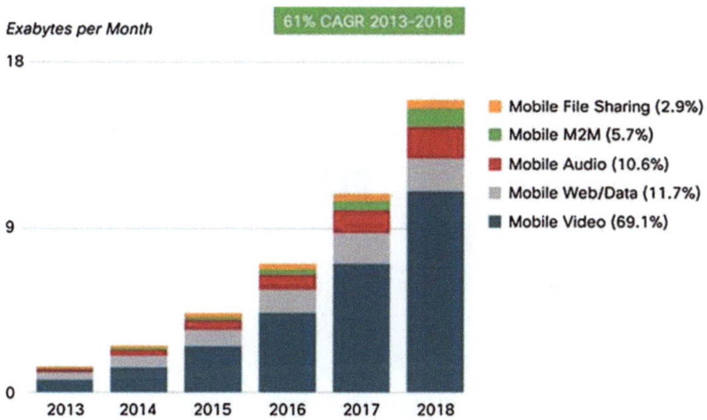

Fig. 2 Global mobile data traffic growth forecast showing >10x increase in 5 years. Figures in parentheses refer to traffic share in 2018. *Source* Cisco VNI Mobile, 2014

costs are escalating per process node (Fig. 4), with a modern day 300 mm fab costing of the order of $4 billion [6]. This rise in costs is itself described by a law called Rock's Law or sometime called Moore's Second Law, which states that the costs of a semiconductor chip fabrication plant doubles every 4 years [7]. 450 mm facilities are still only at early development stage in both the University of Albany New York (Sematech Global 450 Consortium) and at Imec. GlobalFoundries expects the costs of a full scale 450 mm manufacturing facility will be greater than $10 billion and wouldn't be ready until after 2020.

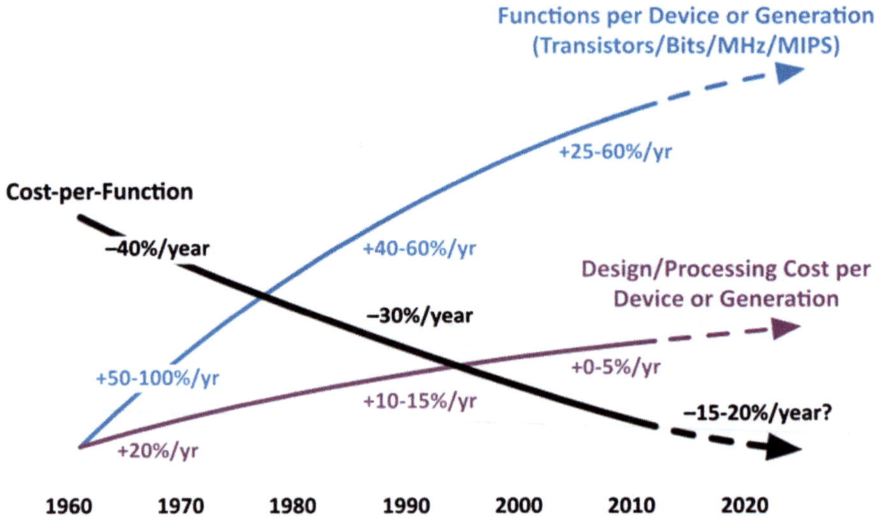

Fig. 3 Opposing economic forces behind growth and innovation in the IC industry. *Source* IC Insights

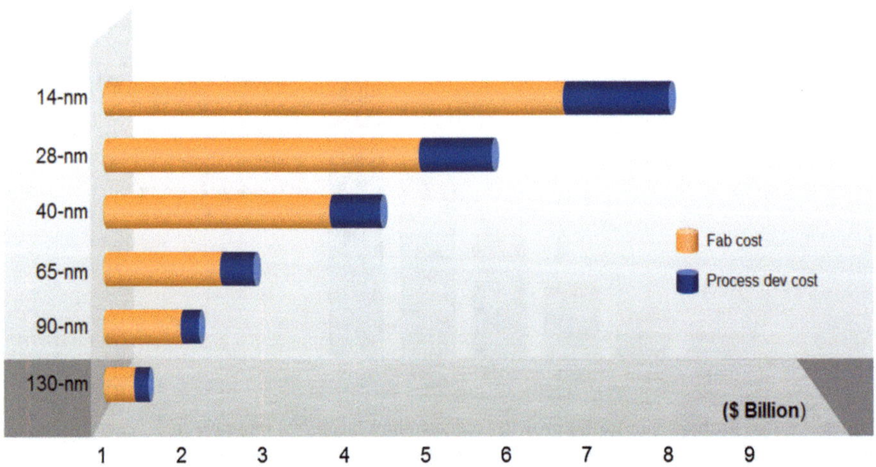

Fig. 4 Rapidly rising fabrication and processing costs of modern CMOS *Source* Global foundries

2 Process Scaling Enablers

Classical CMOS scaling, as described by Dennard [8], which implied simple gate oxide and supply voltage down scaling ($1/\sqrt{2}$ x) with reduced device dimensions ($1/\sqrt{2}$ x per dimension) each new process generation, ended at 130 nm. The SiO2

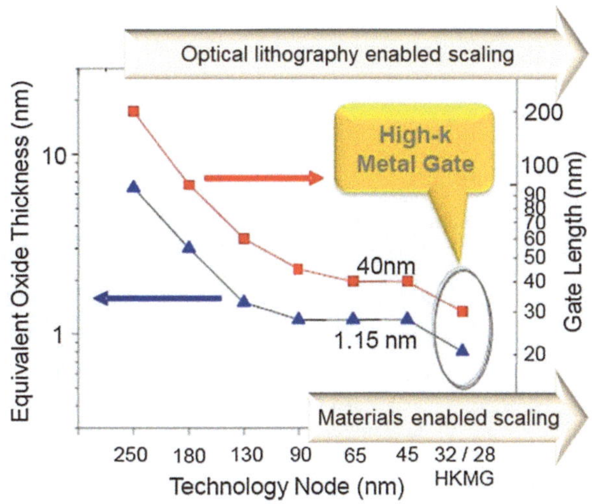

Fig. 5 The history of gate scaling enabled through lithography and advanced materials. *Source* Global Foundries, IMEC

gate dielectric scaling had reached the limit of what was achievable (1–2 nm) before gate leakage arising from carriers tunnelling through the SiO2 insulator created excessive off-state static power loss. To avoid the roadblocks which arose at the 90 nm node, new materials based improvements were made to improve performance with process down-scaling—Fig. 5 [9]. In particular, the increase in substrate doping used to suppress short channel effects, negatively impacted carrier mobility in the transistor channel. Strain engineering materials were introduced in order to compensate for the reduction of drive current with scaling [10]. This approach relies on the fact that carrier mobility in silicon can be improved by stressing the silicon lattice—Fig. 6. Tensile strain improves electron mobility (SiN cap layer for NMOS), while compressive strain enhances hole mobility (SiGe is grown in the recessed Si source/drain regions, which expands to compress the adjacent Si channel resulting in uniaxial stress along the channel in the direction of current flow). Although the embedded SiGe method used for PMOS is much more effective than the cap layer method used for NMOS, the effect is to increase the mobility of PMOS devices almost to the level of NMOS in the most modern processes today. Transistor strain has increased proportionately each subsequent process node.

Intel was the first company to replace the SiO2 with an alternative gate insulator with a higher dielectric constant (5x larger), namely Hafnium Dioxide (HfO2), which they did at their 45 nm node (Fig. 7i) [11]. This reduced gate leakage by 25x which allowed the resumption of equivalent gate oxide scaling at 45 nm and beyond. This materials innovation was paired with the use of metal gates which was needed to remove poly-gate depletion and reduce the impact of mobility degradation [12]. Two different methodologies were introduced (Fig. 7ii) to implement the high-k dielectric and metal gate (HKMG) combination, namely (traditional) *gate first* and (new) *gate last* [13]. TSMC first introduced HKMG a

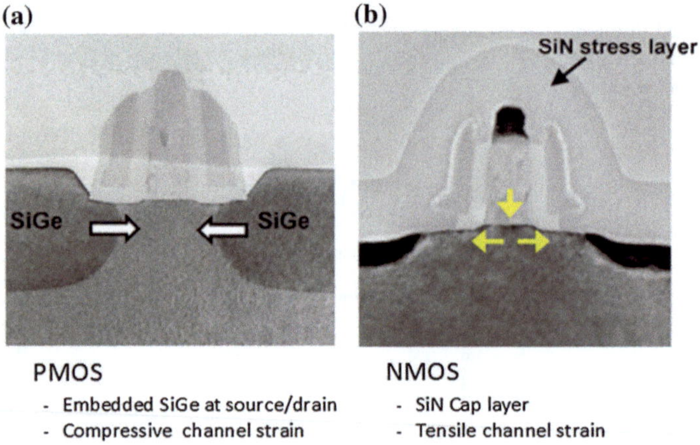

Fig. 6 Strain engineering at 90 nm and beyond. **a** PMOS hole mobility improvement, **b** NMOS electron mobility improvement *Source* Intel

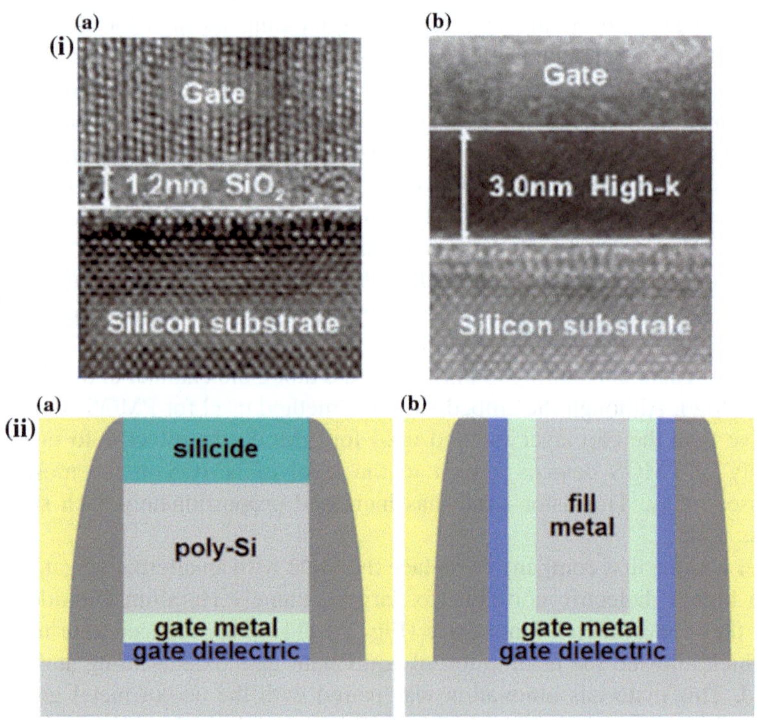

Fig. 7 Solutions to continued scaling after 40 nm, High-k dielectric (*i*) **a** Traditional SiO2, **b** High-k dielectric. *Source* Intel. and Metal gate (*ii*) **a** Gate-first (MIPS), **b** GAte-last. *Source* IBM

generation later than Intel for 28 nm and it too used the *gate last* approach. Samsung and GlobalFoundries have been using the gate first approach in 28 nm. From 20 nm onwards, all foundries have adopted the more reliable *gate last* HKMG method.

Scaling below 28 nm is dominated by what can be achieved by the imaging and patterning equipment. Intel has extended the physical limits of 193 nm immersion lithography to the 22 nm node using single exposure. TSMC transitioned to double patterning at the 20 nm node, while Intel is making the transition at the 14 nm node with its second generation FinFet technology. Product plans for 14/16 nm FinFet technologies are firmly in place and the techniques used to achieve 14/16 nm CMOS will likely be extended to 10 nm through the further use of immersion based optical lithography and multiple mask patterning [14]. Extreme Ultra Violet (EUV), with 13.5 nm wavelength and requiring much fewer mask steps than immersion lithography is on the horizon but is unlikely to be ready until the 7 nm node. The main limitation, beyond the much increased cost of equipment, is the reduced wafer throughput achievable with EUV. As of the beginning of 2014, ASML claim they can get a wafer throughput of 50 wafers per hour (SPIE). This throughput rate would need to more than double and reliability improve before EUV could be committed to high volume manufacturing. Intel has proposed a combined approach [14] using complementary lithography at the 10 nm node, where multiple 193 nm via patterning steps are replaced by EUV patterning steps. Another alternative that is under investigation at the moment is the use of massively parallel E-beam lithography which requires no masks but has a slow throughput rate. It too could be combined with immersion lithography and/or EUV to create a cost-optimised solution at 10 nm and beyond.

3 The FPGA Experience of Process Scaling

With the exclusion of (flash) memories, FPGAs are the largest transistor count integrated circuits available today. The transistor count for the largest Xilinx devices each process generation are plotted out in Fig. 8. The most recent largest device is a 20 nm Virtex Ultrascale XCVU440 with 4.4 million logic cells, which is more than 2x larger than the previous largest 28 nm device (Virtex 2000T) [15]. This is equivalent to 50 million ASIC gates for a total of 20 billion transistors. Up to 28 nm, the transistor count has been increasing at a rate of about 0.4x. However, with the advent of 3D packaging technology (Sect. 4), the maximum number of transistors per generation has taken off at an even steeper rate of >2x. Not only has the size of the IC's increased at a very fast rate but so too has the complexity, with a wide range of different features from thousands of hardened DSP blocks to multiple 28 Gbps transceivers and various analog mixed signal functions.

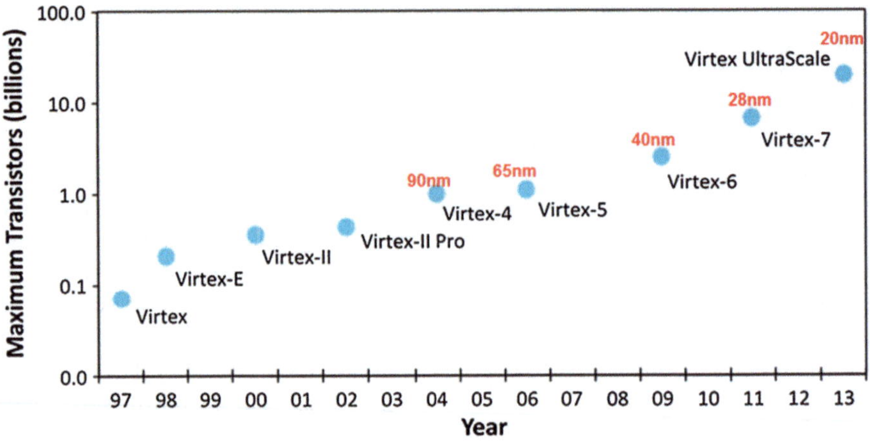

Fig. 8 Xilinx top-end FPGAs maintaining record breaking transistor count per IC generation

3.1 FPGA Design Considerations for Transfer to 28 nm Process Node

Xilinx has moved all new chip developments to TSMC since the 28 nm node. This has been a pivotal process for Xilinx, not just because it was the first process to include HKMG, but because of the range of processes made available and the importance of optimal process choice to achieve both high performance and low power FPGA operation. TSMC offers three processes at 28 nm: the 28LP, 28HP and 28HPL. A lot of analysis was done to work out what the best process choice would be for FPGAs. 28LP is the only more traditional Polysilicon/Silicon Oxy-Nitride (Poly/SiON) process, which is low leakage and low power but also relatively low speed with respect to HPL and HP and more suited to consumer mobile applications. The real choice came down to 28HP or 28HPL for FPGAs, both of which are HKMG, where 28HP has added embedded SiGe source/drains for PMOS strain.

It has been verified that the 28HPL process is the optimal process for FPGA design. The reduced yield and increased leakage associated with the eSiGe in 28HP in the range of operation of the FPGA was considered too risky. The larger allowed voltage headroom in 28HPL compared to 28HP allows for a flexible trade-off in power with performance through the supply voltage [16]. For example, in Fig. 9, high performance mode ($V_{CC} = 1.0$ V) for 28HPL actually gives higher performance than 28HP at lower static power in the range of performance operation for an FPGA. Furthermore, in low-power mode ($V_{CC} = 0.9$ V), 28HPL has 65 % lower static power than 28HP. Dynamic power is also reduced by ~20 % at this voltage. Note, a voltage ID is stored in each FPGA device, so that adaptive voltage scaling can be applied to set the supply voltage at the minimum level needed for that device to still meet its performance targets.

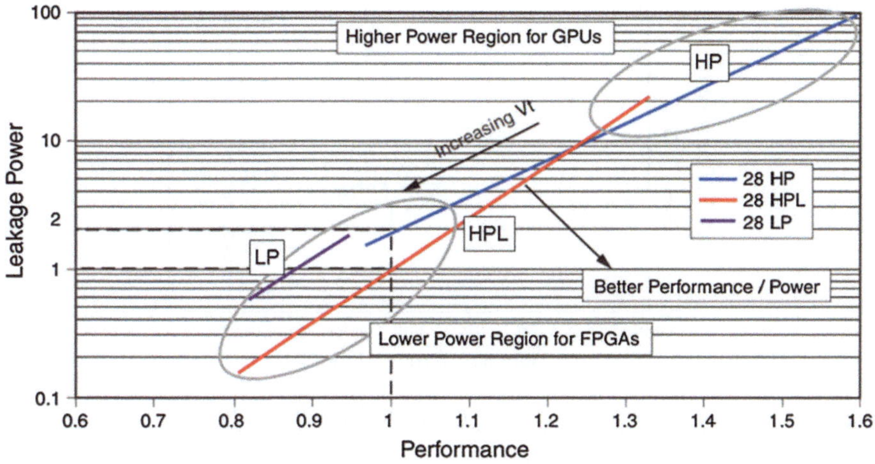

Fig. 9 Leakage power driving process choice for HPL in 28 nm FPGAs

The flexibility in the 28HPL process allows for operation in a low leakage region (Fig. 9), so that the process can be applied to a low power family of FPGAs (Artix-7), without the need to migrate to the 28LP process. This reduces the need for a complete redesign.

FPGAs are generally used as replacements for ASICs (customised) application specific integrated circuits) and ASSPs ((off-the-shelf) application specific standard products). They are more often than not placed in high performance professional equipment, so that keeping static power levels down is critical for customers. Product reliability is dependent on the thermal response, since every 10 °C rise in temperature doubles the chance of failure. The higher static power associated with 28HP means that an incremental change in power consumption can cause the device temperature to exceed the specified operating temperature. This is clear from Fig. 10, where two FPGAs, one in 28HPL (Xilinx) and one in 28HP, are compared at similar activity levels (equivalent number of logic cells for same design implementation). The 28HP FPGA will reach its maximum power limit earlier than the 28HPL FPGA, so that performance would have to be throttled back to maintain sufficient margin for the device not to exceed its allowed junction temperature. Alternatively, the ambient temperature needs to be lower, which isn't attractive in many wired communications applications in rack-style environments, since this means increased cooling is required.

A further factor which arises for ASIC/ASSP designs, when migrating from node to node, is the escalating up-front NRE (non-recurring engineering) costs when doing a design for the first time in a new node (Fig. 11) [17]. For instance, the NRE costs have jumped from an estimated \$85 million in 40 nm to over \$170 million in 28 nm. Such costs include R&D expenses, masks, embedded software (IP licences), test and the yield cost associated with product ramp-up. Extrapolation to the 14/16 nm node shows NRE costs will increase to about \$340

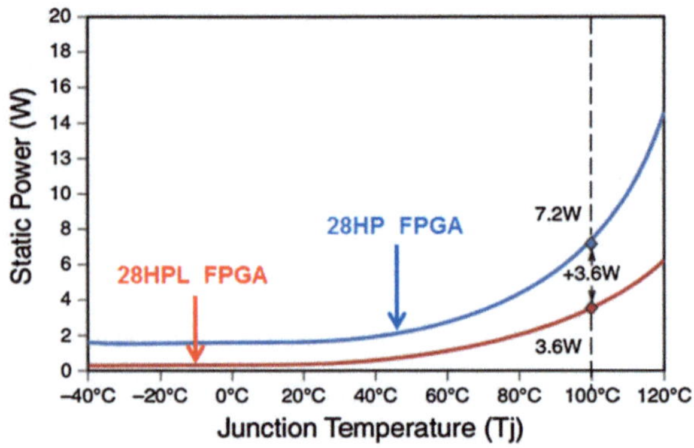

Fig. 10 Max static power as a function of T_j for 28HP and 28HPL FPGAs

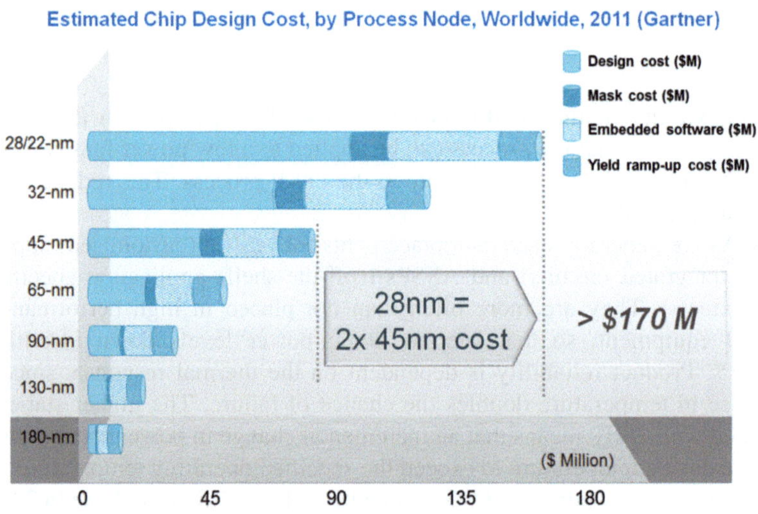

Fig. 11 Escalating NRE costs for ASIC and ASSP designs

million, so that only a small cohort of ASIC and ASSP chip providers, especially those implementing very large scale SoCs (system on chip), can afford to convert to beyond the 28 nm node. The escalating up-front NRE costs are often prohibitive and put considerable pressure on designers to make sure their designs are "first time right" to avoid costly re-spins. FPGA technology, on the other hand, will always benefit from being in the leading edge technology node, encouraging broader adoption among end users who don't have to shell out the NRE costs for their designs and who require a fast time-to-market. FPGA technology, with the inherent benefits of programmable integration and in-the-field flexibility and scalability, can address a wide variety of different applications, while the recent innovation in I/O solutions allows for multi-Terabits per second data transfer rates on and off chip [15].

3.2 The Road to 20 nm and 16 nm FinFet Technologies

3.2.1 Scaling Down to 20 nm

TSMC has chosen to implement double patterning at the 20 nm node and have it in full production by the time 16 nm FinFet technology arrives. Intel, on the other hand, has implemented FinFet technology at the 22 nm node but will wait until the 14 nm node for the implementation of double pattering. In going from 28 to 20 nm, the extra costs are not just associated with more expensive masks but also with the extra number of masks associated with the double patterning and the requirement for local interconnect. For this, 20 nm offers 20 % speed improvement and 30 % power improvement over its predecessor 28 nm, as well as $\sim$2x increase in density.

Increased design challenges occur at 20 nm with respect to 28 nm. For instance, layout dependent effects (LDE) are more pronounced due to the reduced separation distances between devices—one such effect being the well proximity effect (WPE), where the threshold voltage of a transistor is dependent on how near it is to an adjacent well. Another effect is device aging, which is strongly correlated to the degradation over time of the gate dielectric and its interface to the silicon. Since the gate dielectric is scaled down to only a few atoms in equivalent thickness in 20 nm, it is more susceptible to hot carrier injection (HCI) and bias temperature instability (BTI). The result of aging is the increase of the threshold over time and the reduction of the carrier mobility in the transistor channel. These aspects are included in the design rules and layout guidelines for 20 nm.

Xilinx is already shipping 20 nm FPGAs since November 2013. These Ultrascale devices [15] will complement the 28 nm devices and not seek to replace them. In fact this approach will be extended to 16 nm, so that Xilinx will have 3 nodes running concurrently (28/20/16), where devices will be differentiated based on performance and functionality rather than just which process node they have been manufactured in.

3.2.2 Prospects for 16 nm FinFet Technology and Beyond

Planar devices have reached the end of the road at the 20/22 nm process nodes. Scaling the planar device any further doesn't make sense, since it starts to behave more like a resistor than a switch with the drain competing with the gate to control channel conduction. The traditional solution of shrinking down the gate dielectric to increase gate control has reached the limit at 20/22 nm. The jump to FinFet technology represents one of biggest revolutions to affect IC manufacturing in recent years. The raised 3D structure has the gate wrapped around 3 sides of the fully depleted thin vertical silicon fin [18]. The restrictions of the lithography require simple, repeatable, regular patterns and so the implementation on chip resembles a "sea of fins", where the whole chip is patterned with fins of a single pitch. The same lithographic design rules as 20 nm are used, except for the front-end 3D transistor structure. An example of a recently published 6T SRAM cell from TSMC [19] can be seen in Fig. 12 with the smallest published area of only 70 nm^2. SRAM is a key indicator of process performance, particularly the minimum allowed voltage in order to maintain the functions of read, write and hold with process variability.

Data from TSMC (Fig. 13) demonstrates the improvement in digital and analog transistor performance which can be obtained when migrating from 28 nm HKMG to 14 nm FinFet technology [20]. The scalability of 2x increase in logic density is the same as for 20 nm, while the speed improves to >35 %, or equivalently the power reduces to <55 % with respect to 28 nm, and the Ion/Ioff ratio is greatly improved.

One of the most favourable aspects of FinFet technology is its improved analog performance. All key aspects of analog design are improved. Matching is better, transconductance and output resistance have increased, and hence transistor gain, as well as switch off-stage leakage has been reduced. While thermal noise remains the same, 1/F noise has reduced due to the lower dopant concentration in the depleted channel. The effective removal of back-gate effects allows for a much flatter on-resistance switch characteristic and a more linear source follower and consequently more highly accurate track-and-hold operation at higher sampling frequencies with more headroom. The main downside from an analog perspective is the increased extrinsic capacitance [21] associated with the fins which gives rise to an overall degraded F_T, despite the improved transconductance.

While it is difficult to make direct comparisons of analog circuits from one process node to another by reasons of differing power supplies and transistor characteristics requiring differing circuit optimisations, it is still useful to see the high-level trend in porting complex analog circuits. As a design of experiments [22], a current-steering based comparator and a CMOS based comparator, both with pre-amplifiers and ready detectors, were compared between 40 nm planar and 16 nm FinFet technologies, where both were optimised to drive similar loads (Fig. 14). While the delays at the ready outputs in both cases are considerably lower in 16 nm, the most remarkable observation is the reversal of fastest circuit architecture between 40 and 16 nm. The current-steering approach no longer dominates pure CMOS for high-speed at this node, as it did at previous planar process nodes.

Achieving yield in high volume production is a major challenge for FinFet technology. Intel announced recently a delay of one quarter in releasing its 14 nm FinFet because of reliability concerns. TSMC plans full 16 nm production before the end of 2014. The plan for Xilinx is to be among the first to tape out a product in 16 nm TSMC, which TSMC have pegged at a year beyond full production of 20 nm. This again will be about 1.5–2 years ahead of complex ASIC or ASSP based SoCs.

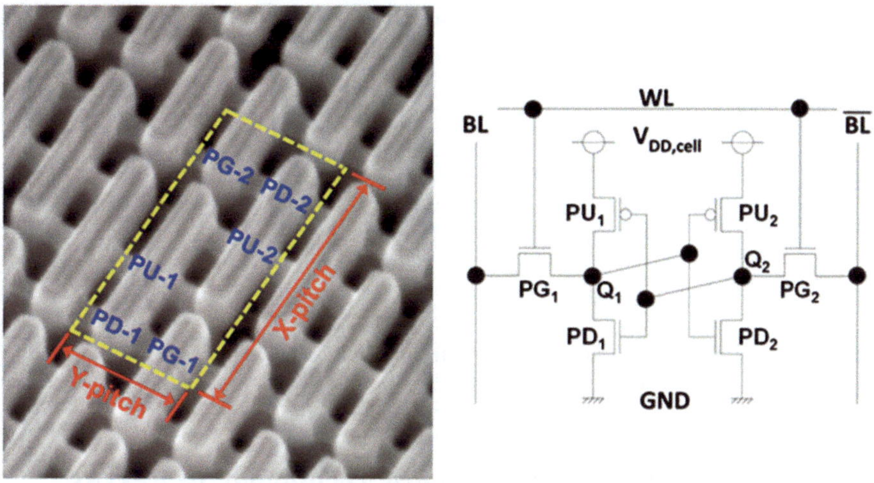

Fig. 12 TSMC 16 nm bulk FinFet 6T SRAM cell in 70 nm^2

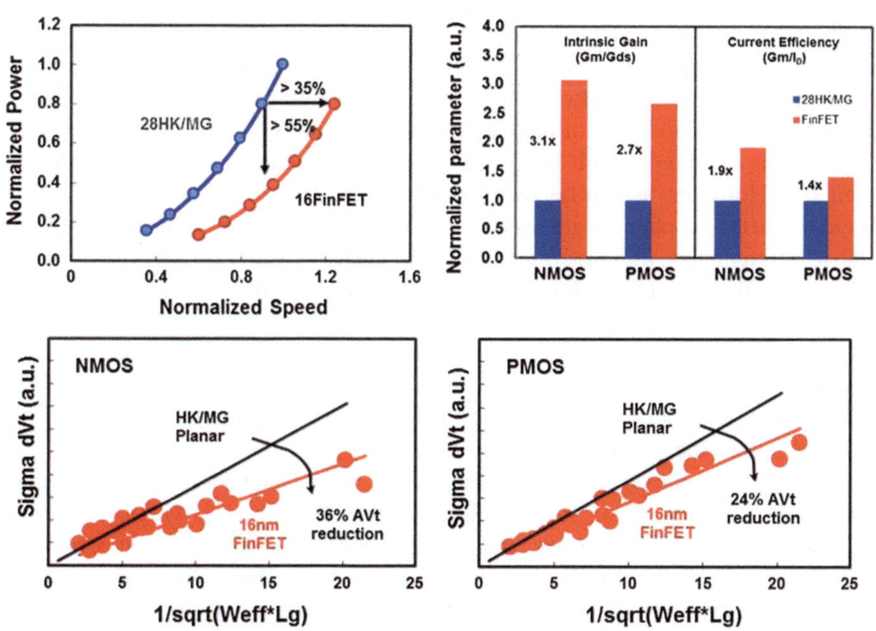

Fig. 13 TSMC 16 nm bulk FinFet Analog performance compared to 28 nm HKMG [20]

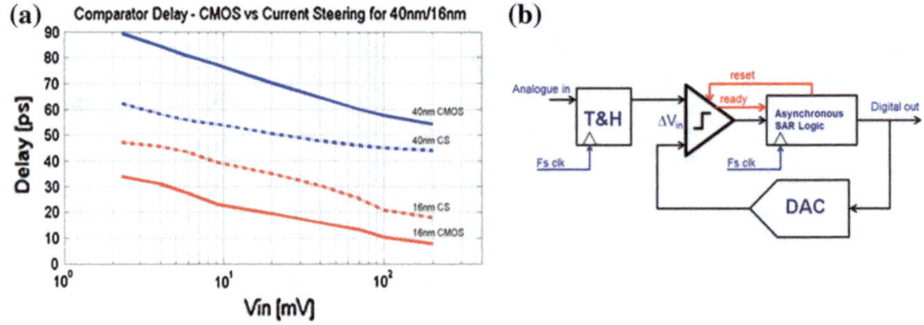

Fig. 14 **a** Dynamic performance of current-steering and CMOS comparators(ready output) in 40 nm planar and ported 16 nm FinFet technologies; **b** SAR ADC application of high-speed comparator

4 Application of More than Moore Technology to FPGAs

The trend designated as More than Moore (MtM) has been acknowledged by ITRS [23] as one of the major trends for roadmapping separate from "More Moore" scaling trends. The MtM approach is being driven by the requirement for more functionality both on chip and in package. It is demonstrated here how MtM methods have been employed to both complement Moore scaling through homogeneous integration but also to extend the reach of integration by combining different functions from different technologies through heterogeneous integration.

The traditional method for increasing system capacity is to connect multiple ICs together on a printed circuit board and to have them interact with each other via their I/O. However, as can be seen from Fig. 15a, the pace at which the internal logic resources on chip are increasing far outpaces the rate of increase of standard I/O to interconnect the various chips. It is clear, in the case of Xilinx FPGAs (Fig. 15b), that the I/O to logic cell ratio has decreased dramatically from generation to generation. While this trend motivates larger size die so as to keep most signals on chip, as chip die size gets larger, yield goes down exponentially and so building very large scale dice becomes very difficult and expensive. Largest devices on the latest process nodes cost disproportionately more and availability is delayed due to the impact of die size on yield.

With the innovative 3D-IC approach, using SSIT (stacked silicon interconnect technology), a number of smaller high yielding dice are laid side-by-side on a silicon interposer in such a way that all dice behave as one single integrated die [17, 24]—Fig. 16. In this way, it is possible to effectively linearize what was originally an exponential curve of die cost (proportional to 1/yield) versus die area. Regular chip metal interconnect is used to route the dice of the SSIT device. As demonstrated in Fig. 18, it makes possible more than 100 times the die connectivity bandwidth per Watt, at a fraction of the latency, and at much less power, than is possible using traditional I/O (either standard parallel or high-speed serial). Known good die are combined on the interposer, so that very high yields can be

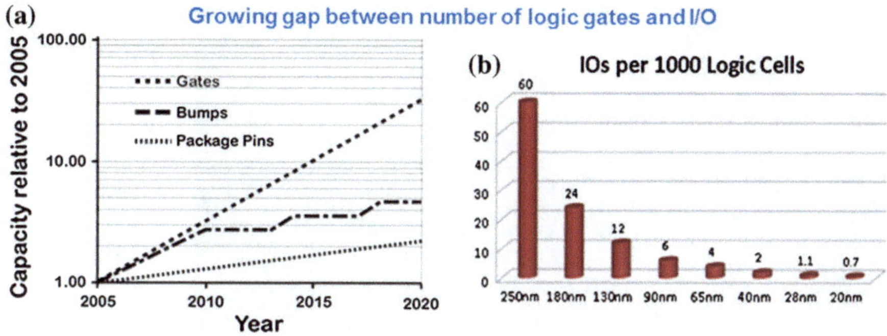

Fig. 15 Chip I/O bottleneck with (**a**) ITRS showing 15x drop in I/O-to-logic ratio by 2020, (**b**) I/O per 1000 logic cells of largest FPGAs per node

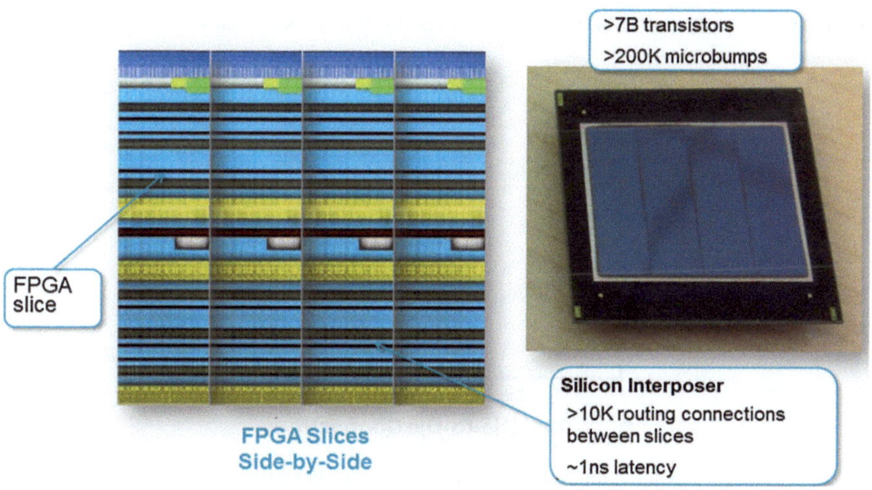

Fig. 16 Virtex 2000T in Stacked Silicon Interconnect Technology (SSIT) using four 28 nm FPGA slices on common 65 nm interposer

achieved for a very large chip, well beyond the yields which could be achieved if such a chip was integrated as one piece of silicon. The elements of the SSIT are evident in Fig. 17, which shows a cross-section from the 28 nm FPGA die down to the balls of the BGA package. The interposer is a large single slab of standard thinned 65 nm silicon with four top metal layers. It is used to carry the die-to-die interconnections but also contains the TSVs which act a bridge to the C4 bumps to allow power/ground and user I/O in and out of the complete system. Die-to-die microbump pitches are 5x smaller than standard chip bumps (C4) giving 25x reduction in aerial density. Another 25x improvement in aerial density is obtained in going from the C4 bumps to the package bumps. This gives an overall improvement of ~600x in aerial density in going from standard package bumps to die level microbumps.

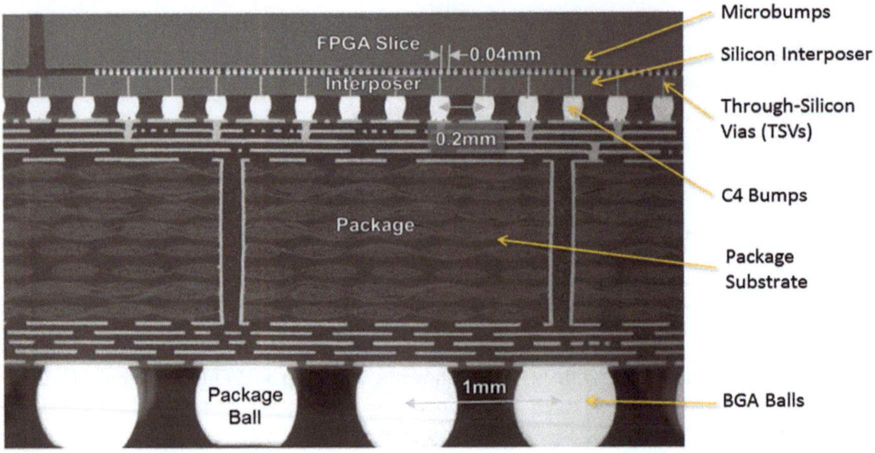

Fig. 17 Cross-section of SSIT from FPGA slices down to package balls

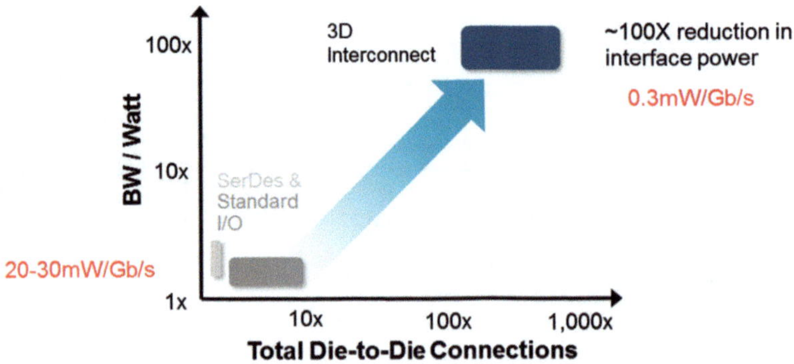

Fig. 18 Up to 100x bandwidth/Watt advantage for 3D-IC I/O versus conventional I/O

Based on this methodology, a Virtex 2000T has been manufactured in 28HPL with double the capacity of what would have been achievable based on Moore's Law scaling alone. It contains 2 million logic cells, commensurate with >7 billion transistors, and there are up to 10,000 routing channels between individual FPGA slices via the interposer.

Putting it all together, the net benefit of both Moore's Law and "More than Moore" integration in system power reduction through advanced integration in 28HPL can be seen in Fig. 19. The power savings in the 3D-IC implementation are principally achieved through elimination of standard I/O power between multiple devices.

The SSIT 3D-IC approach lends itself naturally to the combination of FPGA slices with slices of other technologies to create a mixed-mode SoC. The principal electronic components of logic, memory, CPU and analog are each best suited to

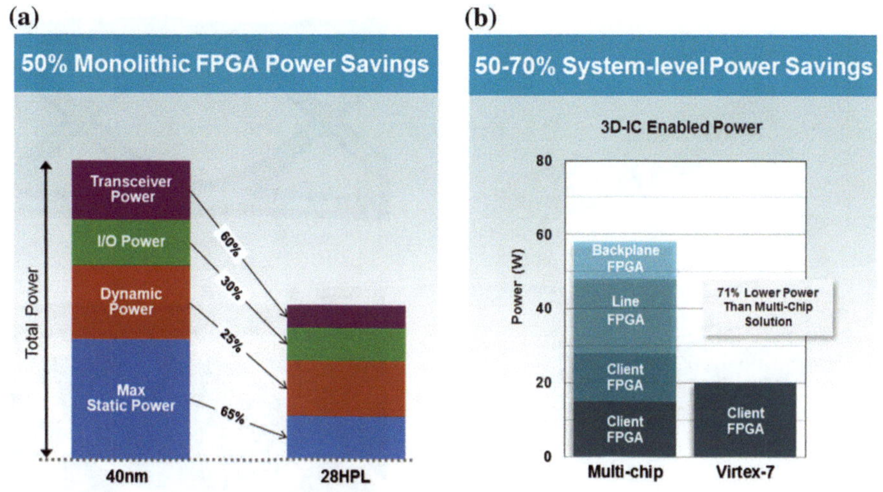

Fig. 19 Power reductions achieved through scaling and innovation for (**a**) 40 nm to 28 nm HPL porting and (**b**) multi-chip to a single SSIT enabled SoC

their own optimized process type. While it might be feasible to combine all such technologies in one leading edge process node, it doesn't make economic sense to do so. Indeed for analog, smaller does not necessarily mean better, since analog performance deteriorates with smaller die areas and reduced power supplies, despite the finer lithographies. The gap between current analog technology nodes and current digital nodes continues to widen each successive new CMOS generation. Most analog designs will stay parked at or above 28 nm for a long time to come. Factors such as double/multi-patterning in finer processes and the analog uncertainty that it brings, as well as the lower supply voltages and expense of all the extra masks, will make it prohibitive for analog migration for quite some time.

A set of heterogeneous FPGAs has been developed at Xilinx based on SSIT [24], the largest of which (XC7VH870T) contains three FPGA slices and two 28 Gbps serial transceiver slices. The total I/O throughput rate is 2.8 Tb/s from one such module and lends itself to Nx100G Ethernet ports and 400-Gbps Ethernet line cards, optical transponders, and base-station and remote radio head applications. An exemplary family member is shown in Fig. 20. The measured 28 Gb/s eye diagram shows excellent correspondence to the simulated case due to the very low jitter (RJ of 230 fs).

One of the other main benefits of SSIT, beyond those discussed above, is the excellent noise isolation which can be achieved between die. The very low noise eye pattern demonstrated in Fig. 20 is obtained despite the digital noise on the large FPGA VLSI. The experiment in Fig. 21 further underlines this fact [25]. Here, two separate 65 nm die, one containing sixteen 125MS/s 13-b ADCs and the other containing sixteen 1.6GS/s 16-b DACs are combined on a single interposer with two 28 nm FPGAs. With the two FPGA slices fully active and using 12 W of dynamic power, cross-talk isolation of less than -92dBc was measured at the DAC output. Hence, this scheme of using high performance analog die, separated from

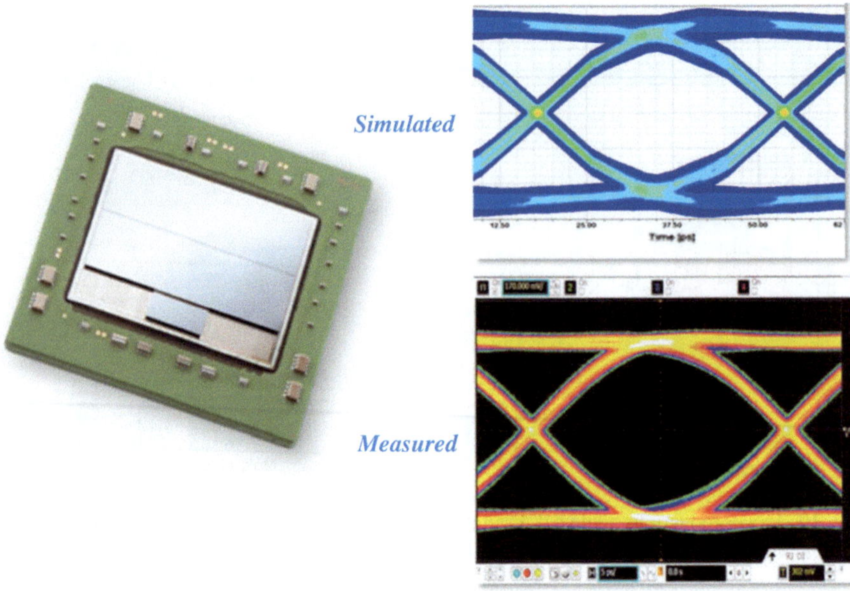

Fig. 20 Dual FPGA slices with 8 × 28 Gb/s Serdes die

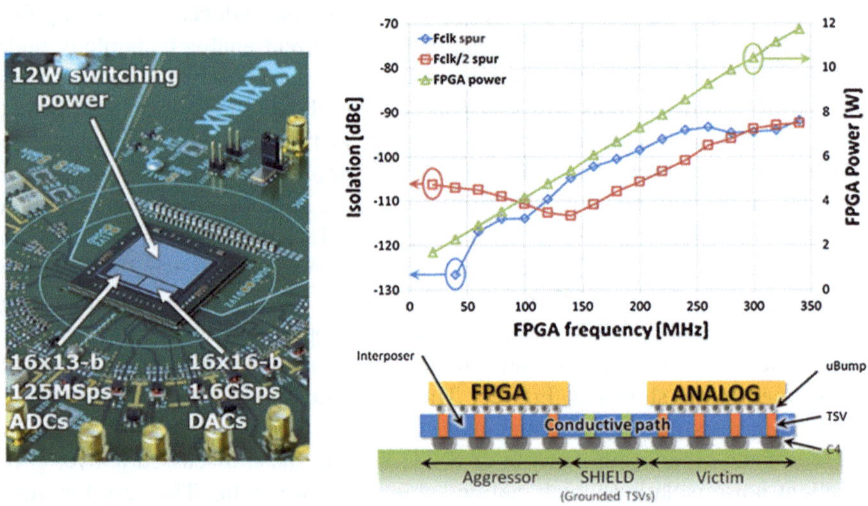

Fig. 21 Excellent noise isolation at DAC output for Heterogeneous SSIT with 2 × 28 nm FPGA slices and 2 × 65 nm analog data converter die

the digital VLSI FPGA but connected via an interposer, provides an excellent means for complete digital to analog cross-talk isolation up to very high frequencies. This is unmatched by any other method.

5 Conclusion

This chapter reviewed the key drivers behind Moore's Law and the important role FPGAs play at the forefront of the IC technology roadmap. Insight was provided into the important design considerations of process selection for a whole family of FPGAs when migrating to a new process node. For instance, judicious choice of the HPL process at the 28 nm node has meant an optimal mix of process high-performance and low power suitable for FPGA applications.

FPGAs have been early adopters of More than Moore technology and practical examples were given of the various benefits. These include two orders of magnitude reduction in digital interface power, reduction of cost and board space through inclusion of separate optimized technologies within one package, and exceptional digital-to-analog cross-talk isolation.

The availability of FinFet technology has given new life to further Moore scaling and should spur on new innovations to take advantage of its very favourable analog and digital performance characteristics. However, the best days of process scaling, with 2x economic gains between process generations, are over. Finding cost effective alternatives to 193 nm immersion lithography is going to dominate the further progress of process scaling at 10 nm and beyond.

References

1. D. Buss, "Technology in the internet age", in Proc. IEEE ISSCC, 2002., pp. 18–21
2. S.Kipp, "Exponential bandwidth growth and cost declines", Network World 2012, http://www.networkworld.com/news/tech/2012/041012-ethernet-alliance-258118.html
3. Cisco Visual Networking Index: Global Mobile Data Traffic Forecast Update, 2013-2018, http://www.cisco.com/c/en/us/solutions/collateral/service-provider/visual-networking-index-vni/white_paper_c11-520862.html
4. G. E. Moore, "Cramming more components onto integrated circuits," Electronics, vol. 38, no. 8, 1965, pp. 114–117.
5. G. E. Moore, "Progress in digital integrated electronics," in Proc. IEEE IEDM, 1975, pp. 11–13
6. A. Manocha, Foundry-driven innovation in the mobility era, Semicon West 2013, keynote
7. C.A Mack, "Fifty years of Moore's Law", IEEE Trans. Semi Manufacturing, Vol 24, No. 2, 2011 pp. 202–207
8. R.H. Dennard "Design of ion-implanted MOSFETs with very small physical dimensions", IEEE JSSC, vol. SC-9, 1974, pp. 256–268
9. M. Heyns, "Nanoelectronics and More-than-Moore at IMEC", US National Institute of Standards and Technology, 2011, http://www.nist.gov/pml/div683/conference/upload/Heyns.pdf
10. T. Ghani, "Challenges and innovations in nano-CMOS transistor scaling", 2009, http://www.intel.com/content/www/us/en/silicon-innovations/nano-cmos-transistor-scaling-presentation.html
11. K. Mistry et al, "A 45 nm Logic Technology with High-k + Metal Gate Transistors, Strained Silicon, 9 Cu Interconnect Layers, 193 nm Dry Patterning, and 100 % Pb-free Packaging", in Proc. IEEE IEDM, 2007, pp. 247–250

12. J. Kavalieros, "Tri-Gate Transistor Architecture with High-k Gate Dielectrics, Metal Gates and Strain Engineering", in Proc. VLSI Symposium, 2006, pp. 50–51
13. M. Frank, "High-k/metal gate innovations enabling continued CMOS scaling", in Proc. ESSCIRC 2011 pp. 50–58
14. R. Schenker, "Foundations for Scaling Beyond 14 nm", in Proc. IEEE CICC, 2013, pp. 1–4
15. Xilinx 20 nm Ultrascale family, http://www.xilinx.com/ultrascale
16. M. Klein et al, "Leveraging power leadership at 28 nm with Xilinx 7 series FPGAs", Xilinx whitepaper, 2013
17. P. Quinn, "FPGA based silicon innovation exploiting "More than Moore" technology, in Proc. IEEE PRIME, 2013, pp. 11–12
18. W. Maszara et al, "FinFETs Technology and Circuit Design Challenges", in Proc. ESSCIRC, 2013, pp. 3–8
19. T. Song et al, "A 14 nm FinFET 128 Mb 6T SRAM with VMIN Enhancement Techniques for Low-Power Applications", in Proc. IEEE ISSCC, 2014, pp. 232–233
20. S-Y Yang et al, "A 16 nm FinFET CMOS Technology for Mobile SoC and Computing Applications", in Proc. IEDM 2013, pp. 9.1.1–9.1.4
21. J. C. Tinoco et al, "Impact of extrinsic capacitance on FinFet RF performance", IEEE Trans. Microwave Thy, 2013, pp. 833–840
22. M. Wagemans, "Design strategies for high-speed single-slice intrinsic 8-bit SAR ADCs in planar and FinFET technologies", M.Sc. report TU Eindhoven, at Xilinx Ireland, Dec. 2013
23. Semiconductor Industry Association, http://www.siaonline.org/cs/papers–publications/statistics
24. L. Madden et al, "Advancing high performance heterogeneous integration through die stacking", in Proc. ESSCIRC, 2012, pp. 18–24
25. C. Erdmann et al, "A Heterogeneous 3D-IC Consisting of Two 28 nm FPGA Die and 32 Reconfigurable High-Performance Data Converters", in Proc. IEEE, ISSCC, 2014, pp. 120–121

The Impact of CMOS Scaling on the Design of Circuits for mm-Wave Frequency Synthesizers

Francesco Svelto, Andrea Ghilioni, Enrico Monaco, Enrico Mammei and Andrea Mazzanti

Abstract Transceivers for wireless communications at millimeter-waves are becoming pervasive in several commercial fields. Taking advantage of a cut-off frequency of hundreds of GHz, CMOS technology is rapidly expanding from Radio Frequency to Millimeter-Waves, thus enabling low-cost compact solutions. The question we raise in this article is whether scaling is just providing advantages at mm-waves or not. We present experimental data of single devices, comparing 65 and 32 nm nodes in a wide-frequency range. In particular, switches used in VCOs for tank components tuning, MOM and AMOS capacitors, inductors. f_T and f_{MAX} increase though slower than in the past, r_{on}*C_{off}, a figure of merit for switches, improves correspondingly. As a consequence, wide-band circuits benefit from scaling to 32 nm. As an example, a frequency divider-by-4, based on differential pairs used as dynamic latches, realized in both technology nodes and able to operate up to 108 GHz, is discussed. On the contrary, passive components do not improve and eventually degrade their performances. As a consequence, a conventional LC VCO, relying on tank quality factor, is not expected to improve. In this work we discuss a new topology for Voltage Controlled Oscillators, based on inductor splitting, showing low noise and wide tuning range in ultra-scaled nodes.

F. Svelto (✉) · A. Ghilioni · A. Mazzanti
Dipartimento di Ingegneria Industriale e dell'Informazione, Università degli Studi di Pavia, Via Ferrata 1, 27100 Pavia, Italy
e-mail: francesco.svelto@unipv.it

E. Monaco · E. Mammei
Studio di Microelettronica, STMicroelectronics, Via Ferrata 4, 27100 Pavia, Italy

P. Harpe et al. (eds.), *High-Performance AD and DA Converters, IC Design in Scaled Technologies, and Time-Domain Signal Processing*,
DOI: 10.1007/978-3-319-07938-7_10,

1 Introduction

Scaling of CMOS devices is motivated by the performance improvement and cost savings of digital circuits. But analog processing is not necessarily receiving just benefits from scaling. The dramatic increase of f_{max} and f_T, obeying to Moore's law, has opened up to analog processing at ever increasing frequency, from RF up to mm–waves and beyond [1], making systems on a single chip possible. On the other hand, transistors mismatches and process variations increase and back-end of line does not favor the realization of high quality passive components, key for wireless signal processing [2]. Moreover, scaled nodes characterized by metal levels thinner and closer to the substrate are expected to provide passives, and capacitors in particular, with poorer characteristics [2, 3]. In this work we present experimental data of single devices, comparing 65 and 32 nm nodes in a wide-frequency range up to mm-waves. In particular, switches used in VCOs for tank components tuning, MOM and AMOS capacitors, inductors. f_T and f_{MAX} increase though slower than in the past, r_{on}*C_{off}, a figure of merit for switches, improves correspondingly. As a consequence, wide-band circuits benefit from scaling to 32 nm. As an example, a frequency divider-by-4, based on differential pairs used as dynamic latches, realized in both technology nodes and able to operate up to 108 GHz, is proposed [4, 5]. 32 nm realized prototypes have been characterized within the 14–70 GHz frequency range, maximum measured frequency limited by available instrumentation. Operation bandwidth is in excess of 60 % and maximum power consumption is 4.8 mW. The inductor-less divider is also extremely compact occupying an active area of only $55 \times 18\ \mu m^2$.

On the contrary, passive components do not improve and eventually degrade their performances. As a consequence, a conventional LC VCO, relying on tank quality factor, is not expected to improve. In this work we discuss a new topology for Voltage Controlled Oscillators, based on inductor splitting, exploiting low noise and wide tuning range in ultra-scaled nodes [6]. Realized prototypes prove the following performances: an oscillation frequency tunable between 33.6 and 46.2 GHz corresponding to a 31.6 % total tuning range, a power dissipation of 9.8 mW, phase noise at 10 MHz offset ranging between -115 and -118dBc/Hz with a $1/f^3$ corner of $\sim$800 kHz, and a resulting FoM between 175 and 180dBc/Hz.

The article is organized as follows: Sect. 2 discusses performances of active devices, switches, capacitors and inductors comparing measured results in 65 nm versus 32 nm. Section 3 presents the inductor-less divider by four, Sect. 4 the LC VCO using inductor splitting and achieving wide tuning range. Section 5 draws conclusions.

2 Components in Scaled CMOS Technologies

Measurements of single components are provided with realizations in different nodes in the literature but almost no experiment on samples below 65 nm [7]. In this section we review performances of active and passive components down to 32 nm.

a. *Transistors*

Several past investigations have provided experimental evidence of the fast increase of f_T and f_{max} with CMOS node scaling [3, 8–11]. An immediate advantage for example in broad-band high frequency blocks derives and opens to processing at ever increasing frequency. However, f_T is increasing at reduced pace in recently scaled nodes with an increase of about +15 % between two subsequent scaling steps [3]. In the design of narrow-band tunable circuits like VCOs, transistors can be used also as switches to allow frequency variation. Transistors f_t nor f_{max} best capture device performances when the device is used as a switch. More insight comes from the on-resistance and the off-capacitance of the device. An ideal switch should realize a short circuit in the on-state ($r_{on} = 0\ \Omega$) and an open circuit in the off-state ($C_{off} = 0$ F). In order to assess the quality of the switch we refer to a figure of merit (FoM_{sw}) given by $r_{on}C_{off}$. Observing that $r_{on} \propto 1/g_m$ and $C_{off} \propto C_{gs}$, FoM_{sw} is inversely proportional to transistor's f_T. We have measured two devices, characterized by 20 μm/65 nm and 19.2 μm/32 nm aspect ratios, respectively. Measured FoM_{sw} are 795 and 550 fs, for the 65 and 32 nm devices respectively, corresponding to about 30 % improvement, in good agreement with the f_T increase reported in [3].

b. *Passive components*

While transistors improve their performances with scaling, this is not necessarily true for passive components, where the impact of the different back-end of line (BEOL) plays a key role [2]. Figure 1 shows a comparison between the stack of metals.

The thickness of the two topmost copper metals is the same in the two technologies, but layers are closer to the substrate, in 32 nm node. The distance is almost 10 % shorter. Lower metal levels, usually employed for higher density devices, do not lend themselves to realizing high frequency components. In fact, in 32 nm they are much thinner and closer to the substrate, thus suffering significant increase of both series resistance and parallel capacitance. The reduction of metal physical size also determines a reduction of contact footprint, which is in turn responsible for a significant increase of device resistance.

Inductors are commonly realized with the topmost metals plus alucap to exploit their minimum square resistance and parasitic capacitance. We have realized two samples having 70 pH inductance in 65 nm and 110 pH in 32 nm, respectively. The metal trace in 65 nm has a width of 12 μm, while that in 32 nm has a width of 5 μm. The radius is 26 μm for both. Measured results are displayed in Fig. 2.

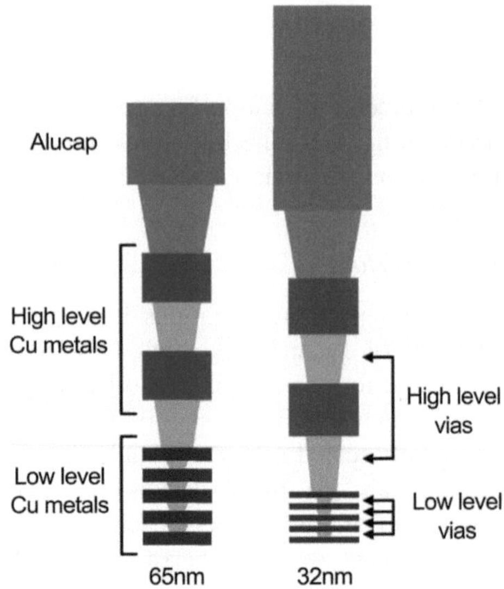

Fig. 1 BEOL comparison between 65 nm (*left*) and 32 nm (*right*)

The quality factors are slightly different but this is due to the different trace width, despite the same thickness. Inductors with same geometrical parameters, simulated in the two different nodes, demonstrate almost the same performances as expected.

Varactors are key components to provide continuous frequency tuning of LC VCOs. Few measurements are available in the literature, and all performed below 20 GHz frequency and with large C_{max}/C_{min} [12–14]. We have measured accumulation mode varactors with C_{max}/C_{min} equal to 1.6, realized in 65 nm and 32 nm. Measured capacitance and quality factor at 40 GHz versus tuning voltage are displayed in Fig. 3. Maximum Q is of the same order, while the minimum is higher at 65 nm, evidencing degradation due to scaling, as observed by other authors at lower frequency [14].

The discrete tuning in VCOs is commonly realized by using switched MOM capacitors, implemented through the lower metal layers to increase the capacitance density, suffering a reduced quality factor due to the higher series resistance of metals. A comparison between capacitors realized in the two different technologies has been carried out analyzing the performance of samples close to 200 fF in both nodes. Figure 4 shows the quality factor versus frequency.

The two capacitors have the same structure in terms of number of metal layers and interconnection of the fingers. As expected the quality factors for the 32 nm MOM is lower, the difference being around $15 \div 20$ %. Even more interesting is evaluating the performance of the switched MOM capacitors. Figure 5 sketches its equivalent circuit, highlighting the parasitics of the switch as well as the MOM: R_{on} and C_{off} are the on resistance and off capacitance of the switch respectively, while R_{mom} models the finite quality factor of the capacitor.

The most important parameter of the switched MOM is the ratio C_{max}/C_{min}, that can be derived as follows:

Fig. 2 Measured quality factor for the 70 pH inductor integrated in 65 nm and the 110 pH one integrated in 32 nm

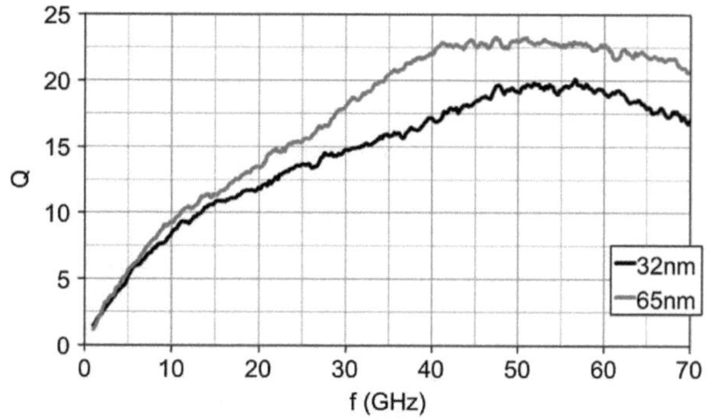

Fig. 3 Measured capacitance (*solid lines*) and quality factor (*dashed lines*) for varactors

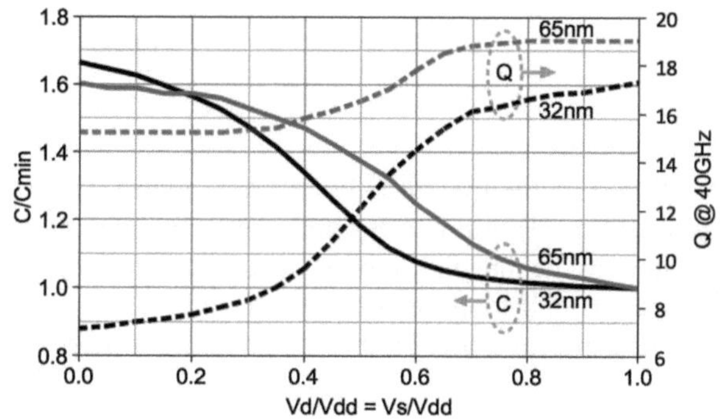

Fig. 4 Measured quality factor of mom capacitors versus frequency for 65 and 32 nm nodes

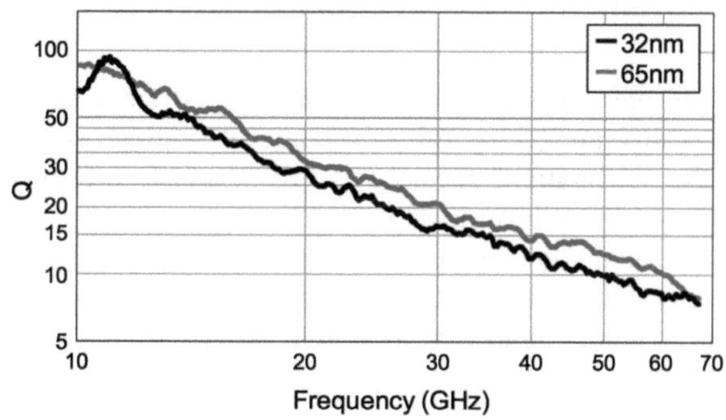

Fig. 5 Equivalent circuit of switched mom

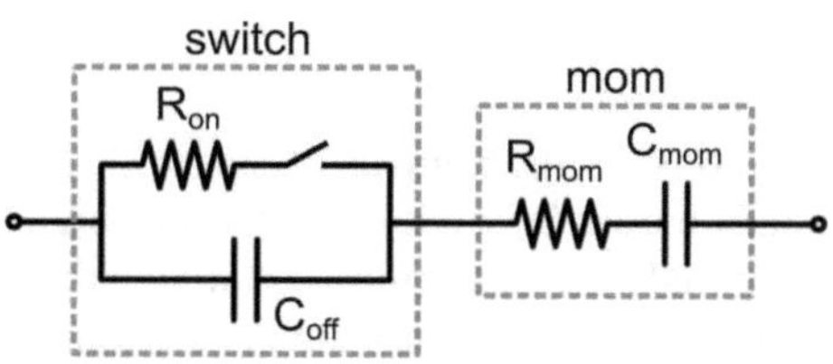

$$\frac{C_{max}}{C_{min}} = 1 + \frac{C_{mom}}{C_{off}} \tag{1}$$

while the overall quality factor is minimum when the switch is closed, and equal to:

$$Q_{on} = \frac{Q_{mom}}{1 + \frac{R_{on}}{R_{mom}}} \tag{2}$$

Combining (2) with (1) and the definition of FoM_{sw}, the minimum quality factor of the switched MOM as a function of C_{max}/C_{min} and FoM_{sw} can be derived as follows:

$$Q_{on} = \frac{Q_{mom}}{\left(\frac{C_{max}}{C_{min}} - 1\right) FoM_{sw}\omega_0 Q_{mom} + 1} \tag{3}$$

Based on the measured FoM_{sw} and quality factor of MOM capacitors, we have plotted Eq. (3) in Fig. 6 for both the technology nodes. No significant difference is evident in the range of interest, meaning that the better switch but the worse 32 nm capacitor tend to compensate each other. As a result, we cannot expect any significant improvement simply by adopting a more scaled node.

3 Inductor-Less Frequency Divider By Four

The presented inductor-less frequency divider is a wideband circuit based on an architecture taking full advantage of scaling, as demonstrated in this paragraph. The schematic is reported in Fig. 7, and it is formed by a cascade of four stages closed in an inverting loop.

The single stage uses clocked differential pairs working as dynamic CML latches and employing transistors only. Without any input injected signal, the divider is an autonomous ring oscillator, and its self-oscillation frequency is calculated as [15]:

$$f_{osc} = \frac{1}{2N_s R_{eq} C_L \ln\left(1 + \frac{A_{osc}}{R_{eq} I_b}\right)} \tag{4}$$

where N_s is the number of stages, R_{eq} is the equivalent resistance of the PMOS loads biased in triode, I_b is the biasing current of the latch and A_{osc} is the signal amplitude. Such amplitude can be calculated as $A_{osc} = (0.84 * R_{eq} * I_b)$ for $N_s = 4$ [15]. Since the load resistance is synthesized by PMOS transistors biased in triode, it is inversely proportional to g_m. Moreover, the load capacitance of each latch is

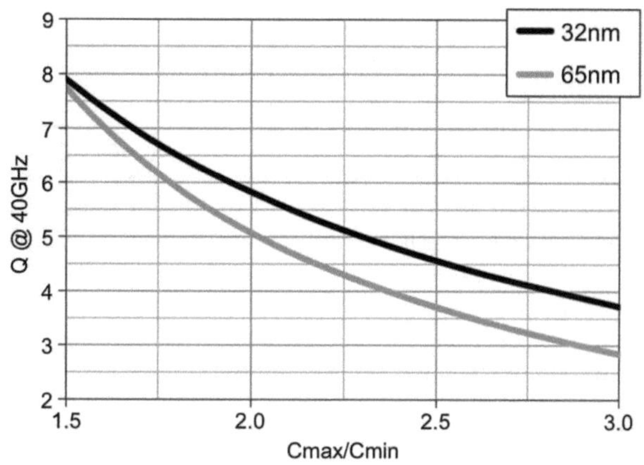

Fig. 6 Minimum quality factor of switched mom capacitors versus C_{max}/C_{min} in 65 and 32 nm based on Eq. (3)

mainly determined by the gate capacitance of the cascaded stage, thus f_{osc} is expected to be proportional to f_T. Measured maximum self oscillation frequencies in 65 nm and 32 nm are 18 and 27 GHz respectively [4, 5], showing a 50 % increment. This value is larger than the increment of device parameters f_T and f_{max} since the buffer stage in 32 nm is scaled down more than proportionally to the device feature scaling. This leads to a smaller capacitance loading the divider and thus to a higher self-resonance frequency as described by Eq. (4).

A detailed analysis of the locking behavior of the divider, as originally developed in [5], is reported here to provide design guidelines and insight in circuit behavior in scaled nodes. As shown in the inset of Fig. 7, the NMOS and PMOS devices of each dynamic latch are driven by complementary clock phases. If the clock amplitude is large enough to switch on both NMOS and PMOS during half the clock period, and switch them completely off during the second half, read and hold phases of the latch are well defined and the description of the behavior of the circuit is intuitive since it resembles the operation of a standard digital frequency divider. To derive equations for the divider dynamics, refer to Fig. 8, showing current and voltage waveforms, both at the input and at the output of one of the latches, when the divider is locked. To simplify the analysis, it is assumed at this point that the PMOS synthesize an infinite equivalent load resistance R_{off} during the hold state, and the clock frequency is low enough to hard switch the differential pair during the read phase.

The output voltage waveforms of the latch are an exact replica of the input ones, delayed by half the clock period. During the hold phase, since the latch is switched off, no current flows. In the second half of the input period, the output differential voltage of each latch has to overcome the voltage threshold V_{sw} required to completely steer the current in the pair before the end of the read phase in t_1, t_5, to ensure a correct operation. If this is not the case, the following stage is not completely switched and the digital state does not correctly propagate across the ring. Assuming the differential pair always in saturation during the read phase (this condition is well approximated in practice), and observing that the output

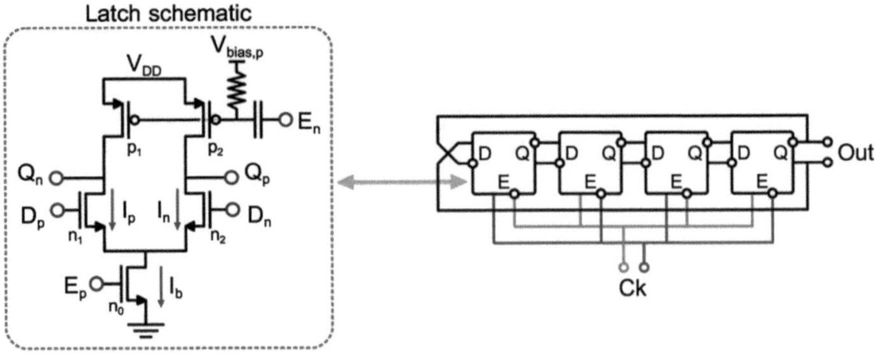

Fig. 7 Synchronous divider by four realized with dynamic load modulated latches

capacitor is charged/discharged with exponential rise/fall, the maximum frequency can be calculated solving the following system of equations:

$$\begin{cases} Q_p(t) = V_{DD} - \left(V_{DD} - Q_p^*\right) e^{-\frac{t-t^*}{R_{on}C_L}} \\ Q_n(t) = Q_n^* - \left(Q_n^* - V_{DD} + R_{on}I_b\right)\left(1 - e^{-\frac{t-t^*}{R_{on}C_L}}\right) \end{cases} \tag{5}$$

$Q_p(t)$ and $Q_n(t)$ are the single ended time domain voltages at the outputs of the latch during the two read phases, as identified in Figs. 7 and 8, and C_L is the parasitic capacitance loading the latch. When the latch is switched on, R_{on} and I_b are the equivalent load resistance and the biasing current respectively. Again referring to Fig. 8, for the first read phase between t_0 and t_1 we have $t^* = t_0$, $Q_p^* = Q_{p0}$ and $Q_n^* = Q_{n0}$, while for the second read phase between t_2 and t_3 we have $t^* = t_2$, $Q_p^* = Q_{p2}$ and $Q_n^* = Q_{n2}$. Due to the differential behavior of the circuit, the operation between t_4 and t_8 is exactly the same as in t_0, t_4 but with opposite sign. The maximum locking frequency can be calculated assuming that $(t-t^*) = 1/(2f_{max})$, i.e. assuming the read phase lasts exactly the minimum time required to meet the threshold voltage $V_{sw} \sim \sqrt{2}V_{ov}$, i.e. $Q_{n2} - Q_{p2} = V_{sw}$, and observing that $Q_{p2} = Q_{p1}$, $Q_{n2} = Q_{n1}$, $Q_{p3} = Q_{n0}$ and $Q_{n3} = Q_{p0}$:

$$f_{max} = \frac{1}{2R_{on}C_L(1.41 + 0.59\gamma)\sqrt{\gamma}} \tag{6}$$

For given $\gamma = V_{sw}/(R_{on}\, I_b)$, i.e. the ratio between the voltage threshold V_{sw} to switch the pair and the asymptotic output amplitude $R_{on}I_b$, a higher fmax is achieved when reducing $R_{on}C_L$. On the other hand, the lower γ the higher fmax, for given time constant $R_{on}C_L$, because the time required to surpass the threshold reduces.

The circuit operates down to DC under the assumption $R_{off} \to \infty$. We release now this condition and inspect the dependence of f_{min} on R_{off} [5]. Modulation of

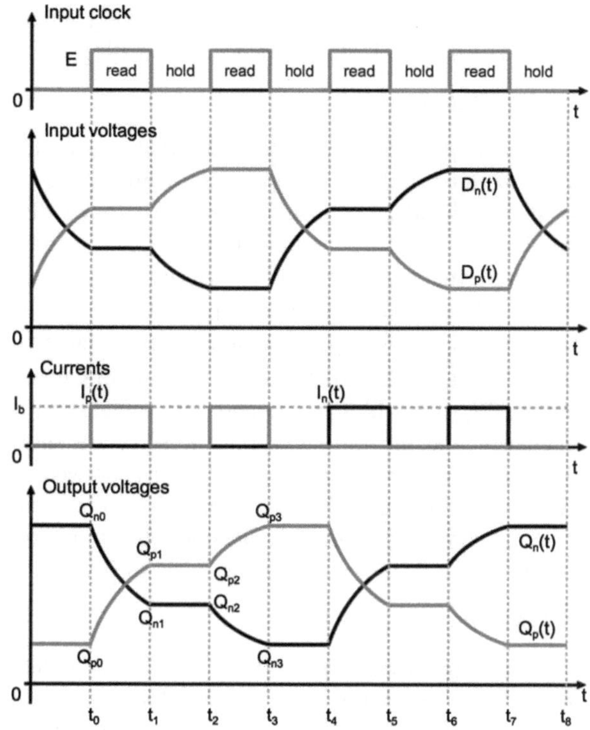

Fig. 8 Voltage and current waveforms in a dynamic load modulated latch with infinite R_{off}, when embedded in a locked divider by 4

the load device renders R_{off} much higher than R_{on}, leading to $f_{min} \ll f_{max}$. Close to f_{min}, during the read mode, the two output voltages of the latch have enough time to settle at V_{dd} and $V_{dd} - R_{on}I_b$ respectively. During the hold mode, the differential voltage should not discharge below V_{sw} in order to retain a correct digital state and ensure the locking of the divider. The maximum hold time is derived as:

$$V_{DD} - \left(V_{DD} - R_{on} I_b e^{-\frac{t_{hold}}{R_{off} C_L}} \right) < V_{sw} \tag{7}$$

Setting $t_{hold} = 1/(2f_{min})$, the minimum locking frequency is found as

$$f_{min} = \frac{1}{2R_{off} C_L \ln \frac{1}{\gamma}} \tag{8}$$

A reverse dependence on the discharging time constant during the hold phase emerges as expected, together with a mild dependence on γ.

It should be noted that f_{min} is designed to be as low as possible, to have the widest locking range, by maximizing the equivalent resistance of the PMOS loads during the hold phase. In this way there is no leakage of the charge stored in the load parasitic capacitors and the state is kept indefinitely. Based on (8), we also conclude that f_{min} is not influenced by scaling.

a. *Circuit design*

The analysis has been carried out assuming an ideal circuit in order to separate fundamental principles of operation from second order effects caused by implementation details. Insights for a well designed divider-by-4 are gained through simulations, adopting the models of a 32 nm technology design kit. Latches have minimum channel length, $W_{n0} = 8\ \mu m$, $W_{p1,2} = 3.2\ \mu m$ while $W_{n1,2}$ are swept between 2.6 and 12.6 μm. Both C_L, directly proportional to $W_{n1,2}$, and γ, inversely proportional to $\sqrt{W_{n1,2}}$, are changed at the same time. The power consumption, which is set by the tail transistor n_0, remains constant to 3.5 mW. In turns, the smaller $W_{n1,2}$ the higher f_{min} and f_{max}, suggesting a trade-off between maximum frequency and the ratio f_{max}/f_{min}. Results are shown in Fig. 9. Final choice of $W_{n1,2}$ is 6 μm, as optimum compromise between high f_{max} and high f_{max}/f_{min}. Moreover, since f_{min} does not depend on the technological features while f_{max} takes advantage of reduced parasitics, the implementation of this divider architecture into a further scaled node leads to increasing both f_{max} and f_{max}/f_{min}.

b. *Experiments*

Test chips have been realized by STMicroelectronics in CMOS32n-LP technology. The input mm-wave signal, clocking the latches, is made differential by means of an on-chip transformer realized with two rectangular spirals of 30 × 70 μm. The divider is then followed by a three stages buffer driving the output pads and the 50 Ω input of the measurement setup. Photomicrograph of the test-chip is shown in Fig. 10. Active area of the divider is extremely low, 18 × 55 μm.

All experiments have been performed with 1 V supply. Figure 11 compares four measured (solid lines) and simulated (dotted-lines) sensitivity curves. The plot evidences a significant extension of the lower divider operating frequency, f_{min}, when the amplitude of the input signal increases. This is because a larger input leads to better switch-off of the PMOS loads, when the latches are in the hold mode, thus raising quickly the equivalent resistance R_{off} and reducing f_{min}. The frequency location of each band can be continuously tuned by varying the biasing DC voltage of the PMOS loads ($V_{bias,p}$) from 430 to 180 mV. The measured bandwidth ranges from 14 to 70 GHz input frequency, limited at low frequency by the high-pass transfer function of the integrated balun and at high frequency by our available instrumentation. The fractional bandwidth of operation is in excess of 90 and 60 % at minimum and maximum frequencies, respectively.

Finally Fig. 12 compares the phase noise measured at the input and at the output when the divider is driven by a 60 GHz signal source. The output phase noise is ~12 dB lower, up to ~10 MHz frequency offset, as theoretically expected due to division by 4, demonstrating a negligible noise degradation introduced by the divider. Experimental results are summarized and compared

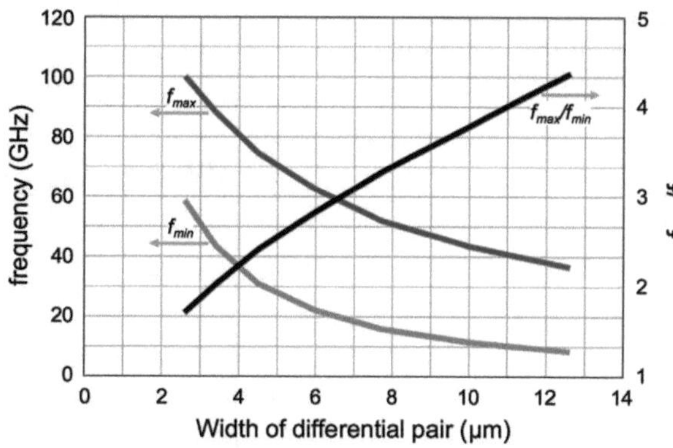

Fig. 9 Simulated maximum and minimum locking frequencies versus the size of the differential pair

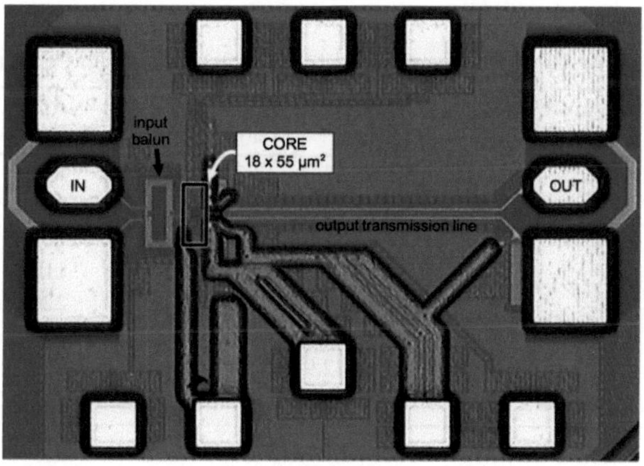

Fig. 10 Frequency divider's chip photomicrograph

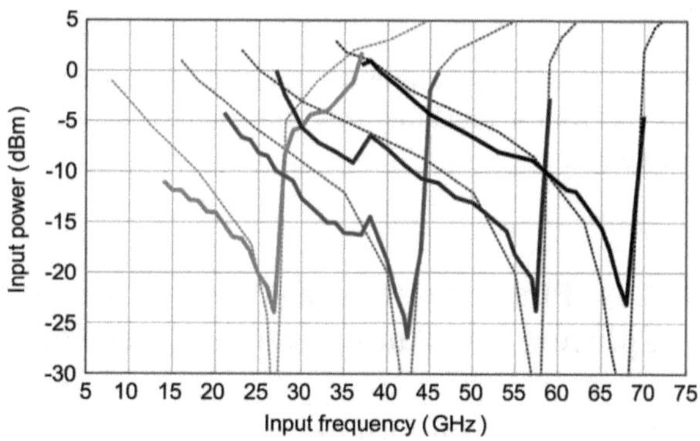

Fig. 11 Measured (*solid*) versus simulated (*dashed*) sensitivity curves for different PMOS bias voltages

against recently reported mm-waves injection-locked dividers providing division factors of 3 and 4 in Table 1. The proposed dynamic CML latch leads also to an outstanding frequency locking range in each sub-band, in excess of 60 %, at

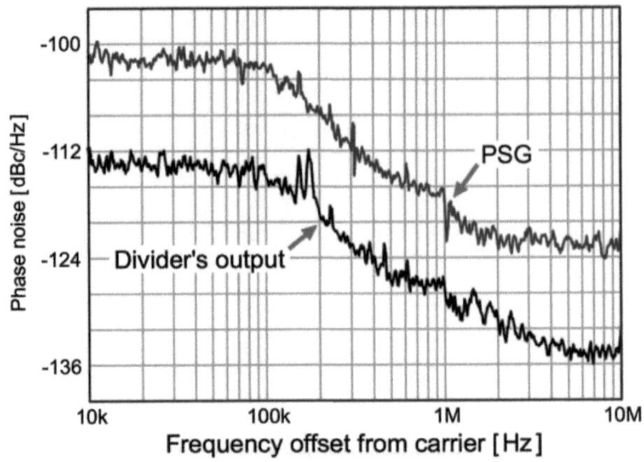

Fig. 12 Measured phase noise at divider's input (*top*) and output (*bottom*)

Table 1 Measurement summary and comparison with the state of the art

References	f_{in}/f_{out}	$f_{min}-f_{max}$ (GHz)	L.R. (%)	P_{diss} (mW)	Area (μm^2)	Tech CMOS (nm)	FoM (GHz/mW)
[16]	3	58.6–67.2	13.7	5.2	170 × 220	65	1.65
[17]	3	48.8–54.6	3.5[a]	3.0	300 × 300	65	0.57
[18]	4	79.7–81.6	2.4	12	106 × 330	65	0.16
[19]	4	62.9–71.6	3.2	2.8	110 × 130	90	0.82
[20]	4	82.5–89.0	7.6	3.0	220 × 290	90	2.17
[21]	4	67.0–72.4	7.7	15.5	870 × 760	90	0.35
[22]	4	58.5–72.9	21.9	2.2	160 × 260	65	6.55
This work	4	14[b]–70[c]	60–90	1.3–4.8	18 × 55	32	6.67–17.5

[a] Estimated from reported sensitivity curves
[b] Minimum input frequency limited by on-chip spiral balun
[c] Maximum input frequency limited by our available instrumentation

limited power dissipation. Getting rid of resonant LC loads, the proposed divider has the widest tunability and the smallest active area and lends itself to further scaled nodes.

4 Low Phase Noise MM-Wave VCO with Inductor Splitting for Tuning-Range Extension

The LC tank Q is limited by capacitors, at mm-waves, preventing better performances just due to scaling. Design of VCOs in this band is further penalized by trade-off between phase noise and frequency tuning range, as confirmed by experimental results reported in the literature, and summarized in Fig. 13. Tuning range encompasses a rapid degradation, as frequency increases. The limits of the

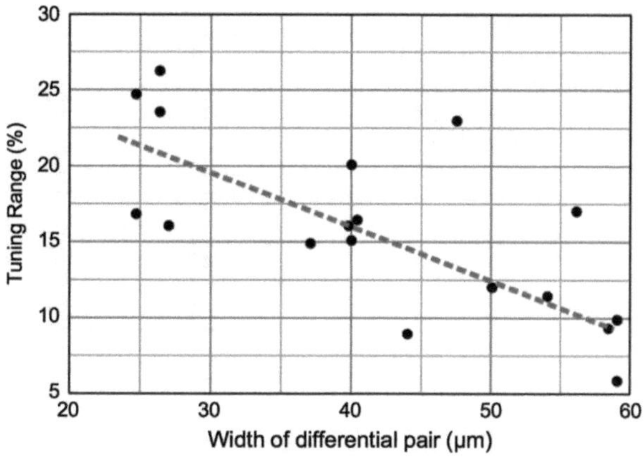

Fig. 13 Tuning range versus center frequency of recently reported mm Wave CMOS oscillators

traditional LC resonator, tuned by switching capacitors in parallel with the inductor, are discussed in this section and an improved topology where the switch is placed in series with the tank inductor is proposed [6]. It is shown that the latter is more robust against switch parasitics and leads to a much wider frequency tuning step without compromising the tank Q.

a. *Switched LC-resonator with wide tuning range*

Figure 14 shows a simplified equivalent circuit of an oscillator with the traditional tank with switched capacitor C_T and the proposed alternative made of a switch in series with inductor L_T. The negative resistance models the active devices while C_{fix}, comparable or even larger than C_T at mm-Waves, represents the fixed capacitance loading the tank, introduced primarily by parasitics of the negative resistance and buffer. When the switch is on, the two tanks have the same resonance frequency equal to $1/(2\pi(L_T(C_{fix} + C_T))^{1/2})$ but when the switch is off, the behavior of the two circuits is remarkably different.

The oscillation frequency f_0 for the top-right circuit in Fig. 14 is:

$$f_0 = \frac{1}{2\pi\sqrt{L_T\left(C_{fix} + \frac{C_T C_{off}}{C_T + C_{off}}\right)}} \tag{9}$$

If the parasitic capacitance of the switch is negligible compared with C_T and C_{fix}, (i.e. $C_{off} \ll C_T, C_{fix}$) the oscillation frequency simplifies to $f_0 \approx \frac{1}{2\pi\sqrt{L_T C_{fix}}}$, i.e. C_{fix} sets an upper bound to the maximum frequency step.

In the bottom-right circuit of Fig. 14, C_{off} appears in series with $C_T + C_{fix}$ instead of C_T only, determining a larger variation of the equivalent tank capacitance and removing the limitation introduced by C_{fix} on the maximum oscillation frequency, given by:

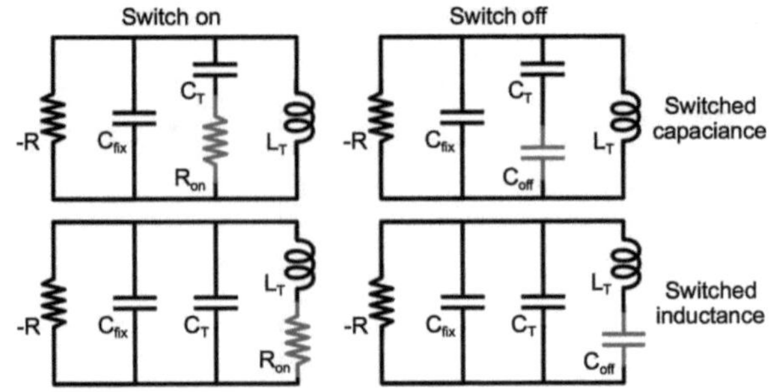

Fig. 14 Simplified equivalent circuit of an LC-tank oscillator with a switch in series to the capacitor and inductor

$$f_0 = \frac{1}{2\pi\sqrt{L_T \frac{(C_T+C_{fix})C_{off}}{(C_T+C_{fix})+C_{off}}}} \tag{10}$$

which, in the limit case of C_{off} much lower than C_{fix} and C_T, simplifies to $f_0 \approx \frac{1}{2\pi\sqrt{L_T C_{off}}}$.

Looked at in an alternative way, for the same relative frequency step, the switch in series with L_T can assume a much higher C_{off}. It can be therefore realized with a larger transistor leading to a lower R_{on} and less penalty to the tank quality factor. As a numerical example, assuming typical values of $L_T = 100$ pH, $C_T \sim C_{fix} = 100$ fF, the resonance frequency with the switches closed is 35.6 GHz. A fractional frequency step of 20 % sets $C_{off} = 50$ fF in the switched capacitor while allowing $C_{off} = 400$ fF when the switch is in series with the inductor. With $C_{off} = 1.27$ fF/μm and $R_{on} = 432$ Ω·μm measured for the 32 nm CMOS technology, the on resistance of the switch comes out to be 11 and 1.37 Ω for the switched capacitor and inductor, respectively. With lossless reactive components, the tank quality factor would be 8.1 and 16.3 respectively, thus showing a remarkable advantage. Simulation results taking into account losses of the tank components are shown in Fig. 15. The plot compares the minimum Q of the traditional tank and the proposed solution at different tuning steps for a center frequency of 40 GHz. For $f_{max}/f_{min} = 1.2$, connection of the switch in series with the inductor improves the tank Q by 50 %, from 3 to 4.5. Simulations reveal advantage increases further when targeting a larger tuning step.

b. *Circuit design*

A wide tuning range VCO based on the proposed LC-tank has been designed in 32 nm CMOS technology. The center frequency is 40 GHz targeting high-speed Gb/s wireless communications at V-band with a sliding-IF receiver architecture [23]. Tuning range in excess of 20 % is required to cover 57 to 66 GHz bandwidth with margin against process variations and poor device and parasitic modeling. The schematic of the VCO is shown in Fig. 16.

Inductor L_T with capacitors C_T and C_{fix} realizes the resonator. Transistor M_{sw} splits the inductor and, by exploiting its parasitic capacitance in off-state, allows a

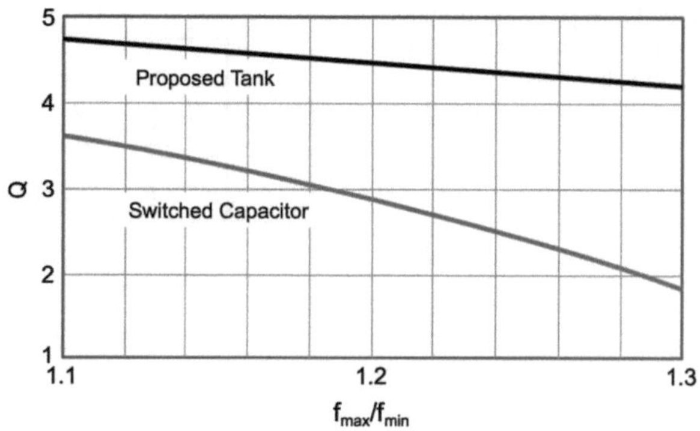

Fig. 15 Simulated minimum Q at 40 GHz versus the tuning step for the traditional and proposed tank

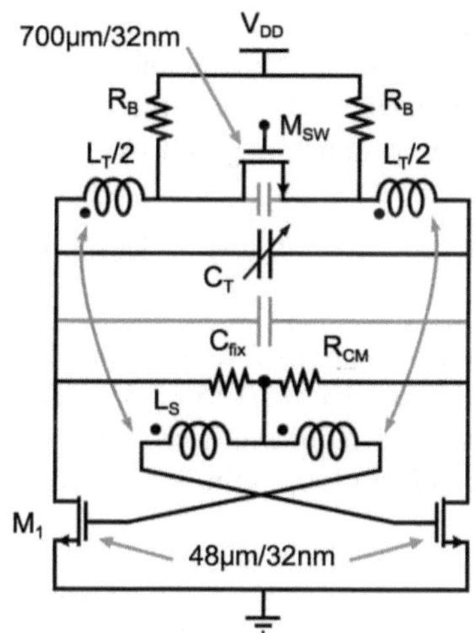

Fig. 16 Schematic of the proposed VCO

coarse tuning-step dividing the total oscillator tuning range in two main sub-bands. Finer tuning is implemented by realizing C_T with a bank of three binary-sized switched MOM capacitors and a small varactor. Being M_{sw} very large, the on resistance has a negligible impact on the overall tank quality factor which varies from 4 to 5.5 in the tuning range. Transistors M_1–M_2, with a current consumption regulated by the two top biasing resistors R_b, compensate resonator losses. The feedback from the tank to the gate terminals of M_1–M_2 is implemented through the secondary coil of a transformer L_S, and the gate biasing is provided by resistors R_{cm} connected to the center tap of L_S. Feedback via the transformer is required to prevent the circuit from latching when M_{sw} is off. It also leads to larger loop gain. To gain insight, Fig. 17 shows the simplified equivalent circuit of the oscillator with M_{sw} on and off.

In the lower sub-band, with M_{sw} on, the tank impedance at resonance frequency f_{low} is $R_{PON} = 2\pi f_{low} L_T Q$ and the loop gain is:

Fig. 17 Equivalent circuits of the oscillator in the two main sub-bands

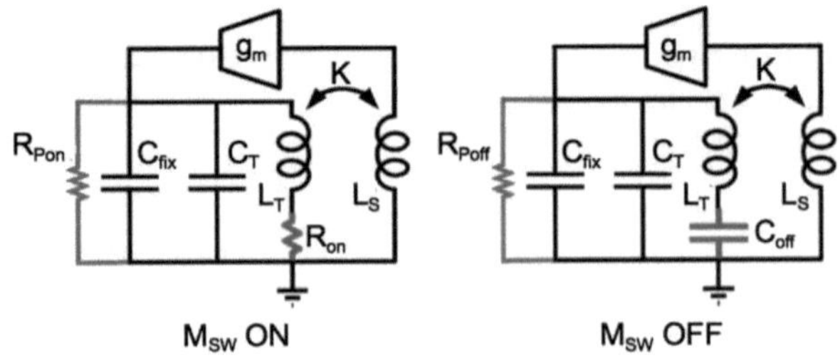

$$G_{loop} = k\sqrt{\frac{L_S}{L_T}} g_m 2\pi f_o L_T Q \qquad (11)$$

Looking now at the right circuit in Fig. 17, which represents the oscillator with M_{sw} off, the tank impedance is infinite at DC. The transformer suppresses the loop gain at DC and avoids the circuit from latching. At resonance frequency f_{hi}, the tank impedance is given by:

$$R_{POFF} = 2\pi f_{hi} L_T Q \alpha^2 \quad \text{with} \quad \alpha = \frac{C_{off}}{C_T + C_{fix} + C_{off}} \qquad (12)$$

The capacitive divider drastically limits the impedance at the drain of the transistors ($\alpha^2 = 0.35{-}0.45$). The transformer thus helps raising the loop gain, otherwise requiring a prohibitively high g_m. By inspection of the right circuit in Fig. 17 the loop gain is given by:

$$G_{loop} = k\sqrt{\frac{L_S}{L_T}} \alpha g_m 2\pi f_o L_T Q \qquad (13)$$

In the realized VCO the estimated impedance magnitude of the tank ranges from 80 to 105 Ω. The transformer has been designed with a primary inductor L_T of 100 pH and secondary L_S of 120 pH. Magnetic coupling k is 0.75. Core active devices M_1–M_2 in Fig. 16 are 48 μm wide over minimum channel leading to a transconductance of 25 mS which ensures a loop gain larger than 1.7.

c. *Experiments*

Test chips have been realized by STMicroelectronics in CMOS32 nm-LP technology. For characterization the VCO drives a buffer, realized with an open-drain differential pair, and a frequency divider-by-four [5]. Both the signal at mm-Waves and a replica scaled in frequency are thus available. A micrograph of the test chip is shown in Fig. 18.

The oscillation frequency is tunable from 33.6 to 40.8 GHz with M_{sw} on and from 38.8 to 46.2 GHz with M_{sw} off, corresponding to 31.6 % total tuning range. The wide overlapping of 2 GHz between the two bands is more than expected. An accurate redesign reducing overlap to a minimum would allow a tuning range of

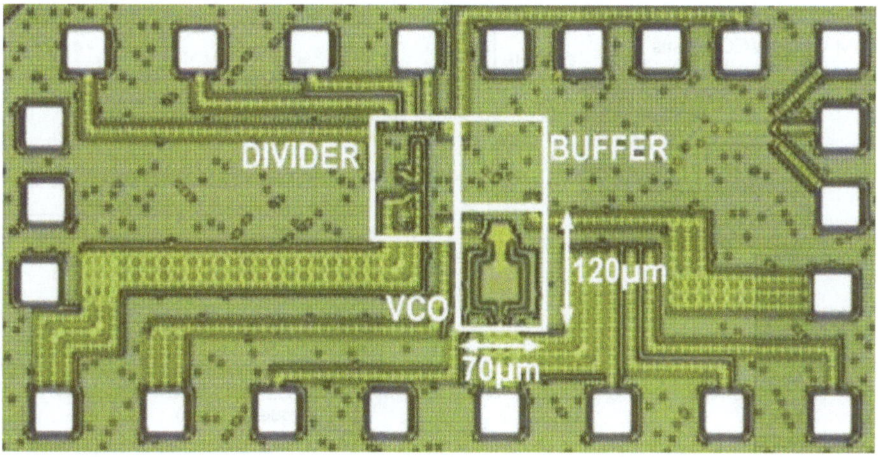

Fig. 18 Die microphotograph of the realized VCO

Fig. 19 Measured phase noise for the VCO oscillating at 40 GHz

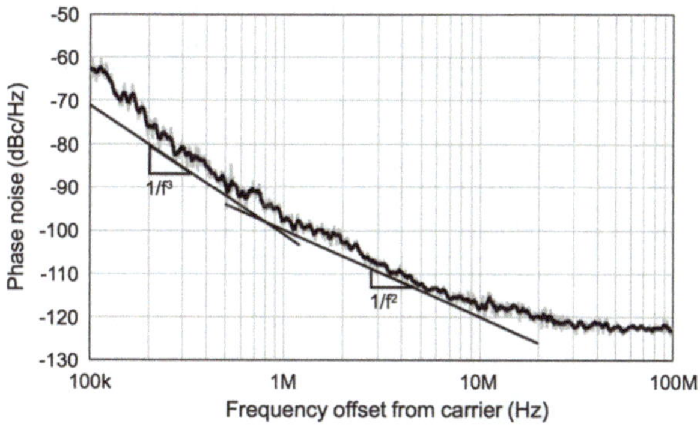

35 %. The power dissipation is 9.8 mW from a 1 V supply. The phase noise has been measured after the frequency divider and reported to the carrier frequency assuming negligible noise introduced by the divider. Figure 19 shows a typical plot, for a carrier frequency of 40 GHz. The phase noise at 10 MHz offset is −118 dBc/Hz with a $1/f^3$ corner frequency of ~800 kHz. Phase noise at 10 MHz offset and the FOM versus the oscillation frequency are shown in Fig. 20. The phase noise ranges from −115.2 to −118 dBc/Hz with a corresponding FOM from 177.5 to 180 dBc/Hz.

Experimental results are summarized and compared to recently reported mm-Wave VCOs in Table 2. VCOs exploiting transformer tuning [24, 25] achieve record tuning ranges, much more than typically required, at the cost of a severe in-band phase noise penalty, as evidenced by the very low minimum FOM. The proposed VCO has a state-of-the-art FOM, despite being realized in an ultra-scaled 32 nm node, comparable with traditional switched capacitors or varactor solutions [26, 27].

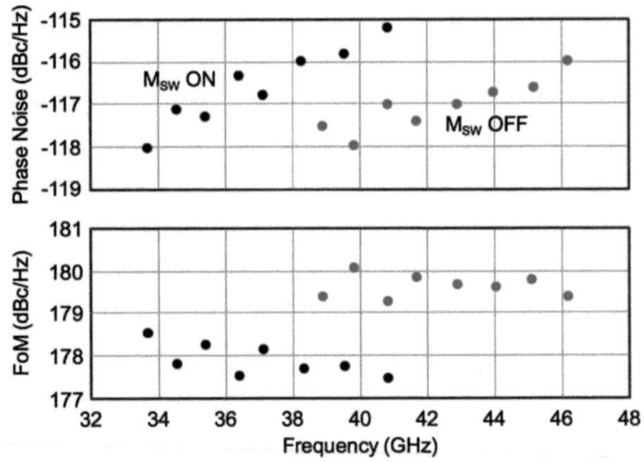

Fig. 20 Phase noise and FoM of the VCO versus frequency

Table 2 VCO performance summary and comparison with the state of the art

Reference	Freq (GHz)	T.R. (%)	P_{diss} (mW)	PN@10 MHz (dBm/Hz)	FoM (dBc/Hz)	Tech CMOS (nm)
[24]	57.5/90.1	44.2	8.4 /10.8	−104.6/−112.2	172/180	65
[25]	34.3/39.9	15.0	14.4	−118/−121	178.4/180	65
[26]	43.2/51.8	22.9	16.0	−117/−119	179/180	65
[27]	21.7/27.8	24.8	12.2	−121	177.5	45
This work	33.6/46.2	31.6	9.8	−115.2/−118.0	177.5/180	32

5 Conclusions

Transistors improve their performances with scaling, but this is not necessarily true for passive components, where the impact of the different back-end of line (BEOL) plays a key role. This article shows data around single devices, comparing 65 nm versus 32 nm nodes in a wide frequency range. As an example of a wide-band circuit taking advantage of faster devices, available in ultra-scaled nodes, we propose an inductor-less frequency divider by four demonstrating state of the art frequency locking range together with lowest occupied area. On the contrary, conventional LC VCOs are penalized by further scaled nodes where passive components degrade. This observation motivates the research of alternative VCOs better suited for LO generation in ultra-scaled nodes. We have proposed a solution with resonator using inductor splitting and achieving a tuning range in excess of 30 % at 40 GHz and a state of the art noise FOM.

References

1. Razavi, B., "Gadgets Gab at 60 Ghz,"Spectrum, *IEEE*, vol. 45, no. 2, pp. 46,58, Feb. 2008
2. Jinglin Shi; Kai Kang; Yong Zhong Xiong; Brinkhoff, J.; Lin, F.; Xiao-Jun Yuan, "Millimeter-Wave Passives in 45-nm Digital CMOS,"*Electron Device Letters, IEEE*, vol. 31, no. 10, pp. 1080,1082, Oct. 2010
3. Jan, C. -H; Agostinelli, M.; Deshpande, H.; El-Tanani, M.A.; Hafez, W.; Jalan, U.; Janbay, L.; Kang, M.; Lakdawala, H.; Lin, J.; Lu, Y.-L.; Mudanai, S.; Park, J.; Rahman, A.; Rizk, J.; Shin, W.-K.; Soumyanath, K.; Tashiro, H.; Tsai, C.; Vandervoorn, P.; Yeh, J.-Y; Bai, P., "RF CMOS technology scaling in High-k/metal gate era for RF SoC (system-on-chip) applications,"*Electron Devices Meeting (IEDM), 2010 IEEE International*, vol., no., pp. 27.2.1,27.2.4, 6–8 Dec. 2010
4. Decanis, U.; Ghilioni, A.; Monaco, E.; Mazzanti, A.; Svelto, F., "A Low-Noise Quadrature VCO Based on Magnetically Coupled Resonators and a Wideband Frequency Divider at Millimeter Waves,"*Solid-State Circuits, IEEE Journal of*, vol. 46, no. 12, pp. 2943,2955, Dec. 2011
5. Ghilioni, A.; Mazzanti, A.; Svelto, F., "Analysis and Design of mm-Wave Frequency Dividers Based on Dynamic Latches With Load Modulation,"*Solid-State Circuits, IEEE Journal of*, vol. 48, no. 8, pp. 1842,1850, Aug. 2013
6. Mammei, E.; Monaco, E.; Mazzanti, A.; Svelto, F., "A 33.6-to-46.2 GHz 32 nm CMOS VCO with 177.5dBc/Hz minimum noise FOM using inductor splitting for tuning extension,"*Solid-State Circuits Conference Digest of Technical Papers (ISSCC), 2013 IEEE International*, vol., no., pp. 350,351, 17–21 Feb. 2013
7. Changhua Cao; Eunyoung Seok; O, K.K., "Millimeter-Wave CMOS Voltage-Controlled Oscillators,"Radio and Wireless Symposium, 2007 IEEE, vol., no., pp. 185,188, 9–11 Jan. 2007
8. Changhua Cao; O, K.K., "Millimeter-wave voltage-controlled oscillators in 0.13-μm CMOS technology,"*Solid-State Circuits, IEEE Journal of*, vol. 41, no. 6, pp. 1297,1304, June 2006
9. Dickson, T.O.; Yau, K. H K; Chalvatzis, T.; Mangan, A.M.; Laskin, E.; Beerkens, R.; Westergaard, P.; Tazlauanu, M.; Ming-Ta Yang; Voinigescu, S.P., "The Invariance of Characteristic Current Densities in Nanoscale MOSFETs and Its Impact on Algorithmic Design Methodologies and Design Porting of Si(Ge) (Bi)CMOS High-Speed Building Blocks,"*Solid-State Circuits, IEEE Journal of*, vol. 41, no. 8, pp. 1830,1845, Aug. 2006
10. Razavi, B., "A 300-GHz Fundamental Oscillator in 65-nm CMOS Technology,"*Solid-State Circuits, IEEE Journal of*, vol. 46, no. 4, pp. 894,903, April 2011
11. Woerlee, Pierre H.; Knitel, M.J.; van Langevelde, R.; Klaassen, D. B M; Tiemeijer, L.F.; Scholten, A.J.; Zegers-van Duijnhoven, A.T.A., "RF-CMOS performance trends," Electron Devices, IEEE Transactions on, vol. 48, no. 8, pp. 1776,1782, Aug 2001
12. Post, I.; Akbar, M.; Curello, G.; Gannavaram, S.; Hafez, W.; Jalan, U.; Komeyli, K.; Lin, J.; Lindert, N.; Park, J.; Rizk, J.; Sacks, G.; Tsai, C.; Yeh, D.; Bai, P.; Jan, C. -H, "A 65 nm CMOS SOC Technology Featuring Strained Silicon Transistors for RF Applications," Electron Devices Meeting, 2006. IEDM '06. International, vol., no., pp. 1,3, 11–13 Dec. 2006
13. Shien-Yang Wu; Chou, C.W.; Lin, C.Y.; Chiang, M.C.; Yang, C.K.; Liu, M.Y.; Hu, L.C.; Chang, C.H.; Wu, P.H.; Chen, H.F.; Chang, S.Y.; Wang, S.H.; Tong, P.Y.; Hsieh, Y.L.; Liaw, J.J.; Pan, K.H.; Hsieh, C.H.; Chen, C.H.; Cheng, J.Y.; Yao, C.H.; Wan, W.K.; Lee, T.L.; Huang, K.T.; Lin, K.C.; Yeh, L.Y.; Ku, K.C.; Chen, S.C.; Lin, H.J.; Jang, S.M.; Lu, Y.C.; Shieh, J.H.; Tsai, M.H.; Song, J.Y.; Chen, K.S.; Chang, V.; Cheng, S.M.; Yang, S.H.; Diaz, C.H.; See, Y.C.; Liang, M.S., "A 32 nm CMOS Low Power SoC Platform Technology for Foundry Applications with Functional High Density SRAM," Electron Devices Meeting, 2007. IEDM 2007. IEEE International, vol., no., pp. 263,266, 10–12 Dec. 2007
14. Dajiang Yang; Yuanli Ding; Huang, S., "A 65-nm High-Frequency Low-Noise CMOS-Based RF SoC Technology," Electron Devices, IEEE Transactions on, vol. 57, no.1, pp. 328,335, Jan. 2010

15. Gangasani, G.R.; Kinget, P.R., “Time-Domain Model for Injection Locking in Nonharmonic Oscillators,”*Circuits and Systems I: Regular Papers, IEEE Transactions on*, vol. 55, no. 6, pp. 1648,1658, July 2008
16. Hsieh-Hung Hsieh; Fu-Lung Hsueh; Chewn-PuJou; Kuo, F.; Chen, S.; Tzu-Jin Yeh; Tan, K.K.-W.; Po-Yi Wu; Yu-Ling Lin; Ming-Hsien Tsai;, “A V-band divide-by-three differential direct injection-locked frequency divider in 65-nm CMOS,”*Custom Integrated Circuits Conference (CICC), 2010 IEEE*, vol., no., pp. 1–4, 19–22 Sept. 2010
17. Xiao Peng Yu; van Roermund, A.; Xiao Lang Yan; Cheema, H.M.; Mahmoudi, R.;, “A 3 mW 54.6 GHz Divide-by-3 Injection Locked Frequency Divider With Resistive Harmonic Enhancement,”*Microwave and Wireless Components Letters, IEEE*, vol. 19, no. 9, pp. 575–577, Sept. 2009
18. Mayr, P.; Weyers, C.; Langmann, U.;, “A 90 GHz 65 nm CMOS Injection-Locked Frequency Divider,”*Solid-State Circuits Conference, 2007. ISSCC 2007. Digest of Technical Papers. IEEE International*, vol., no., pp. 198–596, 11–15 Feb. 2007
19. Yamamoto, K.; Fujishima, M.;, “70 GHz CMOS Harmonic Injection-Locked Divider,”*Solid-State Circuits Conference, 2006. ISSCC 2006. Digest of Technical Papers. IEEE International*, vol., no., pp. 2472–2481, 6–9 Feb. 2006
20. Chung-Chun Chen; Hen-WaiTsao; Huei Wang;, “Design and Analysis of CMOS Frequency Dividers With Wide Input Locking Ranges,”*Microwave Theory and Techniques, IEEE Transactions on*, vol. 57, no. 12, pp. 3060–3069, Dec. 2009
21. Chao-An Yu; Tang-NianLuo; Chen, Y.E.;, “A V-Band Divide-by-Four Frequency Divider With Wide Locking Range and Quadrature Outputs,”*Microwave and Wireless Components Letters, IEEE*, vol. 22, no. 2, pp. 82–84, Feb. 2012
22. Liang Wu; Luong, H.C.;, “A 0.6 V 2.2mW 58-to-73 GHz divide-by-4 injection-locked frequency divider,”*Custom Integrated Circuits Conference (CICC), 2012 IEEE*, vol., no., pp. 1–4, 9–12 Sept. 2012
23. Vecchi, F.; Bozzola, S.; Temporiti, E.; Guermandi, D.; Pozzoni, M.; Repossi, M.; Cusmai, M.; Decanis, U.; Mazzanti, A.; Svelto, F., “A Wideband Receiver for Multi-Gbit/s Communications in 65 nm CMOS,”*Solid-State Circuits, IEEE Journal of*, vol. 46, no. 3, pp. 551,561, March 2011
24. Jun Yin; Luong, H.C., “A 57.5-to-90.1 GHz magnetically-tuned multi-mode CMOS VCO,”*Custom Integrated Circuits Conference (CICC), 2012 IEEE*, vol., no., pp. 1,4, 9–12 Sept. 2012
25. Nariman, M.; Rofougaran, R.; De Flaviis, F., “A switched-capacitor mm-wave VCO in 65 nm digital CMOS,”*Radio Frequency Integrated Circuits Symposium (RFIC), 2010 IEEE*, vol., no., pp. 157,160, 23–25 May 2010
26. Murphy, D.; Gu, Q.J.; Yi-Cheng Wu; Heng-Yu Jian; Xu, Z.; Tang, A.; Wang, F.; Chang, M.-C.F., “A Low Phase Noise, Wideband and Compact CMOS PLL for Use in a Heterodyne 802.15.3c Transceiver,"Solid-State Circuits, IEEE Journal of, vol. 46, no. 7, pp. 1606,1617, July 2011
27. Osorio, J.F.; Vaucher, C.S.; Huff, B.; v.d.Heijden, E.; Anton de Graauw, “A 21.7-to-27.8 GHz 2.6-degrees-rms 40Mw frequency synthesizer in 45 nm CMOS for mm-Wave communication applications,"Solid-State Circuits Conference Digest of Technical Papers (ISSCC), 2011 IEEE International, vol., no., pp. 278,280, 20–24 Feb. 2011

Digital Enhanced Transmitter Concepts for Nanometer-CMOS Technologies

M. Fulde and F. Kuttner

Abstract This chapter shows novel digital enhanced transmitter concepts based on RF-D/A converters as example for innovative circuit solutions mitigating the technology scaling related drawbacks for analog and RF design. Moving analog requirements from voltage to time domain allows to fully benefit from high switching speed in scaled CMOS. RFDACs are key building blocks for digital TX architectures and have to provide very high dynamic range at high clock frequencies and high power efficiency. Innovative digital RFDAC concepts are presented based on current mode and capacitive operation. Pre-distortion, calibration, distributed mixers and novel decoding schemes are employed to fulfill tough cellular and co-existence specifications and to allow multi-mode operation in future digital TX architectures.

1 Introduction

The increasing demand for global mobile computing and internet access (any standard, any place, any time) drives the continuous integration of analog and RF functionality into deeply scaled digital CMOS technologies. Highly integrated system-on-chip solutions enable lower cost and reduced power consumption. However, the classical analog device performance like g_m, g_{ds} or noise tends to suffer from technology scaling, e.g. due to pronounced short channel effects and reduced supply headroom.

This work shows kind of a top down approach for RF IC design in scaled technologies. A digital transmitter concept serves as example for scaling friendly system partitioning and scaling friendly architectures that are combined with

M. Fulde (✉) · F. Kuttner
Intel Mobile Communications Austria, Villach, Austria
e-mail: michael.fulde@intel.com

P. Harpe et al. (eds.), *High-Performance AD and DA Converters, IC Design in Scaled Technologies, and Time-Domain Signal Processing*,
DOI: 10.1007/978-3-319-07938-7_11,

circuit techniques like calibration or pre-distortion to deal with analog imperfections in scaled CMOS technologies. The focus is on the design of the RFDAC as typical mixed-signal circuit suffering from analog imperfections and benefiting from higher switching speed.

The chapter is organized as follows: Sect. 2 describes some key system considerations for the digital transmitter and derives specifications for the RFDAC. In Sect. 3 the design of a current-mode RFDAC in 28 nm CMOS is shown in detail with special focus on technology related design issues. Besides implementation aspects also measurement results from a test IC are presented. Section 4 shows an even more digital and scaling friendly concept for a capacitive RFDAC which may be suited for transceiver solutions in technologies beyond 28 nm.

The focus in this work is on cellular applications; however, the concepts could be easily extended to other radio access technologies (RAT) like WLAN or Bluetooth.

2 System Level Aspects of Digital TX

To derive the specifications for the RFDAC some system level requirements for the digital TX need to be discussed. Flexibility is one of the key aspects in today's cellular transceiver solutions. The support of multiple standards and modes like 2G (GSM/EDGE), 3G (WCDMA/HSPA+) and 4G (LTE/LTE-CA) is mandatory. In addition a variety of frequency bands (>40 actually defined in 3GPP) needs to be covered and co-existence with other connectivity technologies like WLAN, Bluetooth or GNSS needs to be supported.

Obviously low cost and low power consumption are key requirements. Low cost translates typically in small chip area, high level of integration, low pin count and low amount of external components. Small chip area and high level of integration can be achieved by moving RF functionality as much as possible in the digital domain. Here technology scaling also helps to reduce power consumption. However, the remaining RF building blocks need to be suitable for deep submicron CMOS. Avoiding the use of external TX filters is another measure reducing costs but increasing the requirements on the dynamic range of the transmitter. The need for very high dynamic range at the TX output is caused by today's frequency division duplex systems and co-existence scenarios. TX and RX (same or different RAT) are active at the same time but on different frequencies. Consequently the TX noise that falls into the RX frequency band needs to be very low. The following example illustrates this: in order to not de-sensitize the RX the injected noise from TX should be much less than the thermal noise from the 50 Ω input impedance, e.g. −180 dBm/Hz. Assuming a duplexer attenuation of about 47 dB and a maximum output power of 24 dBm the total TX noise in the RX band should not exceed −157 dBc/Hz at maximum power.

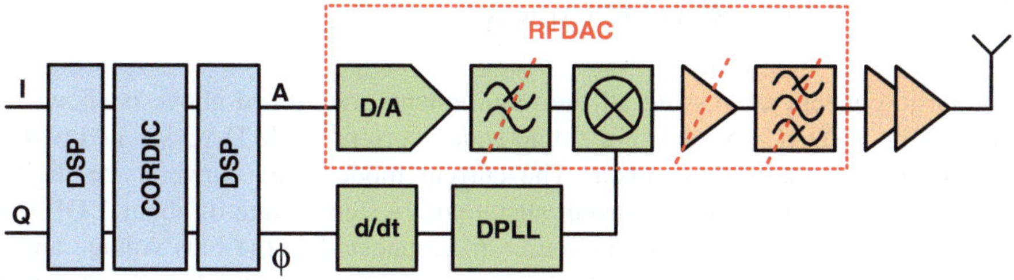

Fig. 1 High-level overview of digital polar TX showing functionality of RFDAC

The digital polar transmitter concept based on RFDAC [1] as shown in Fig. 1 addresses these tough requirements. The digital IQ data from baseband is converted into polar data (amplitude and phase) in the digital front-end using a CORDIC algorithm and digital filtering/interpolation. The derivative of the phase signal (=modulation frequency) is fed to a digital PLL producing the phase-modulated LO signal. The RFDAC block combines almost all the remaining analog functionality in the amplitude path of the transmitter.

Instead of an analog reconstruction filter a D/A converter with very high sample rate is used. The D/A converter is combined/stacked with the up-conversion mixer (fed by phase modulated LO signal) to re-use the DAC current and to avoid a dedicated g_m cell. The RFDAC is providing enough output power to directly drive the external power amplifier, i.e. no additional analog/RF sensitive driver is needed. Finally the external SAW filter can be removed if the RFDAC provides very high dynamic range. Apart from the external power amplifier and the digital controlled oscillator core in the DPLL the RFDAC is the only remaining analog/ RF block in the signal chain.

Obviously the combination of different analog/RF functions in one block yields very tough requirements for the RFDAC:

- High-speed (∼1–2 GHz): to avoid an analog reconstruction filter the sample rate of the DAC needs to be high enough to shift the repetition images far away from RX bands which can mean up to 1 GHz. Additional co-existence restrictions may enforce even higher sample rates up to 2 GHz. Thus, timing is very critical in the decoding logic and the control of the DAC switches. However, scaled technologies and high switching speed help in this case.
- High-power (∼12 dBm): without dedicated PA driver the RFDAC needs to provide about 12 dBm peak power which translates into signal currents of up to 100 mA. Therefore power efficient structures are mandatory, e.g. class-B operation, where the DAC current consumption follows the output signal. Class-B operation is equivalent to a single-ended DAC (for current mode).
- High-resolution (∼100 dB/17 bit): since the digital TX chain does not provide additional filtering the DAC needs very high resolution and dynamic range. The sum of quantization and thermal noise contributed by the DAC should not exceed −166 dBc/Hz in a 4 MHz bandwidth (RX band) in worst case which translates into approximately 17bit effective resolution.

3 Current Mode RFDAC Concept

The basic concept addressing all the requirements mentioned above is shown in Fig. 2. The RFDAC consists of a single-ended current mode DAC stacked with a simple current commuting mixer. The current mode DAC converts the digital input word into a (baseband) current signal with a sample rate of about 1 GHz. It comprises basically just a current source fix connected to the bias voltage and a switch controlled by the digital amplitude information. To avoid spur generation the clock signal for the DAC is directly derived from the LO signal (e.g. LO/2). Since the baseband frequency is typically much lower a significant oversampling ratio is guaranteed. The bias voltage of the DAC is generated via a current mirror that is fed by another (slow) D/A converter. In this way the DAC gain can be controlled over wide range in digital way. The signal current is fed to a single-balanced mixer pair doing the up-conversion to LO rate. To isolate the thin oxide devices of mixer and DAC from the high voltage swing at the output an additional cascode stage is inserted (not shown). Finally a LC balun (trafo) converts the differential current signal into a single-ended voltage signal and filters out the high-order LO components.

In order to achieve the required resolution a segmented approach for the 14bit DAC is chosen [2]. Since the far-off noise/resolution requirements are determined by the differential non-linearity (DNL) of the DAC the thermometer coded segment comprises 10 MSBs resulting in 1024 unit cells. The 4 LSBs in the binary coded part add another 4 binary scaled cells. To avoid the digital complexity of a 10bit binary-to-thermometer decoder, the decoding scheme is split into two parts. The 1024 cells are arranged in an array of 32 $\times$ 32 cells controlled by 32 thermometer coded line and 32 thermometer coded column signals derived from 5 binary bits each. To select the active cells a local decoder in each unit cell is added.

3.1 Current Mode RFDAC Design Details

The sizing of the current source transistor is mainly determined by impedance, noise and matching requirements.

- Impedance: the finite output impedance of the current source yields a certain deterministic integral non-linearity (INL) according to [3]. This kind of deterministic, low-order INL shape only affects the in-band resolution or signal quality (system level spec = error vector magnitude, EVM) and the close to carrier harmonics (system level spec = adjacent channel leakage ratio, ACLR). Typical values for EVM ($\sim$3 %) and ACLR (45 dB) result in quite relaxed numbers for INL and current source impedance. Moreover the switch in series helps to get sufficient impedance.

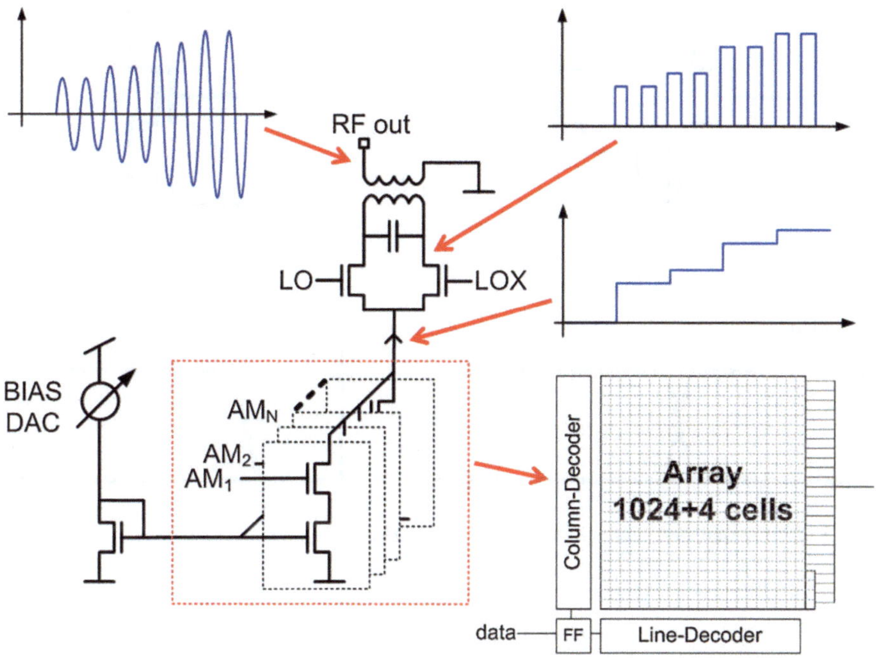

Fig. 2 Basic concept of stacked current mode RFDAC and mixer

- Noise: the thermal noise contribution of the current source is given by

$$\overline{I_n^2} = 4kT\gamma g_m. \tag{1}$$

The tolerable thermal noise level can be calculated from the total noise budget of −166 dBc/Hz for the DAC comprising thermal and quantization noise. To achieve low thermal noise a low g_m/I_d ratio is desirable which translates into high overdrive voltage and large channel length of the current source transistor. Flicker noise (which is typically pronounced in scaled CMOS technologies) is no concern here since the frequency offset of the RX bands is in most cases several tens of MHz and thus much higher than the 1/f noise corner frequency.

- Matching: the requirements for the current source mismatch σ(I)/I can be calculated out of the DNL specifications of the DAC. The DNL is dominating the high-order harmonics and the "far from carrier" quantization noise behavior and needs to fulfill the above mentioned 100 dB of SNR. The DNL of a N bit segmented DAC with B binary and N–B thermometer coded bits can be approximated with [4]

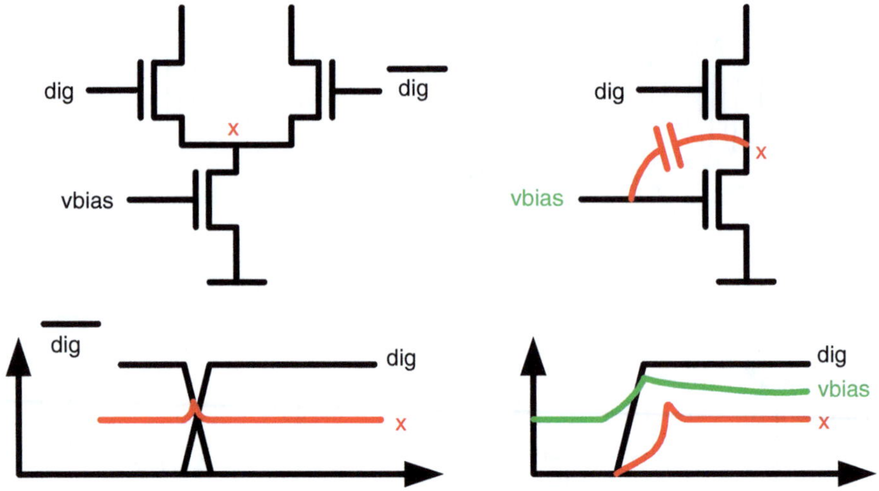

Fig. 3 Differential (*left*) and single-ended (*right*) DAC cells

$$DNL \approx \sqrt{2^{B+1} - 1}\frac{\sigma(I)}{I}LSB. \tag{2}$$

The current source mismatch σ(I)/I finally yields the minimum transistor dimensions according to [5]

$$(WL_{\min}) = \frac{1}{2}\left[A_\beta^2 + \frac{4A_{VT}^2}{(V_{GS} - V_T)^2}\right] \Big/ \left(\frac{\sigma_I}{I}\right)^2 \tag{3}$$

where A_{VT} and A_β are technology related mismatch parameters. Interestingly these parameters improve with technology scaling down to 45 nm and at least do not worsen even in nanometer scaled CMOS [6].

Besides the static design considerations discussed above the single-ended current mode DAC has some critical drawbacks compared to its differential counterpart as shown in Fig. 3. In contrast to the differential version where the current source is always on an the current is just steered between positive and negative branch the on/off switching in the single-ended version really means a complete charging/discharging of the current source drain node *x* since the bias of the current source is not disconnected. This charging/discharging affects especially the dynamic DAC performance:

- To charge the parasitic capacitances at node *x* first some current is needed which is missing in the actual signal current. This effect is quite non-linear and leads to harmonic distortion in the output current.

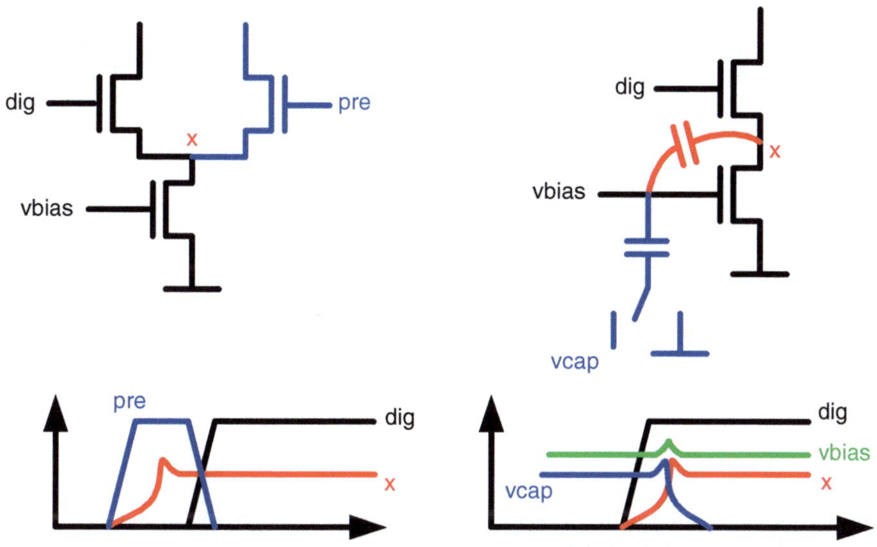

Fig. 4 Single ended DAC cell with pre-charge (*left*) and H2 compensation (*right*)

- Switching on/off the cells yields large voltage jumps at node x causing a significant feedback on the bias voltage via the gate-drain capacitor of the current source. This so-called H2 effect is also non-linear and degrades the dynamic performance of the DAC.

To achieve sufficient dynamic performance also in the single-ended DAC two additional compensation schemes are introduced here.

To compensate for the missing signal current during cell activation a pre-charge switch is added, see Fig. 4. The pre-charge switch is closed for half a clock-cycle before the cell is actually switched on. During this half clock cycle node x is charged to approximately the desired value, so no current is missing when it's really needed. Of course the current flowing through the pre-charge switch results in some current consumption overhead. However, as long as the oversampling ratio is high enough, there is no significant power penalty.

To compensate the H2 effect a dummy capacitor and two further switches are added. In off-state this capacitor is charged to V_{cap}. When the current cell is activated the capacitor is switched to ground at the same time and charge is subtracted from the bias node. This charge is intended to compensate for the unwanted charge coupled on V_{bias} via the gate-drain capacitor in order to keep the voltage flat and avoid the H2 effect as shown in Fig. 4.

The effectiveness of the proposed techniques is shown in Fig. 5.

The simulation compares the spectral performance of a single-ended DAC output current with and without pre-charge and H2 compensation for a 1 MHz sine-wave digital input. Especially the critical second harmonic is significantly improved. However, the additional switching activity increases the noise floor slightly.

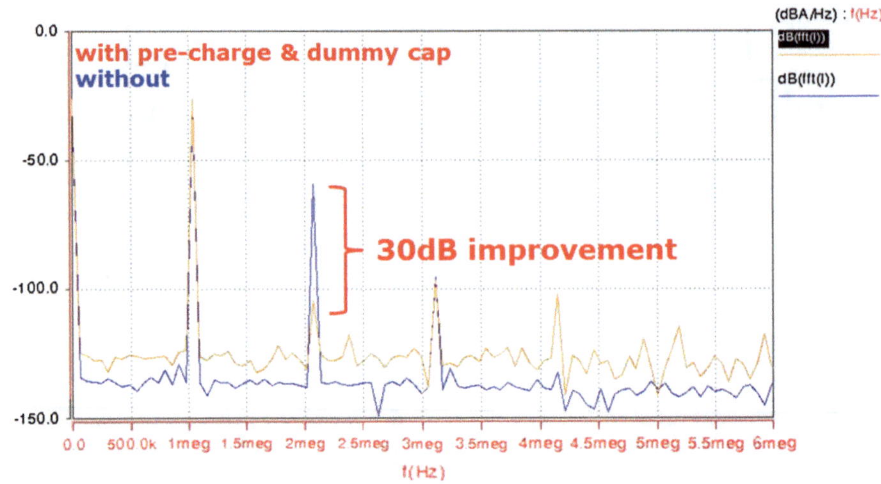

Fig. 5 Simulation result of single-ended DAC with and without pre-charge and H2 compensation for 1 MHz sine-wave input

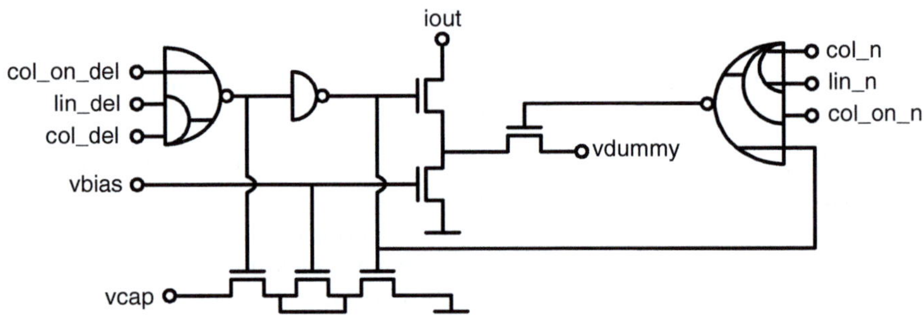

Fig. 6 Schematic of DAC unit cell including pre-charge and H2 compensation

The complete schematic of the current source including pre-charge and H2 compensation capacitor is shown in Fig. 6. The actual switch of the current source is controlled by the local decoder processing the line/column information and is driven by full-swing logic signals. These "cell on/off" logic signals also control the charging and discharging of the dummy capacitor for the H2 compensation that is realized as MOS-cap.

The pre-charge switch is connected to an additional bias voltage V_{dummy} (approximately set to the desired voltage for node x) and controlled by a second local decoder. This additional decoder is controlled by the inverted logic signals for columns and lines which are triggered half a clock cycle before the non-inverted logic signals to generate the pre-charge pulse.

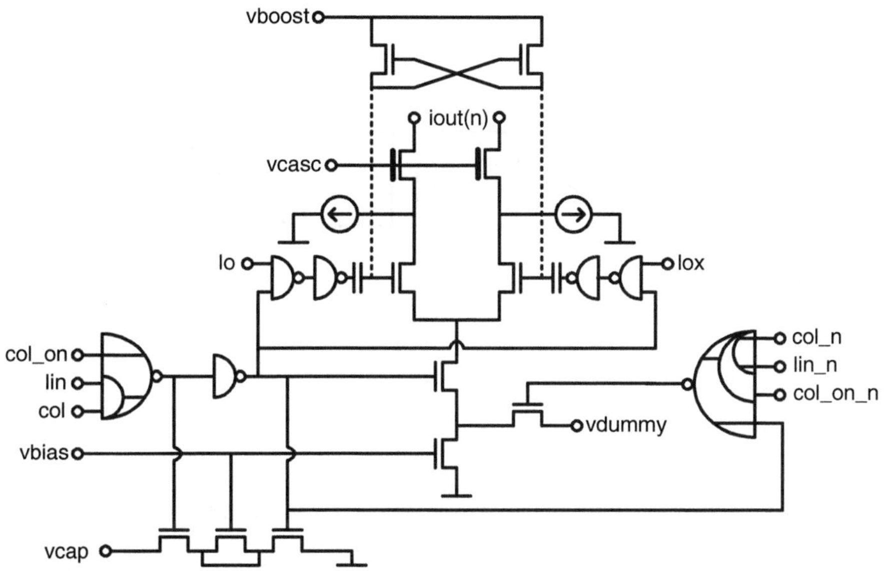

Fig. 7 Schematic of DAC unit cell including distributed mixer, LO boosting, cascode, pre-charge and H2 compensation

To better deal with the analog limitations in scaled technologies it has been decided to distribute the mixer and include it in the DAC unit cell. This has several advantages: there is no voltage swing at the mixer tail node (=DAC output) so the supply headroom limitations are relaxed. This is quite important since the current source is operated with high overdrive and needs to be kept in saturation. Furthermore the big wiring parasitics at the DAC output that degrade the mixer performance are minimized when the mixer is distributed and placed locally in the cell. Also the sizing of the mixer devices is relaxed since the current density no longer changing so drastically. On the other hand some additional functionality needs to be included in the cell. The LO is gated locally by the "cell-on" signal with a NAND gate to reduce the LO leakage to the output. A local LO signal boosting circuitry is introduced. Thus the mixer devices still can be driven with full swing signals and the bias voltage for the mixer tail node can be selected independently. The boosting consists of an AC coupling (MOS-cap) and a cross-coupled NMOS pair that is setting the DC point. The schematic of the complete DAC unit cell including cascode and mixer is shown in Fig. 7.

Another problem is related to the H2 compensation. The H2 compensation circuit only works only effectively if the charges cancel each other. The main issue here is to select an appropriate voltage for the dummy capacitor. Unfortunately a direct replica of the current source gate-drain capacitor cannot be used for area reasons. Instead a MOS-cap is used. Consequently the capacitor values will not match especially over process, voltage and temperature variations. In order to select the correct voltage for the dummy capacitor a calibration circuit is

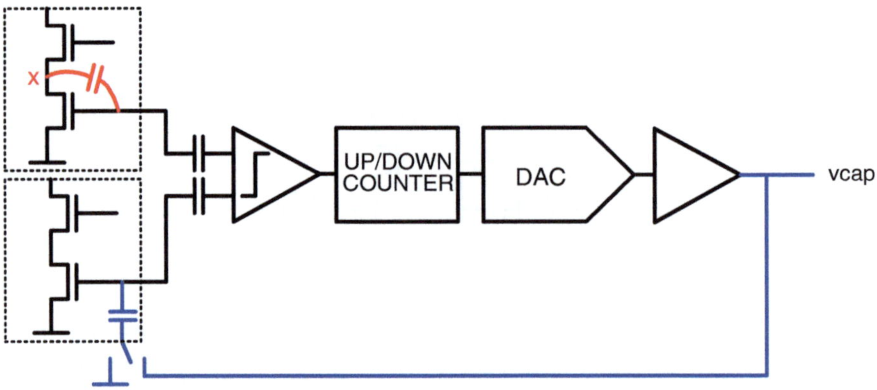

Fig. 8 Simplified schematic of H2 voltage calibration circuit

introduced as shown simplified in Fig. 8. In this concept two kinds of dummy DAC cells are used. One part is always coupling on a virtual bias node via the current source gate-drain capacitor whereas the other part is always coupling on another virtual bias node via the dummy capacitor. The amount of coupling on both nodes is now compared via an offset compensated comparator. The comparator result controls an up–down counter which again sets a simple very slow D/A converter generating the voltage for the dummy capacitor. The voltage is changed via the control loop until the coupling on both nodes is the same. As soon as the coupling is equal the correct H2 compensation voltage is found and used in the array. The calibration can be operated in the background from time to time since typical PVT variations occur quite slowly.

3.2 Measurement Results

The digital polar TX based on current-mode RFDAC has been implemented on a test IC in 65 and 28 nm digital CMOS. Supply voltages are 1.1 and 2.5 V for digital & DAC-array and balun center-tap, respectively. The active chip area for the DAC including balun, biasing and LDO is about 0.2 mm^2 in 28 nm.

Since the RFDAC is covering almost all functionality in the amplitude chain a stand-alone characterization does not make much sense. Instead the measured performance of the complete TX is briefly discussed here.

The measured spectra for 3G modes in 65 nm are shown in Fig. 9. The linearity and in-band signal quality is reflected in the ACLR. In both cases very good linearity is achieved due to DAC pre-charge and H2 compensation even without digital pre-distortion.

As mentioned above the far-off noise is considered as much more critical than linearity. One example from 65 nm test IC for 3G high-band operation at

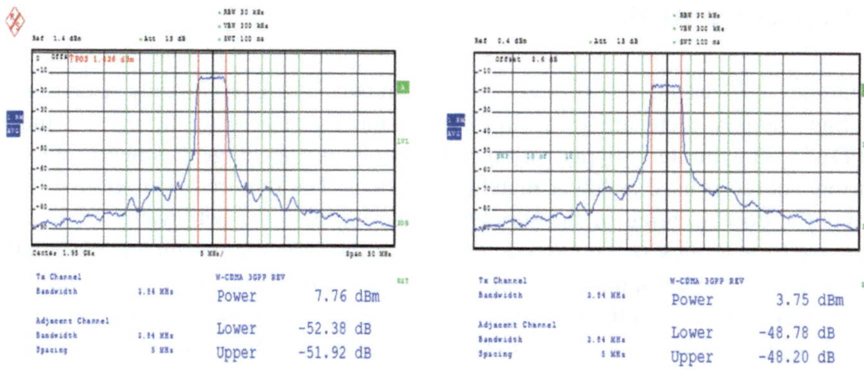

Fig. 9 Measured 3G spectra for WCDMA at 8 dBm (*left*) and HSUPA at 4 dBm (*right*)

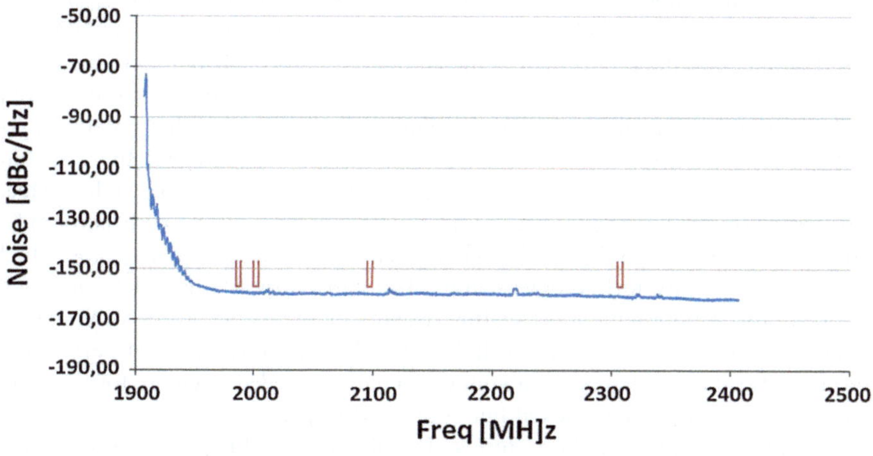

Fig. 10 Measured far-off noise in 3G with different duplex requirements

1907 MHz and 6 dBm output power is given in Fig. 10 where also the different RX frequency bands are indicated. The noise floor is quite flat and achieves the required values of −157 dBc/Hz (at chip level) with sufficient margin. It's worth mentioning that the actual TX signal is only partly visible in the measurement. The reason is the limited dynamic range of the measurement equipment. As mentioned above about 100 dB dynamic range is required. The dynamic range of commercially available spectrum analyzers does not reach these values. To circumvent this limitation a second chain with tuneable notch and LNA is used. The notch filters out the carrier signal and the LNA amplifies the remaining noise floor. Of course careful calibration is needed for this signal chain.

4 Capacitive RFDAC

Although the digital transmitter concept shown above is already technology scaling friendly there are some limitations, mostly related to the current mode RFDAC. The limited supply voltage headroom is the biggest concern since a current source is used as linear/matching element. The required saturation voltage limits the down-scaling of the supply voltage but also the achievable efficiency. The need for a thick oxide cascode device complicates the design and especially the layout of the complex DAC unit cell. Finally the actual signal bandwidth is limited by the bandwidth of the auxiliary circuitry for compensation, calibration and biasing. To show a possible path for further technology scaling a novel RFDAC concept is briefly introduced here. Based on the idea of switched-capacitor power amplifiers [7] a capacitive RFDAC has been developed and tested in 28 nm CMOS. The basic concept is shown in Fig. 11. The DAC comprises an array of matched capacitors that are driven by the phase modulated LO signal. The amplitude modulation is added by a digital mixer which is just a NAND gate. Based on the amplitude information the amount of cells contributing to the output signal is controlled, the off-cells form a capacitive divider. Similar to the current mode approach a matching network is required to filter out higher order components. The capacitive RFDAC concept has several advantages: the structure behaves like a class-D amplifier where high power efficiency is achievable. Instead of a current source a capacitor acts a linear/matching element. So no headroom for saturation is wasted and the resolution is given only by capacitor matching which tends to improve by technology scaling. The capacitors are driven by inverters which are by definition the most scaling friendly elements in CMOS technologies: supply voltage can be reduced well below 1 V and active area is very small. Another advantage is that no high voltage domain is needed anymore for the balun since the matching is not stacked but connected in parallel.

Despite of the clear benefits in terms of scaling the C-RFDAC obviously suffers from some imperfections. The main challenge is related to the supply of the inverters which is now acting also as DAC reference. The current consumption follows the amplitude, so any response of the supply voltage on signal current e.g. via IR drop or imperfect load regulation in the LDO creates non-linear effects. Also the thermal noise of the supply needs to be very low. The non-ideal behavior of the inverters furthermore degrades the performance of the DAC. The AM information (=n) modulates the effective load and the rise-fall time of the inverters, see Fig. 12. This effect creates some amplitude dependent phase shift (AM–PM) resulting in harmonic distortion. Interestingly the phase-shift follows the amplitude in an almost linear manner. So a simple digital pre-distortion can be used to compensate for the effect. The finite, non-linear MOS resistance is also modulating the actual current flowing into the load (AM–AM). Further non-ideal effects can be found in [7].

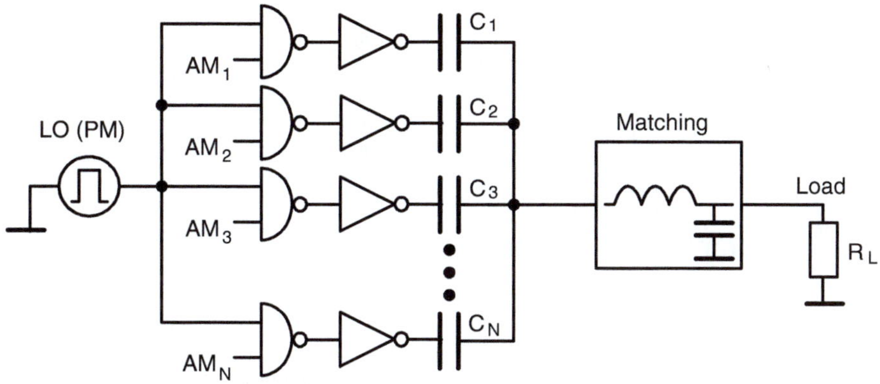

Fig. 11 Basic concept of capacitive RFDAC

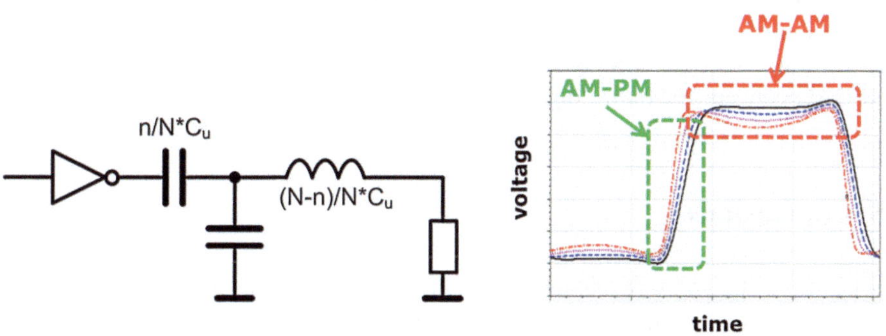

Fig. 12 Load modulation of inverters and resulting non-ideal switching wave-form

Finally the LO signal distribution and the timing alignment is very critical to achieve the tough far-off noise specifications. However this shift from voltage to time domain requirements enables to benefit from faster switching in scaled CMOS. The sizing of inverter and capacitors is not discussed in detail but can be derived in a similar way as shown in Sect. 3 from linearity, output power and resolution requirements.

Also the capacitive RFDAC has been implemented as part of a digital TX chain in 28 nm CMOS. To achieve similar resolution the DAC structure and segmentation is identical to the current mode approach, just the unit cell is replaced and the decoder adapted. A LUT based digital pre-distortion is included. The supply voltage is 1.1 V for digital and analog, no high-voltage domain is needed. The active chip area for the DAC including matching network and LDO is about 0.14 mm^2.

Some key measurement results are shown in Fig. 13.

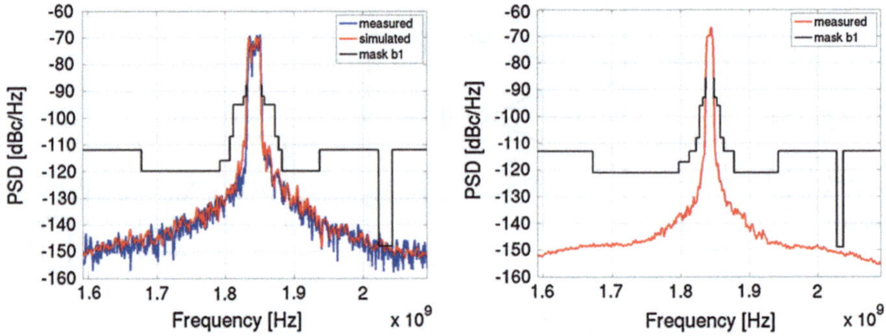

Fig. 13 Measured and simulated spectrum for LTE20 (*left*) and measured spectrum for LTE10 (*right*) at 0 dBm output power

The measured linearity and effective resolution are very promising since the tough out-of-band emission mask is fulfilled even for wideband standards like LTE20 and LTE10. The measured performance proves the feasibility of this RFDAC concept for future digital TX applications.

5 Conclusions

It has been shown by two concrete design examples that technology scaling is no road block for integration of analog/RF functionality. Appropriate system partitioning and scaling friendly architectures combined with innovative circuit techniques enable high resolution analog circuitry even in nanometer CMOS. Moving analog functionality in the digital domain and moving voltage resolution requirements into time resolution requirements further supports scaling and integration.

References

1. Z. Boos et al., "A fully digital multimode polar transmitter employing 17b RF DAC in 3G mode", IEEE International Solid-State Circuits Conference, 2011, pp. 376–378.
2. C.H. Lin and K. Bult, "A 10-b 500-MSample/s CMOS DAC in 0.6 mm^2", IEEE JSSC, Vol. 33, No. 12, December 1998, pp. 1948–1958.
3. B. Razavi, "Principles of Data Conversion System Design", IEEE Press, 1995.
4. A. van den Bosch, M. Borremans, M. Steyaert, and W. Sansen, "A 10-bit 1-GSample/s Nyquist current-steering CMOS D/A converter," IEEE JSSC, Vol. 36, No. 3, 2001, pp. 315–324.
5. J. Bastos, A. Marques, M. Steyaert, and W. Sansen, "A 12bit Intrinsic Accuracy High-Speed CMOS DAC", IEEE JSSC, Vol. 33, No. 12, December 1998, pp. 1959–1969.

6. S. Decoutere et al., "Technologies for (sub-) 45 nm Analog/RF CMOS - Circuit Design Opportunities and Challenges,", Proccedings of IEEE Custom Integrated Circuits, CICC, pp. 679–686, 2006.
7. S. M. Yoo et al., "A Switched-Capacitor Power Amplifier for EER/Polar Transmitters", IEEE International Solid-State Circuits Conference, 2011, pp. 428–430.

Design of a DC/DC Controller IP in 28 nm

Roberto Pelliconi, Tim O'Connor, Gavin Lacy, Noel O'Riordan and Vincent Callaghan

Abstract Power management IPs become more and more attractive in deep sub-micron technologies in order to optimize the power efficiency of today's complex chips demanding Amps of current. The development of a 28 nm analog buck DC/DC controller is presented here, showing the main design challenges and architectural choices. The main idea is to take advantage of technology scaling, so the controlling engine is purely digital, generating the discrete PWM driving signals to the off-chip power FETs. The analog circuits implement the voltage comparisons to set the output voltage level and the PWM signal buffering, which has to be realized in high voltage and with the correct timing. One of the main challenges of the analog design is presented by the requirement of a single input power supply with a wide range from 1.8 V − 5 % up to 3.3 V + 5 %, but just using 1.8 V compliant devices. Since the PWM modulator has to provide a full rail-to-rail output driving to the external FETs, an intense design phase and reliability analysis has been performed, accessing the foundry information and identifying the proper external low-gate-drive FETs. The analog layout implementation in 28 nm has to satisfy stringent rules with proper structure to guarantee that the device models match with the silicon performance, in particular for circuits where a precise signal sensing is critical. Measured results are available.

1 Introduction

The importance of on-chip Power Management Units (PMU) is increasing with the technology scaling, especially (but not exclusively) in battery equipped devices. The main reasons are:

R. Pelliconi (✉) · T. O'Connor · G. Lacy · N. O'Riordan · V. Callaghan
Silicon and Software Systems, Dublin, Ireland
e-mail: roberto.pelliconi@s3group.com
URL: www.s3group.com

P. Harpe et al. (eds.), *High-Performance AD and DA Converters, IC Design in Scaled Technologies, and Time-Domain Signal Processing*,
DOI: 10.1007/978-3-319-07938-7_12,

- Generation of multiple supply domains on a single chip with different specifications of noise, ripple rejection, load/line regulation, also depending on the load current profiles.
- Dynamic control of each supply value to optimize the performance of the analog/digital circuitry. For example, a supply scaling in a digital circuit can trade between speed and power consumption, depending on the application requirements.
- Application dependent customized PMU control. The more the Power Unit is integrated on-chip, the more complex algorithms can be implemented.
- Reduction of the number of external ICs and components to limit the size and the cost of the application board.

Following this industrial demand, a big effort has been put into implementing effective and efficient solutions.

The 2 main architectures proposed to achieve a supply regulation are:

- Linear Drop-Out Regulator (LDO).
- DC/DC Converter.

The choice of the best regulator architecture for each application depends on the desired performances in terms of Output Load Current profile (I_{LOAD}), Efficiency (η), Power Supply Rejection Ratio (PSRR), Ripple Requirements, Quiescent Current (I_Q, the current consumed by the regulator when 0 load current is required), Area, etc.

A quick description of the circuit basics and of the advantages/disadvantages of the 2 options is described.

1.1 Linear Drop-Out

The basic architecture of an LDO [1] with external compensation capacitance is shown in Fig. 1.

A PMOS (or NMOS) transistor, with one node connected to the input supply, modulates its conductivity (ON resistance) in order to guarantee that the output voltage stays constant and at the desired value. This is generally achieved through a feedback control loop which senses the output voltage and tunes the gate voltage of the output switch. More advanced control methodologies with feed-forward techniques can be implemented to improve the circuit stability and performance.

Among the several possible LDO architectures, 2 families can be mentioned:

1. *External compensation capacitor LDOs*: the stability is guaranteed by an external capacitor (value in the order of μF, which cannot be integrated on chip). The main advantages of this architecture are that it is quite simple and can achieve high performances in terms of PSRR, line/load regulation, noise, load step response and quiescent current.

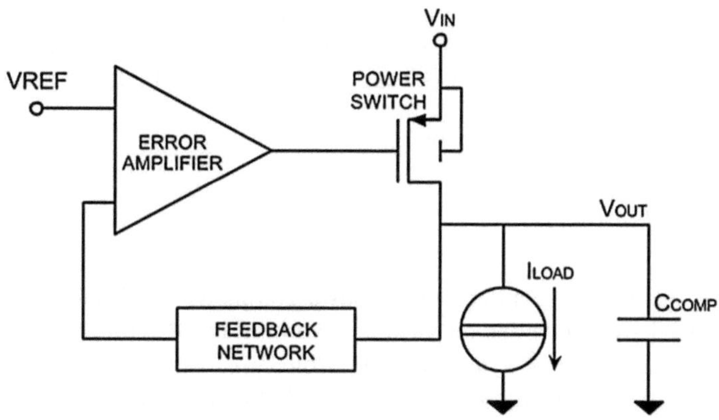

Fig. 1 LDO architecture

2. *Capless LDOs* [2]: these don't require an external big compensation capacitor, but the stability is achieved with a different technique which allows the integration of the main compensation capacitor on-chip (due to a much smaller area required for the capacitor itself, in the order of tens or hundreds of picoFarads). This is a big advantage when a reduced number of external components is a key specification; as a side effect, this architecture has some reduced performance in PSRR, noise, load step response dynamic, area and quiescent current, compared to the external compensation capacitor based LDO.

In general a fully integrated LDO has the big benefit of requiring a limited number of external components (1 external compensation capacitor if the architecture requires it), providing a ripple free regulated voltage, with good PSRR and noise performance, and a low quiescent current. The main disadvantage is that the efficiency is limited by the voltage drop across the output switch, according to the formulae:

$$P_{OUT} = V_{OUT} * I_{OUT}$$
$$P_{IN} = V_{IN} * I_{IN}$$

In the case where $I_{IN} - I_{OUT} << I_{IN}$ (large output currents), the efficiency can be approximated as follows:

$$\eta = \frac{P_{OUT}}{P_{IN}} \cong \frac{V_{OUT}}{V_{IN}}$$

It is clear that the voltage drop across the output switch is the strong limit of the efficiency. Also the dropout voltage (defined as $V_{IN} - V_{OUT}$) has a minimum value around 0.2 V in order to keep the output switch saturated with a reasonable size for a required output current. Quite often, in real industrial designs, the dropout voltage is set by the input I/O supply used as V_{IN} and the core supply used as V_{OUT}, so it is not unlikely to have an efficiency <50 % (for example in many 65 nm process options where the I/O supply is 2.5 V and the LDO generated core supply is 1.2 V).

1.2 DC/DC Converter

The DC/DC Converter architecture [3] has a completely different approach with respect to the LDOs. It relies on switches working in ON/OFF mode, controlled with PWM to generate a V_{OUT} different from V_{IN}. Depending on the way the switches (and the output filters) are connected, V_{OUT} can also be regulated at a value higher than V_{IN} (which isn't possible with the LDO architecture). 3 types of DC/DC can be identified:

1. *Buck Converters*: $V_{OUT} < V_{IN}$ (functionality comparable with LDOs)
2. *Boost Converters*: $V_{OUT} > V_{IN}$ (functionality not possible with LDOs)
3. *Buck/Boost Converters*: V_{OUT} can be set to a value either lower or higher than V_{IN}, depending on the circuit timing (functionality not possible with LDOs).

From now on, the analysis will focus only on Buck Converters. As previously mentioned, the DC/DC converters have a switching behaviour: there are time intervals where a charge (current) flows from the input to the output voltage, and time intervals with charge flowing from the output node to ground. Naming T_C and T_D as the charge and discharge intervals respectively, the regulated output voltage value can be expressed, to a first order approximation, as:

$$V_{OUT} = V_{IN} * \frac{T_C}{T_C + T_D}$$

where the Duty Cycle is defined as:

$$D = \frac{T_C}{T_C + T_D}$$

hence:

$$V_{OUT} = V_{IN} * D$$

As can be seen in Fig. 2, the DC/DC output is a filtered version of the switches output node (V_{SW}). The most efficient filtering is implemented with a LC network, which guarantees minimum losses (only the parasitic effects of the inductors and capacitors); depending on the current load specifications, the switching frequencies and the controller architecture, both L and C can have a wide range of values, however in "general purpose" applications the inductor can have a value around

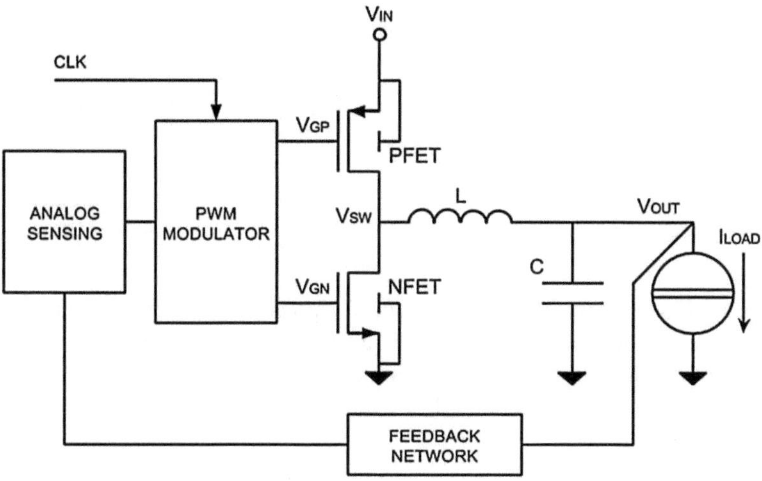

Fig. 2 Buck DC/DC architecture

1 μH and the capacitor can start from tens to hundreds of μF, so both of them are generally external components.

The output power switches can be either integrated on-chip, or placed off-chip, depending on the output current specification.

One of the main advantages of a Buck DC/DC Converter (as opposed to an LDO) is that the efficiency is a lot higher and it is not limited by the $V_{IN} - V_{OUT}$ difference, but it is just related to the ON resistance of the switches, the parasitics of the passive filter components (L and C) and on the power consumed by the analog sensing circuits and PWM controller. For large output current (>200 mA) the efficiency peak (the efficiency is current dependent) can be measured as high as 85/90 % (strongly affected by the power switch limitations), showing a big improvement with respect to the LDOs, that is rarely above 70 % as mentioned previously.

On the other hand, due to its intrinsic switching architecture, the DC/DC output shows some ripple, which can be unacceptable when the supply is used in high precision analog applications. Other disadvantages are that the quiescent current of the entire DC/DC is generally higher, the PSRR is lower and the step response slower than in LDOs, which means that the 2 different solutions can be used for different applications.

1.3 Complex PMU

Nowadays complicated Power Management Units (PMU) can embed both DC/DC Converters and LDOs, quite often connected with the following scheme of Fig. 3.

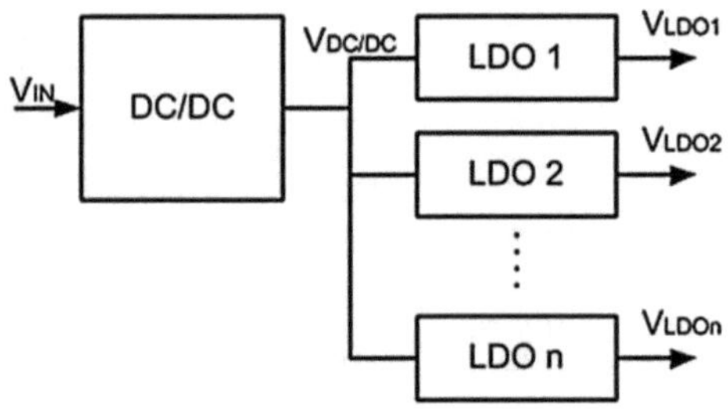

Fig. 3 PMU example

This approach allows the best trade-off for power efficiency and output voltage quality because the main voltage step is absorbed by the high efficiency DC/DC, while the fine voltage regulation with ripple attenuation is achieved with the LDOs without paying a large efficiency drop (the drop-out voltage is kept to the minimum acceptable value).

2 DC/DC Design

2.1 Main Requirement Specifications

The main specifications of the DC/DC solution presented in this chapter are the following:

- Process: 28 nm with 1.0 V core and 1.8 V I/O devices available.
- Power Supply: single, variable value from 1.8 V (−5 %) up to 3.3 V (+5 %) without any possibility to predict or preset the voltage level (so no specialized setup possible).
- Output Voltage Range: from 0.8 to 1.2 V.
- Output Current Range: 0–3 A.
- Output Voltage Accuracy: 3 % (across temperature, current steps of 2.8 A and including any bandgap drift).
- Other features: In-Rush Current Limitation, Over-Current Protection.

2.2 Power Switch Selection

Looking at the Output Current requirement (up to 3 A), it is clear that the switches can't be integrated on-chip: a reasonable limit for switches integration is around 1 A of max load current. The selection of the proper external FET components has been quite complicated due to the following requirements:

1. Current capability up to 3 A.
2. Low ON resistance for V_{GS} down to 1.7 V.
3. Low Gate Charge required for switching.

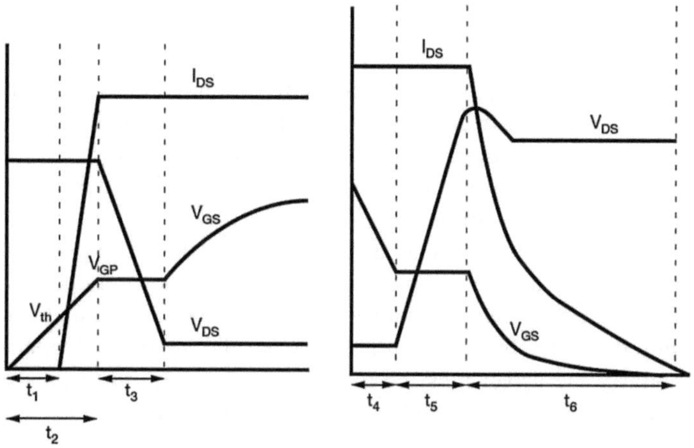

Fig. 4 External FET turn ON (*left*) and OFF (*right*) voltages/current

In order to reach a 3 A driving capability with low V_{GS}, FETs with a large form factor (W, just to compare with MOSFET parameters) are required, and this means large gate capacitance. However, the larger the gate capacitance is, the stronger the driving capability to the gate must be, to achieve an acceptable transition time OFF/ON and ON/OFF. The consequence of a large input capacitance is that a strong on-chip driver stage needs to be designed, and as can be noted, points 2 and 3 conflict (it is very difficult to have a high conductivity and a low gate capacitance at the same time).

In addition to that, it is important to consider that the external power FETs show a particular transient of V_{GS}, V_{DS} and I_{DS}, [4] that drifts quite a lot from the expected transient behaviour of an integrated MOSFET. This is due to the physical structure of the power FETs, which generally present a large Miller capacitance (C_{GD}) affecting the switching transient waveforms as shown in Fig. 4:

Looking at the OFF/ON transition (left of the picture), it can be noted that: V_{GS} starts increasing crossing the threshold voltage (t_1); after that, due to a further increase, the transistor begins to turn-ON, but its V_{DS} is still at the maximum value for quite some time ($t_2 - t_1$). At some point V_{DS} starts dropping (ideally going to 0 V, in reality to $R_{ON}*I_{DS}$), and during such a time interval the phenomena, known as "Miller Plateau", appears blocking the V_{GS} increase (t_3). The reason is that at the Miller Plateau all the charge provided by the driver is all conveyed to the Miller capacitance, to reduce the drain voltage. After time t_3, V_{GS} starts increasing again, eventually reaching the final value.

The ON/OFF transition has a similar behaviour, with the addition of an overshoot at V_{DS}, which effectively couples to V_{GS} (not shown in the picture), creating a kick-back stress to the on-chip driver.

Due to the physics of the external FETs, the Data Sheet description of their characteristics is quite different, compared to what is used to describe an integrated

MOSFET. The main differences (except the maximum absolute ratings to avoid device breakdown) are:

- $R_{DS(ON)}$ (@ different V_{GS} values): allows the estimation of the conductivity of the devices. Such a parameter affects both the DC/DC stability and efficiency. A high $R_{DS(ON)}$ means large power losses, so low efficiency.
- P_D (Maximum Power Dissipation): used to determine if the device is capable of providing the required current, ($P_D = R_{DS(ON)} * I_{DS}^2$).
- I_{DS} (Continuous Drain Current at a certain Junction Temperature): further information on the device driving capability for the desired application.
- Q_G (Total Gate Charge, @ a certain V_{GS}, V_{DS}): allows an estimation of the load effect caused by the FET. The higher Q_G is, the stronger (bigger and more power hungry) the driver must be to achieve the same transition speed. Since the amount of charge is dependent on the voltage applied ($Q = C*V$), those values are expressed defining the voltage levels.
- $t_{D(ON)}$, $t_{D(OFF)}$ (turn-ON/OFF delay times) and t_R, t_F (rise and fall times): describe the dynamic of the transitions. These numbers help to define the maximum clock operating frequency of the DC/DC using those external FETs.

Once the devices are identified, it is important to embed a model (often released by the device vendor) of them in the DC/DC simulations (more details in the driver section). Unfortunately it is not unusual that the devices models aren't properly compatible with the simulation tools, preventing the simulation analysis, so adding risk to the design.

2.3 Controller Overview

The control engine is a key part of the DC/DC block. Like in any feedback controlled systems, it receives information from the output (in this case the output voltage) and defines the next action to take in order to maintain the controlled parameter (output) at the desired value.

Compared to LDOs (whose intrinsic continuous time architecture recalls a classic analog continuous time control loop with poles and zeroes), the DC/DC perfectly fits in a discrete time control methodology, due to the output of the controller being a PWM digital signal. This is not excluding the continuous time methodology [5], and there are several examples of pure analog control techniques for DC/DC, but at some point a PWM signal has to be applied to the external switches.

The presented DC/DC follows a classical digital control technique, depicted in Fig. 5.

The output voltage is measured and converted to the digital domain, then fed to the control state machine, which selects the proper Pulse Width that will be applied to the external switches at next clock cycle (f_{SW}). This means that the regulation acts on the switch Duty Cycle in order to keep the output voltage at the desired

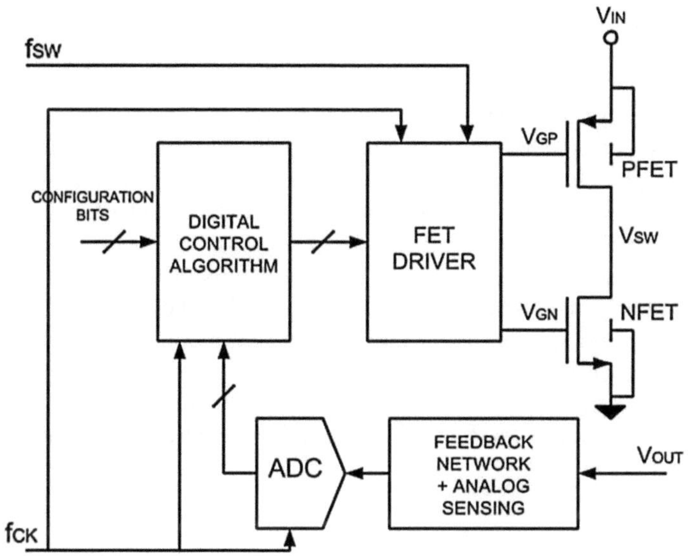

Fig. 5 Digital control loop scheme

value. The sampling (and conversion to a digital domain) of the output voltage happens at a frequency N times higher (usually $N = 2^n$, where "n" can be any integer number, normally between 3 and 5) than the switching frequency f_{SW} and named internal clock frequency (f_{CK}); the pulses applied to the external FETs have a discrete time resolution, with a single step size of $1/f_{CK}$. The definition of the ratio between f_{CK} and f_{SW} leaves some freedom, during the early design phases, to find the best trade-off between regulation and power performance. Selecting a high N (for example 64, or even 128) reduces the approximations in the Duty Cycle applied to the FETs, at the cost of a higher conversion performance required for the ADC (extra power consumption), which has to provide a new data at each T_{CK} ($T_{CK} = 1/f_{CK}$). To partially solve the resolution issue, the controller can calculate the average Duty Cycle of multiple switching periods.

The described functionality assumes that for each switching cycle T_{SW} both PFET and NFET are turned-ON, one at the time, and it doesn't permit that they're driven both OFF (so the possible modes are just PFET = ON, NFET = OFF or PFET = OFF, NFET = ON): this is true when the load current is sufficiently high to keep the value of the current flowing through the inductor always above 0 A. When this is not possible (cases where the load current is low), the controller defines some time interval where both PFET and NFET are switches OFF, the current through the inductor is 0 A and the output voltage is stabilized by the charge stored in the external capacitor.

The digital control state machine implements a well proven and robust algorithm with some parameter configurability in order to guarantee the stability in a wide range of applications. A behavioural model of the state machine, including the information of the external components (FETs, L, C) and of the current load profile is extensively used to finalize the controller setup for stability and regulation performance when different patterns are applied. This numerical approach is necessary because of the strong non-linear nature of the control loop causing the fact that, otherwise, the stability would require a complicated and exhaustive analytical approach; also, since the stability is affected by the f_{SW} setting and the corner values of all the external components, a large set of behavioural test-benches is simulated to cover all the possible cases.

One advantage of this approach is that the details of the analog circuits don't need to be considered at this level of analysis, leaving those design challenges to a separate study and splitting them into smaller and more easily manageable tasks (with resource partitioning, allowing the reduction of the design latency).

In summary, the choice of a digital control methodology brings some advantages and disadvantages (with respect to the analog controller approach), which are summarized below:

- *Advantages*
  - State machine scaling with technology node (both power and area). Getting more and more effective in deep submicron technologies.
  - Process independent, therefore easily portable across different foundries with improved time-to-market performances.
  - Robust and proven algorithm.
- *Disadvantages*
  - Non linear control loop that doesn't lend itself to a compact analytical description of the system.
  - Not optimized as a specific analog control loop can be (as an effect, the output ripple can be larger in the digital control implementation for a given set of external components like FETs, L and C).

Looking at the design trade-offs, the benefits of portability/process independency/robustness, excellent time-to-market capability, have been considered key aspects, so the digital approach has been chosen.

Another important parameter during the DC/DC controller design phase is the selection of the f_{SW} frequency. The general principle is that the higher f_{SW} is, the faster the controller loop is, improving the dynamic of the response to a load step and reducing the output voltage ripple. On the other hand, a faster f_{SW} requires faster FETs, faster internal circuitry (ADC) and costs in terms of overall power efficiency since at each switch clock period both external FETs are turned ON and OFF, charging/discharging their Gate terminals.

In this design the default frequency for f_{SW} is 1.56 MHz, which has been considered the best trade-off between regulation performance and efficiency. Some configuration options have been implemented to either increase or decrease such default a value.

2.4 Analog Design Overview

Looking back at the main requirements, the first challenge faced by the analog design can be immediately noted: the available I/O devices (used in the design) can support 1.8 V, while the only input supply available can vary randomly between 1.8 V (−5 %) and 3.3 V (+5 %), and there isn't any configuration setting which can help the DC/DC to predict the rough value of the external supply; in addition to that, the supply level can vary within the full specified range during the ON time of the DC/DC, so a dynamic control is somehow required.

The design approach to overcome such a constraint has been the partitioning of the DC/DC controller into different supply domains.

The main building blocks and their supplies are:

- *Digital Blocks* (controller and configuration): 1.0 V supply.
- *Analog/Voltage Sensing Blocks*: 1.8 V supply (which can go as low as 1.6 V when the input voltage is 1.71 V). This choice is useful in order to use the I/O devices for any analog operation, without worrying about cascoding each circuitry (more complicated design and verification).
- *Output Driver Block* (for the external FETs): staying at VIN (the main supply voltage level varying between 1.71–3.46 V). This is mandatory in order to properly drive the external FETs (which are connected to VIN).

From the list presented, it is clear that the DC/DC controller works with 3 different supply domains: the digital and analog domain voltages are generated with internal LDOs (capless), properly designed with cascode structures, and able to work with the required V_{IN} variation. The current load of the 2 LDOs can be well estimated with individual blocks simulations during the design phase, and the verification, to prevent voltage stress in the LDOs, is limited to this subsection of the full DC/DC controller. Also, a specialized analysis has been performed during the power-up/enable control, as the LDOs require an external reference voltage to safely startup: for this reason the real DC/DC controller startup is subject to the correct sensing of the following 3 events:

- Sufficient value of the external supply (>1.71 V).
- Valid reference voltage.
- Valid bias current.

Another challenging requirement is the Output Voltage Accuracy within 3 %, including all the possible real mismatch effects, including bandgap uncertainty, any PVT drift of the sensing/comparing circuits, and the worst case current load profiles (instantaneous current steps with positive/negative variations up to 2.8 A in some configurations).

To satisfy such a requirement, all the analog mismatch/process variation effects are corrected with an intense use of trimming and calibration, even if both of them are not tracking the temperature variation (to correct the temperature drifting). Also, a particular care has been taken during the design and the layout implementation to maximize the ADC performance (both in speed/power and precision); the ADC has been implemented with a Flash architecture, to minimize the data latency, which could affect the overall stability loop.

The programmable feedback reference network (to program the target output voltage to selected values between 0.8 and 1.2 V) has been designed with a current steering architecture in order to minimize the temperature dependency, trading the good performance with some area and power consumption penalty.

As described in the controller section, the load dependent transient response has been numerically analysed with extensive simulations of the DC/DC behavioural model, converging to a set of controller parameters and external component values (for inductor and capacitor) and after the selection of the external FET switches. In the most demanding DC/DC operating mode, the values are:

- L = 1.2 uH
- C = 300 μF
  with the external FETs having the following ON resistance:
- $R_{DS(ON)\_PFET} = 0.09\ \Omega$@ Vgs = 1.8 V (typ)
- $R_{DS(ON)\_NFET} = 0.065\ \Omega$@ Vgs = 1.8 V (typ).

2.5 FET Driver

The design of the drivers for the external FET has required a large design effort, due to the challenging specifications. Intuitively, this sub-block should be really simple, because the only feature that it has to provide is to buffer a pair (both NFET and PFET) of digital signals provided by the digital controller making them strong enough to properly drive the external switches; however, in this particular design, this simple task had to deal with the following requirements:

- Pre-elaborate the digital inputs to correctly drive the FETs.
- Translate the 1.0 V modified digital signals to rail-to-rail 1.71 up to 3.46 V ones. This has to be achieved using 1.8 V compliant devices only.
- Be able to drive several nanoCoulombs of charge with fast transition times, to off-chip.
- Keep control of the consumed power to avoid efficiency drop.

The listed points, and the design solutions, will be described in the following sub-sections.

2.6 Digital Pre-elaboration

The digital controller provides the PWM pulse shapes to be applied to the external FETs, depending on the algorithm and the sensed output. As described in the controller section, for each switching period (T_{SW}), both PFET and NFET are turned-ON for a certain time interval and interleaved, in order to keep the average value of the current in the inductor equal to the output load current. Having both FETs ON at the same time is not allowed, as it would create a low resistive path from the input supply V_{IN} to V_{SS}, enabling a massive current flowing which can destroy both FETs in a quite short time interval (currents a lot higher than the devices maximum ratings). This burn-out effect does not happen if the overlapping ON time of the FETs is extremely short; however in this case the peak of current flowing from supply to ground is wasted and not propagated to the output node, causing a dramatic loss of efficiency. This phenomena is known as "shoot-through current" and, as previously described, must be avoided for both reliability and efficiency reasons. Figure 6 shows the PFET/NFET driving signals and currents in a correct configuration and also in a shoot-through condition.

Due to the intrinsic nature of the external FETs, the amount of charge that needs to be added or removed to turn them ON/OFF, their slow response (transition time in the order of tens, up to hundreds of nanoseconds, which is order of magnitudes slower than an integrated logic gate), and the overlapped control signals provided by the digital state machine, (forcing a simultaneous switching of both FETs) can enable at least one shoot-through event for each switching period.

To avoid such a situation, a modulation of the pgate/ngate pulses with a de-overlapping logic before applying them to the FETs is implemented.

The de-overlapping circuitry can follow two different principles:

- Fix delay: this defines the delays based on an accurate simulation analysis. It is the simplest approach, but with a non-optimum performance, since the de-overlapping operation isn't strictly dependent on the real switching of the FETs. The addition of some configuration bits adds some flexibility, thereby permitting a finer optimization during the silicon/board evaluation and allows the use of different external parts (with different transition times).
- Feedback/Sensing loop controlled: this method measures some voltage/current on the FETs (to detect the real turn-OFF event) and defines the exact time for the switching delay, at the price of higher circuit complexity.

The actual implementation in this DC/DC converter is a mix of the two methodologies, as shown if Fig. 7. For both FETs, during the turn-OFF event, a simple circuit (implemented with an inverter) measures the value of the PGATE/NGATE voltage; when a real transition is detected (so not just the control bit provided by the digital controller), an extra (programmable) delay is added, before providing the effective turn-ON command to the other FET. Such a solution doesn't sense the real turn-OFF of the switches, but indicates that the turn-OFF is happening, even if not completed yet; the advantage of sensing PGATE/NGATE is that any delay occurring in the driver circuit is no longer estimated, but really measured and the addition of an extra (programmable, but fixed) delay finalizes the protection.

It must be noted that the amount of de-overlapping has to be sufficient, but potentially not too large: the reason is explained in Fig. 8.

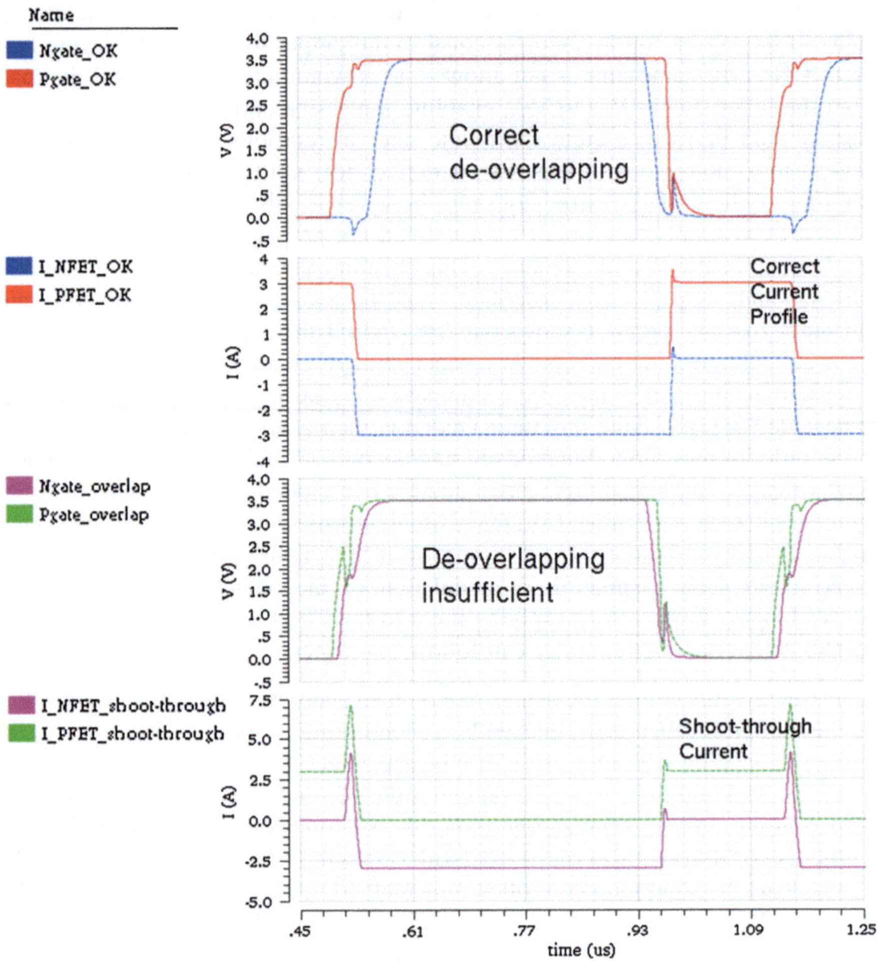

Fig. 6 Correct de-overlapping vesus shoot-through current

When a non-zero current flows through an inductor, such a current can't suddenly drop to zero because of the inertial behaviour of the coil:

$$v_L = L * \frac{di_L}{dt}$$

so when both FETs are OFF, there must be some current source to feed the inductor. The only source is the ground node, through the NFET inversion diode. During this time interval V_{SW} goes negative, and equal to $-V_{THNDIODE}$ (about 0.8/1.0 V, depending on the FET and the current flowing); under such a condition the power consumed is $V_{THNDIODE} * I_L$, which is a lot higher than the normal power dissipation across a FET turned ON, penalizing the power efficiency.

Figure 11, represents the two logic cases (input ‘0’ and ‘1’) also considering the practical implementation where the value of V_{LOGIC} has been chosen to be equal to $V_{IN\text{-}MAX}/2 = 1.75$ V and $V_{IN} = V_{IN\text{-}MAX}$.

Such a fully complementary structure guarantees reliability even during the logic transitions, if both P-side and N-side steps are applied simultaneously and the load capacitors (at nodes V_{OUT}, V_{ON} and V_{OP}) aren’t order of magnitudes larger than the ones belonging to the MOS transistors; in this particular design, the final driver stage has to charge/discharge the large FETs gate capacitance, also going through the chip pads and bondwires (inductors), so an intense simulation analysis has been performed to quantify the amount of voltage stress on the on-chip transistors, as will be shown later. The selection of $V_{LOGIC} = 1.75$ V guarantees that it stays quite constant and close to the target value while $V_{IN} > 2.0$ V; below such a voltage level, the real circuits generating the P/N logic driving and the cascode voltages reach saturation, reducing the effective V_{LOGIC} value down to 1.55 V when $V_{IN} = 1.71$ V. This partially affects the driving capability, but the transition time (which is generally defined as the time needed to go from the applied input variation to 90 % of the full logic swing of the output) is not really degrading, as the voltage step is smaller and so the charge to be provided to the external FETs.

Figure 12 shows the transition of both PFET and NFET drivers, in the real case with PFET/NFET models, maximum output current (3 A), and bond wires inductors representing the chip package effects. The supply value is again rounded to 3.5 V (above the specified value), with T = 105 C and “ss” models (worst case scenario).

As can be noted, during these transient events there are time intervals where the voltage across some device terminals exceeds the allowed limits, which translates into potential reliability issues, mainly related to:

- Oxide breakdown: too large voltage applied across the transistor gate, which physically destroys the oxide layer.
- Hot-carrier Injection: leakage current, due to hot electrons with a sufficient energy to tunnel through the oxide gate or substrate. This phenomena can change the physical characteristics of the transistors, generating a so called “hot carrier degradation” of the device performances.

In order to estimate the effect of the overshoots, the silicon foundry provides a web based tool containing a calculator showing which level of extra voltages can be tolerated for a limited amount of the lifetime (expressed with a Duty Cycle number); such an extra tolerance allowance is a function of the device type, its channel length and gate area.

The results in this particular case are:

- nmos_1v8 : 2.53 V d.c. (Oxide breakdown, 100 degC), (Hot carrier-injection): 2.49 V transient 0.5 % duty cycle (0 degC)
- pmos_1v8 : 2.97 V d.c. (Oxide breakdown, 100 degC), (Hot carrier-injection): 3.18 V transient 0.5 % duty cycle (0 degC)

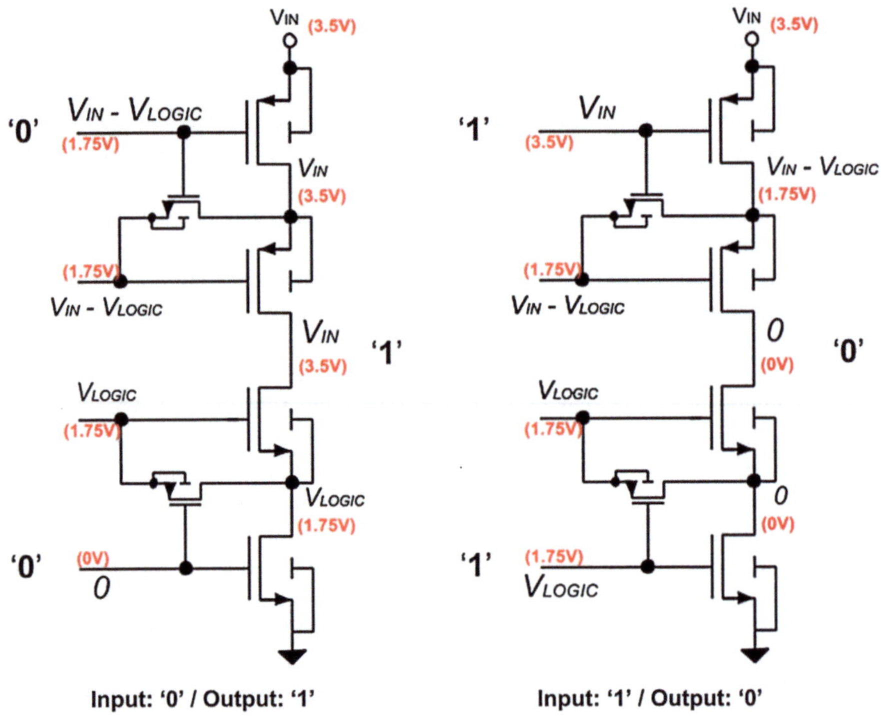

Fig. 11 Output driver stage, logic example

Comparing those numbers with the waveforms of Fig. 12, it can be noted that the circuit will not suffer failures for the expected lifetime, proving the robustness of the design.

3.2 Core Level Shifting

The Core Level Shifting block receives the digital inputs with the driving waveforms from the digital controller and generates the proper signals to drive the Final Driver Stage. In other words,

- Input: 1.0 V domain digital signal
- Outputs: 2 digital signals
  - N-Driving: range in 0/V_{LOGIC}
  - P-Driving: range (V_{IN}–V_{LOGIC})/V_{IN}

One of the main requirements is that both N/P-Driving signals have to work concurrently to properly drive the Final Driver Stage and avoid devices stress, and this is not straightforward, because while the N-Driving is a simple level shift of

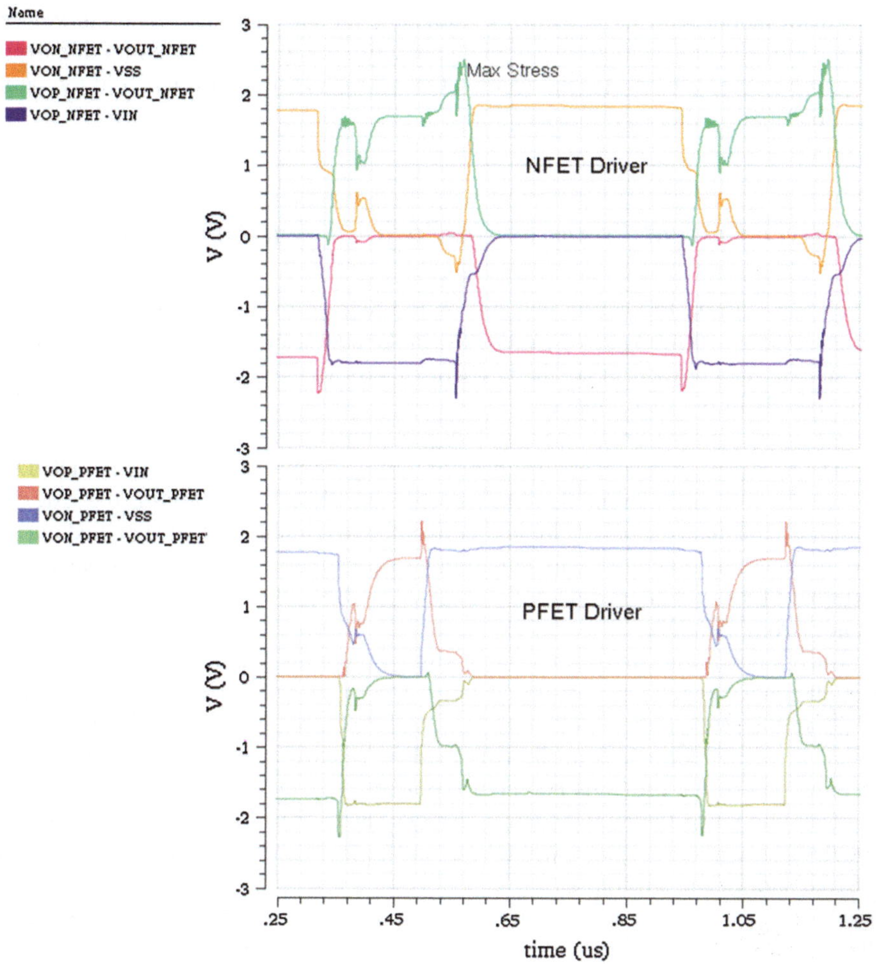

Fig. 12 Output driver stage, voltage stress waveforms

the supply (from 1.0 V to V_{LOGIC}), the P-Driving has to shift the "ground" to ($V_{IN} - V_{LOGIC}$) and the 1.0 V to V_{IN}.

Guaranteeing such a functionality across PVT corners has required an intense design effort and validation.

One important detail to be added is the Power-Down Mode management. As said before, during Power-Down both output FETs have to be kept OFF (so applying V_{IN} to the PFET gate and 0 V to the NFET gate). To achieve that a proper driving has to be generated and applied to the Final Driver Stage, even considering that most of the analog circuits are generally OFF (to minimize the power consumption). To solve this issue, it has been assumed that during Power-Down $V_{LOGIC} = (V_{IN} - V_{LOGIC}) = V_{IN}/2$, and that can easily implemented with a resistor ladder with large resistors (to reduce the DC current); since the Power-Down mode is only static (no switching activity), the maximum amount of current required to keep all the biasing effectively working is 5 μA (also to avoid too large area occupation due to the resistors).

3.3 Over-Current Protection/In-Rush Current Limitation

The Over-Current Protection is based on the sensing of the voltage drop across a small resistor (15 mΩ) placed in series with the output FETs. The resistor is placed as an external component on the board with very low tolerance, in order to avoid big spreads in the functionality.

As a result of the OCD sensing, a digital output flags the event.

The In-Rush Current event can happen during the initial start-up of the DC/DC, where the average current flowing through the FETs and charging the external filtering capacitance can be a lot higher than the maximum current specified during the normal functionality. This situation can potentially cause a burn-out of the external FETs, and an unexpected voltage drop of V_{IN} if the external power source is not strong enough.

Such a condition can be controlled with a proper algorithm implemented in the digital control engine; at start-up the target Duty Cycle (defined by the ratio V_{OUT}/V_{IN}) isn't applied immediately to the external FETs, but it is achieved with gradual small increases starting from 0. This method ensures a lower current peak, preventing any of the previous potential problems.

The amount of tolerated In-Rush Current value affects the start-up time of the DC/DC, for a given selection of FETs, L and C: the higher the In-Rush Current can be, the shorter the start-up is.

A proper control of the start-up routine also helps to prevent any overshoot in the output voltage regulation at the end of the start-up interval, avoiding reliability problems in the load circuitry.

4 Toplevel Integration and Simulation

The full DC/DC Controller layout snapshot is shown in Fig. 13.

The small box the top left of the figure shows the digital controller: as can be noted, its area is almost negligible with respect to the full circuit, also thanks to the 28 nm process used for the design. The FET Driver section is placed at the right hand side of the layout.

The rectangular form factor of the layout has been driven by customer integration requirements.

Due to the intrinsic analog/digital multi-disciplinary nature of the design (with several internal supply domains) and the simulation time intervals which have to

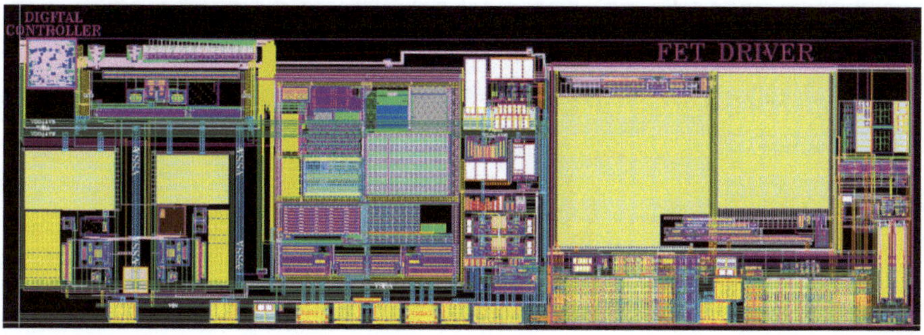

Fig. 13 DC/DC controller toplevel layout

be analyzed to achieve a reasonable design confidence (several milliseconds), the classic analog tools are not capable to provide results in a reasonable amount of time and computing resources.

In order to overcome this limitation, the VerilogAMS methodology has been implemented. The analog blocks have been modeled with the intention of replicating their functionality, but without deeply reproducing their exact dynamics (non-linearity, etc.). Careful attention has been put in the modeling of the supply and biasing dependencies in order to validate the correct startup of the circuitries.

The digital state machine has been simulated using its RTL code, while the output FETs have been described with the models provided by the manufacturers.

Such a level of abstraction is definitely more granular than the initial behavioral simulation of the controller (which uses some limited information of the internal analog circuits and external components), but less precise than a full transistor level analysis (still not possible for a large analysis).

Figure 14 shows the example of the simulation output: a variable current (with steps) is applied to the DC/DC and the plot of the output voltage shows the circuit response, with visual information of the negative/positive voltage dips, and output voltage variations (to double check that they satisfy the requirement specifications).

5 Measurement Results

A test-chip, including a bandgap, a Power-On Reset and the described DC/DC has been fabricated in a 28 nm process. A board with the chip and the selected external components (Power FETs, inductor and capacitor) has been developed and extensively tested across different supply, temperature, and with thermal accelerated reliability test (at maximum supply). Figures 13 and 14 show the measured response of the DC/DC output to an external current step, with a current reduction (Fig. 15) and increase (Fig. 16).

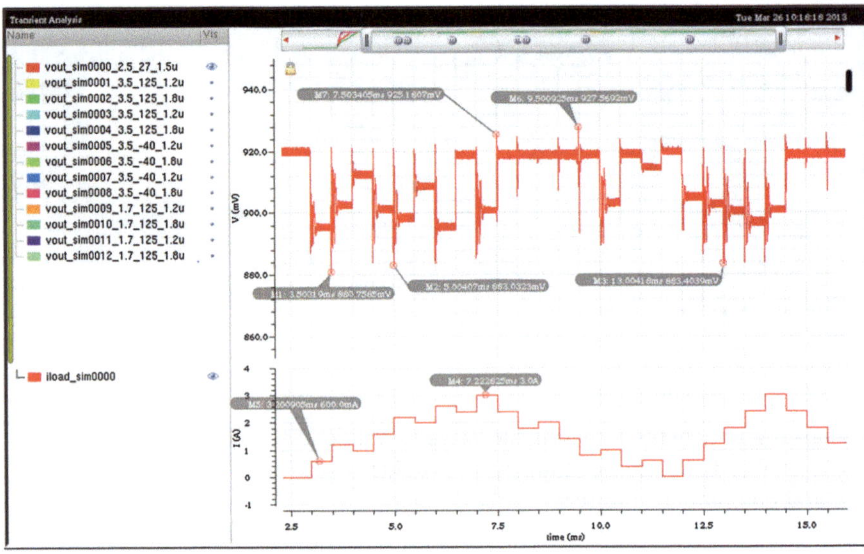

Fig. 14 Toplevel AMS simulation with variable load current

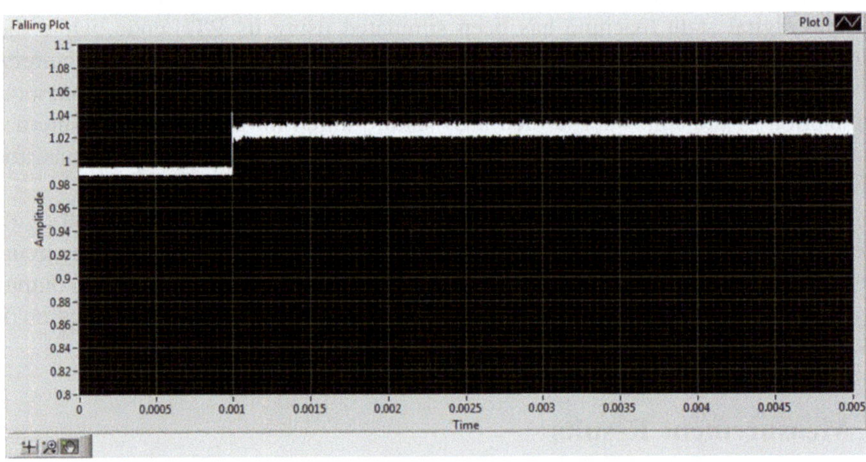

Fig. 15 Measurement: plot of V_{OUT} response to a negative current step

As expected, the first effect of any current step is an instantaneous variation of the output voltage, limited by the external compensation capacitor: after the initial step, the controller starts taking actions to restore the target voltage level, even if it can be noted that it might require some tens (up to hundreds) of f_{SW} cycles before reaching a steady regulation. The transient settling shown in Fig. 16 is a normal behavior of the non-linear digital algorithm; the length of the settling interval can

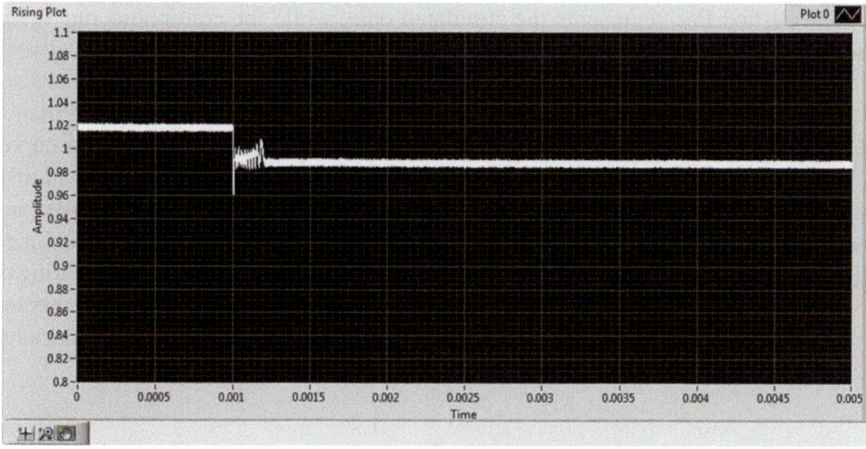

Fig. 16 Measurement: plot of V_{OUT} response to a positive current step

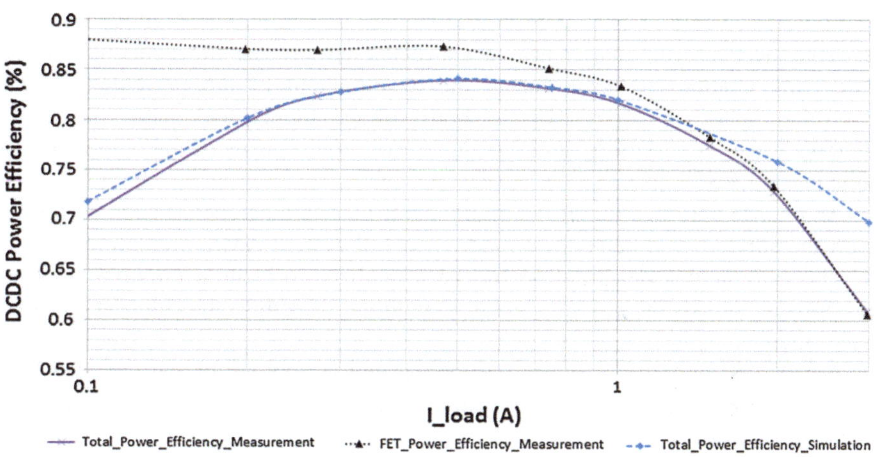

Fig. 17 Efficiency: simulation versus measurements

be reduced with the proper choice of the controller configuration parameters. Due to the controller architecture, the final regulated value is not always the same (even for the same load currents), because the digital state machine implements a sequential algorithm (the output depends on the load current and on the previous state). Nevertheless the final voltage level is always kept within a voltage window defined by the specification requirements.

One of the main figures of merit of a DC/DC converter is the power efficiency. The analysis and the performance achieved with such a design is summarized in Fig. 17.

The dashed line represents the simulated data, while the continuous one is the measured performance. As can be noted, there is a very nice matching between simulation and measurements up to 1.2 A (efficiency >80 % in the interval between 0.2–1.2 A). When the current load increases above 1.5 A the measured performance starts degrading quite quickly. This can be explained looking at the dotted curve, which represents the power efficiency of the power switches only; it is evident that, for large currents, the limiting factor (to efficiency) are the switches, which dissipate too much power with respect to the one estimated during the simulation and claimed in the components data sheet. A possible explanation could be the thermal heating of the FETs, which causes an increase of the ON resistance, increasing even further the heat (power) dissipation. This limitation can be overcome with a better package selection for the FETs, at a price of extra cost in US$.

6 Conclusions

The aggressive time-to-market requirements of the IC industry with the need of power efficient, small area and low cost solutions enables the demand of integrated and power scalable DC/DC controllers. The use of a digital controlling algorithm to drive the PWM modulator provides a robust and process-independent solution. The main challenges faced during the development of the presented DC/DC controller have been driven by the high voltage supply with large and unpredictable range of variability and the aggressive requirements of the output voltage tolerance (3 %).

An intense optimization of the digital controller parameters has been performed with the support of a behavioral model and full system AMS simulations.

The main analog design effort has been finalized to identify circuital solutions to prevent transistor voltage stresses without affecting the highest possible performance.

A first time right test-chip has been fabricated in a 28 nm process and has been fully characterized; the DC/DC controller is now available for integration into the customer's products.

Acknowledgements The authors would like to thank the whole team of Mixed-Signal Designers of Silicon & Software Systems, who worked hard and with high commitment to successfully design, deliver to foundry and measure the DC/DC Controller, adding another successful story to the Company track record.

References

1. Michael Day, Texas Instruments Understanding Low Drop Out (LDO) Regulators.
2. Robert J. Milliken, Jose Silva-Martínez and Edgar Sánchez-Sinencio "Full On-Chip CMOS Low-Dropout Voltage Regulator", IEEE TRANSACTIONS ON CIRCUITS AND SYSTEMS—I: REGULAR PAPERS, VOL. 54, NO. 9, SEPTEMBER 2007, pp. 1879–1890.

3. George Patounakis, Yee William Li and Kenneth L. Shepard "A Fully Integrated On-Chip DC–DC Conversion and Power Management System", IEEE JOURNAL OF SOLID-STATE CIRCUITS, VOL. 39, NO. 3, MARCH 2004, pp. 443–451.
4. Power MOSFET Basics: Understanding Gate Charge and Using it to Assess Switching Performance, Vishay Siliconix.
5. Sujata Verma, S.K Singh and A.G. Rao, "Overview of control Techniques for DC-DC converters", Research Journal of Engineering Sciences, Vol. 2(8), 18–21, August (2013).

Part III
Time-Domain Signal Processing

Kofi Makinwa

As the supply voltage of scaled CMOS processes continues to decrease, the use of time-domain techniques to process analog signals has attracted increasing interest. Apart from their obvious application to phase-locked-loops (PLLs) and time-to-digital converters (TDCs), time-domain techniques have also applied to amplitude-domain analog-to-digital converters (ADCs) and even to linear amplifiers. This section of the book presents six different designs, each of which represents various applications of time-domain signal processing in analog systems.

High Speed Time-Domain Imaging by Shingo Mandai and Edoardo Charbon (TU Delft) describes an array of CMOS digital silicon photomultipliers, intended for use in positron emission tomography (PET). PET is a medical imaging technique in which optical sensors detect pairs of gamma rays emitted indirectly by a positron-emitting tracer, by measuring the arrival times of individual photons. The 4 x 4 array of sensors and their associated TDCs are capable of capturing and digitizing up to 32 million individual photon times-of-arrival per second, with an overall timing resolution of 179 ps.

Fine-Time Resolution Measurements for High Energy Physics Experiments by Lukas Perktold and Jørgen Christiansen (CERN) describes a multi-channel TDC intended for digitizing the arrival times of subatomic particles in high-energy physics experiments. The TDC achieves fine timing resolution by combining a delay-locked-loop (DLL) with a resistive interpolation scheme. The size of its least-significant-bit (LSB) is controlled by a reference clock and so can be continuously adjusted from 5 to 20 ps. When realized in 130 nm CMOS, a prototype TDC achieves a single-shot resolution of less than 2.5 ps (rms).

Time-Domain Techniques for mm-Wave Frequency Generation by Wanghua Wu and Bogdan Staszewski (TU Delft) focuses on the design of a mm-wave frequency synthesizer with a digitally-intensive architecture based on time-domain circuit and calibration techniques. A prototype of a 60-GHz all-digital phase--locked loop (ADPLL) transmitter was implemented in 65-nm CMOS, and achieves excellent phase noise (-75 dBc/Hz at 10 kHz offset), fast locking (3 ls), low reference spurs (-74 dBc), and linear frequency modulation over a 1 GHz range.

In A Deterministic Background Calibration Technique for Voltage Controlled Oscillator Based ADC Sachin Rao (Oregon State University) and Pavan Hanumolu (University of Illinois) describe recent architectures aimed at improving the performance of voltage-controlled-oscillator (VCO)-based ADCs. Such ADCs employ many traditionally digital building blocks and so are amenable to CMOS scaling. Despite this advantage, however, they have only found limited application due to their poor linearity. The authors present a deterministic background calibration method which improves the linearity of VCO-based ADCs from 46 dB to more than 73 dB in a 5 MHz bandwidth.

In Towards Energy-Efficient CMOS Integrated Sensor-to-Digital Interface Circuits Jelle van Rethy et al. (KU Leuven) discuss the issue of whether time--domain sensor interface circuitry is fundamentally more energy-efficient than traditional voltage-domain interface circuitry. To answer this question, theoretical limits are derived for both types of circuitry, which show that the achievable energy efficiency is mainly determined by the sensor itself. Despite this, the author's observe that time-domain interfaces can be scalable, area-efficient and capable of operating from low supply voltages. These observations are illustrated with several practical designs.

In The Ring Amplifier: Scalable Amplification with Ring Oscillators of this section, Benjamin Hershberg and Un-Ku Moon (Oregon State University) introduce the concept of ring amplification, a technique for performing efficient amplification in nanoscale CMOS processes. By using a cascade of dynamically stabilized inverter stages to perform accurate amplification, ring amplifiers are able to leverage the key benefits of technology scaling, resulting in excellent efficiency and performance. After an explanation of the theory behind ring amplifiers, a survey of existing ring amplifier implementations is presented.

High Speed Time-Domain Imaging

Shingo Mandai and Edoardo Charbon

Abstract The chapter focuses on positron emission tomography (PET), a medical imaging technique that combines high speed with high complexity data processing. In PET, optical sensors detect gamma events generated by the annihilation of an electron and a positron, by measuring hundreds of individual photon times-of-arrival. These measurements are carefully analyzed, every 6.4 μs, by ultra-fast networks and distributed reconstruction algorithms, all running in parallel at several gigabit-per-second. In this context, we present an array of 4 × 4 digital silicon photomultipliers (MD-SiPMs) integrated in standard CMOS, capable of capturing and digitizing up to 32 million individual photon times-of-arrival per second, for up to 0.6 million gamma events per second. For each gamma event the equivalent energy is also computed to help the reconstruction algorithms screen out noise. The sensor is the core of the world's first endoscopic digital PET, a tool with unprecedented levels of contrast and detail for early and accurate cancer diagnostics.

1 Introduction

1.1 Medical Imaging

In nuclear medicine, small amounts of radio-isotopes are used to diagnose and localize several ailments, including cancer, heart disease, gastrointestinal, endocrine, and neurological disorders, and other abnormalities found in our organism. There exist a variety of diagnostic technique in nuclear medicine. Scintigraphy is a

S. Mandai (✉) · E. Charbon
Faculty of Electrical Engineering, TU Delft, Delft University of Technology, Mekelweg 4, 2628 CD Delft, The Netherlands
e-mail: s.mandai@tudelft.nl

P. Harpe et al. (eds.), *High-Performance AD and DA Converters, IC Design in Scaled Technologies, and Time-Domain Signal Processing*,
DOI: 10.1007/978-3-319-07938-7_13,

two-dimensional (2-D) imaging technique, while Single Photon Emission Computed Tomography (SPECT) and Positron emission tomography (PET) are three-dimensional (3-D) tomographic techniques. The most important advantage of PET imaging over SPECT is that it can obtain images with a much higher sensitivity (by approximately two to three orders of magnitude); i.e. due to the capability of collecting a higher percentage of the emitted events [1, 2]. Before conducting a PET scan, a radiopharmaceutical, which is a short-lived radioactive tracer isotope, is injected into a patient. The tracer is combined into a biologically active molecule that is absorbed selectively in tissues of interest. As the radioisotope undergoes positron emission decay (also known as positive beta decay), it emits a positron, an antiparticle of the electron with opposite charge. The emitted positron travels in the tissue for a short distance, which is typically less than 1 mm, before it annihilates with an electron available in the surrounding medium. After the annihilation, a pair of gamma photons is produced with approximately opposite directions (180°) as shown in Fig. 1. The coincident gamma photons are detected when they reach a scintillator in the detector ring, creating a burst of visible light which is detected by photomultiplier tubes (PMTs) or silicon based photon sensors. After collecting tens of thousands of coincidence events along with straight lines, known as lines of response (LORs), it is possible to localize their source using appropriate reconstruction algorithms. Photons that are not detected in coincidence (i.e. within a coincident timing window of a few nanoseconds) are ignored.

1.2 Radiation Coincident Detection and Timing Resolution

Radiation detection systems are a key component of any imaging system. The radiation detection system is composed of a scintillation material (scintillator) and photo sensors as shown in Fig. 1. The scintillation material converts high-energy photons into visible light which can be detectable with a conventional photo sensor. The integral of the visible photons is proportional to the total energy deposited in the detector by the radiation. The timing resolution of a PET detector corresponds to the statistical timing fluctuations or uncertainty due to the timing characteristics of scintillator and photo sensor. Figure 2 shows the coincident detection of two detectors. The output from each detector is discriminated by a certain threshold from detector noise or scattered gamma events which have low energy, and sent to a data acquisition system (DAQ). Since the timing resolution represents the variability in the signal arrival times (time-of-arrival, TOA) for different events, it needs to be properly measured for detecting coincident events to distinguish true events from false events.

The accuracy of the coincidence detection is defined as coincidence time resolution (CTR), as shown in Fig. 2. Good timing resolution of a PET detector, besides helping reduce the number of random coincidences, can also be used to estimate the annihilation point between the two detectors by measuring the arrival time

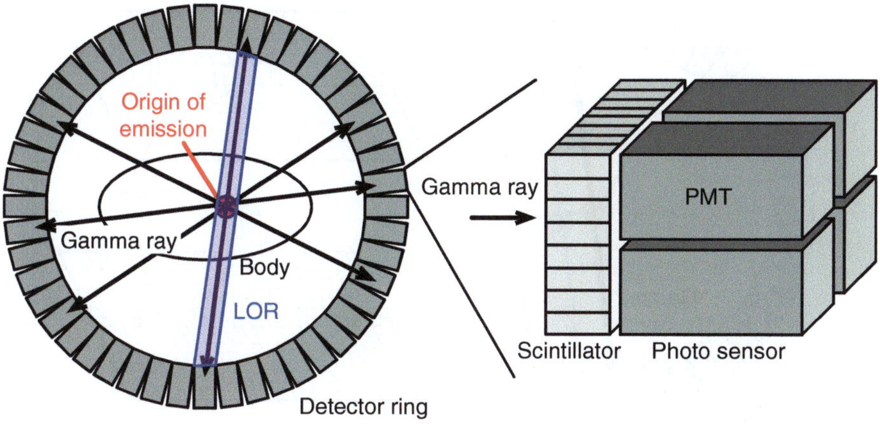

Fig. 1 Structure of PET scanner

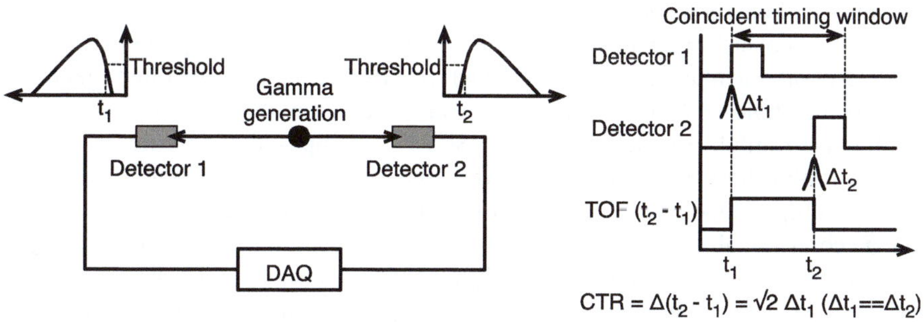

Fig. 2 Detecting coincident events in two detectors

difference of the two gamma photons. This PET scanner is called time-of-flight PET (TOF PET) [3–5]. The advantage of estimating the location of the annihilation point is the improved signal-to-noise ratio (SNR) obtained in the acquired image, arising from a reduction in noise propagation during the image reconstruction process. Figure 3 shows the comparison between non TOF PET and TOF PET. The constructed picture by TOF PET will be sharper with higher contrast.

1.3 Photo Sensors

Photo sensors are coupled to a scintillator using a glue that maximizes the detectability of visible photons generated in the scintillator upon a gamma event. The goal is to collect as many photons as possible with the accurate TOA

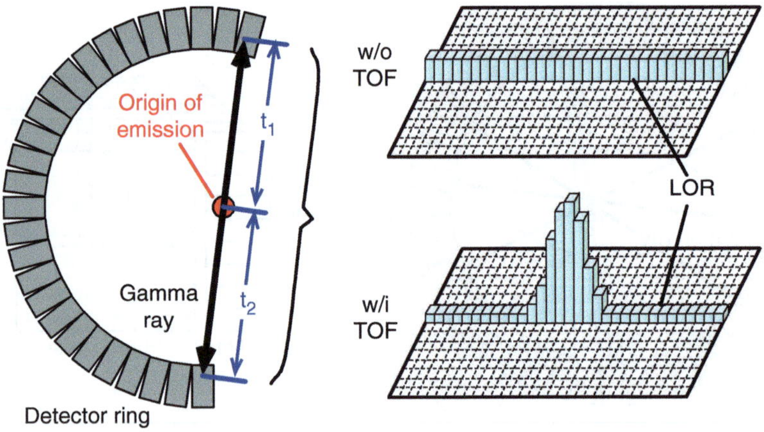

Fig. 3 Advantage of TOF PET

acquisition of gamma photons. SiPMs are a valid solid-state alternative to PMTs because of their robustness to magnetic fields, compactness, and low bias voltage [6]. SiPMs consist of an array of avalanche photodiodes operating in Geiger mode (single-photon avalanche diodes, SPADs). In SPADs, the absorbed light generates an electron-hole that may trigger an avalanche. Two flavors exist for SiPMs: analog and digital. An analog SiPM (A-SiPM) is composed of an array of SPADs, whose avalanche currents are summed in one node, and the output is processed with off-chip components as shown in Fig. 4a [6–12]. In digital SiPMs (D-SiPMs) on the contrary, all of the SPAD digital outputs are combined together by means of a digital OR, and the output is directly routed to an on-chip time-to-digital converter (TDC) to reduce external components and temporal noise as shown in Fig. 4b [13–17].

2 Analysis of Timing Resolution and Proposal for a Sensor Architecture

2.1 Modeling Scintillations

For the emitted photons from a LYSO scintillator, we can assume that detection occurs at time, θ as shown in Fig. 5. The time information of each photon can be considered as statistically independent and identically distributed (i.i.d.) following a probability density function (p.d.f.), which has been modeled as a double-exponential with rise time t_r and decay time t_d [18] $f(t|\theta) = (\exp(-\frac{t-\theta}{t_d}) - \exp(-\frac{t-\theta}{t_r}))/(t_d - t_r)$ when t $>$ θ, otherwise, $f(t|\theta) = 0$. Upon photon impingement, the SPAD jitter and an electrical jitter are convolved with the scintillator-based p.d.f., $f_{emi}(t|\theta)$. The dark counts follow an exponential probability distribution with

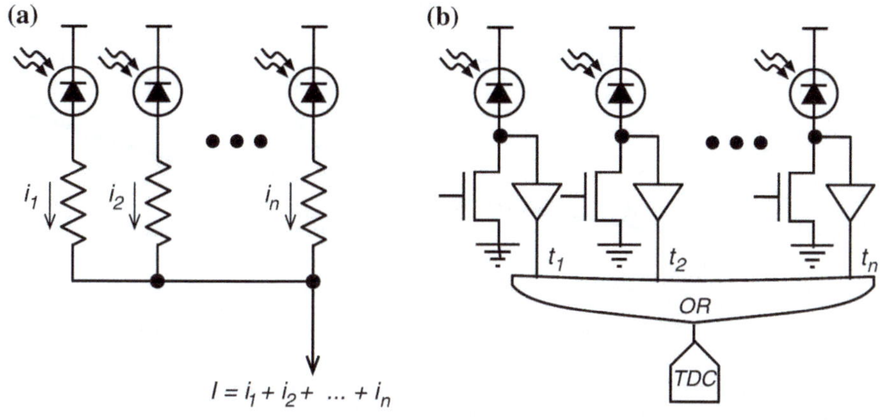

Fig. 4 Configuration of **a** A-SiPMs and **b** D-SiPMs

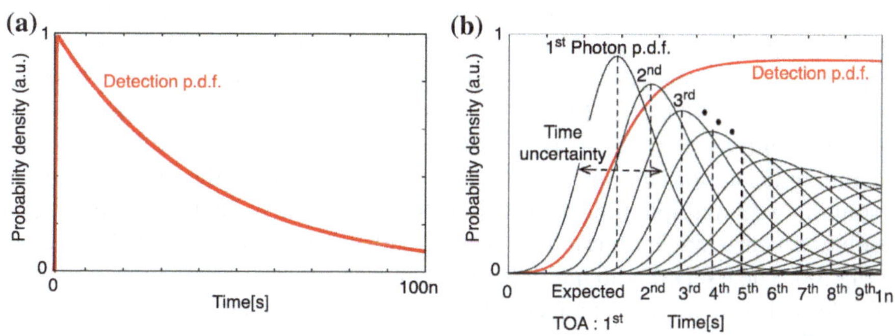

Fig. 5 **a** Detection p.d.f. **b** p.d.f. of the k-th primary photon detected after the first detected photon

event rate, λ, also known as dark count rate (DCR), and reset time, t_r, as $f(t) = \lambda \exp(-\lambda(t - t_r))$ when $t > t_r$, otherwise, $f(t) = 0$. The p.d.f. of the dark counts should also be convolved with an electrical jitter to be $f_{dcr}(t|t_r)$. The detection cycle, or frame, starts at the earliest before θ and it lasts a frame period, T. Thus the dark count p.d.f. is summed up for each reset time and then normalized. The scintillator-based p.d.f. and the dark count p.d.f. are mixed with mixing ratio $\alpha : (1 - \alpha)$ where α is defined by the percentage of photons emitted from scintillator, N, out of total detectable events, $N + \lambda T$, as,

$$f_{emi+dcr}(t|\theta) = \alpha f_{emi}(t|\theta) + (1 - \alpha) \frac{\int_{\theta-T}^{\theta} f_{dcr}(t|tr)\, dr}{\int_{\theta-T}^{\theta} \int_{t_r}^{\infty} f_{dcr}(t|tr)\, dt\, dr}.$$

Finally, the mixed p.d.f. is used for calculating the Fisher information [19] for the rth-order statistics p.d.f. or the joint p.d.f. for the first r-order statistics, then the Crámer-Rao lower bound for the unbiased estimator, θ, is calculated.

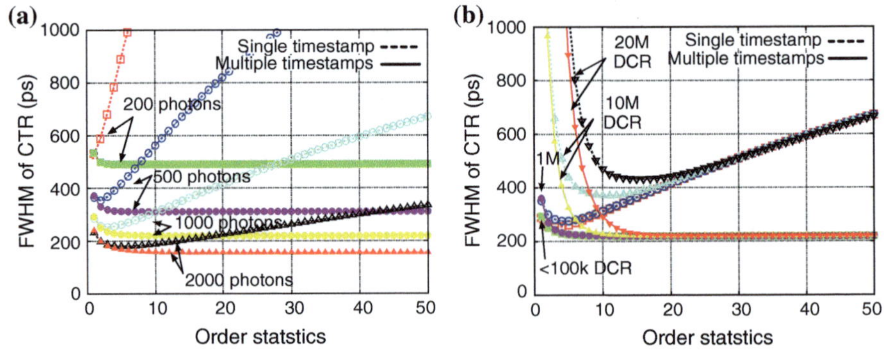

Fig. 6 Order statistics with a single timestamp or multiple timestamps v.s. FWHM of timing resolution: **a** various number of detected photons, 200, 500, 1000, and 2000 at 1 Hz DCR (which is almost negligible), **b** various values of DCR, less than 1,10, 20 MHz at 1000 detected photons

In our SPADs, we assumed normal jitter distributions with a standard deviation of 100 ps, the rise and decay times of a LYSO scintillator are 200 ps and 40 ns, respectively, while the number of detected photons is varied from 100 to 5,000, and DCR varied from 1 to 100 MHz. Figure 6 shows the relation between order statistics and full-width-at-half-max (FWHM) timing resolution. Figure 6a shows that the timing resolution improves with multiple timestamps. Furthermore, the timing resolution with multiple timestamps doesn't degrade due to DCR while the timing resolution with a single timestamp degrades with certain amount of DCR, as shown in Fig. 6b. The FWHM with multiple timestamps improves 13 % if compared to the FWHM with a single timestamp at less than 100 kHz DCR, however, the FWHM is 20 and 40 % better at 1 and 10 MHz DCR, respectively. From this work, it is clear that D-SiPMs capable of providing multiple timestamps are useful not only to improve timing resolution but especially to ensure tolerance to DCR and independence from a threshold and from the energy of the gamma event.

2.2 Multi-Channel Digital Silicon Photomultipliers

The ideal D-SiPM, as shown in Fig. 7a, gives minimum single-photon timing resolution (SPTR) by minimizing the skew, and achieves the statistical approach to estimate TOA by detecting TOAs of multiple photons in a single gamma event. However, the main drawback of the approach proposed in [20] is a low fill factor due to the need for significant silicon real estate to implement per-pixel functionality. In order to keep high PDE, 3-D integration is a promising technique. However, current 3D IC technologies are still expensive and hard to get. To achieve both high fill factor and acquisition of multiple timestamps, a new type of D-SiPM has been proposed [21–23]. Sharing one TDC with several pixels has the advantage of increasing the fill factor while still enabling somewhat independent

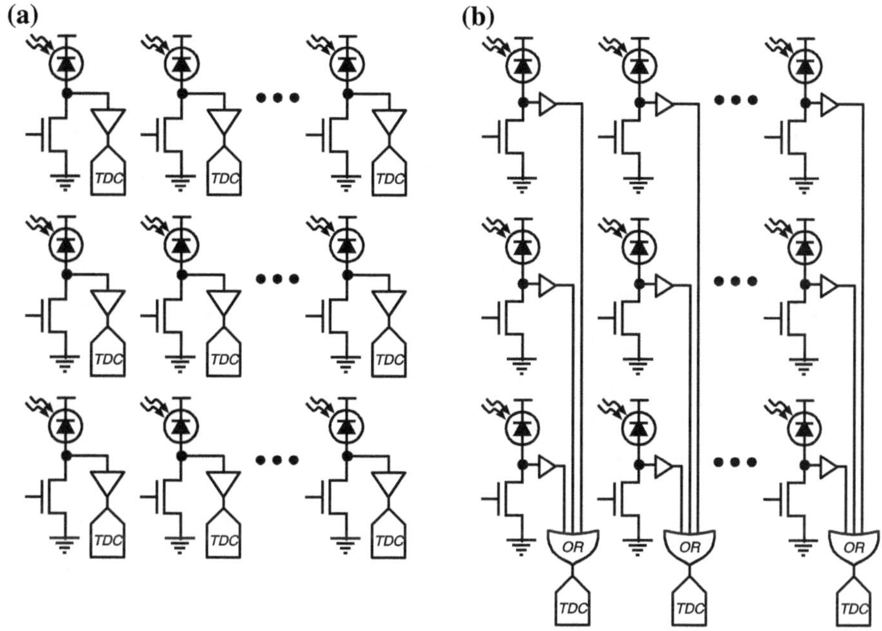

Fig. 7 **a** Ideal D-SiPM. **b** Proposed MD-SiPM

photon TOA evaluation, as shown in Fig. 7b. The skew problem is also improved when compared to conventional D-SiPMs for single-photon detection, and the multiple timing information can be utilized in a statistical approach for multiple-photon detection. This type of SiPMs is called multi-channel digital SiPM, or MD-SiPM. The MD-SiPM can achieve both high fill factor and enough multiple timestamps to improve timing resolution using the same principles outlines earlier.

3 Proposed SiPM Architecture

3.1 SiPM Configuration

Figure 8 shows the proposed MD-SiPM array configuration. Each SiPM in the array comprises 416 photo-detecting cells and measures 800 $\times$ 780 μm^2, adapted to the crystal dimensions. A 20 m gap enables adequate glue reflow in the pixelation process. Each pixel measures 50 $\times$ 30 μm; it generates a sharp pulse in correspondence to a photon detection that is routed directly to a TDC. Adjacent pixels are routed to independent TDCs by triples (every three pixels, the TDC is reused); this approach prevents misses in closely striking photons, thereby reducing local saturation. There are 48 TDCs per SiPM column, each operating

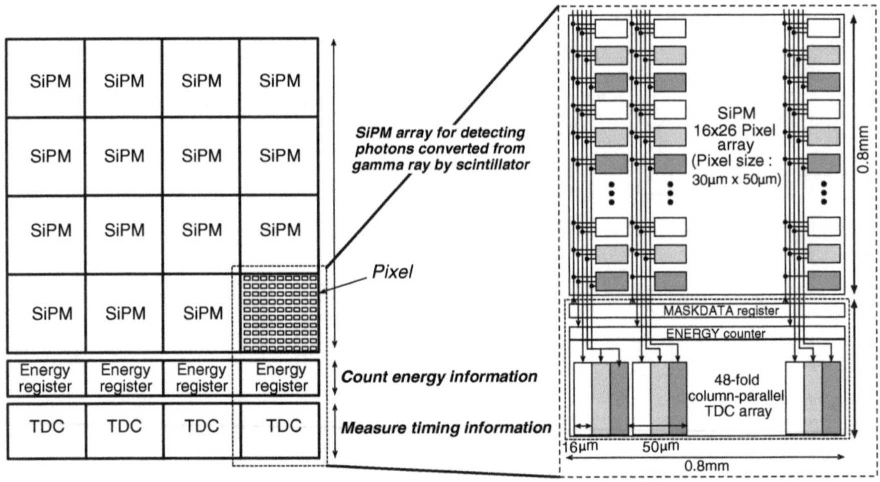

Fig. 8 Block diagram of the proposed sensor capable of detecting a large number of photons and their times-of-arrival with picosecond accuracy

simultaneously with a LSB duration of 44 ps. The number of photons, and thus the energy of the gamma event, is proportional to the number of triggered pixels. Thus, after each event the triggered pixels are counted; the corresponding digital code (1 or 0) is read out along with the complete statistical profiles of the projected photons, and summed up to calculate the total number of photons in one SiPM. Even though the number of cells is 416, it is possible to utilize saturation correction to count more than the number of cells [24]. To minimize photon misses, the array was designed with a fill factor up to 57 %. In our application, the pitch of the SiPMs is 800 μm, so as to match the $800 \times 800\ \mu m^2$ section of the individual crystal in a pixelated scintillator. So the overall fill factor is equivalent to the fill factor of one pixel. To minimize dark counts in a SiPM, a masking procedure is used. Masking an SiPM consists of depriving it from a certain number of SPADs whose activity exceeds a threshold, also known as screamers. The MASKDATA register is used for disabling those pixels with DCR exceeding a threshold, so as to minimize spurious TDC activation. The ENERGY register is used for reading out the number of pixels that detected at least a photon.

3.2 Pixel Architecture

Figure 9a shows the schematic of a pixel; it comprises the SPAD, a 1-bit counter for energy estimation, a masking circuit, and a pulse shaper /column driver. The frame will start after RST resets the SPAD. When a shower of photons is generated in the scintillator, one of them may hit the SPAD in the pixel, generating a digital

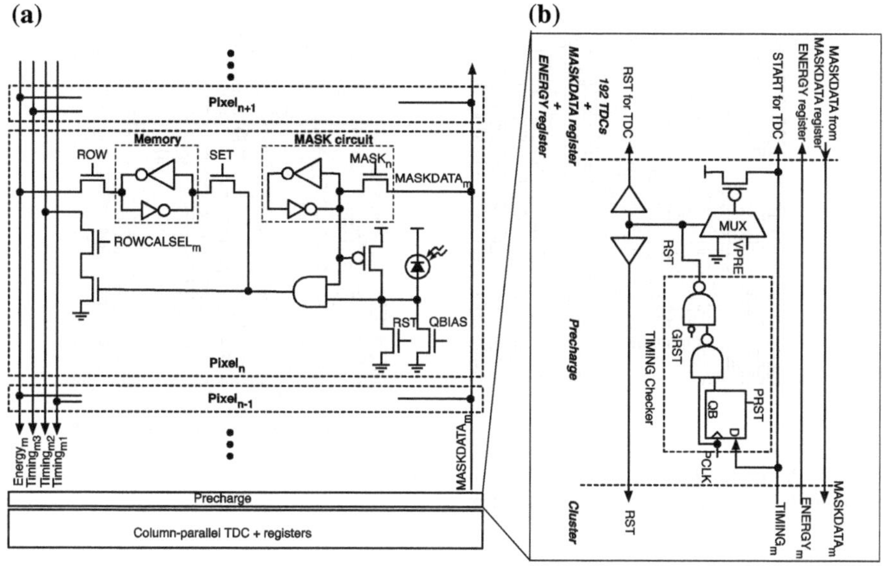

Fig. 9 **a** The pixel architecture comprises a SPAD, a 1-bit counter for energy estimation, a memory for masking, active and passive quenching, and a column driver circuit. **b** Detail of the pre-charge structure implemented at the bottom of the array

pulse to pull-down TIMING$_m$. QBIAS controls the quenching resistance of the SPAD and controls the digital pulse width to be more than the frame period. The event triggers the 1-bit counter, and it is read out as ENERGY$_m$ by ROW after latching the 1-bit counter value by SET. Masking is carried out in advance row-by-row, using signals MASKDATA$_m$ and ROWCALSEL, by bringing the SPAD bias below breakdown and by disabling the signal generated at its anode.

3.3 Readout Architecture from MD-SiPMs to Column-Parallel TDC Bank

Figure 9b shows the schematic of the pre-charge circuit used at the bottom of each column, which is terminated by a TDC. This circuit is used to prepare a TDC for a photon hit by employing signal GRST. TIMING$_m$ is pre-charged during RST for pixels and TDCs; this signal is asserted every 6.4 μs via signal GRST. PCLK is also causing a reset for pixels and TDCs when the number of firing TDCs exceeds a threshold within a pre-determined time, say 100 ns. This early reset mechanism, known as *smart reset* is performed so as to prevent gamma event misses when TDCs are occupied recording previous background photons or noise. VPRE is used to control the pull-up resistance for TIMING$_m$, and TIMING$_m$ used as

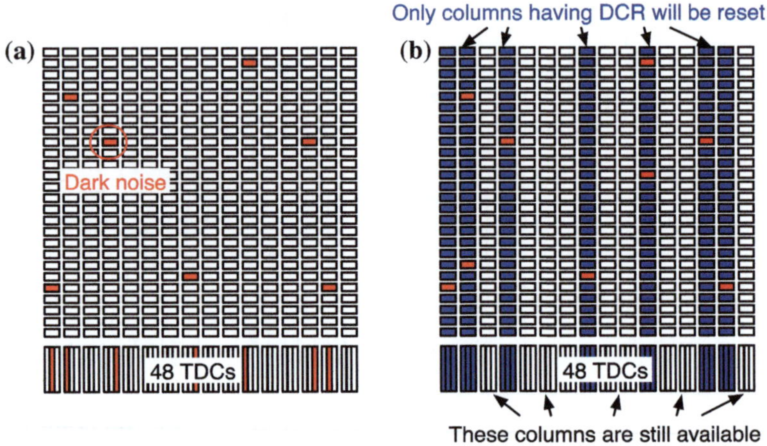

Fig. 10 Principle of smart reset. **a** Pixel and TDC occupation due to dark count noise as an example. **b** The columns that will be reset by *smart reset* in the example

START signals for TDC. Figure 10 shows how *smart reset* is carried out. Figure 10a shows the pixels who had dark count noise and TDCs consequently activated. During the *smart reset*, the blue colored column pixels and TDCs fired due to dark count noise are reset as shown in Fig. 10b, the rest of the column pixels and TDCs are thus available for a gamma event. Conventionally, SiPMs employ an event driven readout because the gamma event is classified as random coincidence. However, D-SiPMs take more than 600 ns to correct and read out the data [17], which is assumed as the dead time, while analog SiPMs take much shorter time. By employing the frame based readout, the data can be read out in the background to reduce dead time. Figure 11 shows the timing diagram of 6.4 s frame based readout. At the beginning of each frame, all the SPADs in every MD-SiPM are reset globally by GRST, and then every 60 ns, the *smart reset* is applied. The number of firing TDCs starts increasing due to dark counts, even in the absence of gamma events. By *smart reset*, pixels and TDCs are reset by PCLK. Upon occurrence of a gamma event, many TDCs fire in a short period ($\leq$100 ns) and the number of firing TDCs exceeds a threshold, thus preventing assertion of PCLK until the end of the frame. At the end of the frame, the pixel data are sent to the pixel memories and the TDC data are latched to TDC registers by SET. These data are read out, during which time the MD-SiPMs start the next frame.

3.4 Column-Parallel TDC Bank

In this MD-SiPM array, a column-parallel TDC utilizing a common VCO has been proposed to mitigate LSB variations. Figure 12 shows the architecture and a simple timing diagram of the proposed column-parallel TDC with a common VCO.

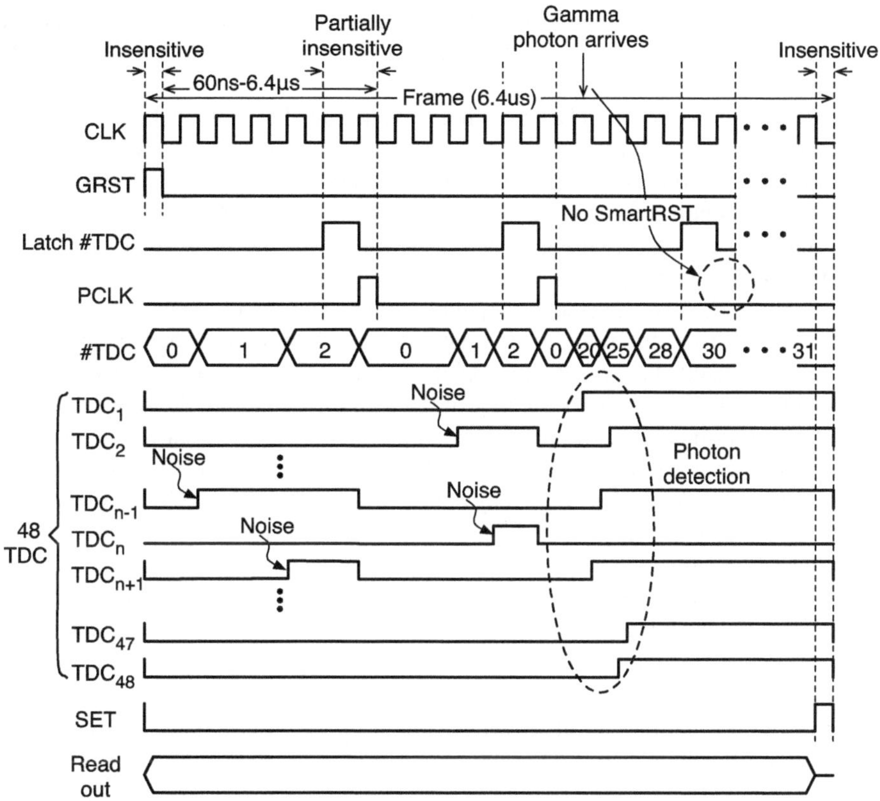

Fig. 11 Timing diagram for frame based readout

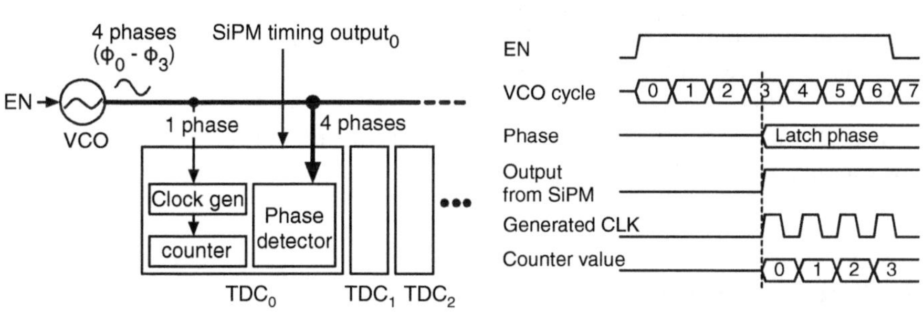

Fig. 12 Architecture of the proposed column-parallel TDC with a common oscillator. On the *right* the logic control timing diagram is shown

A TDC consists of a phase detector, a clock generator, and a counter. The VCO starts to propagate 4 phases, $\phi_0 - \phi_3$ (45 degrees of phase), after EN is asserted high. When the TDC receives a trigger from an MD-SiPM, the phase is latched by the phase detector while the clock for the counter is enabled by the clock generator. 192 TDCs have been implemented in the first version of the column-parallel TDC utilizing a common VCO.

Figure 13a shows one TDC schematic including the phase detector as a fine conversion and a counter as a coarse conversion. The coarse conversion is achieved by counting the clock cycles from the assertion of START (SPAD firing on the corresponding column) to the end of the frame (STOP signal) using the 12-bit counter; the LSB of this conversion is 1.4 ns. In the phase detection, upon START assertion, the phases from the VCO are latched by two-phase detectors with slightly skewed signals, ENB_1 and ENB_2. The small skew is realized by implementing two different inverter chains biased differently and calibrated to have an optimized skew for the best DNL. Figure 13b shows the VCO schematic; the VCO is activated at the beginning of the frame by enabling the ring oscillation and stopped at the end of the frame to save power. Figure 13c shows the bandgap voltage reference circuit for the VCO and the inverter chain in the TDC to ensure stable frequency generation and delay control on the chip. The VCO frequency and delay of the inverter chain in the TDC may also be conveniently programmed.

Figure 14a shows the timing diagram after EN becomes high. The phase starts to propagate while EN is high, and it is latched when EN becomes low. Figure 15b shows the phase detection of the proposed phase detector. Conventionally, the phase detector looks only at ϕ_0 and $\overline{\phi_0}$, ϕ_1 and $\overline{\phi_1}$, ϕ_2 and $\overline{\phi_2}$, and ϕ_3 and $\overline{\phi_3}$, resulting in 8 phases in one oscillation cycle. The LSB corresponds to one over eight of the oscillation period. However, the proposed phase detector employs an interpolation technique to double the phase resolution, thus halving the LSB. By expanding the comparison to ϕ_0 and $\overline{\phi_3}$, ϕ_0 and $\overline{\phi_1}$, ϕ_1 and $\overline{\phi_2}$, and ϕ_2 and $\overline{\phi_3}$, the interpolated phase is detectable, as shown in Fig. 14b. The proposed phase detector utilizes 8 comparator outputs and results in 4 bits resolution. Only four extra comparators and memory are required for the proposed phase detector. Decoding the output of the phase detector is carried out on a host PC after counter and phase values are read out outside the chip. By summing $PHVAL_1$ and $PHVAL_2$, a fine conversion resolution of 5 bits is obtained (corresponding to a LSB of $1.4\ \text{ns}/2^5 = 44$ ps), which, added to the 12 bits of the coarse conversion, corresponds to a total of 17 bits.

4 Characterization

4.1 Chip Fabrication

The sensor chip was fabricated on a 0.35 μm CMOS process, the die size is $4.22 \times 5.24\ \text{mm}^2$. The 4×4 MD-SiPM array occupies $3.2 \times 3.2\ \text{mm}^2$ and each TDC occupies $16 \times 840\ \mu\text{m}^2$ including the readout circuit. A photomicrograph of the MD-SiPM array chip is shown in Fig. 15. The power supply voltage is 3.3 V

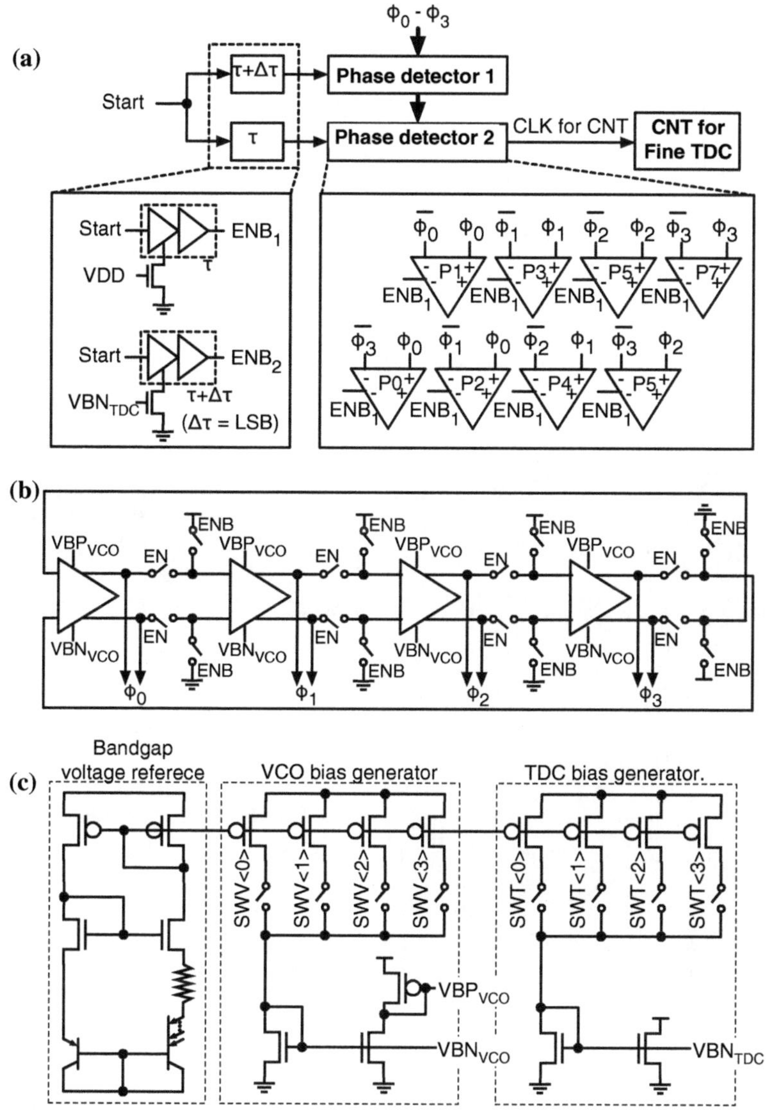

Fig. 13 **a** Schematic of each TDC employing the phase interpolation technique. **b** VCO. **c** Bandgap voltage reference circuit

and the high voltage for SPADs is 22–23 V. The VCO and the bandgap voltage reference consume 80 mA in total, the digital logic 30 mA. The current drawn by each TDC is less than 570 μA, while the MD-SiPM array consumption is 2 mA in the dark. Each SiPM and its fill factor are shown in the figure., along with a denomination. A detail of the SiPM 'D15' is shown in the figure, along with the

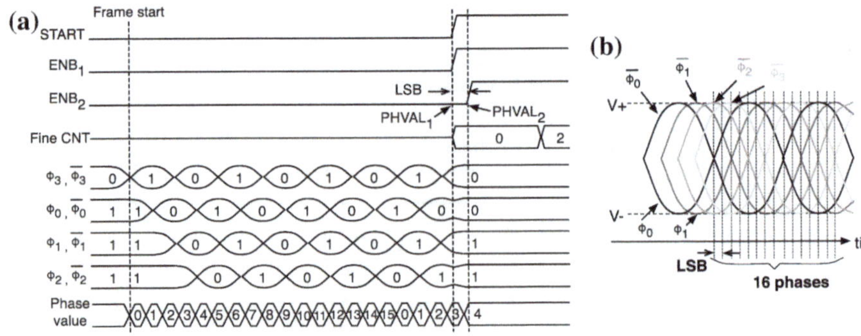

Fig. 14 **a** Timing diagram of the fine TDC. **b** Concept of our proposed phase detector

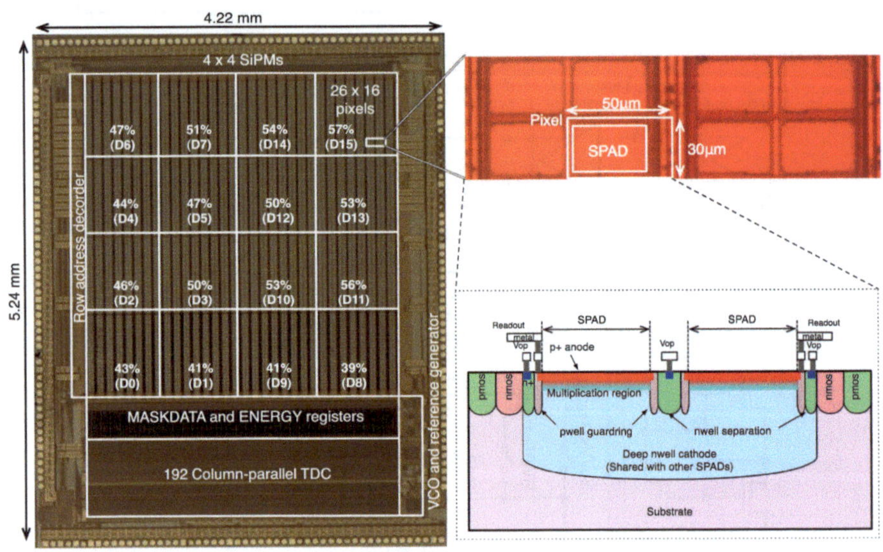

Fig. 15 Chip microphotograph

dimensions of the pixel that achieves a fill factor of 57 %. To maximize fill factor, the electronics was placed at a distance of twice the pitch and implemented in a mirrored fashion.

4.2 Noise and Sensitivity Characterization

Figure 16a shows the cumulative DCR plot for 'D15' SiPM showing the DCR distribution of 416 SPADs for several excess bias voltages and temperatures.

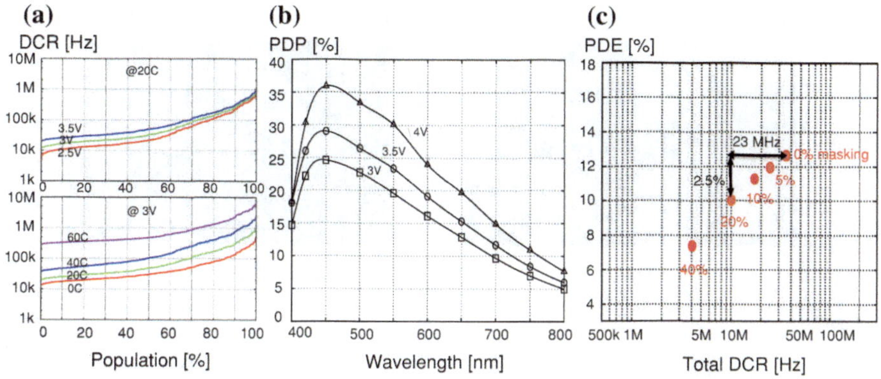

Fig. 16 **a** Cumulative DCR plot for various excess bias voltage and temperature conditions for the 'D15' SiPM. **b** PDP versus wavelength at various excess bias. **c** Relation between DCR and PDE for various SiPMs at 3 V excess bias and 20 °C for several masking levels

The PDP of SPADs is also characterized. We only activate a single SPAD to measure PDP and the dead time of the SPAD is set to be long by using high quenching resistor. Figure 16b shows PDP of a single pixel as a function of wavelength with different excess bias. The temperature dependency of PDP is negligible in the spectral range of interest. The PDP is about 30 % at 4 V excess bias at 420–430 nm which is of interest for TOF PET applications with LYSO scintillators. It means that at most 17.1 % PDE can be achievable by accepting high DCR or by cooling the device. PDE can be calculated based on the PDP measurement results and fill factor by $PDP \times FF$, where FF the fill factor. Masking pixels reduces both DCR and fill factor, and thus PDE. However, the reduction is not linear due to the small percentage of highly noisy pixels (screamers). Thus, small masking levels reduce total DCR in one SiPM faster than PDE, while larger masking has a larger impact on PDE and a smaller impact on total DCR. This mechanism for 'D15' at 3 V excess bias and 20 C can be seen in Fig. 16c.

By using the masking circuitry, optical and electrical crosstalk is measured. Figure 17 shows the DCR map at 4 V excess bias, 30 °C temperature in a section of the MD-SiPM that contains a noisy pixel before (a) and after (b) masking of pixels whose DCR is higher than 1 MHz. Figure 17 also shows the DCR reduction in pixels surrounding the noisy SPAD in 4 × 4 pixels. By turning off the noisy pixel, the DCR decreases approximately 10 % in adjacent pixels as well.

4.3 Column-Parallel TDC Timing Characterization

The TDCs were fully characterized using an electrical input, yielding a single-shot timing uncertainty of 60 ps (FWHM). Figure 18a and b show DNL and INL of a

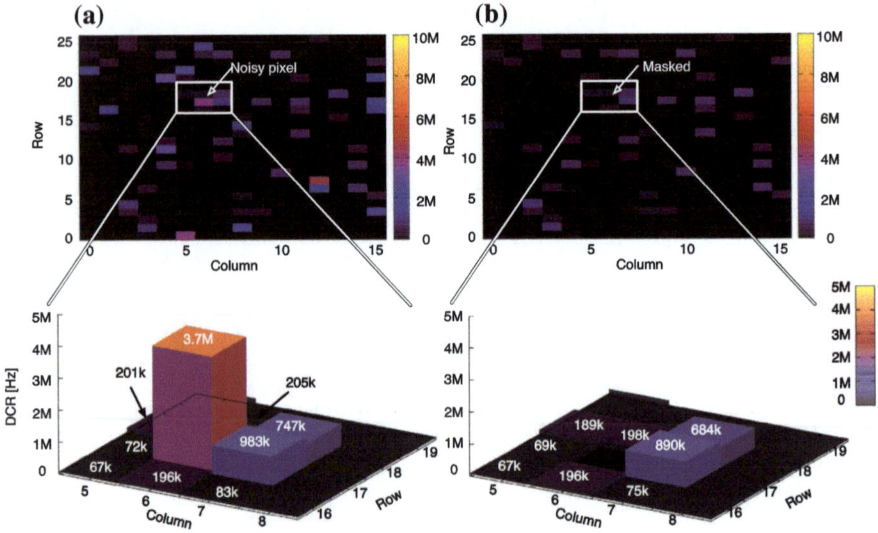

Fig. 17 DCR map (**a**) before masking a noisy SPAD and (**b**) after masking a noisy SPAD. The values in the bottom pictures show the DCR

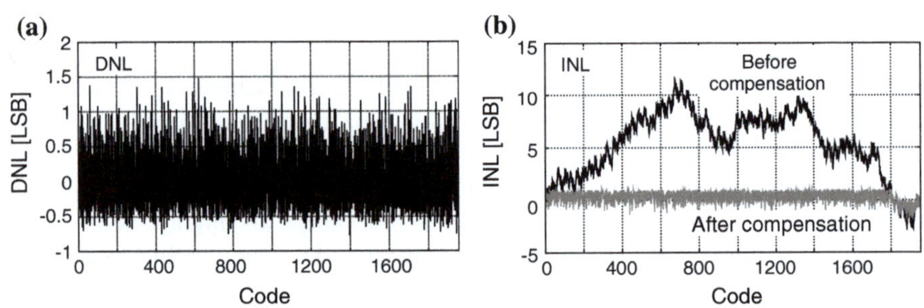

Fig. 18 **a** DNL. **b** INL

typical TDC in the array. Since the INL is mainly caused by the frequency shift of the VCO over the detection period, it can be largely compensated for by means of a lookup table (LUT). Figure 19a summarizes the INL variation for all 192 TDCs after the INL compensation. Figure 19b shows the LSB shift of the TDCs due to temperature and power supply fluctuations. The TDCs suffer from a 6 to 9 % LSB shift in the ±30 °C range and ±10 % power supply variation.

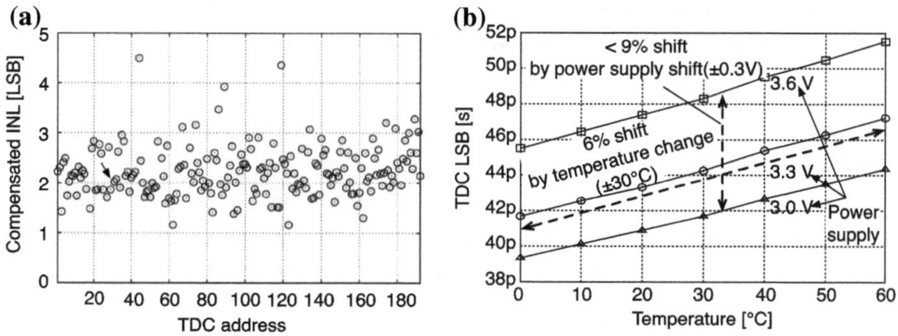

Fig. 19 **a** INL variation after compensation in a TDC when all 192 TDCs are in operation. **b** LSB shift for the TDC due to temperature and power supply fluctuation

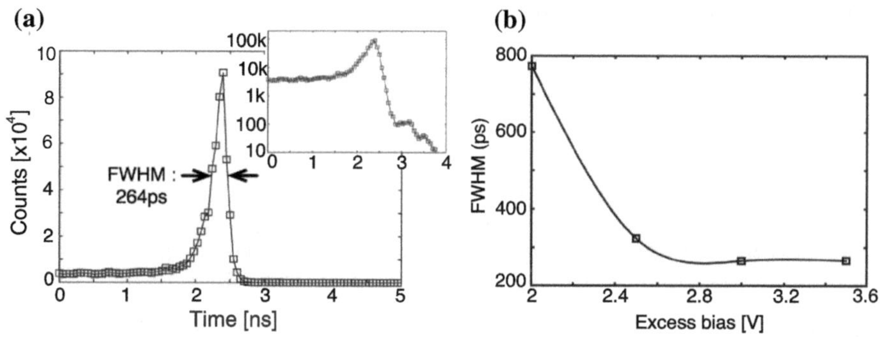

Fig. 20 **a** Single-photon FWHM timing resolution for a single SPAD using a TDC. **b** Single-photon FWHM timing resolution for the complete SiPM at various excess bias voltages

4.4 MD-SiPM Timing Characterization

The timing resolution of 21 % fill factor MD-SiPM was established optically in a TCSPC experiment using a 250 mW, 405 nm laser source (ALDS GmBH) with 40 ps pulse width. Figure 20a shows SPTR measurements of the MD-SiPM. The SPTR of the MD-SiPM is measured with a single-photon level intensity obtained from a laser on average. Asymmetric shape of the count level before and after the laser pulse is due to the fact that a laser intensity is close to one photon per a laser pulse and the relatively high DCR will trigger the MD-SiPM before but not after the laser pulse. The SPTR consists of mainly SPAD jitter, TDC intrinsic jitter, pixel-to-pixel skew of the TDC input lines, and laser jitter. The FWHM timing resolution is 264 ps, including the SPAD jitter, 93.2 ps TDC intrinsic jitter and the pixel-to-pixel skew, when the photon source exhibits 34 ps FWHM jitter. The sensor is operated at 3 V excess bias. Figure 20b shows the FWHM timing resolution as a function of excess bias. The SPTR improves by increasing excess bias for SPADs, because the SPAD jitter dramatically decreases when the excess bias is high.

5 Conclusions

In a PET application, high speed imaging with precise timing information acquisition is required for the gamma detection We have proposed a 4 $\times$ 4 array of D-SiPMs capable of timestamping up to 48 photons, denominated multi-channel D-SiPM or MD-SiPM. We have shown the advantage of generating multiple timestamps in the context of PET. The MD-SiPMs have a pitch of 800 m and comprise 416 pixels each; the timing resolution achieved by the SiPMs is 264 ps FWHM, while each pixel has a fill factor of up to 57 % and a single-photon timing resolution of 114 ps. The sensor is the prototype of the core detector of the worlds's first endoscopic digital PET.

References

1. P. E. Valk, D. L. B. D. Delbeke, D. W. Townsend, and M. N. Maisey, *Positron Emission Tomography*. Springer, 2004.
2. M. E. Phelps, *PET Physics, Instrumentation, and Scanners*. Springer, 2006.
3. W. W. Moses, "Time of flight in PET revisited," *IEEE Trans. Nucl. Sci.*, vol. 50, no. 5, pp. 1325–1330, Oct. 2003.
4. M. Conti, "Effect of randoms on singal-to-noise ratio in TOF PET," *IEEE Trans. Nucl. Sci.*, vol. 53, no. 3, pp. 1188–1193, Jun. 2006.
5. M. Conti, "Focus on time-of-flight PET: the benefits of improved time resolution," *Eur. J. Nucl. Med. Mol. I*, vol. 38, no. 6, pp. 1147–1157, Jun. 2011.
6. HAMAMATSU, "MPPC," http://jp.hamamatsu.com, 2013.
7. T. Nagano, K. Sato, A. Ishida, T. Baba, R. Tsuchiya, and K. Yamamoto, "Timing resolution improvement of MPPC for TOF-PET imaging," in *Proc. IEEE Nuclear Science Symp. Conf.*, 2012, pp. 1577–1580.
8. P. Buzhan, B. Dolgoshein, L. Filatov, A. Ilyin, V. Kantzerov, V. Kaplin, A. Karakash, F. Kayumov, S. Klemin, E. Popova, and S. Smirnov, "Silicon photomultiplier and its possible applications," *Nucl. Instrum. Methods Phys. Res. A*, vol. 504, no. 1–3, pp. 48–52, May 2003.
9. A. G. Stewart, V. Saveliev, S. J. Bellis, D. J. Herbert, P. J. Hughes, and J. C. Jackson, "Performance of 1-mm^2 silicon phopomultiplier," *IEEE J. Quantum Electron.*, vol. 44, no. 2, pp. 157–164, Feb. 2008.
10. N. Zorzi, M. Melchiorri, A. Piazza, C. Piemonte, and A. Tarolli, ''Development of large-area silicon photomultiplier detectors for PET applications at FBK,'' *Nucl. Instrum. Methods Phys. Res. A*, vol. 636, no. 1, pp. 208–213, Apr. 2011.
11. M. Mazzillo, G. Condorelli, D. Sanfilippo, G. Valvo, B. Carbone, A. Piana, G. Fallica, A. Ronzhin, M. Demarteau, S. Los, and E. Ramberg, "Timing performances of large area silicon photomultipliers fabricated at STMicroelectronics," *IEEE Trans. Nucl. Sci.*, vol. 57, no. 4, pp. 2273–2279, Aug. 2010.

12. M. McClish, P. Dokhale, J. Christian, C. Stapels, E. Johnson, R. Robertson, and K. S. Shah, "Performance measurements of CMOS position sensitive solid-state photomultipliers," *IEEE Trans. Nucl. Sci.*, vol. 57, no. 4, pp. 2280–2286, Aug. 2010.
13. T. Frach, G. Prescher, C. Degenhardt, R. Gruyter, A. Schmitz, and R. Ballizany, "The digital silicon photomultiplier— principle of operation and intrinsic detector performance," in *Proc. IEEE Nuclear Science Symp. Conf.*, 2009, pp. 1959–1965.
14. T. Frach, G. Prescher, C. Degenhardt, and B. Zwaans, "The digital silicon photomultiplier— system architecture and performance evaluation," in *Proc. IEEE Nuclear Science Symp. Conf.*, 2010, pp. 1722–1727.
15. Y. Haemisch, T. Frach, C. Degenhardt, and A. Thon, "Fully digital arrays of silicon photomultipliers (dSiPM)—a scalable alternative to vacuum photomultiplier tubes (PMT)," in *Proc. of TIPP*, vol. 37, 2011, pp. 1546–1560.
16. D. Tyndall, B. Rae, D. Li, J. Richardson, J. Arlt, and R. Henderson, "A 100 M photon/s time-resolved mini-silicon photomultiplier with on-chip fluorescence lifetime estimation in 0.13 μm CMOS imaging technology," in *IEEE ISSCC Dig. Tech. Papers*, 2012, pp. 122–124.
17. L. Braga, L. Gasparini, L. Grant, R. Henderson, N. Massari, M. Perenzoni, D. Stoppa, and R. Walker, "An 8x16-pixel 92 k SPAD time-resolved sensor with on-pixel 64 ps 12b TDC and 100Ms/s real-time energy histogramming in 0.13 μm CIS technology for PET/MRI applications," in *IEEE ISSCC Dig. Tech. Papers*, 2013, pp. 486–487.
18. J. Glodo, W. W. Moses, W. M. Higgins, E. V. D. van Loef, P. Wong, S. E. Derenzo, M. J. Weber, and K. S. Shah, "Effects of Ce concentration on scintillation properties of $LaBr_3$:Ce," *IEEE Trans. Nucl. Sci.*, vol. 52, no. 5, pp. 1805–1808, Oct. 2005.
19. S. Park, "On the asymptotic fisher information in order statistics," *Metrika*, vol. 57, no. 1, pp. 71–80, 2003.
20. C. Veerappan, J. Richardson, R. Walker, D. U. Li, M. W. Fishburn, Y. Maruyama, D. Stoppa, F. Borghetti, M. Gersbach, R. K. Henderson, and E. Charbon, "A 160 × 128 single-photon image sensor with on-pixel 55 ps 10b time-to-digital converter," in *IEEE ISSCC Dig. Tech. Papers*, 2011, pp. 312–314.
21. S. Mandai and E. Charbon, "Timing optimization of a H-tree based digital silicon photomultiplier," *Journal of Instrum.*, vol. 8, no. 9, pp. 1–7, 2013.
22. S. Mandai and E. Charbon, "A multi-channel digital silicon photomultiplier array for nuclear medical imaging systems based on PET-MRI," *Proc. IISW*, 2013.
23. S. Mandai, V. Jain, and E. Charbon, "A 780 × 800 μm^2 multi-channel digital silicon photomultiplier with column-parallel time-to-digital converter and basic characterization," *IEEE Trans. Nucl. Sci.*, vol. 52, no. 10, pp. 1805–1808, Oct. 2014.
24. C. Degenhardt, G. Prescher, T. Frach, A. Thon, R. de Gruyter, A. Schmitz, and R. Ballizany, "The digital silicon photomultiplier—a novel sensor for the detection of scintillation light," in *Proc. IEEE Nuclear Science Symp. Conf.*, 2009, pp. 2383–2386.

Fine-Time Resolution Measurements for High Energy Physics Experiments

Lukas Perktold and Jørgen Christiansen

Abstract Fine-time resolution measurements are attracting increasing attention in the high-energy-physics (HEP) community, where a large number of measurement channels must often be realized with a single ASIC. In this contribution, a multi-channel time-to-digital converter (TDC) architecture with a delay-locked-loop (DLL) in its first stage and a resistive interpolation scheme in its second stage is presented. The size of the TDC's least-significant-bit (LSB) is controlled by a reference clock and so can be continuously adjusted from 5 to 20 ps. A global calibration scheme that avoids the need to calibrate each channel separately is also used. Critical design aspects like device mismatch, supply noise sensitivity and process-voltage and temperature (PVT) variation are discussed. When realized in a 130 nm technology, the prototype ASIC achieved a single-shot resolution of better than 2.5 ps-rms. The measured integral-non-linearity (INL) and differential-non-linearity (DNL) were found to be ±1.4 LSB and ±0.9 LSB respectively.

1 Introduction

Many physical quantities like crossing time or energy of a particle can be determined on the basis of time measurements. Recently, high-energy-physics (HEP) detectors have emerged that achieve sub 10 ps-rms resolution, see [1, 2]. To realize their full potential, time measurements in the ps-rms resolution domain are required.

L. Perktold (✉)
ams International AG, Rietstrasse 4, 8640 Rapperswil, Switzerland
e-mail: lukas.perktold@ams.com

J. Christiansen
CERN, 1211 Geneva 23, Switzerland

P. Harpe et al. (eds.), *High-Performance AD and DA Converters, IC Design in Scaled Technologies, and Time-Domain Signal Processing*,
DOI: 10.1007/978-3-319-07938-7_14,

This chapter presents the design of a multichannel, fine-time resolution TDC ASIC aimed at fulfilling the requirements of upcoming HEP detectors like the ATLAS AFP/CMS HPS [3] or LHCb TORCH [4] and others. Accurate absolute time measurements are then required, since it is often necessary to correlate measurements. For this purpose, a time reference is distributed across the experiment to synchronize the generated time stamps. To meet the requirements of many different experiments, a flexible TDC architecture that does not put any constraints on the measurement signal is desired. This rules out the use of START-STOP or noise-shaping TDC architectures.

Today, device mismatch together with jitter due to power supply and thermal noise represent the limiting factors in state-of-the-art TDC design. In recent designs, single-channel time resolutions in the order of 1 ps have been achieved, see [5] or [6]. However, this then necessitates the use of calibration techniques to compensate for device mismatch. Especially in a multichannel environment, calibration on a per-channel basis can lead to considerable circuit overhead and represent a time-intensive task during production.

In this work, a multi-channel TDC architecture is presented. It employs a global calibration scheme that avoids the need for per-channel calibration. Important design trade-offs such as device mismatch and power supply noise susceptibility are discussed, and measurement results on a prototype implemented in a 130 nm technology are presented.

2 Time Measurements in HEP

In HEP and other closely related fields, e.g. positron emission tomography (PET) [7] or fluorescence lifetime imaging microscopy (FLIM) [8] to mention just a few, it is crucial to measure the characteristics of the physical particles that encounter a detector. Quantities such as position, crossing time, momentum or energy of a particle need to be measured with high precision. Whereas crossing point location can be resolved by dividing the sensor into smaller units, crossing time as well as energy can be derived from time measurements.

The block diagram of a single-channel time measurement chain often employed in HEP detector designs is shown in Fig. 1. Often several hundreds or thousands of channels are required to cover the physical area of interest. The charge induced in a sensor is collected and amplified by a charge preamplifier, thus generating an analog signal. By discriminating the signal's amplitude, the crossing time and time-of-arrival of a particle can be extracted. The width of the resulting pulse, which is still continuous in time, can then be digitized using a TDC. In contrast to the traditional approach in which the amplitude of the analog signal is digitized using an analog-to-digital converter (ADC), employing a TDC-based scheme allows system complexity to be drastically reduced, as well as the amount of data generated. As the signal induced in the sensor may be non-ideal, a signal pre-processing stage might be added to suppress glitches and/or define a minimum

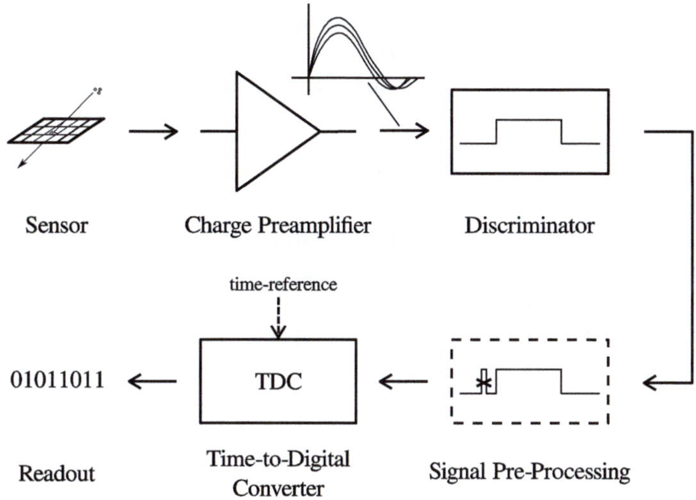

Fig. 1 Time-measurement chain in HEP

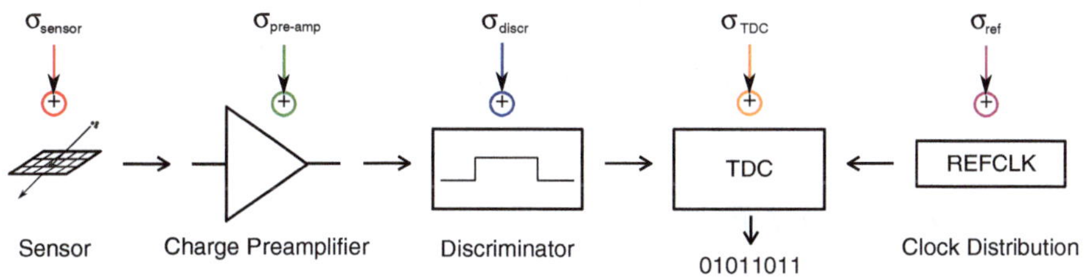

Fig. 2 Timing uncertainties of a typical detector design

pulse width and gap. Depending on the rate at which particles encounter the detector, a great amount of data might be generated, so a dedicated readout scheme to selectively access and retrieve this data is required. As illustrated in Fig. 2, the TDC's timing uncertainty represents only one of many timing uncertainties in the system. As a result, the TDC's timing uncertainty (σ_{TDC}) is often required to be negligible, thus making its design rather challenging.

3 Fine-Time Resolution TDC Design

The time resolution of an ideal TDC is only limited by the size of its LSB. For a uniformly distributed sequence of events, the standard deviation of the error, a function more often referred to as the TDC's quantization noise, can be calculated from Eq. 1. At some point, however, other error sources will become dominant, making further reductions in LSB size unnecessary. As a rule of thumb, the LSB size should be chosen to match the required rms-time resolution. This allows the quantization error to be kept well below the desired resolution.

$$\sigma_q = \frac{LSB}{\sqrt{12}} \tag{1}$$

In a real implementation, the resolution will be degraded by device mismatch as well as by the jitter introduced by power supply and thermal noise. In the case of uncorrelated error sources, which usually is the case in a real application, since the errors originate from different sources, the expected time resolution can be written as

$$\sigma_{TDC} = \sqrt{\sigma_{qDNL}^2 + \sigma_{wINL}^2 + \sigma_{noise}^2 + \sigma_{ref}^2}. \tag{2}$$

Where the rms sum of the quantization error of a specific bin is represented by σ_{qDNL}, the effect of integral non-linearity is denoted by σ_{wINL} and the rms jitter due to thermal and power supply noise is denoted by σ_{noise}. Although jitter introduced by the reference signal is referred to as σ_{ref}, it does not represent an inherent contribution of the TDC itself, nevertheless, due to measurement restrictions, it is often directly included in the TDC's resolution.

3.1 Device Mismatch

Due to device mismatch, the size of each LSB will vary and manifest itself in integral-non-linearity (INL) errors as well as differential-non-linearity (DNL) errors. If measured in advance, however, INL errors can be corrected off-line whereas DNL errors cannot. For uniformly distributed events, larger bins, due to their higher probability of being hit, will collect more events. This will lead to a degradation in time resolution, as larger bins also suffer from larger quantization error. To account for this behavior, the standard deviation of the quantization error needs to be calculated based on the real value of LSB size taking into account their relative probability p_i to receive a hit, defined as LSB_i/T_{ref}. This relation is more clearly expressed by

$$\sigma_{qDNL} = \sqrt{\sum_{i=0}^{N-1} \left(\frac{LSB_i}{\sqrt{12}}\right)^2 \cdot p_i}. \tag{3}$$

where N represents the number of TDC bins. In a similar manner the standard deviation of the INL can be represented by the rms sum of the INL error of each specific bin also weighed by its relative probability to get hit p_i, expressed as

$$\sigma_{wINL} = \sqrt{\sum_{i=0}^{N-1} \left(INL_i - \overline{INL}\right)^2 \cdot p_i}, \quad \text{with} \quad INL_i = \frac{\Delta LSB_i}{2} + INL_{i-1}, \tag{4}$$

where $\overline{INL}$ represents the mean INL error, ΔLSB_i represents the deviation from the mean LSB size and INL_i the accumulated time error with $INL_{-1} = 0\,\text{ps}$.

If the variation of the LSB size is small compared to its nominal value, the non-weighted standard deviation of the DNL and INL across all bins represents a good estimate. However, in the case of large device mismatch, a weighted representation of the TDC's nonlinearities leads to a more precise estimate of the expected rms-time resolution. From Monte-Carlo simulations, the effect of device mismatch can be well estimated early in the design phase.

3.2 Jitter

Timing variations due to power supply and circuit noise modulate the switching time of the circuit, leading to timing errors. As illustrated in Fig. 3, voltage variations can be well modeled as threshold voltage shifts in the vicinity of a switching point as denoted by σ_{vth}. Thereby, the generated timing error is proportional to the slew-rate (SR) of the signal. From small signal analysis, voltage variations in the vicinity of switching point due to circuit noise as well as power supply noise can be calculated. A periodic-steady-state (PSS) analysis following a periodic-noise (PNoise) simulation as proposed by [9] can be used to calculate σ_t as expressed by Eq. (5).

$$\sigma_t = \sigma_{\text{vth}} \cdot \frac{1}{SR} \tag{5}$$

Mathematically, jitter can be expressed on a per bin basis by

$$\sigma_{t_i} = \lim_{M \to \infty} \sqrt{\frac{1}{M} \sum_{n=0}^{M-1} \left(t_n^{ideal} - t_n^{real}\right)^2} \tag{6}$$

where t_n represents the nth crossing of a signal and M the number of cycles. More often such a definition of jitter is referred to as time-interval-error (TIE). If all error sources are assumed to be uncorrelated, the standard deviation of jitter across all bins, denoted by N, can be expressed by Eq. 7

$$\sigma_{noise} = \sqrt{\frac{1}{N} \sum_{i=0}^{N-1} \sigma_{t_i}^2}. \tag{7}$$

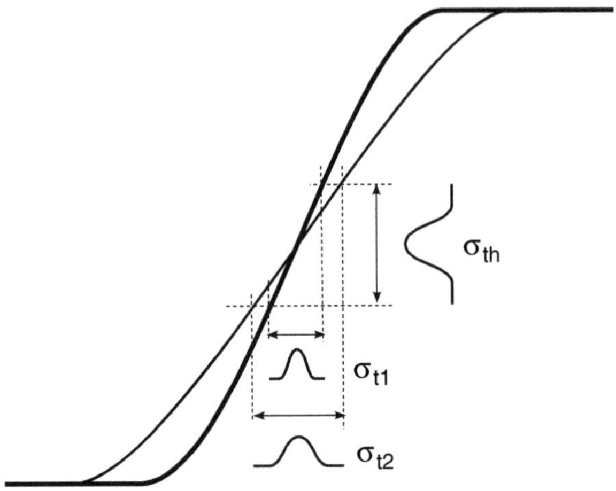

Fig. 3 Timing error dependence on signal slew-rate

Unfortunately, for low-frequency power-supply noise variations, the jitter of adjacent bins will not be truly uncorrelated, making a mathematical representation difficult. Depending on the exact nature of power supply noise, the timing error might not manifest itself as jitter at all. Power supply noise synchronous with the system clock will cause the LSB of the different bins to change by the same amount each clock cycle. This error will finally show up as a non-linearity errors causing an increase of σ_{qDNL} and σ_{wINL}.

Whereas jitter due to thermal noise can be well approximated during design time, timing variations caused by power supply noise are much more difficult to assess early in the design process. To limit the negative effect of power supply noise, it is good practice to minimize propagation delay as well as to employ fast signal edges of timing critical signals.

4 TDC Architecture

To reach ps-rms resolutions, an LSB size of 5 ps is envisaged. A block diagram of the proposed architecture is shown in Fig. 4 as reported in [10]. All measurements are referred to a reference signal allowing multiple TDCs to be synchronized to one common timebase. In a large system where multiple TDCs are operated in parallel, such a synchronization approach is vital. The reference clock signal serves as the time-base of the TDC and is connected to the fine-time interpolator and counter block respectively. Both the fine-time interpolator as well as the coarse counter are shared across all the channels, and so only need to be implemented once per ASIC. The fine-time interpolator generates a set of uniformly distributed signals, here referred to as the fine-time code of the TDC. A counter,

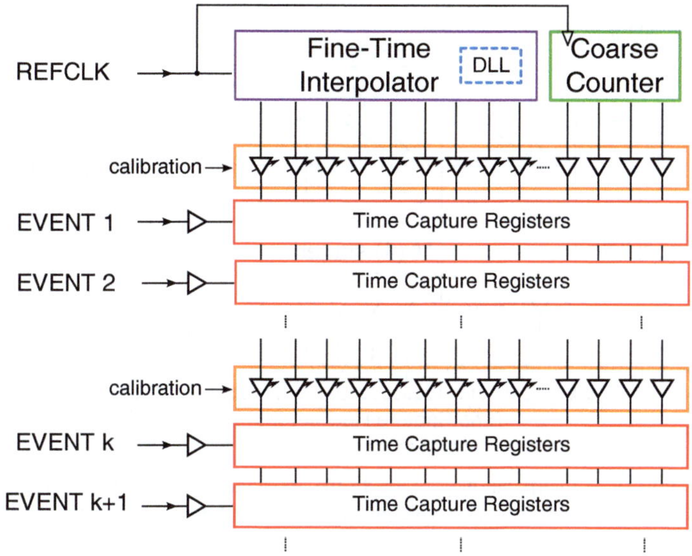

Fig. 4 Proposed multi-channel TDC architecture

which tracks the number of completed clock cycles, is added to efficiently extend the dynamic range of the interpolator by an amount that is only limited by the number of bits of the counter.

The TDC's finest LSB size is generated at the level of the fine-time interpolator. No local interpolation on a per-channel basis is required, thus reducing the complexity of a single channel to a minimum. The generated fine-time as well as the counter code are connected to the respective channels by so-called distributed buffers. To sustain sharp signal edges throughout the channel matrix, several channels are grouped into segments served by a dedicated set of buffers. This efficiently compensates for the RC-delay of the long wires. To compensate for the device mismatch introduced by the fine-time interpolator as well as by the distribution buffers, a calibration feature is integrated at the level of the distribution buffers. No calibration is required on a per-channel basis. PVT variations are compensated by a DLL implemented at the level of the fine-time interpolator block. By controlling the LSB size down to its smallest interpolation level, the LSB size can be adjusted solely by the reference clock frequency. This feature is highly appreciated by the HEP community as it allows time resolution to be traded off against power consumption and to precisely match the TDC's performance to the system requirements.

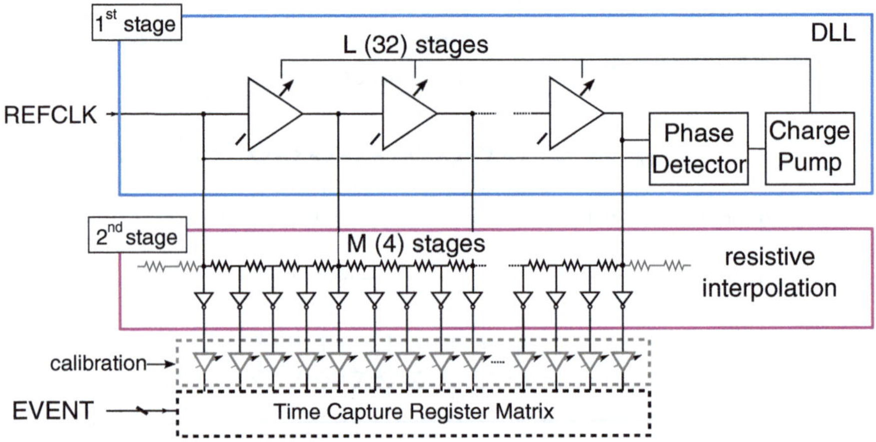

Fig. 5 A multi-stage interpolation concept. In the 1st-stage a DLL is used to reach gate-delay LSB size. In a 2nd-stage resistive voltage division is employed to achieve sub-gate delay resolution

4.1 Fine-Time Interpolator

To achieve sub-gate delay resolutions in conjunction with a large dynamic range, a multistage approach, as has been presented in [11], is pursued. In total $L \cdot M$ timing signals are generated, where L represents the interpolation ratio in the first stage and M the interpolation ratio in the second stage. In the first stage a DLL is employed to generate L uniformly distributed signals. Thereby, the delay of the delay-line elements is adjusted by the feedback loop so that the sum of all elements within the loop precisely match one reference clock period T_{ref}. In the locked condition, the delay of each delay element is equal to T_{ref}/L and is only controlled by the input frequency of the DLL. The second stage is based on resistive voltage division. Signals generated in the first stage serve as inputs of the second stage to derive finer delayed signals to overcome the propagation delay limitations of the technology. This allows LSB sizes as small as T_{ref}/LM to be achieved (Fig. 5).

To reduce the amount of propagation delay within the loop, it is advantageous to run the DLL at high frequencies. The higher this is, the less delay stages are required to achieve a given LSB size. Running the DLL at high frequencies also minimizes the effect of device mismatch and decreases the DLL's noise sensitivity. However, for practical purposes, to reduce potential difficulties with handling high frequencies off-chip, the DLL clock is preferably kept below 2 GHz. In a 130 nm technology, the gate delay is in the order of 20 ps. To ease digital coding later on in the system, integer interpolation factors of base two are preferred. To achieve the envisaged 5 ps LSBs, choosing $L = 32$ and $M = 4$ represents a good choice. Such a combination requires a DLL clock frequency of 1.5625 GHz.

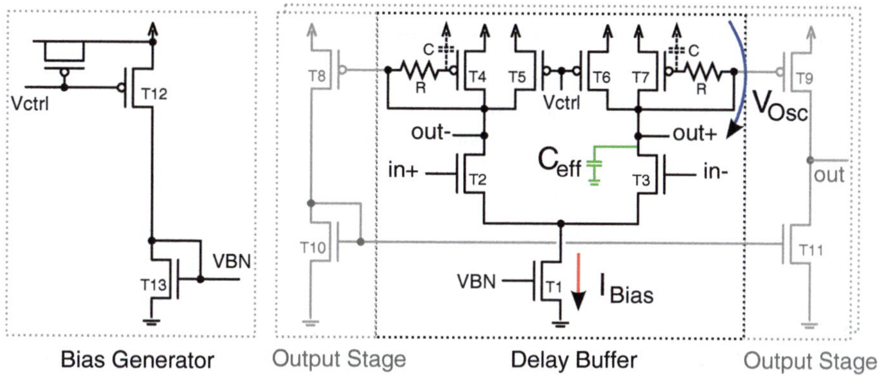

Fig. 6 Bias circuit and fast delay cell element of the DLL. An additional zero is added in the signal path to speed up the cell

4.2 Delay-Cell Element

Careful design of the DLL's delay-cell element is needed to achieve short propagation delays, and to reduce the complexity of the 2nd stage. A modified version of the fully-differential delay buffer presented in [12] is employed. A schematic diagram of the delay buffer together with its bias generation circuit is shown in Fig. 6. The inputs of the delay buffer *in+* and *in−* are connected to the outputs *out+* and *out−* of the succeeding cell. The propagation delay of the cell can be adjusted by means of V_{ctrl} which is adjusted to the desired delay by the feedback mechanism of the DLL. An operational transconductance amplifier (OTA) is employed to generate a single-ended signal that distributes the fine-time code across the channel matrix. Such an approach turns out to be feasible as the propagation delay introduced by the distribution buffers is small compared to the delay introduced by the delay-line itself. To increase the cells robustness against power supply disturbances, large signal swings are preferred.

The propagation delay of the cell can be very roughly estimated by $\tau \approx \frac{V_{osc} \cdot C_{eff}}{2 \cdot I_D}$ where V_{osc} represents the oscillation voltage, C_{eff} the total capacitive load at the output and I_D represents the current defined by *VBN*. To additionally speed up the cell, a zero is added to the signal path to hide the gate capacitance of the top PMOS diode by using resistive peaking [13]. From the equivalent half circuit of the cell, neglecting channel length modulation effects, an output impedance of

$$Z_{out} = \frac{1}{gm_2} \cdot \frac{1 + sRC}{1 + s\frac{C}{gm_2}} \tag{8}$$

can be derived. For $\frac{1}{RC} \ll s \ll \frac{gm_2}{C}$ the output resistance reduces to $\frac{sRC}{gm_2}$ making the load look like an inductor. This can be thought of as hiding the diode-connected load C during the switching cycle. With resistive peaking, an additional reduction

Table 1 Device dimensions of the circuit shown in Fig. 6

Device	Width (μm)	Length (μm)
T1 and T13	18	0.6
T2 and T3	4	0.12
T5 and T6	4.5	0.12
T4, T7, T8 and T9	3	0.12
T10 and T11	1.36	0.12
T12	7.5	0.12
R	11 kΩ	

in propagation delay of 9 % is achieved. The differential-to-single ended (DE/SE) converter is designed to provide similar current drive strength as in the main cell and is sized to reduce the propagation delay employing small length devices.

With the interpolation factors proposed in Sect. 4.1 the maximum propagation delay of the delay cell must not exceed 20 ps. With dimensions as given in Table 1 propagation delays as low as 16 ps can be achieved in a 130 nm technology when the cell is operated with a 1.2 V supply. Across process voltage and temperature, the delay is expected to be vary from 12 ps to 23 ps. To compensate for slow process corners, the supply voltage can be increased up to 1.5 V.

4.3 Resistive Time Interpolation

To overcome the propagation delay limitation of the technology used, a resistive division interpolation concept is employed [14]. Low power consumption as well as high robustness against device mismatches represent the most attractive arguments of a resistive interpolation concept. The basic structure of the concept is depicted in Fig. 7. A resistive voltage divider is connected across the outputs of the DLL. Due to the voltage drop across the resistive ladder finer time delays can be generated. The number of DLL elements involved during the switching cycle depends on the slope of the signals propagating down the ladder. To a first approximation, the number of elements involved in the process can be estimated from the signal slope as expressed by $\frac{t_{slope}}{LSB}$. Fast signal slopes require stronger drivers, whereas on the other hand, slower signal slopes are more sensitive to power supply noise and require more elements at the beginning and the end of the DLL to reach uniformity. To compensate for delays introduced by the RC delay characteristics of the scheme, a non-linear resistive divider has been implemented employing resistor values varying from 28 to 48 Ω. A good trade-off has been found for signal slopes of 120 ps. This involves approximately 6 delay elements or equivalently 24 LSB codes. To profit from the stronger NMOS driving capability, the negative edge is used to distribute the fine-time code across the channel matrix.

A nice feature of the passive interpolator structure is its device-mismatch filtering capability. In a similar manner as reported by [15], due to resistive coupling,

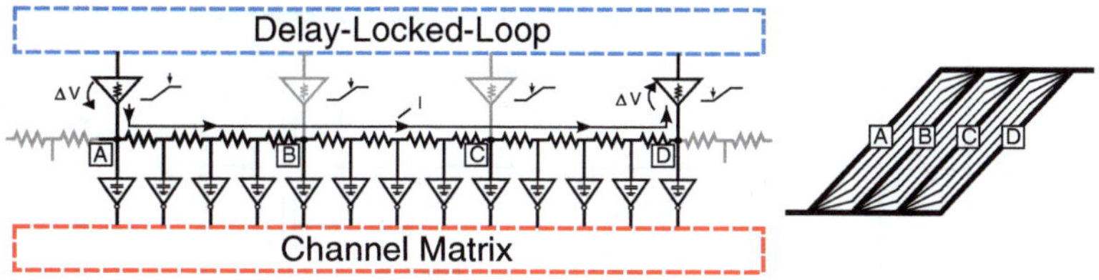

Fig. 7 Resistive voltage division principle

delay variations of adjacent cells are effectively averaged. To a first approximation, the expected 1-sigma standard deviation is scaled by $1/\sqrt{K}$ where K represents the number of elements involved in the signal transition. Due to the long tails of the transitions, a high coupling between adjacent cells is achieved. For the specific implementation, a reduction by more than a factor 5 could be achieved in the mismatch of the resulting falling edges. However, due to the small output buffers used to buffer these timing signals, additional mismatch is added. This finally leads to an improvement of "only" a factor of 1.6.

4.4 Calibration

To compensate for device mismatches introduced by the fine-time interpolator and distribution buffers, an adjustment feature has been included at the level of the distribution buffers. From Monte-Carlo simulations the expected standard deviation of the fine-time interpolator including the distribution buffers across all bins has been estimated to be 1.8 ps-rms. To not only allow DNL errors to be corrected but also INL errors of up to 6.4 LSB, a 5 bit adjustment feature has been implemented, see Fig. 8. Capacitive loading is employed to adjust the propagation of the timing signals. To avoid placing a large capacitive load at the output of the buffer, the adjustment feature is implemented after the first buffering stage. In total up to 64 fF in 2 fF steps can be added, allowing signals to be delayed by up to 32 ps in 1 ps steps. Due to the additional capacitive loading, the power consumption is increased by approximately 25 % compared to a cell without the calibration feature.

4.5 Time Capturing

To capture the time-of-arrival of an event, the actual state of the fine-time code as well as the counter value are stored in so called time capture registers (TCRs). As illustrated in Fig. 9 the event signal is connected to the CLK-input of the TCRs whereas the fine-time code is connected to the D-input of the registers. From the latched code (i.e. the 1 to 0 transition) the exact time-of-arrival can be resolved

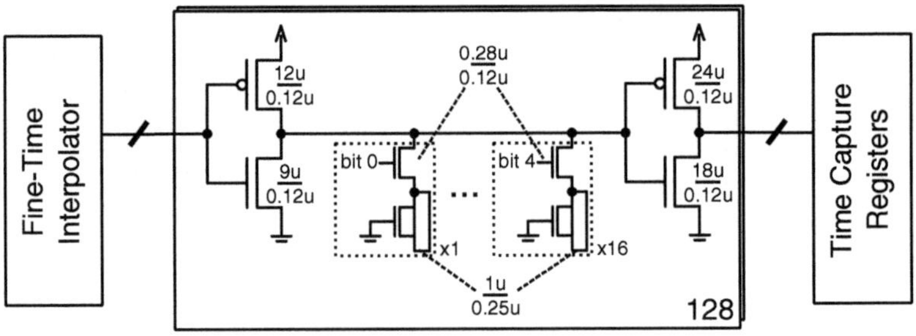

Fig. 8 Schematic diagram of the distribution buffers to distribute the fine-time code. Capacitive loading is used to adjust for device mismatches

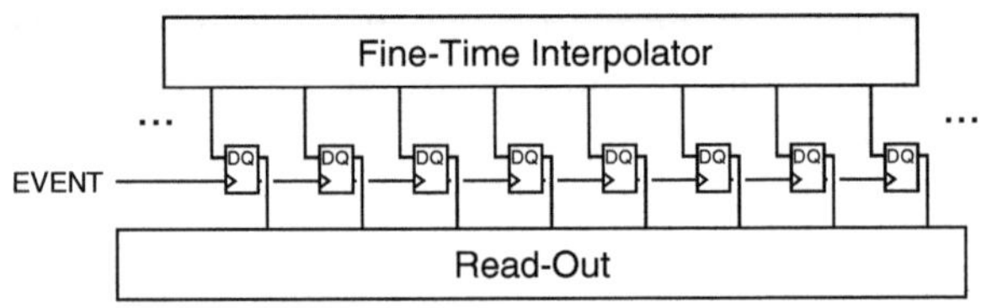

Fig. 9 Illustration of the time capturing concept. The time-of-arrival of an event is captured by sampling the state of the fine-time interpolator

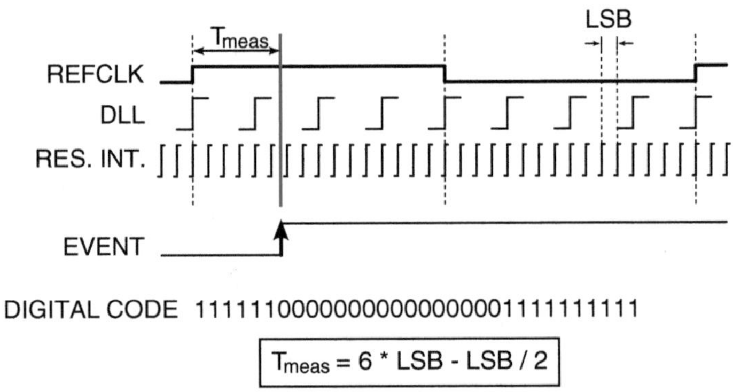

Fig. 10 Timing diagram of the time capturing process

(Fig. 10). For the event to be captured, the first latch of the TCRs needs to be kept transparent to allow the latch to follow the state of the fine-time code signals. This causes the 1st latch to switch with the frequency of the reference input, causing additional power to be consumed.

To balance power consumption and device mismatches of the registers, two different versions of the TCRs have been implemented. One register has been taken from the standard cell library together with its layout, whereas the other

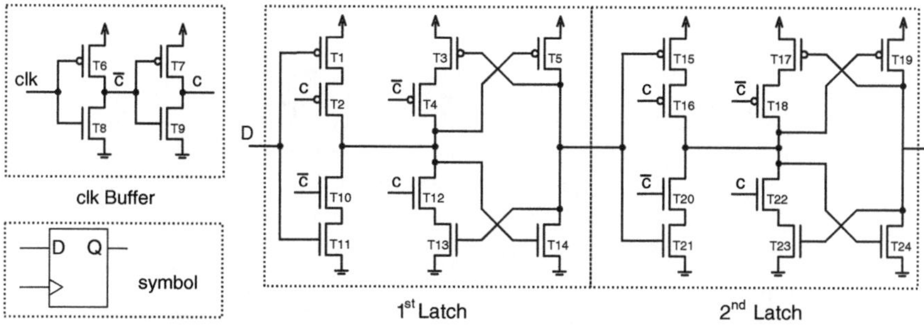

Fig. 11 Schematic diagram of the TCRs. The 1st latch is optimized for timing

Table 2 Device dimensions of the circuit shown in Fig. 11

Device	Standard cell TCR width (μm)	Custom TCR width (μm)	Length (μm)
T1–T7	1	3	0.12
T8–T14	0.5	1.5	0.12
T15–T19	1	1	0.12
T20–T24	0.5	1	0.12

Two different versions have been implemented to compare matching performance

register has been custom designed and laid-out to improve timing performance and reduce parasitic capacitances. The schematic diagram of the employed TCR is depicted in Fig. 11. The dimensions of both versions are listed in Table 2. In the custom TCR, the 1st latch has been optimized for good matching. From Monte-Carlo simulations, the expected 1-sigma variation of the time capturing point is about 2.4 ps-rms for the standard TCR and 1.3 ps-rms for the custom TCR respectively. Whereas, the custom TCR has been designed to achieve sufficient timing accuracy for a 5 ps LSB TDC the standard TCR is expected to be suitable only for a 10 ps LSB TDC. With either register, distinct channel pairs have been implemented to compare simulation results against measurements.

5 Experimental Results

A prototype has been designed and fabricated in a commercial 130 nm technology. In Fig. 12 a micro-photograph of the constructed ASIC wire-bonded to its carrier board is shown. The demonstrator consists of 8 channels, together with the fine-time interpolator as well as the distribution buffers and their bin adjustment feature. Different channel configurations were implemented to investigate the effect of device mismatch, input buffer architecture as well as time capturing concept. With the constructed test setup, the effect of different input buffer architectures was found to be negligible. The performance of the channels making use of the

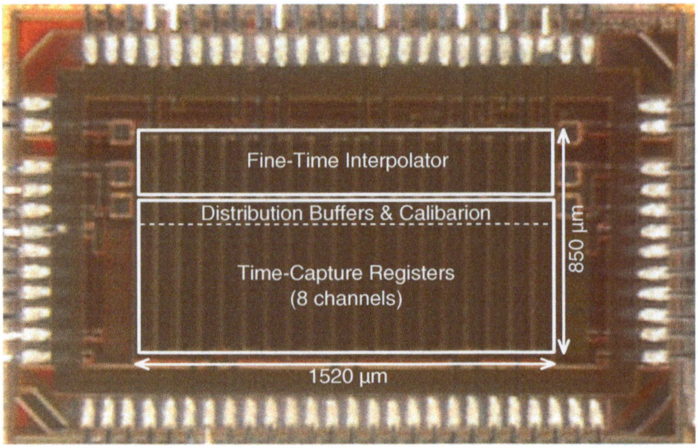

Fig. 12 Micro-photograph of the demonstrator ASIC

Table 3 Performance summary of the TDC for the standard cell and custom TCR

Technology	130 nm	
Supply voltage	1.3 V	
Chip area	1.2 mm^2	
# of channels	8	
	Custom TCR	*Standard cell TCR*
Ref. frequency	1.5625 GHz	781.25 MHz
LSB size	5 ps	10 ps
DNL	±0.9 LSB	±0.9 LSB
INL	±1.4 LSB	±0.7 LSB
Single shot precision	<2.5 ps	~5 ps
Power consumption[a]	36.5 mW	13.3 mW
Dynamic range	640 ps (on chip)	1,280 ps (on chip)

[a] Per channel power consumption excluding shared components

time capturing concept described in Sect. 4.5 was also investigated. The TDC's time reference was supplied by a low jitter clock generator with a measured period jitter of <1 ps-rms. A summary of the resulting performance is listed in Table 3.

5.1 Non-linearities

To extract the transfer characteristic of the TDC, a uniformly distributed sequence of events was generated. From the number of samples collected by each respective bin, the actual LSB size can be approximated by $LSB_i = \frac{\#Events_i}{\#TotalEvents} \cdot T_{ref}$. A sequence of 10^5 events was found to approximate the LSB size with satisfying accuracy.

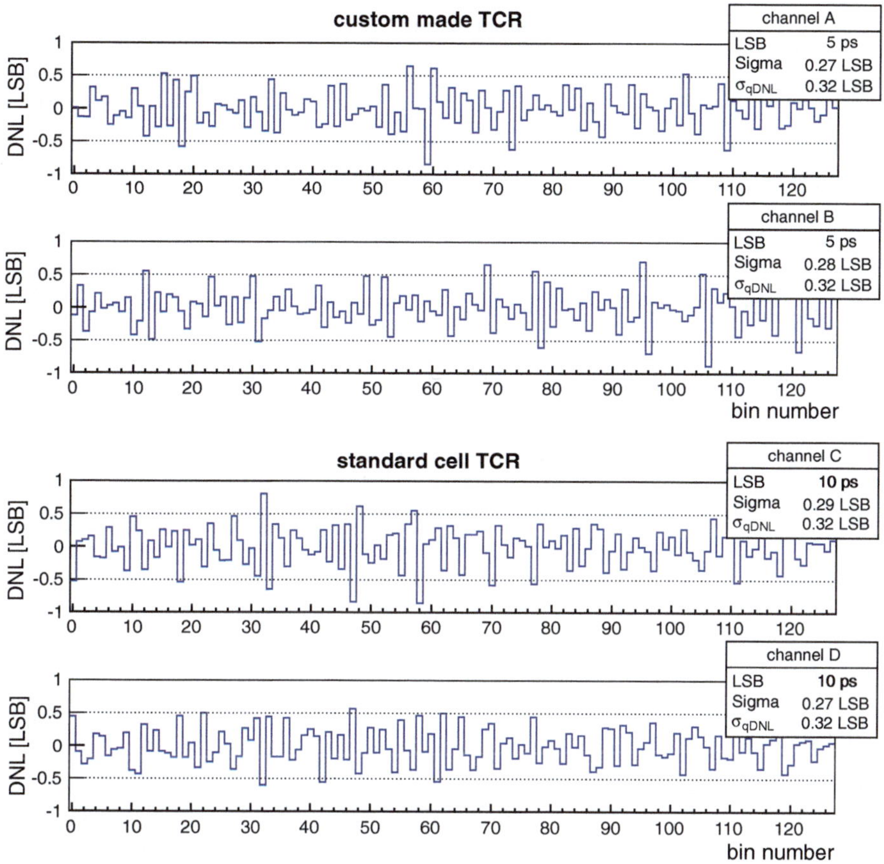

Fig. 13 Measured DNL of all 128 bins of the interpolator for *channel A–D*. The *boxes* list the 1-sigma distribution as well as the weighted DNL error calculated after Eq. (3)

In Figs. 13 and 14 receptively, the measured DNL and INL of channels A–D after global calibration has been applied is shown. For the channels employing the standard TCR (i.e. C and D) the reference clock period has been reduced to 781.25 MHz to account for the lower timing precision of the standard TCR. Across the whole dynamic range of the TDC no missing codes were observed. As a first estimate, the expected single-shot precision can be calculated by $\sigma_{TDC} \approx \sqrt{\sigma_{qDNL}^2 + \sigma_{INL}^2}$. This accounts for timing variations resulting from the TDC's quantization noise and non-linearities. From the measurement results, the expected time resolution for channels using the standard cell TCR and the custom-made TCR of 4 and 2 ps-rms respectively was calculated. From Monte-Carlo simulations, the achievable time resolution has been estimated to be 4.8 and 2.9 ps-rms respectively, well in line with measurements.

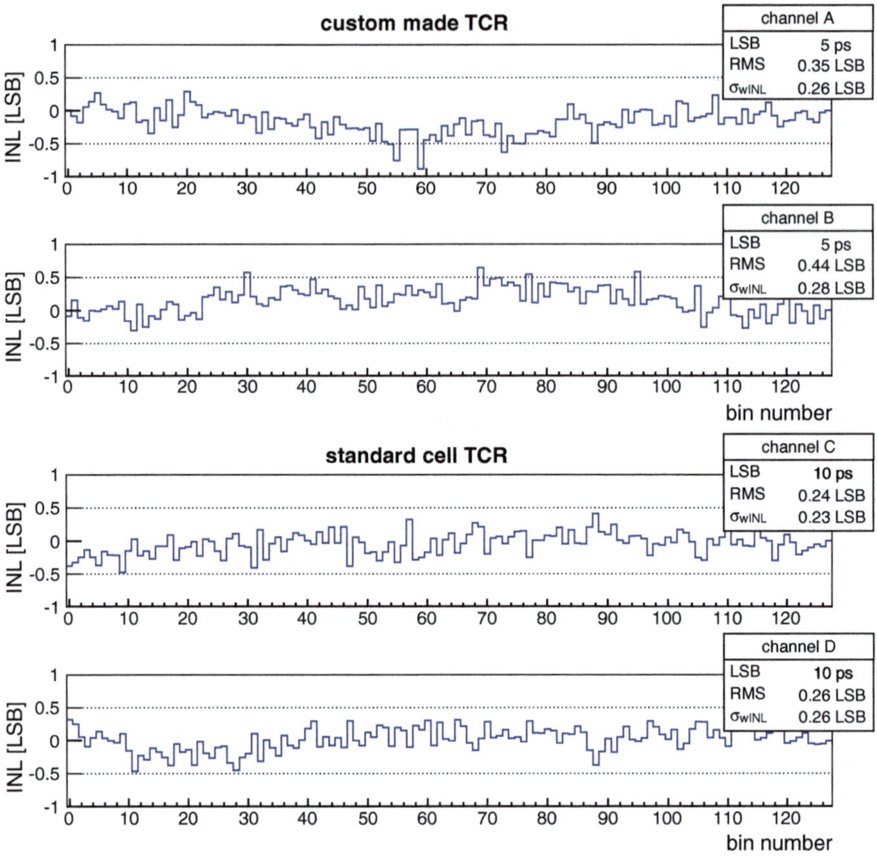

Fig. 14 Measured INL of all 128 bins of the interpolator for *channel A–D*. The boxes list the 1-sigma distribution as well as the weighted INL error calculated after Eq. (4)

5.2 Single-Shot Precision

To measure the actual rms-time resolution often also referred to as the single-shot precision of the TDC, a uniformly distributed sequence of events was generated and sent to two distinct channels. A fixed length of wire was added to one of the channels to generate a constant propagation delay between the two channels. This allows the generation of delay differences that only depend on the length of wire and so are robust to voltage and temperature variations. To generate two copies from a single event, a resistive power splitter is employed. From the collected samples, the bin number difference between the two channels is histogramed and the underlying Gaussian distribution is extracted. Its standard deviation is used to estimate the single-shot precision of the TDC. As two edges are involved in the measurement the standard deviation has to be scaled by the $1/\sqrt{2}$ as expressed by

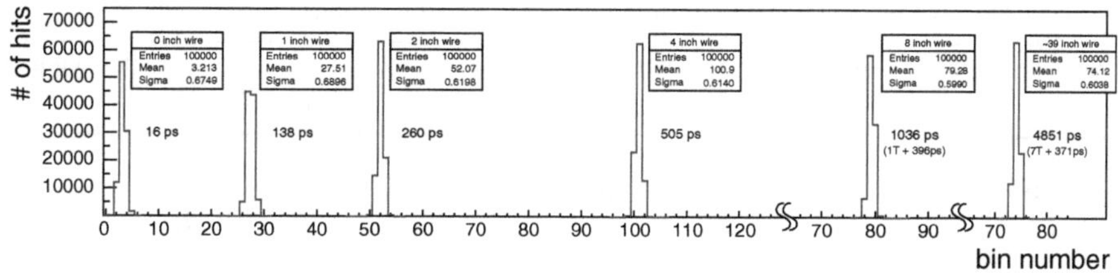

Fig. 15 Measured delay difference of channel pair A and B for a 4 in. length wire

Eq. (9). Using a time difference measurement, only timing contributions coming from the TDC itself as well as jitter resulting from the time reference signal are recorded by the measurement. Any contribution resulting from the event signal itself are excluded from the measurement.

$$\sigma_{TDC} = \frac{\sigma_{\Delta bin} \cdot LSB}{\sqrt{2}} \tag{9}$$

In Fig. 15 the recorded time difference of channel pair A & B employing the custom TCR for different wire delays is shown. For the custom TCR, the TDC reference clock was set to 1.5625 GHz resulting in a 5 ps LSB size TDC. In the case of the standard TCR, the TDC was operated with a 781.25 MHz generating 10 ps LSB sizes. Different time delay differences were generated to record the TDC's precision across its dynamic range. The delay difference between the two channels has been adjusted so that the event signal of the 2nd channel arrives (a) within one clock cycle, (b) one clock cycle later and (c) multiple clock cycles later. However, no notable influence of the wire length could be observed. After calibration, a single-shot precision of 2.5 ps-rms for the channels using the custom made TCR and 5 ps-rms for the channels employing the standard cell TCR has been achieved. This result is well in line with the expected time resolution derived earlier in Sect. 5.1.

5.3 Power Consumption

The power consumption of the TDC is extracted by means of a current-voltage measurement. For the architecture to capture an event, the first latch of the TCR has to be transparent. Lower power is consumed if acquisition is not running. When operated with a 1.5625 GHz reference clock an equivalent power of 34 mW to 42 mW per channel is consumed by the demonstrator. If a higher number of channels is implemented the contribution of shared components can be reduced. In Table 4 the power consumption contribution of a single channel (including 1.3 mW for the I/O buffer) is estimated from simulation to be 36.8 mW and 13.3 mW per channel, for channels employing the custom and standard cell TCR

Table 4 Performance comparison of different channel configurations

Channel	LSB (ps)	Single-shot (ps-rms)	Power per ch.		# ch. per segment
			Acqu. on (mW)	Acqu. off (mW)	
A and B	5	<2.5	36.8	23.8	9
A and B	10	~4	19.3	13.0	9
A and B	20	~7	10.9	7.6	9
C and D	10	~5	13.3	10.5	11

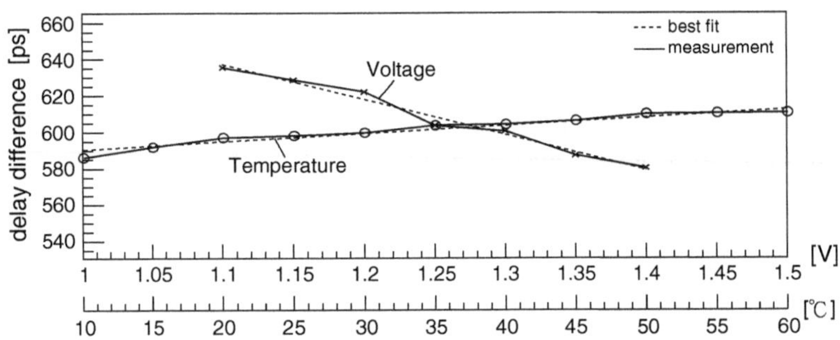

Fig. 16 Measured delay variations due to voltage and temperature shifts

respectively. Less power is consumed at lower reference clock frequencies, i.e. larger LSB size. For the estimate, the number of channels per segment has been chosen to equally load the output of the distribution buffers.

5.4 PVT Sensitivity

Any propagation delay not kept stable across temperature and voltage will suffer from propagation delay variation effects. In the proposed architecture, the LSB size is solely determined by the reference signal's frequency and is thus held stable across PVT variations. However, the event and reference clock I/O buffers as well as the distribution buffers are not adjusted by the loop. Any delay difference introduced in those two paths will show up as a constant time shift in the measurement. Depending on the absolute change in propagation delay, the sign can either be positive or negative.

To characterize the circuit's sensitivity to voltage and temperature variations, an event with a fixed phase relationship with respect to the reference clock was generated and recorded across voltage and temperature. As depicted in Fig. 16, a voltage sensitivity of −0.2 ps/V and a temperature variation of 0.4 ps/°C were measured. These variations were found to be negligible compared to the variations introduced by other building blocks within the measurement chain (e.g. charge preamplifier [16]).

Table 5 Performance comparison

References	Method	LSB (ps)	Single shot	Power[a]	Channels	Robustness[b]
[7]	Time amp.[c]	8.9	–	–	128	~/+
[17]	RC-delay	24.4	15.8 ps-rms	125 mW	8	~/~
[18]	WaveUnion	3.7[d]	2.5 ps-rms	–	10	+/~
[19]	Cap. scaling	12.2	13 ps-rms	20 mW	2	+/~
[14]	Pas. interpl.[c]	4.7	3.3 ps-rms	3.6 mW[e]	1	~/+
[5]	Time amp.[c]	1.25	0.6 ps-rms	3 mW[f]	1	–/~
This	Pas. interpl.	5	2.5 ps-rms	43 mW	8	+/+

[a] Per channel
[b] PVT and mismatch /power supply noise
[c] START-STOP measurement
[d] Equivalent LSB size
[e] 180 MHz sampling frequency
[f] 10 MHz sampling frequency

6 Conclusion

A multi-channel TDC architecture precisely matched to the requirements of next generation HEP experiments has been presented. When operated with a 1.5625 GHz reference clock, 5 ps LSB sizes are generated. After calibration, a DNL and INL of ±0.9 LSB and ±1.4 LSB have been achieved. The single-shot precision of the TDC has been evaluated by means of a time-difference measurement for different wire length differences. For the better matching channels, a single-shot precision of better than 2.5 ps-rms has been demonstrated. The full prototype consumes between 34 mW/channel to 42 mW/channel. Lowering the input clock frequency to 781 MHz (=10 ps LSB sizes), the power consumption can be reduced to 21 mW/channel to 26 mW/channel respectively. The architecture exhibits a time shift in presence of voltage variations of −0.19 ps/mV and experiences a temperature dependence of 0.44 ps/deg. With the measurement precision of the test setup, inter-channel crosstalk between two neighboring channels has been evaluated to be below ±1 LSB.

In Table 5 a comparison with previously published TDCs is given. Compared to other high channel count TDCs, the presented architecture achieves very fine resolution with a high degree of robustness to power supply noise and PVT variations. The time measurements of the architecture are referred to a reference clock that can be common to many devices. In HEP experiments the event signal can arrive at any time, and so the TDC is required to run continuously as well as to offer a large dynamic range using a counter. The reader should observe that in such a continuously running environment, in contrast to START-STOP architectures, the interpolator cannot be disabled to reduce power consumption.

References

1. S. White, M. Chiu, M. Diwan, G. Atoyan, and V. Issakov, "Design of a 10 picosecond Time of Flight Detector using Avalanche Photodiodes," 2009.
2. R. Forty, and M. Charles, "Torch: a novel time-of-flight detector concept," CERN, Geneva, Tech. Rep. LHCb-PUB-2009-030. CERN-LHCb-PUB-2009-030, Nov 2009.
3. L. Adamczyk, "AFP: A proposal to install proton detectors at 220 m around ATLAS to complement the ATLAS high luminosity physics program," 2011. [Online]. Available: http://atlas-project-lumi-fphys.web.cern.ch/atlas-project-lumi-fphys/default.html
4. N. Harnew, "TORCH: A large-area detector for precision time-of-flight measurements at LHCb," *Physics Procedia*, vol. 37, no. 0, pp. 626 – 633, 2012, Proceedings of the 2nd International Conference on Technology and Instrumentation in Particle Physics (TIPP 2011). [Online]. Available: http://www.sciencedirect.com/science/article/pii/S1875389212017427
5. M. Lee, and A. Abidi, "A 9 b, 1.25 ps resolution coarse-fine time-to-digital converter in 90 nm CMOS that amplifies a time residue," *IEEE Journal of Solid-State Circuits*, vol. 43, no. 4, pp. 769 –777, April 2008.
6. P. Keranen, K. Maatta, and J. Kostamovaara, "Wide-range time-to-digital converter with 1-ps single-shot precision," *IEEE Transactions on Instrumentation and Measurement*, vol. 60, no. 9, pp. 3162 –3172, Sept 2011.
7. S. Mandai, and E. Charbon, "A 128-channel, 8.9-ps LSB, column-parallel two-stage TDC based on time difference amplification for time-resolved imaging," *IEEE Transactions on Nuclear Science*, vol. 59, no. 5, pp. 2463–2470, 2012.
8. D. Schwartz, E. Charbon, and K. Shepard, "A single-photon avalanche diode array for fluorescence lifetime imaging microscopy, " *IEEE Journal of Solid-State Circuits*, vol. 43, no. 11, pp. 2546 –2557, Nov 2008.
9. K. Kundert, "Modeling jitter in PLL-based frequency synthesizers," 2003.
10. L. Perktold and J. Christiansen, "A high time-resolution (<3 ps-rms) time-to-digital converter for highly integrated designs," in *Instrumentation and Measurement Technology Conference (I2MTC), 2013 IEEE International*, 2013.
11. L. Perktold and J. Christiansen, "A flexible 5 ps bin-width timing core for next generation high-energy-physics time-to-digital converter applications," in *2012 8th Conference on Ph.D. Research in Microelectronics and Electronics (PRIME)*, 2012, pp. 1–4.
12. J. Maneatis, "Low-jitter and process independent DLL and PLL based on self biased techniques," in 1996 *IEEE International Solid-State Circuits Conference, 1996. Digest of Technical Papers. 42nd ISSCC.*, Feb 1996, pp. 130–131, 430.
13. S. Anand and B. Razavi, "A CMOS clock recovery circuit for 2.5-Gb/s NRZ data," *IEEE Journal of Solid-State Circuits*, vol. 36, no. 3, pp. 432 –439, Mar 2001.
14. S. Henzler, S. Koeppe, D. Lorenz, W. Kamp, R. Kuenemund, and D. Schmitt-Landsiedel, "A local passive time interpolation concept for variation-tolerant high-resolution time-to-digital conversion," *IEEE Journal of Solid-State Circuits*, vol. 43, no. 7, pp. 1666 –1676, July 2008.
15. M.-W. Chen, D. Su, and S. Mehta, "A calibration-free 800 MHZ fractional-n digital PLL with embedded TDC," *IEEE Journal of Solid-State Circuits*, vol. 45, no. 12, pp. 2819–2827, Dec 2010.
16. E. Martin et al., "The 5 ns peaking time transimpedance front end amplifier for the silicon pixel detector in the NA62 Gigatracker," in *Nuclear Science Symposium Conference Record (NSS/MIC), 2009 IEEE*, 2009, pp. 381–388.
17. J. Christiansen, "Manual: HPTDC—high performance time to digital converter," 2004. [Online]. Available: http://tdc.web.cern.ch/tdc/hptdc/docs/hptdc_manual_ver2.2.pdf

18. E. Bayer, P. Zipf, and M. Traxler, "a multichannel high-resolution (5 ps RMS between two channels) time-to-digital converter (TDC) implemented in a field programmable gate array (FPGA)", in *Nuclear Science Symposium and Medical Imaging Conference (NSS/MIC), 2011 IEEE*, 2011, pp. 876–879.
19. J.-P. Jansson, A. Mantyniemi, and J. Kostamovaara, "A CMOS time-to-digital converter with better than 10 ps single-shot precision," *IEEE Journal of Solid-State Circuits*, vol. 41, no. 6, pp. 1286–1296, 2006.

Time-Domain Techniques for mm-Wave Frequency Generation

Wanghua Wu, Robert Bogdan Staszewski and John R. Long

Abstract The demand for higher integration level and lower production cost has driven mm-wave electronics, which have traditionally been implemented in III-V technologies for better RF performance, to be also implemented in CMOS. This motivates the digitization of the mm-wave systems for improved RF performance. This paper focuses on a digitally intensive architecture and time-domain circuit and calibration techniques for mm-wave frequency synthesizer. A 60-GHz all-digital phase-locked loop (ADPLL) transmitter prototype, implemented in 65-nm CMOS, achieves excellent phase noise (−75 dBc/Hz at 10 kHz offset), fast locking (3 μs), low reference spurs (−74 dBc), and linear frequency modulation up to 1 GHz in range.

1 Introduction

The millimeter-wave (mm-wave) industry has been dominated historically by high-performance technologies intended for low-volume production in communication, security and defense applications. However, transistor scaling extends the capabilities of CMOS circuits and systems into the mm-wave range, where the integration density and cost/volume advantages demonstrated by CMOS SoCs for cellular and WLAN may be applied to mm-wave applications. With 65-nm bulk CMOS technologies in production offering peak transit frequency (f_T) and maximum frequency of oscillation (f_{max}) close to 200 GHz [1], several experimental

W. Wu (✉)
Marvell Semiconductor Inc., Central Engineering, 1 Vista Montana Apt 5316, Santa Clara, CA 95134, USA
e-mail: wanghuawu@gmail.com

W. Wu · R. B. Staszewski · J. R. Long
Electronics Research Laboratory/DIMES, Delft University of Technology, Delft
The Netherlands

P. Harpe et al. (eds.), *High-Performance AD and DA Converters, IC Design in Scaled Technologies, and Time-Domain Signal Processing*,
DOI: 10.1007/978-3-319-07938-7_15,

60-GHz transceivers achieving above 4-Gb/s data rate over a 2-m link have been reported [2–4]. These 60-GHz prototypes employing analog transceiver architectures have demonstrated the potential to use deep-submicron CMOS for RF/baseband co-integration. More importantly, there is room for improved RF performance (e.g., lower phase noise, higher output power, etc.), and reduced power dissipation and chip area, since the analog RF circuits cannot fully share the benefits of CMOS scaling.

To maintain reliability when scaling MOS devices, the supply voltage has been reduced to <1 V, while threshold voltage remains almost constant (to suppress leakage current). This reduces the available voltage headroom when transistors are intended to operate as current sources. Moreover, backend capacitors of tens of picofarads are usually required to integrate the baseband filters and loop filters in charge-pump PLLs on-chip, which occupy considerable chip area.

On the other hand, digital gate density doubles and the basic gate delay improves linearly with continued scaling of CMOS technologies (i.e., from 90 to 65-nm, then to 40-nm, and so on). The fast switching characteristics of CMOS logic (rising/falling time of 20 ps in 40-nm CMOS) enables high-speed clocks and fine control of timing transitions. The high density of digital logic (1 Mgates/mm^2) and SRAM (4 Mb/mm^2) allows implementation for many programmable digital functions SOC applications. Thus, the time-domain signal processing is more amendable to IC implementation in the future compared to voltage-domain [5], which motivates the digitization of RF to fit into the deep sub-micron CMOS paradigm.

More digital than analog RF design (i.e., Digital-RF) digitizes the majority of the system with an emphasis on using time-domain techniques to obtain scalability and higher system performance. An analog-to-digital conversion usually takes place in the very early stage of the system in order to benefit the most from the digital signal processing. The Digital-RF increases the reconfigurability and the testability of the RF system, reduces the turnaround cycles by using automated digital implementation tools and flows, is easier to migrate between technology nodes, and may lead to a smaller silicon area and less power consumption.

The following sections focus on time-domain and digitally intensive approaches for mm-wave frequency generation. The all-digital phase-locked loop (ADPLL) is introduced and compared to an analog charge-pump PLL in Sect. 2. Then, a multi-rate ADPLL architecture for mm-wave frequency synthesis and wideband frequency modulation is presented in Sect. 3. Critical time-domain circuits for mm-wave ADPLL are elaborated in Sect. 4, which are high-resolution mm-wave digitally-controlled oscillators (DCOs), and a time-to-digital converter (TDC). On-chip digital calibration techniques are highlighted in Sect. 5. Experimental results of a 60-GHz multi-rate ADPLL prototype are presented, which employs the time-domain techniques presented in this paper and demonstrates both high spectral purity and wideband frequency modulation (FM) capability in Sect. 6.

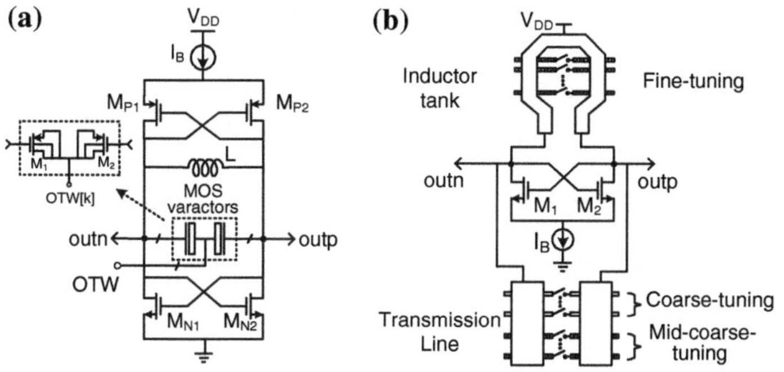

Fig. 3 **a** Schematic of a conventional gigahertz DCO, **b** schematic of a mm-wave DCO with distributed switched-metal capacitors

capacitor (e.g., $\Delta C_0 = 35$ aF in 65-nm CMOS) is ~15 MHz for a 60-GHz carrier, which contributes significant quantization noise when integrated in a PLL (i.e., $f_0 = \frac{1}{2\pi\sqrt{L_0 C_0}}$, assuming $\frac{\partial f_0}{\partial C_0} \approx \frac{\Delta f_0}{\Delta C_0} = -\frac{f_0}{2C_0}$, thus $\Delta f_0 = \frac{60\,\mathrm{GHz}}{2 \cdot 70\mathrm{fF}} \cdot (35\,\mathrm{aF}) = 15\,\mathrm{MHz}$).

The MOS-varactor-based DCO is not suitable for mm-wave DCOs. We propose a DCO tank topology based on distributed switched-metal capacitors, as shown in Fig. 3b to achieve wider tuning range and better PN [19]. The distributed LC-tank consists of a transmission line (TL), inductor or transformer (e.g., Fig. 3b use TL and inductor), and pairs of metal shield strips located beneath the resonator and distributed along its major dimension in various metal layers. These metal strips form digital tuning banks that are distributed along the length of the resonator. Each metal shield strip pair is connected to a MOS switch driven by a digital tuning signal. Activating the switch varies the capacitive load on the resonator and introduces a distinct phase shift in the DCO loop that varies the oscillator frequency. Moreover, the phase shift introduced by each metal strip pair varies with its position along the resonator, the metal layer used for implementation, and its physical dimension (width and spacing). This attribute is further exploited to form coarse- and fine-tuning banks, and thus to optimize the tuning range and frequency resolution, simultaneously.

For coarse-tuning, a digitally controlled TL is used (see Fig. 3b). Its 3-D view is depicted in Fig. 4a. The oscillation signal (e.g., at 60 GHz) runs along the TL in thick top metal to reduce conductor loss and loss from the silicon substrate. Shorting metal strips beneath the differential TL via NMOS switches increases the capacitance per unit length, thus reducing the wavelength $\left(\lambda = \frac{1}{f\sqrt{LC}}\right)$ of the RF signal as described in [20, 21]. This increases the phase shift along the TL and reduces the tank resonant frequency. The tuning elements are connected via the signal path of the resonator in top metal without any additional interconnecting wires, which eliminates complex wiring scheme required for conventional varactor tuning and reduces the wiring capacitance. The design of the NMOS switch

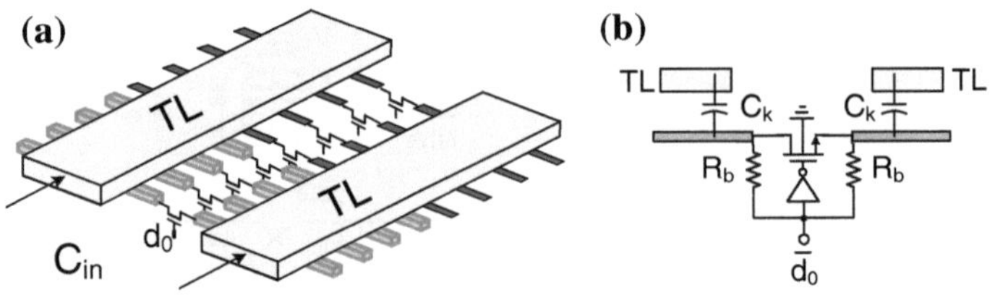

Fig. 4 **a** 3-D view of a digitally controlled TL for coarse tuning, **b** schematic of the NMOS switch

(see Fig. 4b) involves a trade-off between tuning range and Q-factor, which is further exacerbated by the increased operating frequency [22].

In order to obtain smaller discrete tuning steps, two fine-tuning techniques incorporating a distributed switched-metal capacitor bank into either an inductor or transformer are introduced. Both techniques achieve ~2 MHz frequency resolution for a 60-GHz carrier without employing minimum-size components so that interconnection parasitics do not limit the frequency step size and uniformity.

A. *Inductor-Based Fine-Tuning of a mm-Wave DCO*

A 3-D view of the inductor-based fine-tuning bank (FB) is shown in Fig. 5a. Equal width metal shield strips are placed beneath the inductor, which act as a distributed capacitive load (C_L) along its entire length. When the switch in series with C_L (i.e., each metal strip pair) is ON, the inductor sees an increased capacitive load which increases the differential input impedance (Z_{in}) of the tank. Although the C_L array is unit-weighted, the sensitivity of Z_{in} to changes in C_L (i.e., $\partial Z_{in}/\partial C_L$) depends upon the position of the switched-capacitor in the array. Placing C_L close to the center-tap introduces less phase shift compared to directly loading the inductor at the (differential) terminals connected to the oscillator core. The measured frequency tuning step of each FB bit from a 60-GHz DCO testchip is plotted in Fig. 5b [19]. The frequency tuning step for the first bit in FB (i.e., farthest from the center-tap) is 22.5 MHz, which reduces gradually to only 160 kHz for the last bit. The width and the spacing of the metal shield strip pairs can be modified to form a binary weighted fine-tuning bank to simplify the decoding when used in an ADPLL. The simulated Q-factor of the FB is 20 in the 60-GHz band, and varies by only ±0.025 across the 7-GHz DCO tuning range.

B. *Transformer-Coupled Fine-Tuning of a mm-Wave DCO*

For some applications (e.g., linear FMCW radar), it is desirable to have a uniform fine-tuning bank. This can be achieved using the transformer-coupled technique, as shown in Fig. 6. The primary coil of the transformer (L_p) is connected to the MOS cross-coupled pair, while the secondary coil (L_s) is connected to a variable capacitor load (C_L). The mutual coupling factor between L_p and L_s is k_m. Varying C_L changes L_{in} as seen from the primary coil inputs. Compared to loading the primary coil inputs

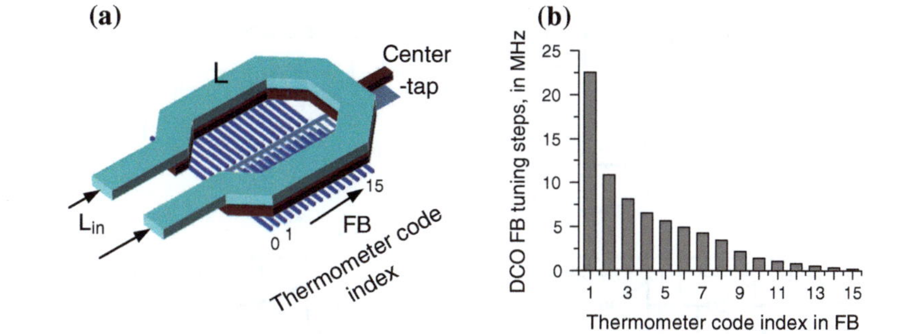

Fig. 5 **a** 3-D view of an inductor-based fine-tuning bank, **b** measured 60-GHz DCO fine-tuning frequency steps with respect to the input control

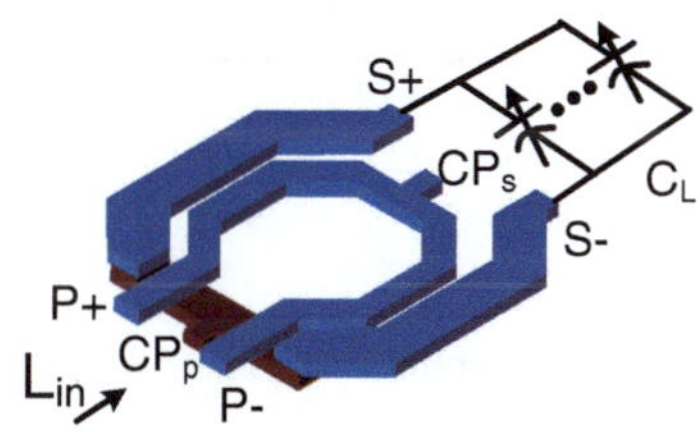

Fig. 6 Principle of transformer-coupled fine-tuning

by C_L directly, the capacitive loading effect on L_{in} is reduced by a factor $k_m^2 L_s/L_p$. The capacitance loading factor can be increased by either reducing the ratio of secondary to primary inductance (L_s/L_p), or by reducing the coupling coefficient k_m. A capacitance reduction factor larger than 10 can be easily achieved when a weakly coupled transformer (e.g., $k_m = 0.25$) is constructed.

A linear FB can be formed by replacing C_v in Fig. 6 by a unit-weighted capacitor array (e.g., a digitally controlled TL). Note that the self-resonance frequency of the transformer with capacitive load C_v should be much larger than the operating frequency (ω_0) to maintain the uniformity in frequency tuning steps, i.e., $\omega^2 L_s C_L < <1$ [19].

The coarse and fine-tuning techniques are based on distributed switched-metal capacitor banks, which exhibit the same temperature coefficient. Thus, the calibration procedure is simplified when used in an ADPLL. Electromagnetic (EM) simulation of the entire resonator structure (including coarse and fine-tuning) captures the distributed LC effect, unwanted capacitive coupling between adjacent tuning elements, and the loss of the tank accurately at mm-wave frequencies.

4.2 Time-to-Digital Converter

A TDC is used as the phase/frequency detector and charge-pump replacement in an ADPLL synthesizer, as shown in Fig. 1. The created variable phase signal is a fixed-point digital word in which the fractional part is measured with a resolution of an

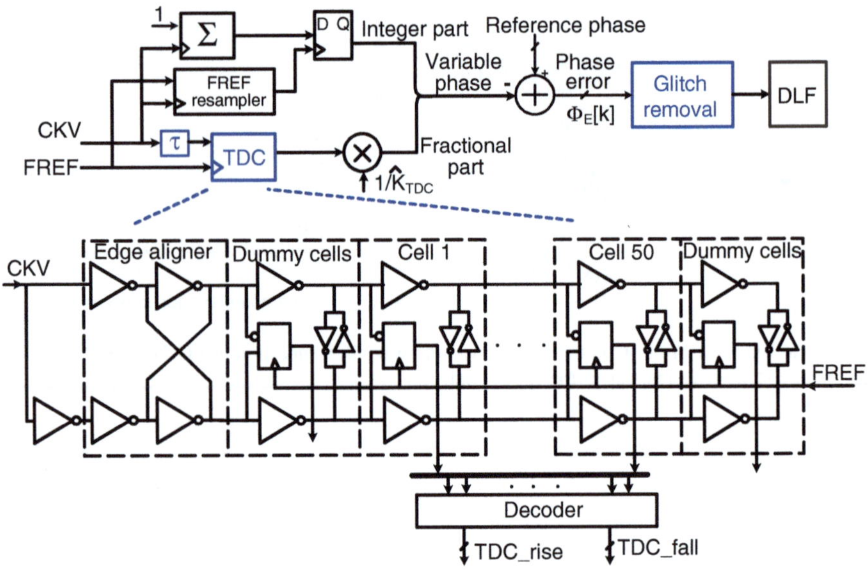

Fig. 7 Simplified schematic of an inverter-chain-based TDC core

inverter delay ($\sim$12 ps in 65-nm CMOS) by means of the TDC core, as shown in Fig. 7. The DCO clock (CKV) gets delayed by a string of inverters whose outputs are sampled at the rising edge of FREF. A pseudo-differential delay chain is adopted to avoid mismatch between rising and falling edge transitions due to differing strengths of the NMOS and PMOS transistors. The 50-bit output obtained from the TDC core forms a pseudo-thermometer code, which is then converted to binary and normalized to the CKV period, T_V. The number of inverters is set to cover one T_V. To increase the dynamic range arbitrarily, an edge counter (CKV) with sufficient word length is added, thus contributing the integer part of the variable phase.

The fixed-point TDC output timestamp (i.e., variable phase) consists of the sampled CKV edge count (integer part) and the normalized delay from CKV to FREF (fractional part). The FREF clock provides triggering moments which sample both the counter and TDC outputs. These different sampling instants could have a timing misalignment τ, indicated in Fig. 7, and thus cause glitches in the phase error when the counter and TDC outputs are combined. Instead of correcting these glitches [23], they are removed by digital signal processing. Digital logic first detects a glitch by comparing the current phase error to that in the previous clock cycle. If the difference is larger than a threshold (e.g., 0.5), the input is assumed to contain a glitch. The phase error is frozen for this clock cycle by disregarding the current phase error to obtain a glitch-free output. In addition, the same logic can be re-used as a lock indicator, or to generate a clock quality monitoring signal by setting a different comparator threshold.

In PLL applications, the reference edge locations are predictable, so significant power is saved by gating the TDC activity off during 95 % of the time between the reference edges. An FREF resampler retimes frequency reference by the DCO clock (CKV), which allows the ADPLL to work in a clock-synchronous manner. To avoid metastability in FREF retiming, FREF is oversampled by both rising and falling edges of CKV, simultaneously, and an edge-selection signal derived from the TDC delay chain chooses the path farther away from the metastable region [5]. Due to the full digital nature of the phase error correction, sophisticated control algorithms via a dynamic change of the loop filter parameters could also be employed. For example, gear shifting of the ADPLL bandwidth to speed-up the frequency settling [24] and change the loop characteristics, such as switching from type I to type II PLL after settling is complete. These types of reconfigurations are difficult to implement in the analog-intensive architectures.

5 Calibration Techniques for High RF Performance

A modulation range of several GHz is required for many mm-wave applications, such as FMCW radar and high-date rate communications. Ideally, a single tuning bank with constant K_{DCO} across the modulation range would be achieved. However, the DCO tuning must be segmented into coarse- (CB) and fine-tuning (FB) banks (i.e., each with different K_{DCO}) to realize both high resolution and a wide tuning range in practice, as discussed in Sect. 4.1. The measured tuning step mismatches for a 60-GHz transformer-coupled DCO prototype is $\sim$15 % in CB (implemented as digitally controlled TL), which is much larger than in FB since dummy cells are not employed there due to the limited LC budget when oscillating at 60 GHz [15]. Moreover, the K_{DCO} in FB varies with the CB tuning word due to the wide coarse-tuning range (e.g., 7 GHz). When the capacitance increases by ΔC, the oscillation frequency (f_o) will decrease by Δf, or approximately $\frac{1}{2}\frac{\Delta C}{C_o}f_o$ (i.e., $\frac{\partial f_0}{\partial C_0} \approx \frac{\Delta f_0}{\Delta C_0} = -\frac{f_0}{2C_0}$). Therefore, Δf will vary with f_o even for the same ΔC, which is the case for modulation frequency range up to a few GHz. As discussed in Sect. 3, the two-point modulation scheme relies on accurate DCO gain (K_{DCO}). Thus, DCO gain calibration and linearization are essential for linear FM sweeping across several GHz in range.

5.1 DCO Gain Calibration and Linearization

The oscillator gain (K_{DCO}) dependence on PVT and frequency makes it necessary to estimate it on an as-needed basis within the actual operating environment. A digital normalization algorithm that measures the phase error present in the loop can be used to calibrate the K_{DCO} of a linear FB [5]. Alternatively, adaptive gain compensation by a sign-LMS loop [25, 26] calibrates the K_{DCO} in the background without interrupting normal frequency modulation. However, they are difficult to apply to a mm-wave DCO for an FMCW application as both CB and FB are used

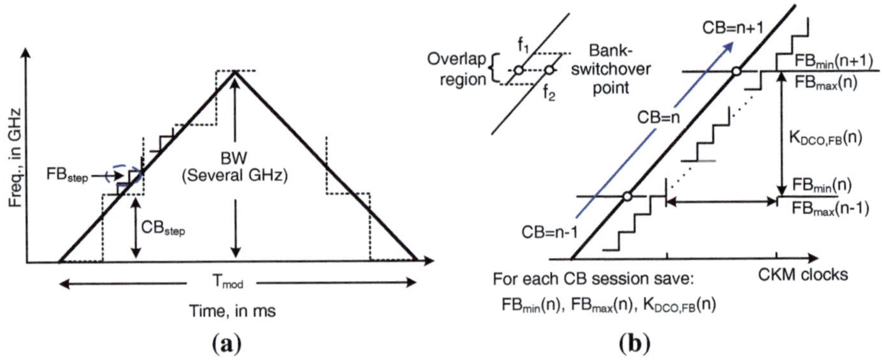

Fig. 8 **a** *Triangular* modulation in an FMCW radar, **b** principle of the closed-loop K_{DCO} linearization algorithm for FMCW generation

for modulation. Each bit in CB has a different K_{DCO}, and the K_{DCO} in FB also varies with CB settings as explained above. It requires over 10 K_{DCO} values to be calibrated in the background, which makes the adaption algorithm too complicated to implement and may not converge to a stable solution. Correcting K_{DCO} via a look-up table for individual bits in each bank employing open-loop calibration algorithms [27] requires a long calibration time (up to hours) and an unacceptably large look-up table for a gigahertz range.

As an alternative, we present a fast, closed-loop DCO gain linearization technique for linear FMCW generation. For a triangular modulation of a slope $k_{mod} = 2BW/T_{mod}$ (BW is the modulation range and T_{mod} is the period of the triangular modulation, as shown in Fig. 8a), the output frequency change within each modulation clock (CKM) is k_{mod}/f_{CKM}, where f_{CKM} is the modulation sampling rate. Instead of finding and storing accurate DCO tuning words (OTWs) for each frequency along the triangular modulation trajectory, accurate OTWs are determined only in the vicinity of the bank-switchover points (see Fig. 8b). Thus, the size of the look-up table is determined by the number of bank-switchover points and independent of the FM rate and range. To ensure monotonic tuning against PVT, the total FB tuning range is set to 1.7 times the frequency step size in CB. The mid-point of the overlap region is a natural choice for a robust switchover. Between the two adjacent bank-switchover points, only FB is used for modulation, which is sufficiently linear, and one normalized K_{DCO} for each sub-range is employed. The above techniques reduce the calibration time at power-up to $\sim$4 s.

5.2 *Mismatch Calibration of Fine-Tuning Bank*

Both inductor and transformer-based fine-tuning techniques described in Sect. 4.1 achieve a raw frequency resolution of $\sim$2 MHz. A $\Sigma\Delta$-modulator with a higher dithering rate may be employed to obtain a resolution on the order of hundreds of

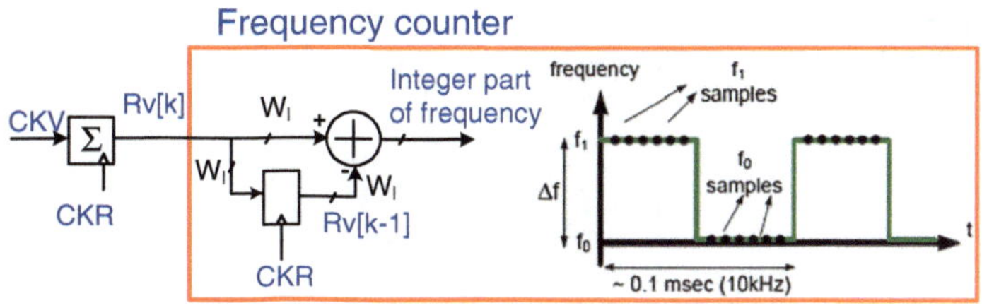

Fig. 9 Open-loop DCO gain calibration based on toggling

Hertz. The fractional bits (dithering bits) and the integer bits of the fine-tuning bank may have mismatches and can introduce nonlinearity in the frequency modulation.

The mismatch of the fractional tuning bits with respect to the average K_{DCO} of the integer bits can be characterized in an open-loop manner by a forced ON/OFF toggling of the fractional bit [27]. Since small capacitance fluctuations in the DCO tank result in proportional frequency fluctuations, changes in capacitance, resulting from on/off switching of the dithering bit, are evaluated by subtracting frequency measurements performed at each of the two states. This open-loop configuration is used since each toggling procedure addresses a specific fine-tuning bit, which could not be done through the normal modulation capability of the ADPLL. The frequency measurements are based on a counter within the ADPLL and multiple readings of the counter (e.g., M readings) are averaged to reduce the quantization error in a single measurement of frequency deviation, especially in the presence of DCO phase noise, as shown in Fig. 9.

The tuning step for this particular bit is Δf_i, and can be calculated by $\Delta f_i = \frac{1}{M}\left[\sum_{k=1}^{M} f_{1_k} - \sum_{k=1}^{M} f_{0_k}\right]$. The number of measurements (N) used is typically on the order of 2^{15} for 1 % fine-tuning step accuracy. The resulted frequency tuning step after averaging is $\Delta f_{avg,i} = \frac{1}{N}\sum_{i=1}^{N} \Delta f_i$. Thus, the normalized tuning step mismatch between the fractional bit and the integer bits of the FB_{Mod} is $\varepsilon = (\Delta f_{avg,i} - \Delta f_{FB})/\Delta f_{FB}$, where Δf_{FB} is the estimated frequency tuning step of the integer bits of FB_{Mod}. The correction factor ε can be then applied to the dithering bit upon modulation.

6 Experimental Results

The 60-GHz ADPLL-based transmitter employing the time-domain circuit and calibration techniques presented in this article was fabricated in TSMC 65-nm bulk CMOS. The die photo is shown in Fig. 10. The ADPLL core occupies 0.5 mm^2 of the 2.2-mm^2 total die area, including bondpads, power amplifier, SRAMs (6 × 2^{13}-bit), and other digital circuitry for debugging. The SRAM is used to take

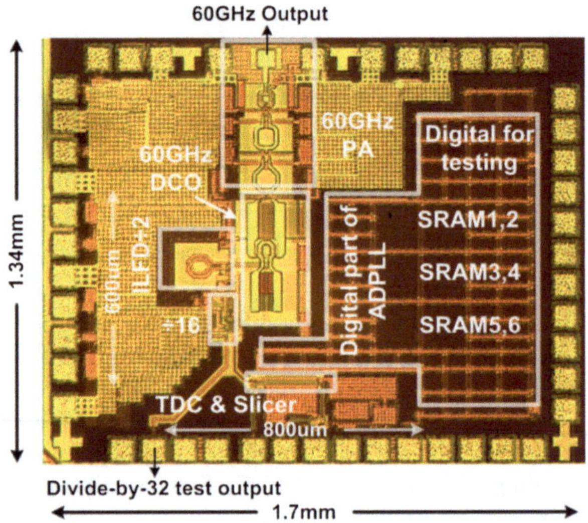

Fig. 10 Die photo of the 60-GHz ADPLL-based transmitter

a system snapshot for debugging purposes, and is also used to store the K_{DCO} calibration data for the GHz range, linear FM. The ADPLL chip consumes 40 mA: 11 mA by the DCO, 23 mA in the frequency prescaler, and 6 mA for the TDC and digital part, while the power amplifier dissipates 34 mA, all from a single 1.2-V supply. The single-ended 60-GHz PA output is measured by probing on die. In addition, a 2-GHz test output (after divide-by-32) is also accessed via the printed circuit board (PCB), providing a convenient way to characterize the 60-GHz ADPLL without on-die probing.

The fractional-N ADPLL can generate arbitrary frequencies ranging from 56.4 to 63.4 GHz. The measured spectrum of the mm-wave output when locked at 60.09987 GHz is plotted in Fig. 11a. A very low reference spur level of −74 dBc is observed, with no other significant spurs detectable. The measured worst case reference spur is −72.4 dBc across the 7-GHz locking range. The out-of-band fractional spurs are filtered out heavily by the type II, 4th order IIR loop filter. For some FCWs (e.g., near integer-N channel), the fractional spurs fall in-band but are always less than −60 dBc.

A spectrum of the 60-GHz carrier close-in is shown in Fig. 11b. It indicates that the PN at 1-MHz frequency offset is −88.5 dBc/Hz ($-41.58 - 10\log_{10}[50\ \text{kHz}]$). Figure 12 compares the PLL output spectrum when free running and locked. It is clear that in the locked spectrum a "shoulder" is formed by the TDC noise in combination with the DCO noise, as set by the PLL bandwidth. The two spectra are very close at larger offsets from the carrier, which indicates the loop operation does not contaminate the DCO PN.

Figure 13 plots the measured PN at various loop bandwidths. It is measured from the divide-by-32 test output (CKV/32), and thus 30.1 dB (i.e., $20\log_{10}[32]$) should be added to refer the PN to the mm-wave (i.e., 60 GHz) output. For a nominal loop bandwidth of 300 kHz, the measured PN is −118 dBc/Hz at 1-MHz offset, which

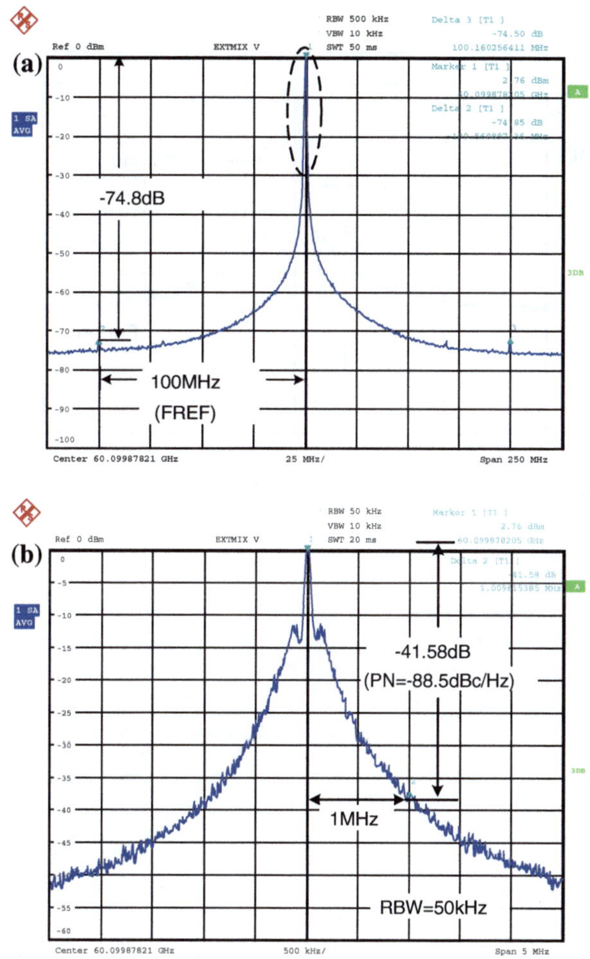

Fig. 11 Measured ADPLL output spectrum when locked at 60.09987 GHz

agrees well with the PN obtained at the PA output shown in Fig. 11b. The integrated PN from 10 kHz to 10 MHz is −45.9 dBc for a loop bandwidth of 300 kHz, which corresponds to an integrated PN of −15.8 dBc (= −45.9 + 20log32) at 61.87-GHz output and rms jitter of 590.2 fs. The measured, integrated PN varies by 1.5 dB across the 7-GHz locking range.

The TDC is self-calibrated in the background to compensate for inverter delay variation with PVT [28]. The measured TDC resolution is 12.2 ps at room temperature, contributing an in-band PN of −80 dBc/Hz at 60-GHz PLL output, theoretically. This agrees well with the measured in-band PN of −78 dBc/Hz at a wide loop bandwidth setting (~1.5 MHz), as shown in Fig. 13. The TDC consumes 4.5 mA, which reduces to only 1.5 mA when edge prediction and power gating are enabled.

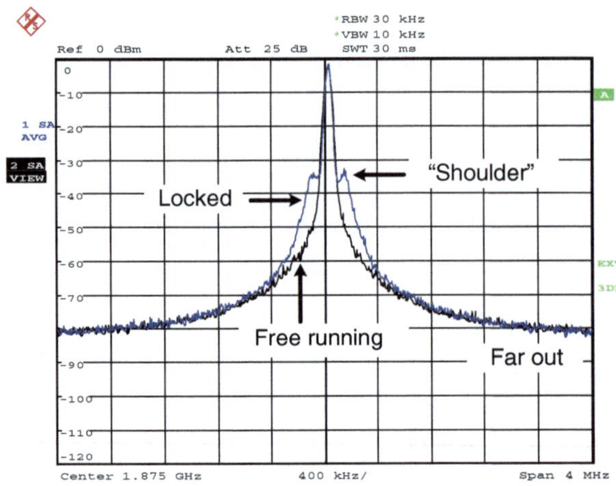

Fig. 12 Measured ADPLL spectrum at divide-by-32 output (free-running and locked)

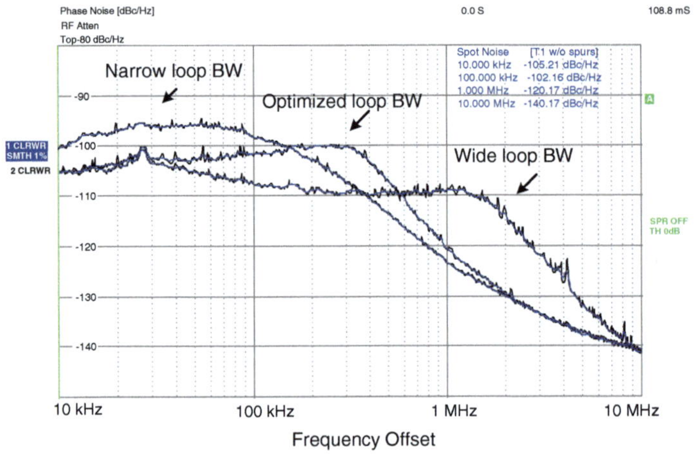

Fig. 13 ADPLL PN at various loop bandwidths (measured at CKV/32 output)

The measured lock-in time is within 3 μs for a frequency step of up to ~10 % of the carrier frequency via the dynamic control on the loop parameters. During frequency acquisition, the loop operates in type I made with a wide bandwidth of 1.5 MHz. It is then switched hitlessly to type II, using a 4th order IIR filter and 300-kHz bandwidth only in tracking mode.

As the first reported ADPLL at 60 GHz, its performance is compared in Table 1 to leading 60-GHz analog PLLs [29–31]. It is the only 60-GHz CMOS PLL capable of the fractional-N synthesis. It exhibits excellent in-band and out-of-band PN performance, fast locking, and an ultra-low reference spur when compared to the prior art. Moreover, it is also capable of a wideband FM, which is demonstrated in Fig. 14 for triangular FMCW generation.

Table 1 Performance comparison of the 60-GHz ADPLL with leading analog PLL counterparts

	This work	ISSCC'09 [29]	JSSC'11 [30]	ISSCC'13 [31]
Center frequency	60 GHz	61.5 GHz	60.5 GHz	63.1 GHz
Architecture	TDC-ADPLL	Charge-pump PLL	20 GHz analog PLLx3	Charge-pump PLL
Type	Fractional-N	Integer-N	Integer-N	Integer-N
Modulation	Yes (2-point)	No	No	No
CMOS	65 nm	45 nm	65 nm	65 nm
Reference freq.	100 MHz	100 MHz	36 MHz	135 MHz
PN $\Delta f = 10$ kHz	−75 dBc/Hz	−70 dBc/Hz	−60 dBc/Hz	−80 dBc/Hz
$\Delta f = 1$ MHz	−90 dBc/Hz	−75 dBc/Hz	−95 dBc/Hz	−91 dBc/Hz
RMS jitter	590.2 fs	NA	NA	238.4 fs
FREF spur	−74 dBc	−42 dBc	−67 ~ −58 dBc @20 GHz	<−45 dBc
Frequency range	7 GHz (11.6 %)	9 GHz hi/lo, 2 VCOs, (14.6 %)[a]	5 GHz (8.3 %)	10.4 GHz (16.5 %)
Locking time	3 μs	NA	NA	NA
Supply	1.2 V	1.1 V	1.2 V	1.2 V
Power consumption	48 mW	78 mW[b]	80 mW[b]	24 mW[b]
Core area	0.48 mm^2	0.82 mm^2 (with pads)	1.68 + 0.8 mm^2 (2 dice, with pads)	0.192 mm^2

[a] PLL locking range is 2.4 GHz, limited by injection locking divider's locking range
[b] Provides quadrature outputs at 60 GHz

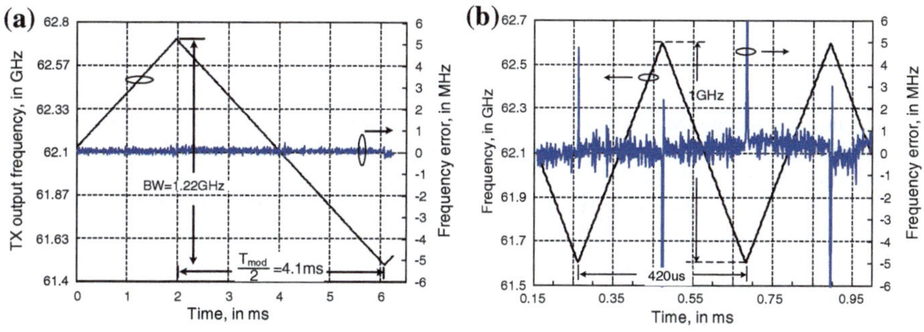

Fig. 14 Measured time-domain frequency characteristics of the FMCW signal: **a** $T_{mod} = 8.2$ ms, BW = 1.22 GHz, **b** $T_{mod} = 0.42$ ms, BW = 1 GHz

Figure 14a plots the instantaneous output frequency of the ADPLL when a triangular modulation across 1.22-GHz in range is applied ($T_{mod} = 8.2$ ms). The frequency error compared to an ideal triangular chirp is also shown in Fig. 14a, with a root-mean-square (rms) value of only 117 kHz. Figure 14b shows the modulation results when the modulation speed is 16 times faster (i.e., 1-GHz change in 210 μs), and the measured frequency error is still smaller than 400 kHz$_{rms}$. Compared to state-of-the-art FMCW generators [32–34], the all-digital architecture achieves wider modulation range for varying modulation slopes, and better phase noise with lower power consumption.

7 Conclusion

The latest time-domain techniques for mm-wave frequency generation have been described. The multi-rate all-digital PLL architecture features fractional-N synthesis with wideband frequency modulation (FM) capability. The digitally intensive nature of the design increases reconfigurability and testability. The mm-wave DCO at the heart of the ADPLL provides wide tuning range and fine tuning steps, simultaneously, through the application of distributed switched-capacitor topology. The weakly coupled transformer attenuates the frequency tuning step by a factor of k_m^2 to obtain tuning steps of ~1-MHz without employing minimum-size components that are difficult to ensure high yield or are susceptible to process variations. Closed-loop and open-loop DCO gain calibration techniques enable the linear FM across a gigahertz modulation range. The 60-GHz ADPLL-based FMCW transmitter prototype demonstrates a 1.2 GHz linear FM sweep with only 117 kHz_{rms} frequency error. It achieves an rms jitter of 590.2 fs, ultra-fast settling (3 μs), and very low reference spur levels (−74 dBc) with no other significant spurs observed, thereby demonstrating the strengths of applying time-domain techniques to mm-wave frequency synthesis.

References

1. Z. Luo, A. Steegen, M. Eller, R. Mann, et al., "High performance and low power transistors integrated in 65 nm bulk CMOS technology," *IEEE Int. Electron Devices Meeting Dig. Tech. Papers*, pp. 661–664, 2004.
2. A. Tomkins, R.A. Aroca, T. Yamamoto, S.T. Nicolson, Y. Doi, and S.P. Voinigescu, "A zero-IF 60 GHz 65 nm CMOS transceiver with direct BPSK modulation demonstrating up to 6 Gb/s data rates over a 2 m wireless link," *IEEE J. Solid-State Circuits*, vol. 44, no. 8, pp. 2085–2099, Aug. 2009.
3. K. Okada, N. Li, K. Matsushita, K. Bunsen, R. Murakami, A. Musa, T. Sato, H. Asada, N. Takayama, S. Ito, W. Chaivipas, R. Minami, T. Yamaguchi, Y. Takeuchi, H. Yamagishi, M. Noda, and A. Matsuzawa, "A 60-GHz 16QAM/8PSK/QPSK/BPSK direct-conversion transceiver for IEEE802.15.3c," *IEEE J. Solid-State Circuits*, vol. 46, no. 12, pp. 2988–3004, Dec. 2011.
4. S. Emami, R.F. Wiser, E. Ali, M.G. Forbes, M.Q. Gordon, X. Guan, S. Lo, P.T. McElwee, J. Parker, J.R. Tani, J.M. Gilbert, and C.H. Doan, "A 60 GHz CMOS phased-array transceiver pair for multi-Gb/s wireless communications," *IEEE Int. Solid-State Circuits Conf. Dig. Tech. Papers*, pp. 164–165, Feb. 2011.
5. R.B. Staszewski, and P.T. Balsara, *All-Digital Frequency Synthesizer in Deep-Submicron CMOS*. WILEY-Interscience, 2006.
6. F.M. Gardner, "Charge-pump phase-locked loops," *IEEE Trans. on Communications*, vol. COMM-28, pp. 1849–1858, Nov. 1980.

7. R.B. Staszewski, K. Muhammad, D. Leipold, C.-M. Hung, Y.-C. Ho, J.L. Wallberg, C. Fernando, K. Maggio, R. Staszewski, T. Jung, J. Koh, S. John, I.Y. Deng, V. Sarda, O. Moreira-Tamayo, V. Mayega, R. Katz, O. Friedman, O.E. Eliezer, E. de-Obaldia, and P.T. Balsara, "All-digital TX frequency synthesizer and discrete-time receiver for Bluetooth radio in 130-nm CMOS," *IEEE J. Solid-State Circuits*, vol. 39, iss. 12, pp. 2278–2291, Dec. 2004.
8. R.B. Staszewski, J. Wallberg, S. Rezeq, C.-M. Hung, O. Eliezer, S. Vemulapalli, C. Fernando, K. Maggio, R. Staszewski, N. Barton, M.-C. Lee, P. Cruise, M. Entezari, K. Muhammad, and D. Leipold, "All-digital PLL and transmitter for mobile phones," *IEEE J. Solid-State Circuits*, vol. 40, iss. 12, pp. 2469–2482, Dec. 2005.
9. L. Vercesi, L. Fanori, F. De Bernardinis, A. Liscidini, and R. Castello, "A dither-less all digital PLL for cellular transmitters," *IEEE J. Solid-State Circuits*, vol. 47, no. 8, pp. 1908–1920, Aug. 2012.
10. R.B. Staszewski, and P.T. Balsara, "Phase-domain all-digital phase-locked loop," *IEEE Trans. on Circuits and Systems II: Express Briefs*, vol. 52, no. 3, pp. 159–163, Mar. 2005.
11. K. Pourvoyeur, R. Feger, S. Schuster, A. Stelzer, and L. Maurer, "Ramp sequence analysis to resolve multi target scenarios for a 77-GHz FMCW radar sensor," *Proc. Int. Conf. on Information Fusion*, pp. 1–7, June 2008.
12. W. Wu, J.R. Long, and R.B. Staszewski, "A digital ultra-fast acquisition linear frequency modulated PLL for mm-wave FMCW radars," *Proc. IEEE Radio Frequency Integration Technolgy Symp.*, pp. 32–35, Dec. 2009.
13. G.M. Brooker, "Understanding millimeter wave FMCW radars," *Proc. Int. Conf. on Sensing Tech.*, pp. 152–157, Nov. 2005.
14. W. Wu, X. Bai, R.B. Staszewski, and J.R. Long, "A 56.4-63.4 GHz spurious free all-digital fractional-N PLL in 65 nm CMOS," *IEEE Int. Solid-State Circuits Conf. Dig. Tech Papers*, pp. 352–353, Feb. 2013.
15. W. Wu, X. Bai, R.B. Staszewski, and J.R. Long, "A mm-wave FMCW radar transmitter based on a multirate ADPLL," *Proc. IEEE Radio Frequency Integrated Circuits Symp.*, pp. 107–110, June 2013.
16. M.E. Heidari, M. Lee, and A.A. Abidi, "All-digital outphasing modulator for a software-defined transmitter," *IEEE J. Solid-State Circuits*, vol. 44, no. 4, pp. 1260–1271, Apr. 2009.
17. R.B. Staszewski, D. Leipold, and P.T. Balsara, "A first multigigahertz digitally controlled oscillator for wireless applications," *Proc. IEEE Radio Frequency Integration Circuit Symp.*, vol. 51, no. 11, pp. 2154–2164, Nov. 2003.
18. J.R. Long, Y. Zhao, W. Wu, M. Spirito, L. Vera, and E. Gordon, "Passive circuit technologies for mm-wave wireless systems on silicon," *IEEE Trans. on Circuits and Systems I: Regular Papers*, vol. 59, no. 8, pp. 1680–1693, Aug. 2012.
19. W. Wu, J.R. Long, and R.B. Staszewski, "High-resolution millimeter-wave digitally-controlled oscillators with reconfigurable passive resonators," *IEEE J. Solid-State Circuits*, vol. 48, no. 11, pp. 2785–2794, Nov. 2013.
20. T.S.D. Cheung, J.R. Long, K. Vaed, R. Volant, A. Chinthakindi, C.M. Schnabel, J. Florkey, Z.X. He, and K. Stein, "Differentially-shielded monolithic inductors," *Proc. IEEE Custom Integrated Circuits Conf.*, pp. 95–98, Sept. 2003.
21. T. LaRocca, S.-W. Tam, D. Huang, Q. Gu, E. Socher, W. Hant, and F. Chang, "Millimeter-wave CMOS digital controlled artificial dielectric differential mode transmission lines for reconfigurable ICs," *IEEE Int. Microwave Symp. Dig.*, pp. 181–184, June 2008.
22. H. Sjöland, "Improved switched tuning of differential CMOS VCOs," *IEEE Trans. on Circuits and Systems II: Analog and Digital Signal Processing*, vol. 49, no. 5, pp. 352–355, May 2002.
23. M. Lee, M.E. Heidari, and A.A. Abidi, "A low-noise wideband digital phase-locked loop based on a coarse–fine time-to-digital converter with subpicosecond resolution," *IEEE J. Solid-State Circuits*, vol. 44, no. 10, pp. 2808–2816, Oct. 2009.
24. R.B. Staszewski, G. Shriki, and P.T. Balsara, "All-digital PLL with ultra fast acquisition," *Proc. of IEEE Asian Solid-State Circuits Conf.*, sec. 11-7, pp. 289–292, Nov. 2005, Taipei, Taiwan.

25. R.B. Staszewski, J. Wallberg, G. Feygin, M. Entezari, and D. Leipold, "LMS-based calibration of an RF digitally controlled oscillator for mobile phones," *IEEE Trans. on Circuits and Systems II: Express Briefs*, vol. 53, no. 3, pp. 225–229, Mar. 2006.
26. G. Marzin, S. Levantino, C. Samori, and A.L. Lacaita, "A 20 Mb/s phase modulator based on a 3.6 GHz digital PLL with −36 dB EVM at 5 mW power," *IEEE J. Solid-State Circuits*, vol. 47, no. 12, pp. 2974–2988, Dec. 2012.
27. O. Eliezer, R.B. Staszewski, J. Mehta, F. Jabbar, and I. Bashir, "Accurate self-characterization of mismatches in a capacitor array of a digitally-controlled oscillator," *Proc. IEEE Dallas Circuits and Systems Workshop*, pp. 1–4, Oct. 2010.
28. R.B. Staszewski, S. Vemulapalli, P. Vallur, J. Wallberg, and P.T. Balsara, "1.3 V 20 ps time-to-digital converter for frequency synthesis in 90-nm CMOS," *IEEE Trans. on Circuits and Systems II: Express Briefs*, vol. 53, no. 3, pp. 220–224, Mar. 2006.
29. K. Scheir, G. Vandersteen, Y. Rolain, and P. Wambacq, "A 57-to-66 GHz quadrature PLL in 45 nm digital CMOS," *IEEE Int. Solid-State Circuits Conf. Dig. Tech. Papers*, pp. 494–495, Feb. 2009.
30. A. Musa, R. Murakami, T. Sato, W. Chaivipas, K. Okada, and A. Matsuzawa, "A low phase noise quadrature injection locked frequency synthesizer for mm-wave applications," *IEEE J. Solid-State Circuits*, vol. 46, no. 11, pp. 2635–2649, Nov. 2011.
31. X. Yi, C.C. Boon, H. Liu, J.F. Lin, J.C. Ong, and W.M. Lim, "A 57.9-to-68.3 GHz 24.6 mW frequency synthesizer with in-phase injection-coupled QVCO in 65 nm CMOS," *IEEE Int. Solid-State Circuits Conf. Dig. Tech. Papers*, pp. 354–355, Feb. 2013.
32. T. Mitomo, N. Ono, H. Hoshino, Y. Yoshihara, O. Watanabe, and I. Seto, "A 77 GHz 90 nm CMOS transceiver for FMCW radar applications," *IEEE J. Solid-State Circuits*, vol. 45, no. 4, pp. 928–937, Apr. 2010.
33. Y.-A. Li, M.-H. Hung, S.-J. Huang, and J. Lee, "A fully integrated 77 GHz FMCW radar system in 65 nm CMOS," *IEEE Int. Solid-State Circuits Conf. Dig. Tech. Papers*, pp. 216–217, Feb. 2010.
34. H. Sakurai, Y. Kobayashi, T. Mitomo, O. Watanabe, and S. Otaka, "A 1.5 GHz-modulation-range 10 ms-modulation-period 180 kHzrms-frequency-error 26 MHz-reference mixed-mode FMCW synthesizer for mm-wave radar application," *IEEE Int. Solid-State Circuits Conf. Dig. Tech. Papers*, pp. 292–293, Feb. 2011.

A Deterministic Background Calibration Technique for Voltage Controlled Oscillator Based ADC

Sachin Rao and Pavan Kumar Hanumolu

Abstract Among different types of time-domain analog-to-digital-converters (ADCs), voltage-controlled-oscillator (VCO) based ADCs have gained prominence. Like other time-based ADCs, VCO based ADCs are amenable to CMOS technology scaling. They provide inherent anti-alias filtering to the input signal and achieve very high resolution by first order noise shaping the quantization error. Despite these advantages, traditional VCO based ADCs have found limited application due to poor linearity. Recently, several new techniques have been developed to improve the overall performance of such ADCs. This chapter briefly discusses such recent advances towards improving the performance of VCO based ADCs. A deterministic background calibration method to improve the linearity of VCO based ADCs is discussed in more detail.

1 Introduction

CMOS technology scaling has led to the development many new time-based analog to digital converter (ADC) architectures that are amenable to technology scaling. Voltage-controlled-oscillator (VCO) based ADC is one such time-based ADC architecture that has recently gained prominence because of their several desirable properties. VCO based ADCs achieve high resolution by first order noise shaping the quantization error and exhibit inherent anti-aliasing. They do not employ analog building blocks such as op-amps, comparators, precision references etc. Therefore

S. Rao (✉)
Qualcomm Technologies Inc, Santa Clara, CA, USA
e-mail: sachinraob@ieee.org

P. K. Hanumolu
University of Illinois, Urbana Champaign, Chicago, IL, USA
e-mail: hanumolu@illinois.edu

P. Harpe et al. (eds.), *High-Performance AD and DA Converters, IC Design in Scaled Technologies, and Time-Domain Signal Processing*,
DOI: 10.1007/978-3-319-07938-7_16,

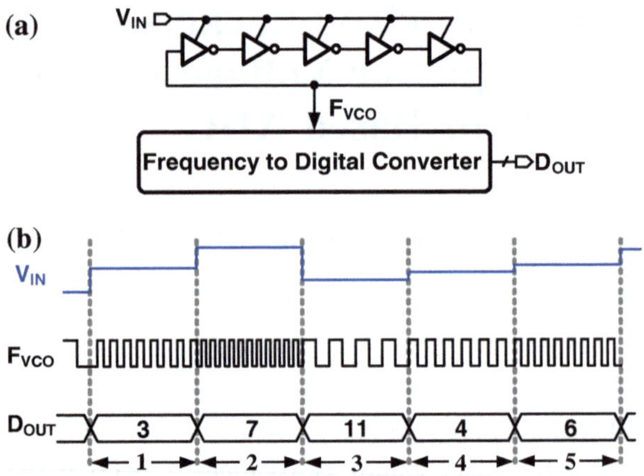

Fig. 1 Simplified VCO-based quantizer. **a** Simplified block diagram. **b** Representative waveforms

they are relatively immune to the reduced supply voltage and poor transistor gain in advanced ultra-deep sub-micron CMOS technologies.

Figure 1 shows a simplified block diagram of a simple VCO based ADCs. It consists of a ring-oscillator wherein the input voltage, V_{IN}, controls the delay of each inverter in the ring-VCO and sets the VCO frequency, F_{VCO}. The digital output, D_{OUT}, is obtained from the frequency to digital converter by measuring the frequency of the VCO output clock waveform. Typically, a counter that determines the total number of rising (or falling) edges of F_{VCO} in a given reference clock period is used as a frequency detector as shown in the waveforms in Fig. 1b. During time period-1, V_{IN} is such that there are 7 rising edges of F_{VCO}. Therefore the digital output at the end of period-1 is 7. During the clock period-2, the input has increased which causes the inverters to be faster and increases the F_{VCO} and resulting in 11 rising edges. In general the ADC output is given by,

$$D_{OUT} \propto F_{VCO} \propto kV_{IN} \tag{1}$$

which indicates that the digital output is proportional to analog input, V_{IN}, if the VCO output frequency is proportional to V_{IN}.

Note that the architecture shown in Fig. 1a employs a 5-inverter stage ring-VCO. However, it uses the output of only one inverter to obtain the digital output. By using all the phases of the VCO, an improved, high resolution VCO based ADC can be realized as shown in Fig. 2 [1, 2]. Here, instead of counting the number of rising edges, the register and counter determine the total number of inverters that undergo transition in one period of the reference clock, F_S. Clearly, since the LSB of the digital output corresponds to one inverter delay rather than one period of F_{VCO}, this architecture achieves better resolution.

The VCO based ADCs shown in Figs. 1 and 2 consist of mostly digital circuits such as inverters, counters, registers etc., therefore they are relatively easy to

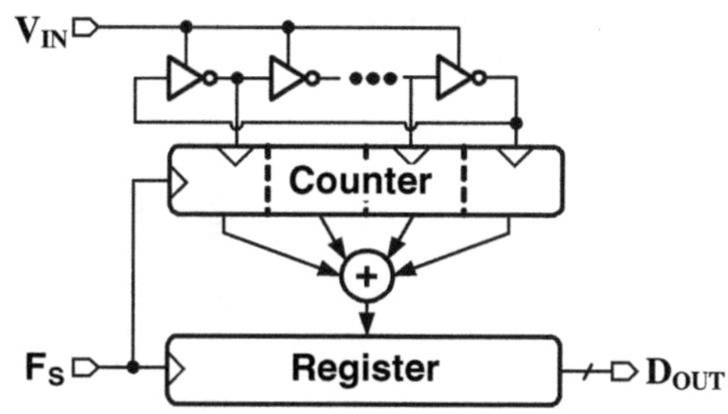

Fig. 2 Improved VCO based ADC with better resolution

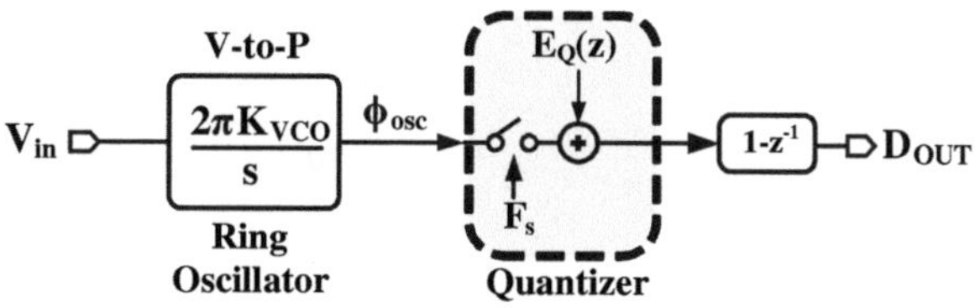

Fig. 3 Equivalent model of VCO based ADC

design. Also, since the delay per stage reduces in advanced CMOS technologies due to faster digital circuits, their resolution improves with technology scaling.

1.1 Equivalent Model

The behavior of VCO based ADCs can be understood with an equivalent model shown in Fig. 3 [3]. The VCO is modeled as a voltage-to-phase (V-to-P) converter that integrates the input voltage, V_{in} to generate an output phase represented by, Φ_{osc}. The phase of the VCO is sampled at a rate F_S, quantized and passed through a digital differentiator to get the digital output, D_{OUT}. The quantizer and digital differentiator effectively model the operation of counter and register in Fig. 2 i.e. the process of determining the total number of inverter transitions in each period of the reference clock.

2 Important Properties of VCO Based ADCs

The equivalent model discussed in the previous section can be directly applied to gain insight into the behavior of VCO based ADCs. Its properties have been thoroughly analyzed by various researchers [1, 3–5]. In this section we will briefly review some of the important properties.

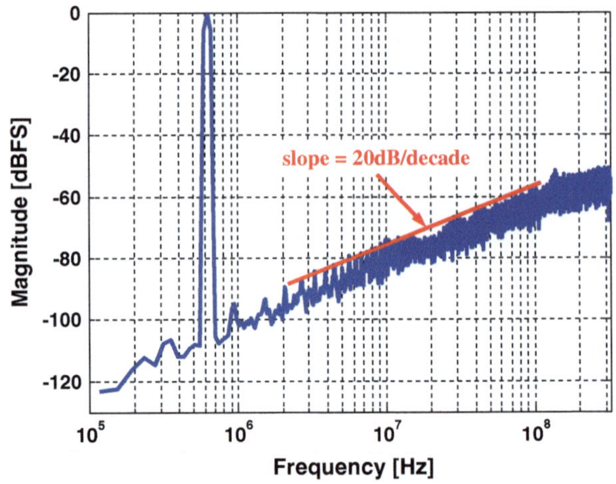

Fig. 4 Ideal spectrum of VCO based ADC

2.1 Resolution

The quantization error in a VCO based ADC is first-order noise shaped. Therefore it achieves high signal to quantization noise ratio (SQNR). This behavior can be readily understood by inspecting the equivalent model as shown in Fig. 3 where the quantization noise, $E_Q(z)$, is assumed to be white. Since $E_Q(z)$ is introduced after the integrator, it is differentiated once before reaching the output. The differentiator first order noise shapes the quantization error. The output spectrum of an 16 delay-stage ideal VCO based ADC, sampled at 640 MHz, is shown in Fig. 4. As expected, the quantization noise is first order shaped.

Note that unlike the quantization noise, the input signal passes through a continuous time integrator and a digital differentiator before reaching the output. Therefore there is no net integration or differentiation, and output contains only a scaled version of the input.

Because the VCO based ADC behaves like a first order $\Delta\Sigma$ modulator, well known equations can be directly applied to find the maximum SQNR. The maximum SQNR, expressed in dB, is given by [4, 5],

$$\mathrm{SQNR}_{\mathrm{max,dB}} = 20\log_{10}(\mathrm{N}) + 30\log_{10}(\mathrm{OSR}) - 3.4 \tag{2}$$

where OSR is the oversampling ratio and N is the maximum possible number of inverter transitions in one period of the reference clock.

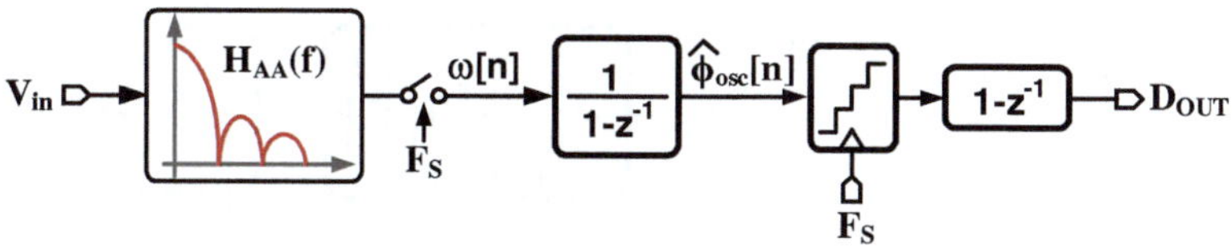

Fig. 5 Model of VCO based ADC depicting anti-alias filter

2.2 Anti-alias Filtering

Consider the model shown in Fig. 5. Here, the input voltage is first passed through a filter, $H_{AA}(f)$, sampled and then passed through a discrete time integrator to get the quantity, $\widehat{\Phi}_{osc}[n]$. This quantity is then quantized and differentiated to obtain the digital output D_{OUT}. It has been shown in [5] that if the filter transfer function is given by,

$$H_{AA}(f) = K_{VCO} e^{-j\pi T_S f} \frac{\sin(\pi T_S f)}{\pi f} \quad (3)$$

then, the systems shown in Figs. 3 and 5 are equivalent. This confirms that the VCO based ADC introduces an inherent first order anti-aliasing filter.

2.3 Sampler Metastability and Offset

For a given resolution, the sampling flip-flops in a VCO based ADC is less susceptible to metastable behavior when compared with a voltage domain quantizer such as the flash ADC. In a VCO based ADC, the sampling flip-flops input voltage swing is close to rail-to-rail and is relatively independent of the ADC resolution. On the contrary, in a flash ADC, as the resolution is increased the minimum voltage swing at the inputs of the sampling flip-flops reduce. This in-turn increases the probability of encountering a metastable behavior. In fact, it is shown in [4] that the flip-flops in a VCO based ADC with M effective number of bits (ENOB) is $2^M - 1$ times less likely to encounter a metastable event.

Offset errors in a flash ADC cause non-linearity and missing codes. A VCO based ADC is to a large extent immune to such errors. This is because, such errors are introduced into the sampled VCO phase which is obtained after integrating the input voltage. Consequently, such errors are high pass shaped by the digital differentiator. On a similar note, the mismatch between the delay cells in the VCO is also first order noise shaped by the differentiator. A thorough analysis of the effect of metastability and offset errors can be found in [4, 6].

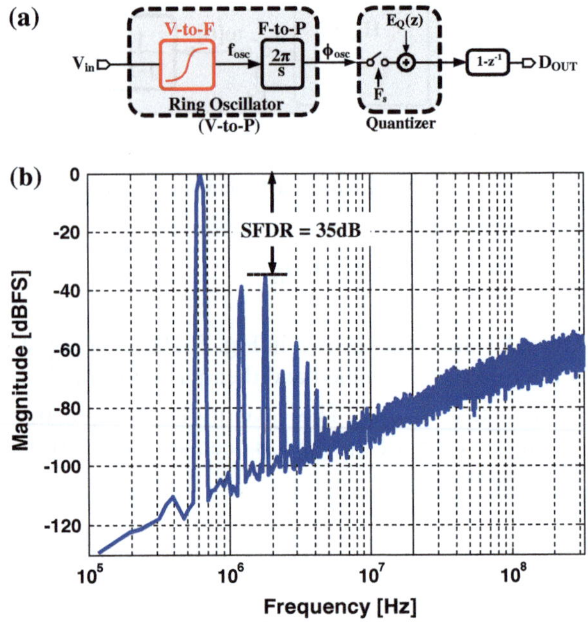

Fig. 6 Real VCO based ADC **a** Equivalent model. **b** Output spectrum

2.4 Effect of VCO Tuning Non-linearity

The ideal VCO based ADC output spectrum, shown in Fig. 4, assumes the tuning characteristics of the VCO is linear. In practice, the tuning characteristics is non-linear and it severely limits the ADC performance. To understand the effect of tuning non-linearity, consider the equivalent model shown in Fig. 6a. Here, the voltage to phase conversion (V-to-P) is explicitly shown in two separate steps namely, voltage-to-frequency (V-to-F) conversion and frequency-to-phase (F-to-P) conversion. Since ring-oscillator based structures are usually employed to realize the VCO in a effective manner, its V-to-F tuning characteristics is highly non-linear. This non-linearity introduces harmonic tones and limits the achievable signal to noise and distortion ratio (SNDR). Figure 6b shows the simulated output spectrum of a real 16 delay-stage VCO based ADC including the effect of tuning non-linearity. It achieves an SNR and SNDR of 77 dB and 33.4 dB, respectively, with an OSR of 64. As shown in the spectrum the SNDR is clearly limited by the harmonic tones. The VCO tuning non-linearity is the biggest hurdle in achieving good SNDR in a VCO based ADC.

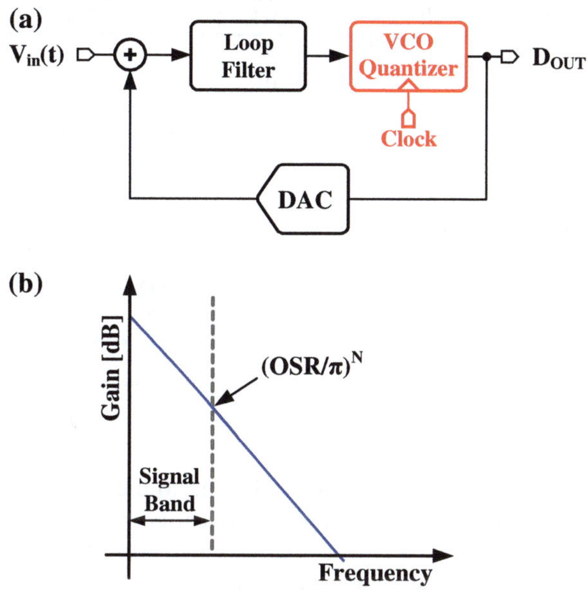

Fig. 7 Non-linearity suppression using negative feedback loop. **a** Block diagram. **b** Typical filter gain versus frequency

3 Prior-Art VCO Based ADC Architectures

To overcome the non-linearity in the tuning characteristic of VCO, several techniques have already been developed. This section briefly reviews some such important techniques.

3.1 Non-linearity Suppression Using Negative Feedback

One of the first techniques that effectively improved the performance of VCO based ADCs employed a negative feedback loop [2]. As shown in Fig. 7a, the VCO based ADC is placed at the last stage of a continuous time $\Delta\Sigma$ loop. Because of the negative feedback loop, the non-linearity is suppressed approximately by the gain of the loop filter. While this technique is very effective in suppressing non-linearity in the tuning characteristics, it requires a high order loop filter to sufficiently suppress the non-linearity. A typical loop filter gain as a function of frequency, shown in Fig. 7b, indicates that the filter gain and hence the non-linearity suppression is minimum at the edge of the signal band. Table 1 summarizes the approximate magnitude of band-edge harmonic tone suppression for different orders of loop filter and OSR. To improve the SFDR of the ADC output spectrum, shown in Fig. 6b, to more than 80 dB a band-edge suppression of about 50 dB is required. Table 1 indicates that a fourth order or a third order loop filter is required with an OSR of 40 and 20, respectively. Since the VCO based ADC itself provides

Table 1 Magnitude of suppression at band-edge

OSR \ N	2 (dB)	3 (dB)	4 (dB)
10	20	30	40
20	32	48	64
40	44	66	88

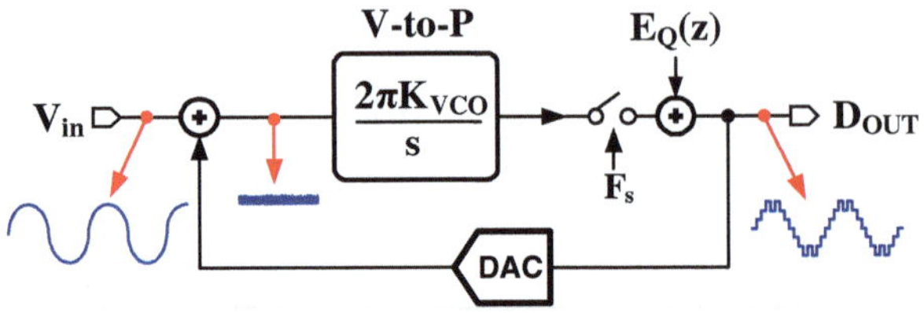

Fig. 8 Non-linearity suppression using VCO as integrator

an additional order of noise shaping to the quantization noise, loop filter order is determined by the desired non-linearity suppression rather than the quantization noise.

To relax the loop filter requirements the VCO based ADC was used as a integrator in [7]. A simplified block diagram of basic idea behind this approach is shown in Fig. 8. Here, the output of the quantizer is not differentiated. Therefore, it represents the VCO phase and the VCO acts as an integrator. Because of the high gain of this integrating quantizer, the voltage swing at the VCO input is minimized by the negative feedback loop as depicted in Fig. 8. Since in this architecture the input exercises a very small section of the non-linear VCO tuning characteristics, it is immune to the VCO tuning non-linearity thereby relaxing the loop filter requirement.

3.2 Non-linearity Suppression Using Feed-Forward

Feed-forward techniques can also be employed to minimize the signal swing at the VCO input [8]. Figure 9 shows a simplified block digram of this approach. To minimize the signal swing at the VCO input, a two step quantizer is used. The output of the loop filter is quantized by a first stage flash-ADC and the quantization error from the first stage is processed by the VCO based quantizer. Since the VCO based quantizer processes only the quantization error of the first stage, this stage is less sensitive to VCO tuning non-linearity. The loop filter shapes the errors due to mismatch in the gains of Path-1 and Path-2, thereby making this architecture robust to such errors. A standalone two step quantization approach has also been proposed in [9, 10]. However, due to the absence of a loop filter and other calibration techniques, the performance is sensitive to PVT variations.

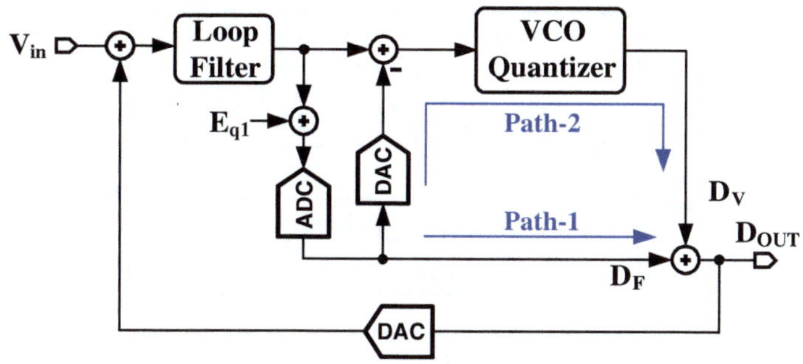

Fig. 9 Non-linearity suppression using residue canceling quantizer

The feedback and the feed-forward techniques discussed above require a highly linear, high-speed digital to analog converter (DAC). Since the DAC is in the feedback path, its linearity should be better than the desired linearity of the entire ADC. Realizing a highly linear, high-speed, multi-bit feedback DAC is challenging.

3.3 Non-linearity Correction

Another approach that seeks to overcome the VCO tuning non-linearity employs calibration wherein non-linearity introduced in the V-to-F conversion is corrected in the digital domain [5, 11, 12]. A simplified block diagram of a correlation based background calibration techniques is shown in Fig. 10 [5, 11]. Here, the VCO's V-to-F tuning curve is assumed to be represented by a third order polynomial. Using pseudo-random sequences $t_1[n]$, $t_2[n]$ and $t_3[n]$, a correlation based calibration engine finds the coefficients of the V-to-F polynomial of the replica VCO ADC. These coefficients are then used to correct the distortion introduced in the signal paths. The first version of this architecture [5] required highly linear voltage to current converter which consumed significant power while the second version overcame this requirement by calibrating even the voltage to current converters non-linearity [11].

3.4 VCO Based ADC Using Two Point Operation

Another technique that overcomes VCO non-linearity problem without using either a multi-bit feed-back DAC or calibration is shown in Fig. 11 [13]. The input analog signal is first converted to a two level signal using a two level modulator. The two level signal is then digitized by the VCO based ADC. Since the input to

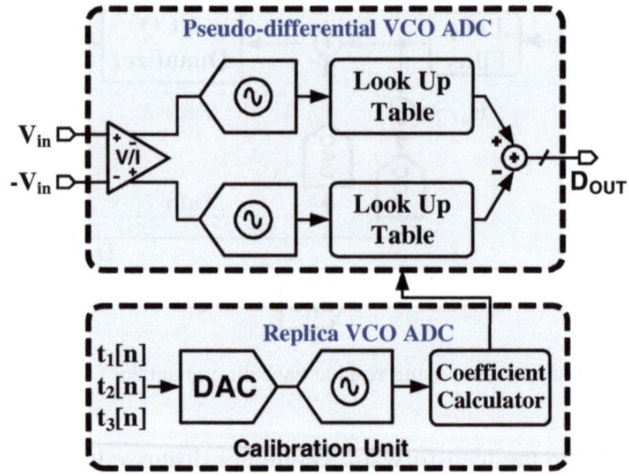

Fig. 10 Non-linearity using correlation based background calibration

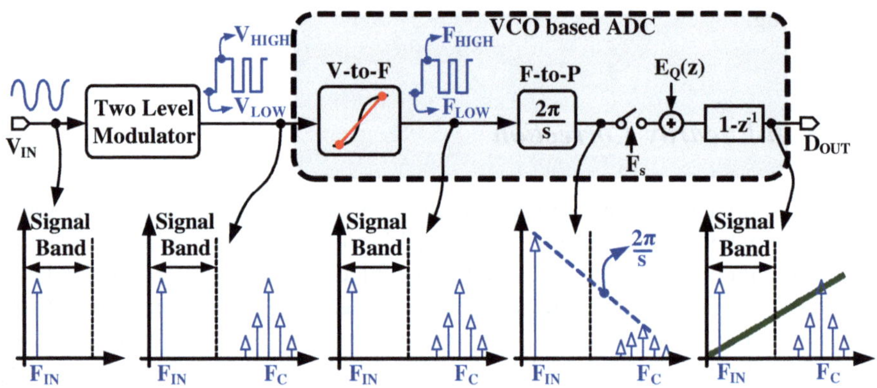

Fig. 11 Block diagram of linear VCO based ADC using two point operation

the VCO is a two level signal, it forces the VCO to operate at only two operating points, thereby ideally making this architecture immune to VCO non-linearity. The representative spectra shown in Fig. 11 confirms the linear behavior.

4 Deterministic Calibration Method

Though the two point architecture shown in Fig. 11 achieves good linearity, it requires a linear front-end two level modulator which consumes considerable power. It is desirable to improve the linearity of VCO based ADC without using such front-end two level modulators or multi-bit DACs. In this section a new

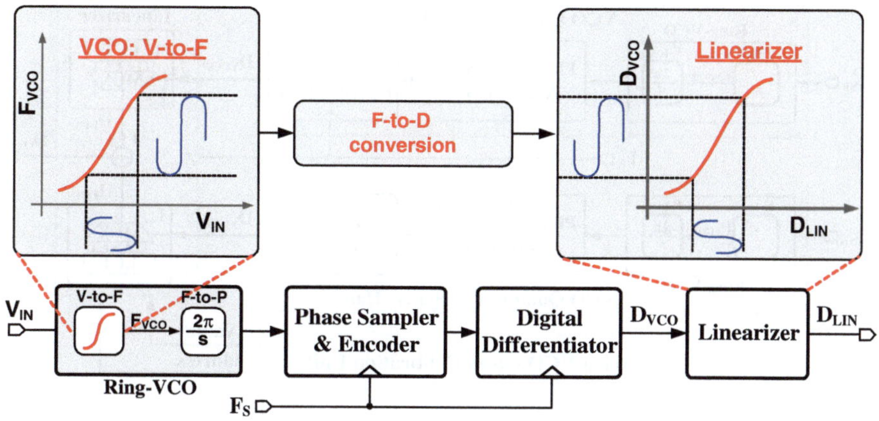

Fig. 12 Basic idea behind the deterministic calibration method

background calibration architecture to linearize the VCO based ADC is discussed. This calibration architecture makes no assumption about the order of V-to-F tuning characteristics and it corrects all the non-linear terms. Additionally it has a simple implementation.

Figure 12 depicts the basic idea behind this calibration architecture [14]. As discussed earlier, within the VCO based ADC the analog input voltage, V_{IN}, follows a non-linear V-to-F tuning curve and the VCO oscillates at a frequency, F_{VCO}. This frequency is converted to a digital word, D_{VCO}, using the phase-sampler, encoder and digital differentiator which effectively realize a frequency-to-digital (F-to-D) converter. The distortion introduced in the V-to-F tuning characteristic is corrected by the linearizer which introduces an inverse V-to-F transfer function in the signal path as shown in Fig. 12. The inverse transfer function is introduced in the linearizer by mapping D_{VCO} to a non-linearity corrected output, D_{LIN}, using a look-up-table. The look-up-table is periodically updated using a replica-VCO based calibration unit (not shown in figure).

4.1 Overall ADC Architecture

A simplified architecture of the overall ADC is shown in Fig. 13. The differential input signals, V_{INP} and V_{INN}, are separately processed by the positive and negative half circuits, respectively. The distorted outputs in each half circuit, D_{VCOP} and D_{VCON} are corrected by linearizers L_P and L_N, respectively. The linearizer outputs are digitally subtracted to get the final ADC output, D_{OUT}. This pseudo-differential architecture cancels the residual even order distortion terms and improves the SNR by 3 dB [5]. The calibration unit runs continuously in the background and uses a replica VCO to find the inverse transfer function. Its implementation details are discussed next.

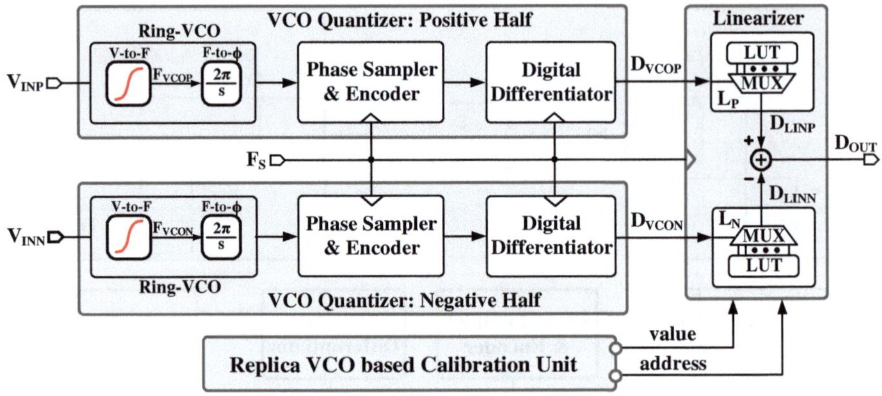

Fig. 13 Overall ADC architecture

4.2 Calibration Unit Details

To find the inverse transfer function, the calibration unit uses the relationship between the digital output of the VCO based ADC and the VCO oscillation frequency. To understand the relationship, consider the N delay-stage VCO quantizer block diagram and the waveforms shown in Fig. 14. Here, the input voltage is a constant which is set such that the digital output has a value equal to 'n' for all sampling periods. This indicates that during each sampling period (=1/F_S), exactly 'n' delay stages undergo transition. Therefore the VCO oscillation frequency is given by,

$$F_{VCO} = \left(\frac{n}{2N}\right)F_S. \tag{4}$$

Using the above equation and substituting 'n' with the VCO digital output, D_{VCO}, the VCO oscillation frequencies for all possible digital outputs is determined. The calibration unit generates the inverse transfer function by finding the input voltages that force the replica VCO to operate at all the frequencies determined from Eq. 4.

Figure 15 shows the simplified block diagram of the calibration unit. It is essentially a highly digital frequency-locked-loop wherein the replica VCO oscillation frequency is set by the digital input code, D_{REF}. The replica VCO oscillation frequency is detected using the frequency detector by counting the total number of rising edges of the divided replica VCO clock in each period of a reference clock, $F_{REF-FLL}$. The accumulator integrates the error between D_{REF} and detected replica VCO frequency, D_{REP}. The high gain of the accumulator along with the negative feedback loop drives the error to zero. Therefore, in steady-state, the replica VCO oscillation frequency is precisely,

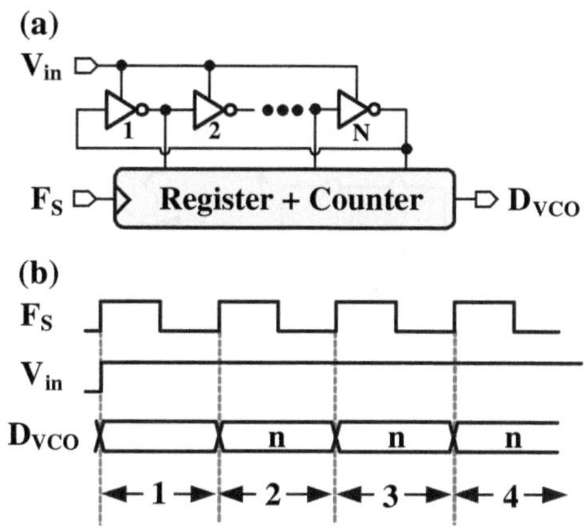

Fig. 14 N delay-stage VCO operation. **a** Simplified block diagram. **b** Timing diagram for a constant input voltage

$$F_{REP-VCO} = D_{REF} N_{DIV} F_{REF-FLL}. \quad (5)$$

The goal of this approach is to force the replica VCO to operate at all the frequencies obtained from Eq. 4 and determine the VCO input voltage. This is easily achieved by properly choosing the parameters of Eq. 5 for a given VCO-based quantizer. For example, in the implementation discussed in [14] a 5-bit VCO quantizer was used. Therefore, with $N_{DIV} = 8$ and $F_{REF-FLL} = F_S/2^{10}$, $D_{REF} = 4x$ where $x = 0, 1, 2, \ldots, 31$ forces the replica-VCO oscillation frequency to all the desired frequencies obtained from Eq. 4. For each of these frequencies, the steady-state accumulator output, D_{ACC}, represents the desired replica VCO input voltage in digital representation. To construct the inverse transfer function, the calibration unit cyclically generates all D_{REF} values cyclically one after the other and writes the steady-state accumulator value for each D_{REF} into the look-up-table.

4.3 Implementation of Linear DAC

For the architecture shown in Fig. 15 to be effective, the accumulator output needs to be converted to an analog quantity with an accurate DAC. In fact, the linearity of the DAC needs to be better than the desired ADC linearity after correction. However, this DAC need not operate at high sampling rates. Therefore, a 1-bit $\Delta\Sigma$ DAC is employed as shown in Fig. 16. The digital $\Delta\Sigma$ modulator converts the 15-bit accumulator to a 1-bit data stream, which is converted to an analog quantity by a 1-bit DAC. The low-pass filter removes the shaped quantization noise. Since a 1-bit DAC has well defined output levels, it is inherently linear. Its sensitivity to dynamic behavior can be minimized by employing RZ-DAC or modified RZ-DACs as discussed in [15].

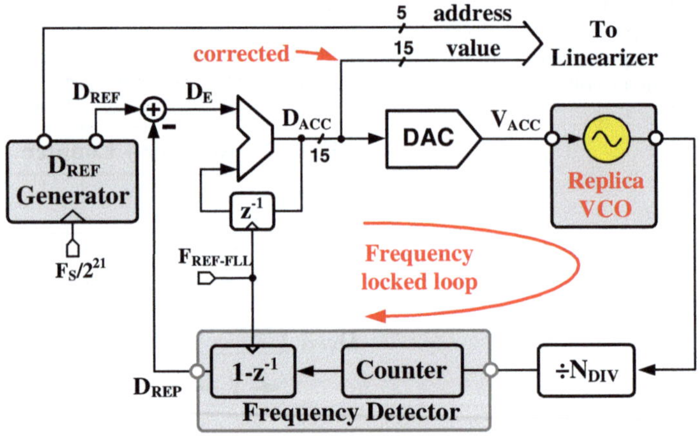

Fig. 15 Simplified block diagram of the calibration unit

Fig. 16 Linear DAC implementation

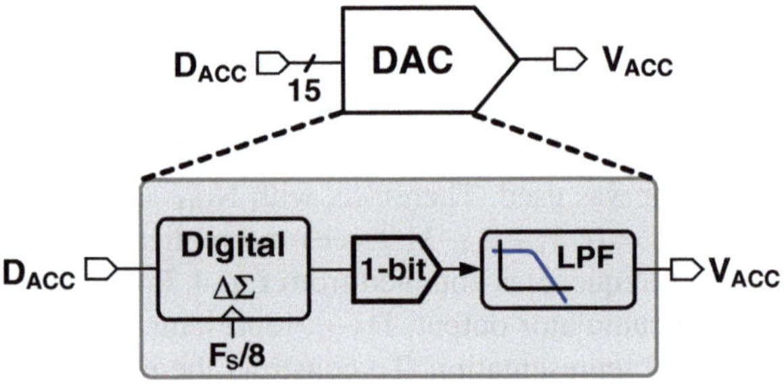

4.4 Circuit Design Details

To validate the above discussed calibration method, a prototype ADC is designed and fabricated in a 90 nm CMOS process [14]. The ADC along with the entire calibration-unit is implemented on-chip. The prototype consists of a 16 delay-stage current starved VCO-based quantizer sampled at 640 MHz. The look-up-table is essentially a 32:1, 15-bit multiplexer. The look-up-table and most of the calibration-unit are digital in nature. They are described in a high level description language and laid out using standard place and route tools. The circuit design of the quantizer, calibration-unit, look-up-table etc., and the effect of various non-ideal circuit behavior are discussed in detail in [15].

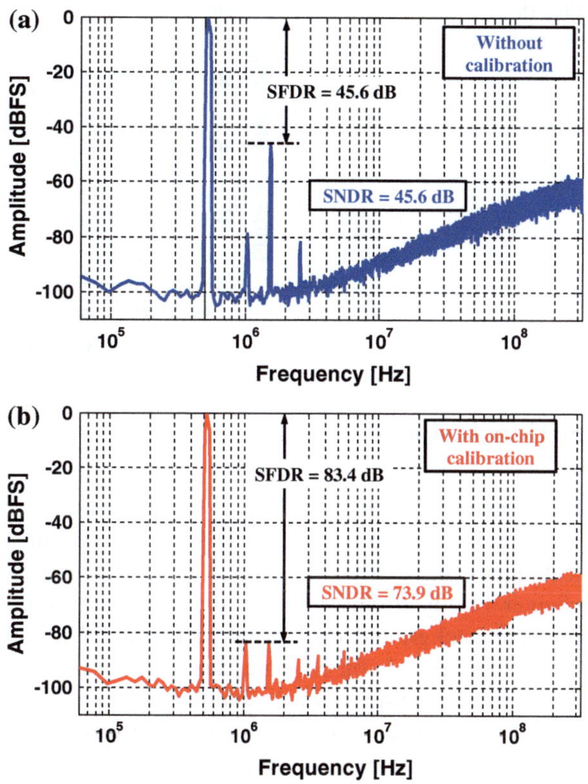

Fig. 17 Measured output spectrum. **a** Without calibration. **b** With on-chip calibration

4.5 Experimental Results

The measured output spectrum of the prototype ADC, for a 500 kHz input signal with and without on-chip calibration is shown in Fig. 17. Without calibration, the SNDR is limited to about 45 dB in 5 MHz signal bandwidth. The SNDR is limited by the third harmonic distortion tone. With on-chip calibration, the harmonic distortion tones are reduced less than −80 dB and the SNDR improves to 73.9 dB. The SNDR, with calibration, is limited by the thermal and quantization noise rather than the harmonic tones. Figure 18 shows the measured SNR and SNDR as a function of different input amplitudes. As expected, both SNR and SNDR increase with input amplitude and follow a 20 dB/decade slope. Overall, the ADC achieves a dynamic range of about 77 dB. The total power consumption of the ADC is around 4.1 mW, with the analog circuits such as the VCO and the DAC consuming 1.2 mW and the remaining circuits such as the calibration unit, phase samplers, encoders etc., consuming 2.9 mW. The ADC achieves a figure-of-merit of 91–110 fJ/conc-step for various input frequencies. These measurement results clearly demonstrate the effectiveness of this technique.

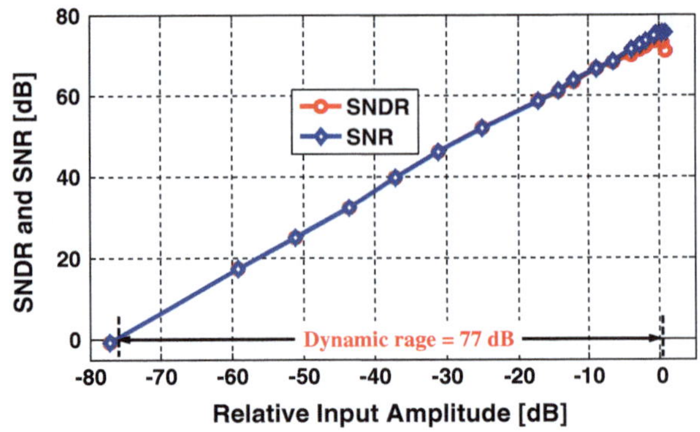

Fig. 18 SNR and ANDR versus input amplitude

5 Conclusions

In summary, VCO based ADCs possess several desirable features but their SNDR is severely limited by the VCO's tuning non-linearity. Recently, several new techniques have been developed to overcome the tuning non-linearity. This chapter briefly discussed such techniques and presented the details of a deterministic background calibration technique. The calibration technique is implemented completely on-chip and has a highly digital implementation. Unlike other calibration techniques, this technique makes no assumption about the order of V-to-F non-linearity. Since a digital frequency locked loop finds the exact input voltage irrespective of the order of V-to-F tuning nonlinearity, this technique can be used as is to correct higher order non-linearity. Because the calibration unit represents the VCO's input voltage with respect to the references of the 1-bit DAC, the calibrated outputs are automatically scaled to the DAC references. This feature can also be used to realize PVT insensitive multi-stage noise shaping architectures which can achieve higher orders of noise shaping. Overall, the ADC achieves state of the art figure-of-merit. Since digital circuits consume significant portion of the total ADC power, its figure-of-merit is expected to improve in more advanced CMOS technologies.

References

1. M. Hovin, A. Olsen, T. Lande, and C. Toumazou, "Delta-sigma modulators using frequency-modulated intermediate values," IEEE J. Solid-State Circuits, vol. 32, no. 1, pp. 13–22, Jan. 1997.
2. M. Straayer and M. Perrott, "A 12-Bit, 10-MHz bandwidth, continuous-time ΣΔ ADC with a 5-Bit, 950-MS/s VCO-based quantizer," *IEEE J. Solid-State Circuits*, vol. 43, no. 4, pp. 805–814, Apr. 2008.
3. U. Wismar, D. Wisland, and P. Andreani, "A 0.2 V 0.44 μW 20 kHz analog to digital ΣΔ modulator with 57 fJ/conversion FoM," in *Proc. ESSIRC*, Sep. 2006, pp. 187–190.

4. M. Straayer, “Noise shaping techniques for analog and time to digital converters using voltage controlled oscillators,” Ph.D. dissertation, Dept. Elect. Eng. Comput. Sci., Mass. Inst. Technol., Cambridge, MA, 2008.
5. G. Taylor and I. Galton, “A mostly-digital variable-rate continuous-time delta-sigma modulator ADC,” *IEEE J. Solid-State Circuits*, vol. 45, no. 12, pp. 2634–2646, Dec. 2010.
6. J. Kim, T. Jang, Y. Yoon, and S. Cho, “Analysis and design of voltage-controlled oscillator based analog-to-digital converter,” *IEEE Trans. Circuits and Syst. I, Reg. Papers*, vol. 57, no. 1, pp. 18–30, Jan. 2010.
7. M. Park and M. Perrott, “A 78 dB SNDR 87 mW 20 MHz bandwidth continuous-time ΔΣ ADC with VCO-based integrator and quantizer implemented in 0.13 μm CMOS,” *IEEE J. Solid-State Circuits*, vol. 44, no. 12, pp. 3344–3358, Dec. 2009.
8. K. Reddy, S. Rao, R. Inti, B. Young, A. Elzhasly, M. Talegaonkar, and P. Hanumolu, “A 16-mw 78-db sndr 10-mhz bw ct ΔΣ adc using residue-cancelling vco-based quantizer,” *IEEE J. Solid-State Circuits*, vol. 47.
9. A. Gupta, K. Nagaraj, and T. Viswanathan, “A two-stage ADC architecture with VCO-based second stage,” *Circuits and Systems II: Express Briefs, IEEE Transactions on*, vol. 58, no. 11, pp. 734–738, 2011.
10. S. Z. Asl, S. Saxena, P. K. Hanumolu, K. Mayaram, and T. S. Fiez, “A 77 dB SNDR, 4 MHz MASH ΔΣ modulator with a second-stage multi-rate VCO-based quantizer,” in *Custom Integrated Circuits Conference (CICC), 2011 IEEE*, Sept. 2011, pp. 1–4.
11. G. Taylor and I. Galton, “A reconfigurable mostly-digital delta-sigma ADC with a worst-case FOM of 160 dB,” *IEEE J. Solid-State Circuits*, vol. 48, no. 4, pp. 983–995, Apr. 2013.
12. J. Daniels, W. Dehaene, M. Steyaert, and A. Wiesbauer, “A 0.02 mm^2 65 nm CMOS 30 MHz BW all-digital differential VCO-based ADC with 64 dB SNDR,” in *Proc. IEEE Symp. VLSI Circuits*, Jun. 2010, pp. 155–156.
13. S. Rao, B. Young, A. Elshazly, W. Yin, N. Sasidhar, and P. Hanumolu, “A 71 dB SFDR open loop VCO-based ADC using 2-level PWM modulation,” in *Proc. IEEE Symp. VLSI Circuits*, Jun. 2011, pp. 270–271.
14. S. Rao, K. Reddy, B. Young, and P. K. Hanumolu, “A 4.1 mW, 12-bit ENOB, 5 MHz BW, VCO-based ADC with on-chip deterministic digital background calibration in 90 nm CMOS,” in *Proc. IEEE Symp. VLSI Circuits*, 2013, pp. C68–C69.
15. S. Rao, K. Reddy, B. Young, and P. K. Hanumolu, “A deterministic digital background calibration technique for VCO-based ADCs,” *IEEE J. Solid-State Circuits*, vol. 49, no. 4, pp. 950–960, Apr. 2014.

Towards Energy-Efficient CMOS Integrated Sensor-to-Digital Interface Circuits

Jelle Van Rethy, Valentijn De Smedt, Wim Dehaene and Georges Gielen

Abstract The ever increasing demand for improved autonomy in wireless sensor devices, drives the search for new energy-efficient sensor interface topologies in CMOS technology. Recently, time-based conversion has gained a lot of interest due to its high potential to implement highly-digital circuitry. While voltage-based analog integrated circuits suffer from the decreased supply voltage and voltage swing in highly-scaled CMOS technologies, time-based processing takes advantage of the increased timing resolution. However, how do these time-based sensor interface circuits compare to their amplitude-based counterparts fundamentally? To answer this question, theoretical limits are derived in this chapter for both implementations, which shows that the sensor itself is actually the dominant factor in limiting the achievable energy efficiency. Time-based topologies, however, enable the implementation of highly-digital interfaces, which are scalable, area-efficient and have low-voltage potential. These observations are illustrated with several practical designs.

1 Introduction

A growing range of sensor application domains, such as Wireless Sensor Networks (WSNs) and Smart Sensors, make sensors ubiquitous in today's smart ambient world and the Internet of Things [1–3]. A major obstacle in realizing these autonomous wireless systems is their power/energy consumption, since these systems can not massively be battery- or grid-powered in practice. Therefore, energy efficiency is a primary objective towards the realization of autonomous

J. Van Rethy (✉) · V. De Smedt · W. Dehaene · G. Gielen
KU Leuven, ESAT—MICAS, Kasteelpark Arenberg 10, 3001 Leuven, Belgium
e-mail: jelle.vanrethy@esat.kuleuven.be

G. Gielen
e-mail: georges.gielen@esat.kuleuven.ac.be

P. Harpe et al. (eds.), *High-Performance AD and DA Converters, IC Design in Scaled Technologies, and Time-Domain Signal Processing*,
DOI: 10.1007/978-3-319-07938-7_17,

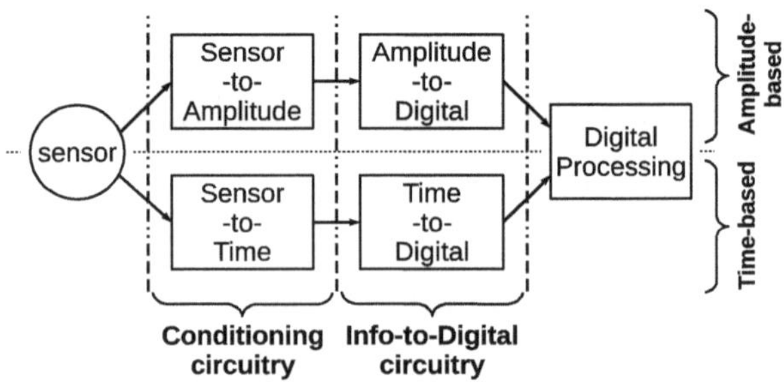

Fig. 1 Amplitude- versus time-based sensor interfacing chain

sensor systems. Recently, time-based interfaces have attracted a lot of attention, since they can be implemented in a digital manner, allowing the use of reduced supply voltages and scaled technologies [4–6]. In highly-scaled CMOS technology, time-to-digital converters can benefit from the increased timing resolution, while voltage-based interfaces suffer from the decreased supply voltage [7]. However, how does this translate to sensor interfaces? This chapter investigates the fundamental difference between these two approaches for both capacitive and resistive sensor interfaces in Sect. 2. Theoretical bounds will be derived and linked to practical implementations, discussed in Sect. 3. Finally, conclusions will be drawn in Sect. 4.

2 Amplitude- Versus Time-Based Sensing

In order to convert an amplitude/time-based signal to the digital domain by means of an analog/time-to-digital converter (ADC/TDC), an amplitude/time signal should exist in which the sensor information is contained. Therefore, a sensor interface can usually be split up into two main parts: a conditioning circuit, which converts sensor information into a convenient format, and a data converter (Fig. 1). This means that for time-based sensor interfaces, the sensor information should first be converted to the time domain. The question here still remains whether this is more energy efficient than its amplitude-based counterpart for the same resolution. Therefore, the theoretical thermal-noise limits will be derived for both the sensor-to-amplitude and the sensor-to-time conditioning circuit from Fig. 1.

2.1 Capacitive Sensor Interfacing

Capacitive sensors are an attractive choice for energy-efficient applications, since they do not consume static power. A capacitive sensor can be modeled as a base capacitance C_0 in parallel with a sensor variation $\pm\Delta C_S$, as depicted in Fig. 2a. Without loss of generality, for the sensor-to-amplitude converter, the well-known

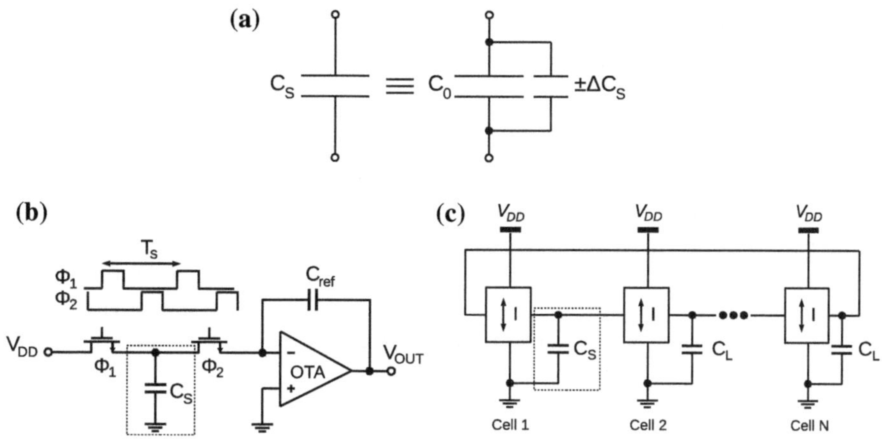

Fig. 2 **a** Modeling of the capacitive sensor with base capacitance C_0 and sensor variation $\pm\Delta C_S$. **b** Amplitude-based conditioning circuit. It is implemented as a Switched-Capacitor-based integrator with the sensor integrated as one capacitive element. **c** Time-based conditioning circuit. It is implemented as a CCO with the sensor integrated in one stage

Switched-Capacitor (SC) integrator is used (Fig. 2b) with an active OTA with feedback capacitor C_{ref} to do the charge-to-voltage conversion [8, 9]. For the time-based implementation, a Capacitance-Controlled (Ring) Oscillator (CCO) is used as a sensor-to-time conversion element (Fig. 2c) [6, 10, 11]. It is important to note that in the following derivation only thermal noise is taken into account in order to determine the theoretical limits. In addition, the results only apply to the conditioning circuitry and are independent of the subsequent ADC/TDC.

2.1.1 Amplitude-Based Switched-Capacitor Integrator

To derive a lower bound for the energy consumption of this conditioning circuit for a given SNR, first a formula for the SNR is derived. In phase ϕ_1 (Fig. 2b), the sensor capacitance is charged to V_{DD}, which results in a total charge of $C_S V_{DD}$. However, as the sensor capacitance consists of both a base capacitance C_0 and a capacitance variation $\pm\Delta C_S$, the peak-to-peak sensor charge information is only $\pm\Delta C_S V_{DD}$. The rms signal charge power is then given by:

$$Q^2_{sig,rms} = \frac{\Delta C_S^2 V_{DD}^2}{2} \tag{1}$$

To find a theoretical lower bound on the thermal noise in the SC circuit, a derivation based on [9] is carried out. All the noise contributions are referred to the equivalent noise charge on the sensor capacitor C_S. In [9] it has been shown that the noise charge associated with the first phase ($\phi_1 = 1$) is due to the thermal noise of the switch. Under the assumption that $\Delta C_S = 0$, the noise charge on the sensor capacitance C_S can be expressed as:

$$q^2_{Cs,\phi 1} = kTC_0 \tag{2}$$

with k the Boltzmann constant and T the absolute temperature in Kelvin. At the beginning of phase 2 ($\phi_1 = 0$ and $\phi_2 = 1$), the charge on capacitor C_S is discharged into the virtual ground created by the OTA and is stored on C_{ref} at the end of phase 2. In phase 2, the total noise charge is the sum of the contributions due to the switch resistance and the OTA (assumed here to be a single stage). However, it can be shown that for the most power-efficient solution, the noise is mainly determined by the noise of the OTA [9], which means that:

$$q^2_{Cs,\phi 2} = \frac{4}{3}kTC_0 \tag{3}$$

Since all noise sources are uncorrelated, the total noise charge power on C_S is equal to the sum of $q^2_{Cs,\phi 1}$ and $q^2_{Cs,\phi 2}$:

$$q^2_{Cs} = kTC_0\left(1 + \frac{4}{3}\right) \tag{4}$$

This leads to an equation for the SNR as the ratio $Q^2_{sig,rms}/q^2_{Cs}$:

$$SNR_{\mathrm{C,SC}} = \frac{1}{4.67kT}\left(\frac{\Delta C_S}{C_0}\right)^2 V^2_{DD}C_0 \tag{5}$$

To link the energy consumption to the SNR, an equation for the energy consumption has to be found. Therefore, a value for the current consumption in the OTA is derived, based on the minimum current required by the OTA to ensure accurate charge transfer within the duration of phase 2 [8]. It can be shown that the minimum g_m required is $g_m = 2C_0K/T_s$ [9], with $K = \ln(2^{\mathrm{ENOB}+1})$ and T_s is the sampling period (Fig. 2b). The energy consumption of the OTA thus equals:

$$E_{OTA} = I_{OTA}V_{DD}T_s = 2C_0V_{DD}K\frac{I_{OTA}}{g_m} \tag{6}$$

where I_{OTA}/g_m is a circuit-topology-dependent constant [8], but one that is nearly technology independent [7]. The total energy consumption, however, is the sum of the energy consumption of the OTA and the energy consumption related to the charging of the capacitor C_S during phase 1 ($\phi_1 = 1$):

$$E_{\mathrm{total}} = C_0V^2_{DD} + 2C_0V_{DD}K\frac{I_{OTA}}{g_m} \tag{7}$$

If we then combine (5) and (7), a measure is obtained for the minimum energy consumption as a function of a given SNR and sensor sensitivity $\Delta C_S/C_0$:

$$E_{C,SC} = 4.67kT\left(\frac{C_0}{\Delta C_S}\right)^2\left(1+2K\frac{I_{OTA}}{g_m V_{DD}}\right)SNR \tag{8}$$

2.1.2 Time-Based Capacitance-Controlled Ring Oscillator

To derive the minimum energy consumption for the time-based implementation of the CCO in Fig. 2c, the following equation for the period of the oscillator is defined:

$$T = 2\frac{C_S V_{DD}}{I} + 2(N-1)\frac{C_L V_{DD}}{I} \tag{9}$$

where N is the number of stages, I is the current to charge and discharge the load capacitor and C_L is the load capacitance of the stages other than the sensor stage. It will be seen that the ratio of C_0 to C_L is important, and therefore a ratio factor $R_C = C_0/C_L$ is introduced. If we consider the sensor from Fig. 2a, (9) can be split into a free-running period T_0 and a variation $\pm\Delta T$ due to the measurand:

$$T_0 = 2\frac{C_0 V_{DD}}{I} + 2(N-1)\frac{C_L V_{DD}}{I} = \frac{2(N-1+R_C)C_0 V_{DD}}{R_C I} \tag{10}$$

$$\Delta T = 2\frac{\Delta C_S V_{DD}}{I} \tag{11}$$

The signal power can thus be defined as:

$$\Delta T^2_{sig,rms} = \left(\frac{\Delta T}{\sqrt{2}}\right)^2 = 2\frac{\Delta C_S^2 V_{DD}^2}{I^2} \tag{12}$$

Based on the theoretical analysis in [12], the minimum achievable jitter (noise) can be derived for the ring oscillator from Fig. 2c. Since the oscillator is not completely symmetrical due to the introduced ratio R_C, the total jitter variance is the sum of the variance of the jitter introduced in every stage:

$$\begin{aligned}\sigma^2_{T_0} &= \left[\frac{2kTC_0}{I^2} + \frac{2(N-1)kTC_L}{I^2}\right]\left(1+\frac{2}{3}\frac{g_{d0}V_{DD}}{I}\right)\\ &= \frac{2(N-1+R_C)kTC_0}{R_C I^2}\left(1+\frac{4}{3}\frac{V_{DD}}{V_{GS}-V_T}\right)\end{aligned} \tag{13}$$

In this expression, the long-channel expression for g_{d0}, which is the device output conductance for zero v_{ds}, is used for simplicity [12]. For the calculation of the minimum jitter, it is assumed that $V_{GS} \approx V_{DD}$ and $V_T \approx V_{DD}/2$. The maximum achievable SNR can then be equated as the ratio $\Delta T^2_{sig,rms}/\sigma^2_{T_0}$:

$$SNR_{\mathrm{C,CCO}} = \frac{1}{3.67kT}\left(\frac{\Delta C_S}{C_0}\right)^2\left(\frac{R_C}{N-1+R_C}\right)V_{DD}^2 C_0 \tag{14}$$

Combining (10) and (14) and given that the energy consumption of an inverter-based ring oscillator can be equated as $E = V_{DD}IT_0$, one can calculate that:

$$E_{\mathrm{C,CCO,inverter}} = 7.33kT\left(\frac{C_0}{\Delta C_S}\right)^2\left(\frac{N-1+R_C}{R_C}\right)^2 SNR \tag{15}$$

and gives the minimum energy consumption for the inverter-based CCO for a given SNR and given sensor sensitivity $\Delta C_S/C_0$. On the other hand, the energy consumption of a biased differential ring oscillator is $E = NV_{DD}IT_0$ and if we assume the same minimum jitter in this case, the minimum energy consumption for a biased differential CCO is:

$$E_{\mathrm{C,CCO,biased}} = 7.33kT\left(\frac{C_0}{\Delta C_S}\right)^2\left(\frac{N-1+R_C}{R_C}\right)^2 SNR \cdot N \tag{16}$$

which is a factor N worse than its inverter-based counterpart. This can be easily understood by noting that all stages consume current, even when they are not contributing to the signal.

2.2 Resistive Sensor Interfacing

A similar derivation can also be done for resistive sensor interfaces. However, it is slightly different because a resistive sensor itself also generates noise. Consider the network in Fig. 3a. It shows a differential resistive sensor with a nominal value R_0 and sensor variation $\pm\Delta R_S$, placed in a voltage divider configuration with an excitation voltage V_{DD}. The output voltage V_S can then be expressed as:

$$V_S = \frac{R_0 \pm \Delta R_S}{2R_0}V_{DD} = \frac{V_{DD}}{2} \pm \frac{\Delta R_S}{2R_0}V_{DD} \tag{17}$$

where $\pm V_{DD}\Delta R_S/(2R_0)$ is the amplitude-based sensor signal. The integrated rms noise generated by this resistive sensor at the output is:

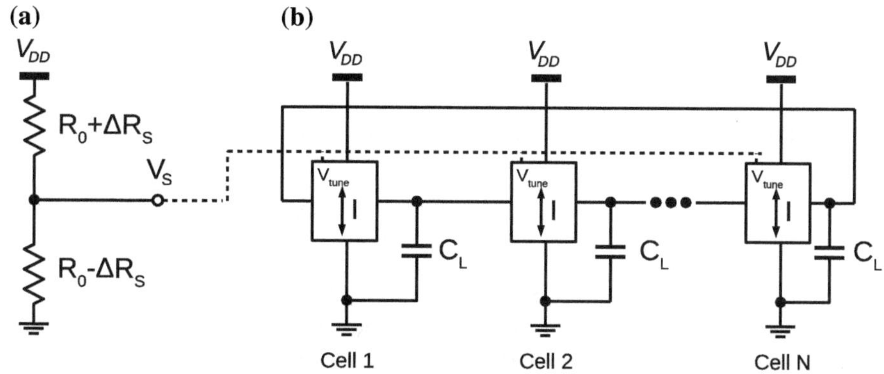

Fig. 3 **a** Differential resistive sensor with nominal resistance R_0 and variation $\pm\Delta R_S$. **b** Sensor-to-period conversion through a voltage-controlled oscillator

$$\sigma_{V_S}^2 = 2kTR_0 f_{BW} \tag{18}$$

where f_{BW} is the cut-off frequency of the captured sensor signal. The SNR of the sensor itself can then be expressed as:

$$SNR_{\mathrm{R,VD}} = \frac{1}{16kTR_0 f_{BW}} \left(\frac{\Delta R_S}{R_0}\right)^2 V_{DD}^2 \tag{19}$$

Since the energy consumption of the voltage divider is equal to $V_{DD}^2/(2R_0 2f_{BW})$, the minimum energy consumption is equal to:

$$E_{\mathrm{R,VD}} = 4kT\left(\frac{R_0}{\Delta R_S}\right)^2 SNR \tag{20}$$

The voltage-controlled oscillator (VCO), however, also adds noise to the output signal. To calculate this, it is assumed that the delay of the oscillator stages is linearly dependent on the applied voltage (e.g. [13]). Similar to (10) and (11), one can write:

$$T = T_0 + \Delta T = 2N\frac{C_L V_{DD}}{I} + 2N\frac{C_L V_{DD}}{I}\frac{\Delta R_S}{R_0} \tag{21}$$

which results in a signal power of:

$$\Delta T_{sig,rms}^2 = 2\left(N\frac{C_L V_{DD}}{I}\right)^2\left(\frac{\Delta R_S}{R_0}\right)^2 \tag{22}$$

Neglecting the noise of the resistive sensor itself, the SNR of the oscillator's output signal is then equal to (using (13) and $V_{GS} - V_T \approx V_{DD}/2$):

$$SNR_{\mathrm{R,RCO}} = \frac{1}{3.67kT}\left(\frac{\Delta R_S}{R_0}\right)^2 V_{DD}^2 N C_L \tag{23}$$

Similarly to (15), the minimum energy consumption can be written as:

$$E_{\mathrm{R,RCO}} = 7.33kT\left(\frac{R_0}{\Delta R_S}\right)^2 SNR \tag{24}$$

and expresses the minimum energy needed to achieve a certain SNR for a given sensor sensitivity $\Delta R_S/R_0$. This means that the maximum SNR and minimum energy consumption is either determined by the voltage divider ((19) and (18)) or by the oscillator ((23) and (24)).

2.3 Discussion and Comparison

2.3.1 SNR

For both the amplitude- and time-based implementations, formulas have been derived that give the maximal theoretical achievable SNR and the minimum theoretical achievable energy consumption for a given SNR, and are summarized in Table 1. It is interesting to see that for all implementations, the SNR is quadratically dependent on the sensor sensitivity $\Delta C_S/C_0$ or $\Delta R_S/R_0$. In addition, for capacitive sensor interfaces, it is linearly dependent on the base capacitance C_0. This is a very important observation, since this means that it is fully determined by the sensor's design/choice and that there are almost no degrees of freedom left. However, for the time-based CCO implementation, the SNR increases with R_C, which is thus a design parameter. This can be explained by the fact that the sensitivity $\Delta T/T_0$ increases with R_C [10]. However, the ratio R_C is limited by the need to ensure the proper functioning of the oscillator [10]. Similar to the SNR for capacitive sensors, the SNR for resistive sensors in a voltage divider is inversely proportional to the base resistance R_0 and is thus also determined by sensor design/choice.

The following discussion mainly focuses on capacitive sensors. However, similar conclusions can be drawn for resistive sensors. In Fig. 4, the SNR (in dB) is plotted as a function of the base capacitance C_0 of capacitive sensors. The sensor sensitivity $\Delta C_S/C_0$ is taken equal to 10 % for both topologies, which is a common value for MEMS capacitive sensors [14]. For the CCO, a 5-stage CCO is considered with a ratio $R_C = 5$, which is feasible to implement [10]. V_{DD} is taken equal to 1 V.

From the figure it is clear that the inherent SNR (without oversampling) is limited to a range of 55–70 dB for both conditioning circuits. It also shows that there is not really a large fundamental difference between the two implementations

Table 1 Theoretical achievable SNR and minimum energy consumption

		SNR	E_{min}
$\Delta C_S/C_0$	SC	$\frac{1}{4.67kT}\left(\frac{\Delta C_S}{C_0}\right)^2 V_{DD}^2 C_0$	$4.67kT\left(\frac{C_0}{\Delta C_S}\right)^2\left(1+2K\frac{I_{OTA}}{g_m V_{DD}}\right)SNR$
	$CCO_{inverter}$	$\frac{1}{3.67kT}\left(\frac{\Delta C_S}{C_0}\right)^2\left(\frac{R_C}{N-1+R_C}\right)V_{DD}^2 C_0$	$7.33kT\left(\frac{C_0}{\Delta C_S}\right)^2\left(\frac{N-1+R_C}{R_C}\right)^2 SNR$
	CCO_{biased}		$7.33kT\left(\frac{C_0}{\Delta C_S}\right)^2\left(\frac{N-1+R_C}{R_C}\right)^2 SNR \cdot N$
$\Delta R_S/R_0$	VD	$\frac{1}{16kTR_0 f_{BW}}\left(\frac{\Delta R_S}{R_0}\right)^2 V_{DD}^2$	$4kT\left(\frac{R_0}{\Delta R_S}\right)^2 SNR$
	RCO	$\frac{1}{3.67kT}\left(\frac{\Delta R_S}{R_0}\right)^2 V_{DD}^2 N C_L$	$7.33kT\left(\frac{R_0}{\Delta R_S}\right)^2 SNR$

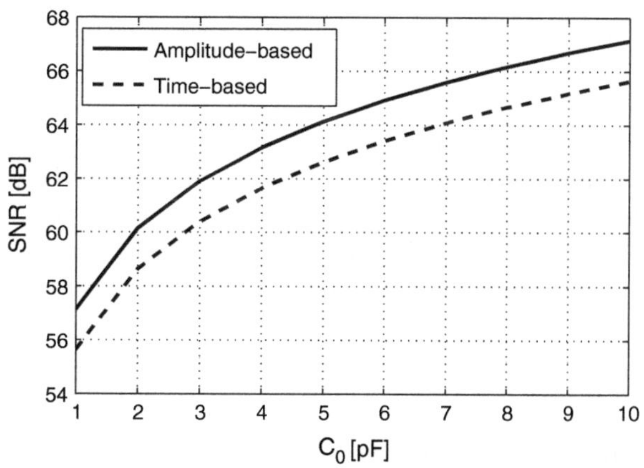

Fig. 4 SNR [dB] as a function of the base capacitance C_0 and a sensor sensitivity $\Delta C_S/C_0$ of 10 % for both the amplitude- and time-based conditioning circuit

(only 2 dB). The time-based solution has a slightly lower SNR limit, which is mainly due to the limited ratio R_C. Although the inherent SNR is thus more or less fixed by the sensor design, it can still be increased by means of oversampling the measured/digitized output (if it is not quantization-noise limited [8]). For every doubling of the oversampling ratio (OSR), there is a doubling in the SNR, as the noise power is halved. This, however, does not influence the minimum energy consumption E_{min}, as doubling the SNR means doubling the oversampling and thus doubling the energy consumption.

2.3.2 Energy Consumption

It can be seen from Table 1 that the minimum energy consumption E_{min} is linearly dependent on the SNR, but quadratically dependent on the inverse of the sensor sensitivity for all implementations. This is an important observation that already was made in [8]. This implies that the theoretical minimum achievable energy efficiency is mainly dependent on the sensor itself.

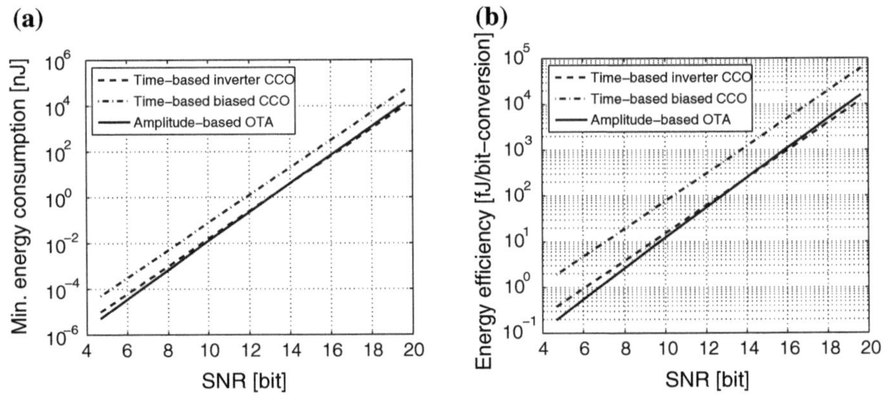

Fig. 5 **a** Theoretical minimum energy consumption as a function of the SNR. **b** Theoretical energy efficiency as a function of the SNR

To compare the energy consumption of both approaches, Fig. 5a plots the minimum energy needed as a function of the SNR (expressed in bits) for the time-based CCOs and amplitude-based SC interfaces, while Fig. 5b plots their corresponding energy efficiency. In literature, the energy efficiency of sensor interfaces is usually expressed in terms of the well-known FoM for A/D converters:

$$\mathrm{FoM}_{\mathrm{A/D}} = \frac{P}{2f_{BW}2^{\mathrm{SNR[bit]}}} \tag{25}$$

where P is the power consumption and f_{BW} is the signal bandwidth of the converter.

For the simulations, all values are taken equal as before ($R_C = 5$, $N = 5$, $V_{DD} = 1$ V and $\Delta C_S/C_0 = 10$ %). For the OTA, the ratio g_m/I_{OTA} is taken equal to 5 V^{-1}, which corresponds to a differential pair in strong inversion [7]. From the figures, it is clear that the energy efficiency of the time-based inverter CCO is more or less equal to that of the amplitude-based OTA. The fundamental limits of these two implementations are therefore comparable and competitive. However, the differential biased CCO is fundamentally a factor N worse in energy efficiency. But it is important to note that differential structures usually outperform their single-ended counterparts in terms of common-mode noise rejection [15], which might justify the use of biased differential structures in some applications.

2.3.3 Quantization-Noise-Limited Designs

If we would compare the theoretical minimum energy efficiency with the state of the art, it is clear that this is not straightforward, as the $\mathrm{FoM}_{\mathrm{A/D}}$ (25) does not take into account the sensor sensitivity, which is of importance for sensor interfaces

(Table 1). Although it is very hard to compare the energy efficiency between different designs, we do see some trends. Some published energy efficiencies are added to the plot in Fig. 6a. Note that the theoretical curves are derived for a sensor sensitivity of 10 %, which might not match the sensor sensitivity of the published sensor interfaces. Nevertheless, from the plot it is clear that the published energy efficiency, usually in the pJ/bit-conversion range, can be several orders of magnitude worse than the theoretical thermal limit. One explanation is that the theoretical thermal noise limit is hard to achieve in practical designs, due to other noise sources than the intrinsic noise. But a more important argument is that so far, only the conditioning circuit from Fig. 1 has been taken into account. This means that the conditioned signal is not yet digitized. In practical designs, the quantization noise of the info-to-digital circuit (Fig. 1) is often the limiting factor, which means that the SQNR is lower than the SNR of the conditioned signal. Therefore, oversampling, etc. is needed to increase the SQNR, which means that the total energy consumption increases. In addition, the energy consumption of the info-to-digital circuit also contributes to the total energy consumption.

To illustrate this, consider a single-bit first-order noise-shaped ΔΣ modulator as an information-to-digital converter (Fig. 1). Oversampling and first-order noise shaping are exploited to increase the SQNR [16]:

$$\mathrm{SQNR}_{\Delta\Sigma,1\mathrm{bit}}[\mathrm{dB}] = 2.55 + 30\log(\mathrm{OSR}) \tag{26}$$

where OSR is the oversampling ratio. To achieve a certain SQNR, the corresponding OSR can be calculated through (26). To calculate the total energy consumption of the sensor-to-digital converter, let us assume a sensor sensitivity of 10 % with $C_0 = 5$ pF. From Fig. 4 can be seen that for $C_0 = 5$ pF, the inherent SNR of the conditioning circuit is ≈63 dB, which equals ≈10 bit of resolution. With this SNR, the corresponding energy/power consumption for the conditioning circuit can then be calculated through the formulas in Table 1. However, without oversampling (OSR = 1), the S(Q)NR at the output of the sensor-to-digital converter is only 2.55 dB (26). This means that we achieve less than 1 bit resolution at the system output, but with an energy consumption in the conditioning circuit that equals 10 bits of resolution. In order to increase the resolution at the output, the OSR should be increased, but, as a result, the energy consumption also is multiplied with the OSR. The resulting energy efficiency as a function of the SNR at the output is plotted in Fig. 6b. To simplify the calculation, we do not consider the power consumption of the ΔΣ modulator itself. From the figure it can be seen that at lower resolutions, the quantization noise is the limiting factor. However, at higher resolutions, the thermal noise of the conditioning circuit becomes the limiting factor and limits the achievable energy efficiency. At the crossing of the theoretical curves, there is an optimum regarding the energy efficiency.

The question then is, how can we obtain thermal-noise-limited designs at lower resolutions, so that we can achieve fJ/bit-conv. energy efficiencies? This can be achieved by lowering the power consumption in the conditioning circuit. However, as the minimum power/energy consumption is mainly determined by the sensor

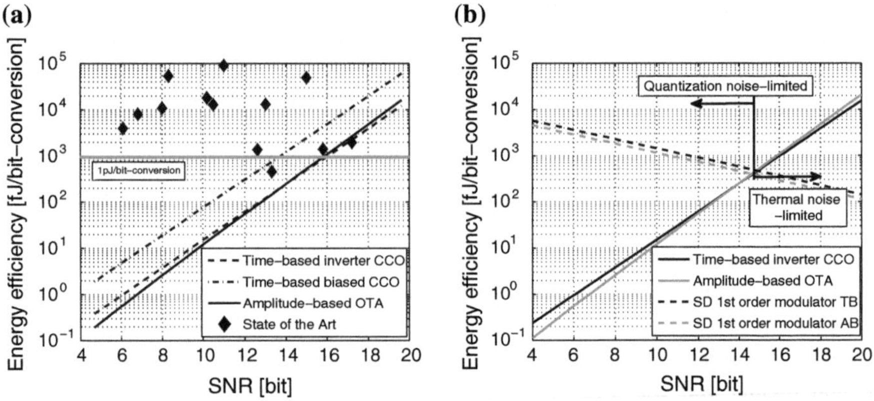

Fig. 6 **a** Theoretical energy efficiency as a function SNR with state of the art added. **b** Theoretical minimum energy efficiency (pJ/bit-conv.) as a function of the SNR for both the amplitude-based and the time-based conditioning circuit and complete sensor-to-digital conversion

sensitivity and the base capacitance C_0, there is no degree of freedom in lowering the power/energy consumption in the conditioning circuit, apart from using a better sensor. This means that at lower resolutions, the performance usually is quantization-noise limited. In oversampled digitizers, the oversampling ratio OSR must therefore be decreased to lower the energy consumption. This can be done with e.g. higher-order modulators, multi-bit quantization, etc. In addition, the power consumption of the modulator itself is not taken into account in this discussion and will also add to the total energy consumption in practical designs. Therefore, the power consumption of the digitizer should be as low as possible.

2.4 Important Conclusions

It has been shown that the theoretical minimum achievable power/energy consumption and the corresponding energy efficiency of a conditioning circuit is mainly determined by the nominal value of the sensor (C_0 and R_0) and by its sensitivity ($\Delta C/C_0$ and $\Delta R_S/R_0$). In addition, it has been shown that the theoretical limits are comparable for both the time-based and the amplitude-based implementations. However, there still seems to be a gap between the theoretical achievable energy efficiency and energy efficiency in practical sensor interface designs, especially for sub-14 bit resolution. This is explained by the fact that these sensor-to-digital converters are usually quantization-noise limited and not thermal-noise limited. Furthermore, the information-to-digital converter itself also adds to the total power consumption of the system, while this power consumption was not taken into account in the derivation. Therefore, to minimize the power consumption, highly-digital implementations seen in time-based interfaces are preferred over analog, biased amplitude-based interfaces. Digital implementations tend to consume less static power, do not need to be biased and are more area-efficient, making these time-based interfaces a primary choice. In addition, they tend to be more scalable and easier to implement than their amplitude-based counterparts.

3 Implementations of Time-Based Sensor-to-Digital Converters

In this section, several design examples will be elaborated and compared with the theoretical results from the previous section. These time-based implementations are based on the use of an on-chip relaxation/ring oscillator as an interfacing element for both capacitive (Sects. 3.1 and 3.3) and resistive sensors (Sects. 3.2 and 3.4). Different methods of implementing the time-to-digital conversion are presented, resulting in the full sensor-to-digital converter (SDC).

3.1 Design 1: PM-Based Open-Loop SDC

The first design example illustrates the use of an on-chip capacitance-controlled oscillator (CCO) as a period-modulation (PM)-based interfacing element for external capacitive MEMS sensors [17]. It employs a 6-stage coupled sawtooth differential biased ring oscillator as a conditioning circuit to perform the capacitance-to-period modulation, as depicted in Fig. 7a [13]. This time/period-based output signal can then be easily digitized by using a simple (reset) counter and decimator, which makes the digitizer entirely digital. Therefore, the power consumption of the TDC is negligible compared to the power consumption of the conditioning circuit (CCO) [17]. The quantization resolution Q of the period-to-digital converter is dependent on the sampling frequency f_{CLK} and on the frequency tuning range Δf_{CCO} of the CCO and it is chosen to be 1 in this design. However, it is interesting to note that by quantizing in this way, the quantization noise is first-order noise-shaped (Fig. 7b), due to the integration of the sampling error [18].

The circuit has been prototyped in 130-nm CMOS technology and takes only 0.05 mm^2 [17]. It is optimized for maximum sensor sensitivity, which means that the ratio $R_C \approx 6$. It consumes 371 μW from a 1.2 V supply voltage. From the measurement results of the CCO in [17], it can be calculated that for an input capacitance range of 3.7–13.7 pF, the corresponding SNR is 59 dB (9.5 bit) for a free-running frequency of 4.59 MHz (bandwidth = 2.3 MHz). This means that the corresponding energy efficiency of the CCO is 110 fJ/bit-conv. However, due to the single-bit quantization in the period-to-digital converter, oversampling is

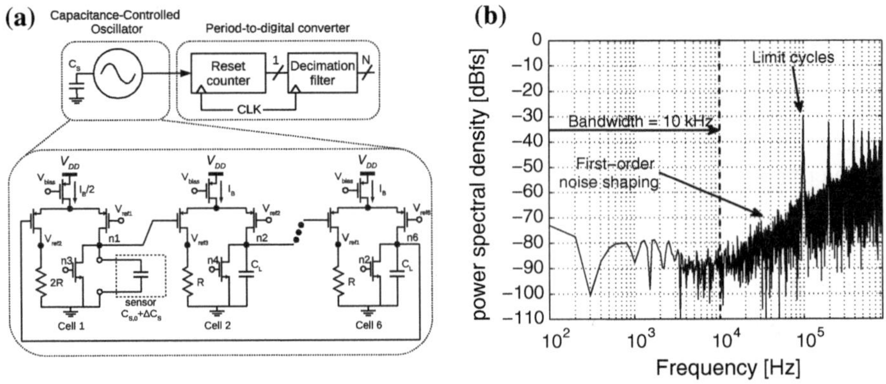

Fig. 7 **a** Block diagram of the PM-based sensor interface and the schematic of the implemented 6-stage coupled sawtooth CCO. **b** Measured spectrum of the digital output of the sensor interface showing the quantization noise shaping

needed to increase the resolution at the digitized output. This results in a SNR of 64.68 dB (10.5 bit) for a bandwidth of 10 kHz (Fig. 7b) [17] and corresponds to an energy efficiency of 12.7 pJ/bit-conv. This means that the energy efficiency is degraded with a factor 115, which confirms the analysis done in Sect. 2. The two FoMs are added in Fig. 11 to compare with the other designs. Although this time-based interface actually only includes a CCO with a very simple TDC, it is operated in an open-loop manner and needs a reference clock f_{CLK}. This makes that it is very vulnerable to temperature and supply-voltage variations. Other implementations in Sects. 3.2 and 3.4 show an improved resilience towards supply-voltage and temperature variations.

3.2 Design 2: PWM-Based Open-Loop SDC

The second implementation shows an open-loop pulse-width-modulation (PWM)-based sensor interface for resistive sensors. Compared to the first design example, it is independent of supply-voltage and temperature variations [19]. This is done by building a ring oscillator of which the stage delay is differentially modulated: each half of the stage delays is increased and decreased simultaneously. By using an output latch, a PWM signal can be extracted of which the duty cycle and not the absolute period is a measure for the sensor value. A block diagram of this oscillator system is shown in Fig. 8. To have high control linearity, the coupled sawtooth oscillator is used [13]. To decrease the power consumption of the oscillator, unused stages are turned off. The interface circuit has been prototyped in 40 nm CMOS technology [19]. The performance of the sensor interface is evaluated by measuring the output spectrum when applying a 100 kHz input

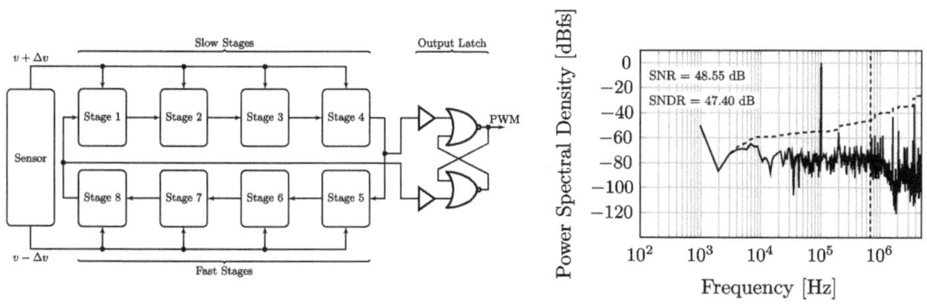

Fig. 8 Block diagram of the PWM-based sensor interface and measured output spectrum of the PWM-based sensor signal with 100 mV input amplitude

signal (Fig. 8). For a 100-mV-amplitude differential input signal, this results in 7.6 bits of resolution at 1 V supply voltage for a bandwidth of 700 kHz. The corresponding energy efficiency is then equal to 66.2 fJ/bit-conv. In order to digitize the PWM sensor signal, the signal is single-bit sampled by a clock of 50 MHz. In [19], it has been calculated that for an accuracy of 8 bits, at least 64 periods of the oscillator need to be sampled by this clock. Under the assumption that 100 periods are needed to accurately digitize the PWM sensor signal, this results in a FoM of the sensor-to-digital conversion of 6.6 pJ/bit-conv. Both FoMs are also added to Fig. 11. To quantify the temperature and supply-voltage resilience, measurements have been performed on the chip [19]. The maximum temperature error is 79 ppm/°C over a −20 to 100° temperature range. The output deviation over a 0.8–1.5 V supply voltage span is below 1.42 %/V.

3.3 Design 3: BBPLL-Based Fully-Digital SDC

The design in Sect. 3.1 is an open-loop architecture, which needs an accurate reference clock f_{CLK}. The proposed design in this section is based on locking a digitally-controlled oscillator (DCO) to the CCO by using a digital Bang-Bang Phase-Locked Loop (BBPLL) (Fig. 9), which means that the conversion is based on the relative frequencies of the two oscillators and not on their absolute frequency [6]. The two oscillators (SCO and DCO in Fig. 9) are 3-stage inverter-based ring oscillators and are operated at only 0.3 V. The D-flip-flop functions as a single-bit phase detector. This design shows the supply-voltage scalability of fully-digitally-implemented time-based sensor interfaces for capacitive sensors [6]. The conditioning circuit in this case is the 3-stage inverter-based ring oscillator with the sensor integrated in one stage (SCO). The time-to-digital converter is the BBPLL-based converter [6, 20]. The interesting feature of this topology is that it resembles a $\Delta\Sigma$ modulator, although it is actually a PLL [20]. This means that the single-bit quantization noise is first-order noise shaped and oversampling can be exploited. The circuit has been prototyped in 130-nm technology [6]. Measurement data of the conditioning circuit (SCO)

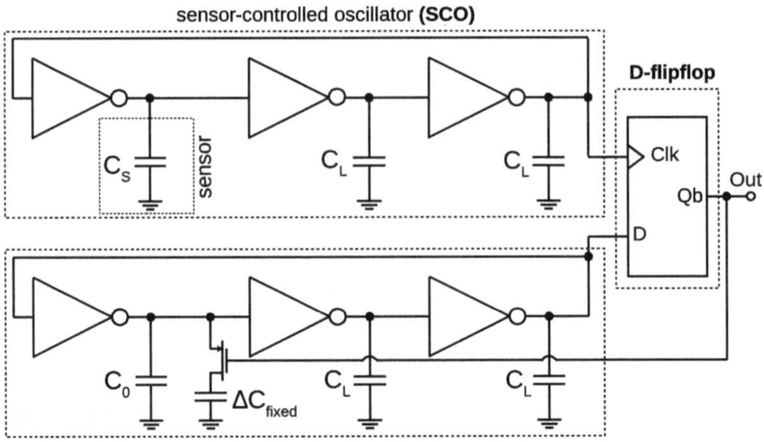

Fig. 9 Circuit implementation of the BBPLL-based capacitive sensor-to-digital interface circuit

alone are unavailable, but the sensor-to-digital conversion achieves a resolution of 6.1 bit for a bandwidth of 500 Hz, while consuming 270 nW from a 0.3 V supply voltage [6]. This results in a FoM of 3.9 pJ/bit-conv., which again is in the range of several pJ instead of fJ (thermal limit) for 6.1 bits of resolution (Fig. 6), which means that this design is quantization-noise limited.

3.4 Design 4: BBPLL-Based Force-Balanced Wheatstone Bridge SDC

The final design shows a second-order BBPLL-based sensor interface for resistive sensors [21]. This interface exploits the feedback nature of the BBPLL-based sensor interface to improve the Power-Supply Rejection Ratio (PSRR) and the temperature insensitivity. In addition, as in the design in Sect. 3.3, the conversion is not based on the absolute frequency of the oscillators, neither is it dependent on the accuracy of an external reference clock f_{CLK}. This makes the interface highly robust. This design proposes a force-balanced Wheatstone bridge technique to improve the PSRR of the entire sensor interface [21]. Furthermore, the sensor-to-digital conversion is incorporated in the circuitry needed to do the force-balancing, which makes extra building blocks unnecessary. Figure 10 shows the architecture of the implementation. In this implementation, the Sensor VCO in Fig. 10 can be seen as the conditioning circuit. It is a 4-stage Maneatis differential VCO with replica bias feedback to bias the delay cells [21]. The time-to-digital conversion is in this case done by the second-order BBPLL, which consists of the D-flip-flop as a single-bit quantizer, the first-order digital filter, the resistive DAC and the Loop VCO in the feedback path (Fig. 10). The Loop VCO is implemented with an

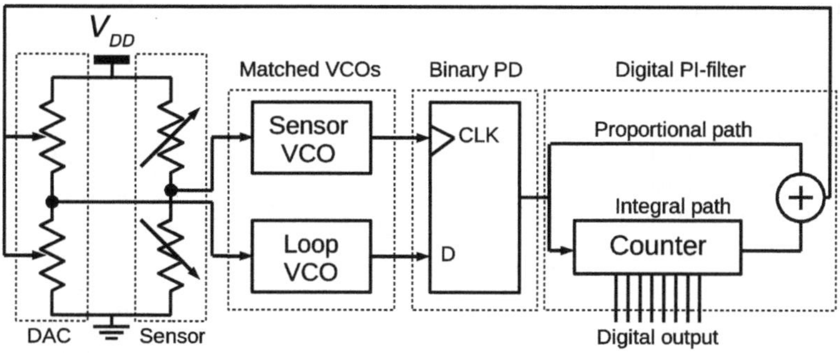

Fig. 10 Time-based architecture of the second-order BBPLL-based force-balanced Wheatstone bridge sensor interface

identical VCO, which means that the two VCOs are matched, which is needed for the cancellation of the temperature and supply-voltage variations. The circuit has been prototyped in 130-nm CMOS technology. Measurements on the Sensor VCO show that the SNR at the output of the conditioning circuit is 40.5 dB (6.4 bit) [21]. The nominal operating frequency of the oscillator is 6.6 MHz with a power consumption of 31.24 μW at 1 V supply voltage. This results in an energy efficiency of only 56 fJ/bit-conv. for the conditioning circuit. The complete sensor interface achieves 64.44 dB SNR (10.4 bit) for a bandwidth of 10 kHz and a total power consumption of 124.5 μW. This results in an energy efficiency of 4.6 pJ/bit-conversion, which again is in the range of pJ and not in the range of fJ. The two FoMs are added to Fig. 11 to compare with the other implementations and theoretical limits. In addition to the state-of-the-art energy efficiency, this interface is also very robust, due to its time-based feedback loop implementation. The measured maximum absolute output error is only 0.7 % in a supply voltage range of 0.85–1.15 V, while the maximum absolute output error in a temperature range of 100 °C is only 0.56 % or 56 ppm/°C drift sensitivity.

3.5 Comparison

In Fig. 11, the performance of the various designs is compared with the theoretical limits. Note that the theoretical limits are derived for a sensor sensitivity of 10 % and a nominal capacitance of 5 pF. These values might differ among the designs, shifting the theoretical curves up or down. The energy efficiency of the conditioning circuit of each design is plotted in the graph with underscript$_1$, while the energy efficiency of the entire sensor-to-digital conversion is plotted with underscript$_2$. It can be seen that the latter is roughly two orders of magnitude higher than the former, due to the quantization noise limitation, as discussed in Sect. 2.3. The arrows in the graph show how the energy efficiency is shifted if no quantization

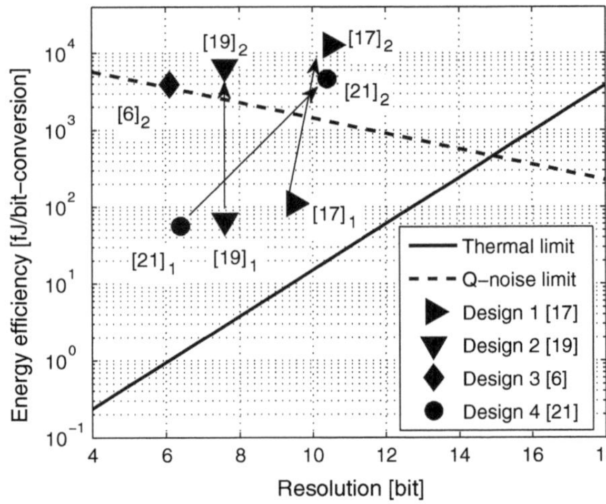

Fig. 11 Comparison of the four presented designs with the theoretical limits. Underscript$_1$ shows the energy efficiency of the conditioning circuit, while the underscript$_2$ plots the energy efficiency of the total sensor-to-digital converter

(underscript$_1$) and quantization (underscript$_2$) is considered. The quantization makes that the state-of-the-art energy efficiency for sensor interfaces still is in the pJ/bit-conv. range and not in the fJ/bit-conv. range.

4 Conclusions

This chapter has described and compared the theoretical minimum achievable SNR and energy efficiency for both amplitude-based and time-based sensor interfaces, for both capacitive as resistive sensors. It shows that the sensor itself (fixed C_0 or R_0 and limited sensor sensitivity) is the dominant factor in limiting the maximum energy efficiency. This means that comparing sensor interfaces is rather difficult, since it mainly depends on the sensor itself. In addition, in practical implementations, the quantization noise is often the limiting factor, which means that the energy consumption of the sensor-to-digital converters is several orders of magnitudes higher than the thermal limit in the conditioning circuitry. This means that the state-of-the-art energy efficiency for sensor interfaces still is in the pJ/bit-conv. range and not in the fJ/bit-conv. range, as state-of-the-art ADCs are. Although the fundamental limits of amplitude- and time-based implementations are comparable, time-based sensor interfaces are an attractive choice, because they can be implemented in a digital manner, enabling scalable, low-area, low-voltage and low-power design. This has been demonstrated with four different time-based implementations. In order to improve on today's state-of-the-art, not only should the energy efficiency of the circuitry be improved, but sensor design optimization is as, or even more, important.

Acknowledgment The work of J. Van Rethy was supported by FWO Flanders.

References

1. G. C. Meijer, *Smart sensor systems*. John Wiley & Sons, 2008.
2. B. Warneke, M. Last, B. Liebowitz, and K. Pister, "Smart dust: Communicating with a cubic-millimeter," *Computer*, vol. 34, pp. 44–51, 2001.
3. F. Lewis, "Wireless sensor networks," *Smart Environments: Technologies, Protocols, and Applications*, 2004.
4. J.-M. Park and S.-I. Jun, "A resistance deviation-to-time interval converter for resistive sensors," in *IEEE SoC Conference*, 2008, pp. 101–104.
5. Z. Tan, S. H. Shalmany, G. C. Meijer, and M. A. Pertijs, "An energy-efficient 15-bit capacitive-sensor interface based on period modulation," *IEEE J. of Solid-State Circuits*, vol. 47, no. 7, pp. 1703–1711, 2012.
6. H. Danneels, K. Coddens, and G. Gielen, "A fully-digital, 0.3 v, 270 nw capacitive sensor interface without external references," in *proceedings of ESSCIRC*, 2011, pp. 287–290.
7. A. Annema, B. Nauta, R. Van Langevelde, and H. Tuinhout, "Analog circuits in ultra-deep-submicron cmos," *IEEE J. of Solid-State Circuits*, vol. 40, no. 1, pp. 132–143, 2005.
8. M. A. Pertijs and Z. Tan, *Energy-Efficient Capacitive Sensor Interfaces*. Springer, 2013.
9. R. Schreier, J. Silva, J. Steensgaard, and G. C. Temes, "Design-oriented estimation of thermal noise in switched-capacitor circuits," *IEEE Trans. on Circuits and Systems-I: Regular Papers*, vol. 52, no. 11, pp. 2358–2368, 2005.
10. J. Van Rethy, H. Danneels, K. Coddens, and G. Gielen, "Capacitance-controlled oscillator optimization for integrated capacitive sensors with time/frequency-based conversion," in *proceedings of EUROSENSORS XXV*, 2011, pp. 1301–1304.
11. M. Shulaker, J. Van Rethy, G. Hills, H. Wei, H.-Y. Chen, G. Gielen, H.-S. P. Wong, and S. Mitra, "Sensor-to-digital interface built entirely with carbon nanotube fets," *IEEE J. of Solid-State Circuits*, vol. 49, no. 1, pp. 190–201, 2014.
12. R. Navid, T. Lee, and R. Dutton, "Minimum achievable phase noise of rc oscillators," *IEEE J. of Solid-State Circuits*, vol. 40, no. 3, pp. 630–637, 2005.
13. S. Gierkink and E. Van Tuij, "A coupled sawtooth oscillator combining low jitter with high control linearity," *IEEE J. of Solid-State Circuits*, vol. 37, no. 6, pp. 702–710, 2002.
14. W. Bracke, P. Merken, R. Puers, and C. Van Hoof, "Ultra-low-power interface chip for autonomous capacitive sensor systems," *IEEE Trans. on Circuits and Systems-I: Regular Papers*, vol. 54, no. 1, pp. 130–140, 2007.
15. A. Hajimiri, S. Limotyrakis, and T. Lee, "Jitter and phase noise in ring oscillators," *IEEE J. of Solid-State Circuits*, vol. 34, no. 6, pp. 790–804, 1999.
16. S. Norsworthy, R. Schreier, and G. Temes, *Delta-sigma data converters: theory, design, and simulation*. New York, 1997.
17. J. Van Rethy and G. Gielen, "An energy-efficient capacitance-controlled oscillator-based sensor interface for mems sensors," in *proceedings of A-SSCC*, 2013, pp. 443–446.
18. J. Kim, T.-K. Jang, Y.-G. Yoon, and S. Cho, "Analysis and design of voltage-controlled oscillator based analog-to-digital converter," *IEEE Trans. on Circuits and Systems-I: Regular Papers*, vol. 57, no. 1, pp. 18–30, 2010.
19. V. De Smedt, G. Gielen, and W. Dehaene, "A 40 nm-cmos, 18 μw, temperature and supply voltage independent sensor interface for rfid tags," in *proceedings of A-SSCC*, 2013, pp. 125–128.
20. J. Van Rethy, H. Danneels, and G. Gielen, "Performance analysis of energy-efficient bbpll-based sensor-to-digital converters," *IEEE Trans. on Circuits and Systems-I: Regular Papers*, vol. 60, no. 8, pp. 2130–2138, 2013.
21. J. Van Rethy, H. Danneels, V. De Smedt, W. Dehaene, and G. Gielen, "Supply-noise-resilient design of a bbpll-based force-balanced wheatstone bridge interface in 130-nm cmos," *IEEE J. of Solid-State Circuits*, vol. 48, no. 11, pp. 2618–2627, Nov 2013.

The Ring Amplifier: Scalable Amplification with Ring Oscillators

Benjamin Hershberg and Un-Ku Moon

Abstract Ring amplification is a technique for performing efficient amplification in nanoscale CMOS technologies. By using a cascade of dynamically stabilized inverter stages to perform accurate amplification, ring amplifiers are able to leverage the key benefits of technology scaling, resulting in excellent efficiency and performance. A generalized view of basic small-signal theory is first presented, followed by a deeper discussion of the time-domain operation of a ringamp in the context of a specific ringamp structure. We conclude with a survey of existing ringamp implementations and techniques reported in literature.

1 Introduction

In ecology, bio-diversity is often a key indicator of the fitness and resilience of an ecosystem. In a similar way, the diversity of viable solutions and approaches available in the world of analog circuits is an indicator of the health of the analog design ecosystem. The range of technical requirements for analog signal processing blocks in practical design applications is as broad and vast as the ways in which their commercial implementations are used to enhance the many facets of society, work, and leisure. There is no one-size-fits-all solution here. Rather, a variety of solutions allow us to select the best tool for the job and leverage technology scaling to the fullest.

For design in nanoscale CMOS, amplifiers are one area where diversity has arguably been lost. Only a small sub-set of the viable amplifier topologies that once ruled in micron and submicron CMOS design remain competitive in nanoscale CMOS. The impact of this is readily observed in the study of ADCs, where

B. Hershberg (✉)
Imec, Leuven, Belgium
e-mail: benjamin@hershberg.com

U.-K. Moon
Oregon State University, Corvallis, OR, USA

P. Harpe et al. (eds.), *High-Performance AD and DA Converters, IC Design in Scaled Technologies, and Time-Domain Signal Processing*,
DOI: 10.1007/978-3-319-07938-7_18, © Springer International Publishing Switzerland 2015

amplifier-less SAR ADC topologies have been able to leverage technology scaling to consistently achieve conversion efficiencies far surpassing ADC topologies that rely heavily on amplification [1, 2].

A common misconception is that this poor scaling performance is a fundamental flaw of amplification itself. In actuality, it is mainly due to an adherence to the old paradigms of how amplifiers should be built. Conventional opamp topologies were conceived of at a time when 2.5 V supplies were considered low-voltage, and the intrinsic properties of transistors were quite different from that of a 14 nm FinFET [3]. Transistors were physically much larger with larger internal parasitic capacitances often approaching even that of the load capacitance being charged. Supply voltages provided much more headroom for stacking devices, and intrinsic device gains were higher. All these factors created a design world where cascoding was preferable to cascading, current biasing was necessary, and small-signal analysis was the prevailing and sufficient design paradigm.

Since those times, a lot has changed. Semiconductor technology has continued to evolve in a direction aimed at improving density, efficiency, and speed for digital logic gates. The ability of a conventional opamp topology to flourish in this new environment is fundamentally limited due to inherent incompatibilities in the underlying approach. Applying additional techniques such as calibration, gain-enhancement, and output-swing enhancement may enable an opamp to function in nanoscale environments, but it won't grant it the ability to scale at the same pace as digital performance improvements. A truly scalable amplifier must operate natively in its environment, in a way that implicitly uses the characteristics of scaled CMOS to its advantage, transforming potential weaknesses into inherent strengths. Since technology scaling is deliberately designed to favor the time-domain world of high-speed digital, viable nanoscale analog techniques are likely to be found in the time-domain realm as well. In order to fully exploit the abilities of a transistor, the biasing and small-signal properties of the device must be viewed as highly coupled, time-dependent variables which can be applied as feedback to each other with respect to time.

Here we explore one such technique: ring amplification. A ring amplifier (ringamp, RAMP) is a small modular amplifier derived from a ring oscillator which naturally embodies all the essential elements of scalability. It can amplify with rail-to-rail output swing, efficiently charge large capacitive loads using slew-based charging, scale well in performance according to process trends, and is simple enough to be quickly constructed from only a handful of inverters, capacitors, and switches.

2 A Small-Signal, Steady-State Perspective

The transient, large-signal, and small-signal operation domains are often much more interdependent in ring amplifier topologies than classical amplifier topologies. Despite this added complexity, as a starting point we can begin with the steady-state small-signal analysis of classical amplifier analysis.

Simply put, a ringamp is a multi-stage amplifier stabilized by a dominant output pole. A generalized three stage ringamp is shown in Fig. 1a, consisting of three inverting gain stages and optional stabilization networks. For now, we will consider the simple case where the stabilization networks have a steady-state, frequency-independent gain of 1 and each gain stage is a single-pole system. This results in Fig. 1b: a cascade of three inverting stages.

The open-loop gain of this three pole system is given by

$$H(s) = \frac{g_{m1}r_{o1} \cdot g_{m2}r_{o2} \cdot g_{m3}r_{o3}}{(1 + sr_{o1}C_{p1})(1 + sr_{o2}C_{p2})(1 + sr_{o3}(C_{p3} + C_L))} \quad (1)$$

where r_{ox} is the impedance seen at pole/node X, g_{mx} is the trans-conductance of inverter stage X, and C_{px} is the total capacitance seen at pole/node X.

Recalling basic stabilization theory, in order to transform this structure from an unstable ring oscillator into a stable ring amplifier, we must create a sufficiently large ratio between the lowest frequency pole in the system and the higher frequency poles. Given that an external load capacitance is required for any practical switched-capacitor design scenario, it is then relatively simple to prove that the strategy of stabilizing with a dominant output pole (p_3) will always yield the maximum amplifier bandwidth and highest efficiency.

The optimal output pole location can be created by first placing p_1 and p_2 at the highest frequencies possible and then adjusting the location of p_3 until it is at a sufficiently low, stabilizing frequency. Critically, for a three-stage opamp using conventional techniques and current biasing, this would not yield a very practical solution. The internal poles would still be relatively large, which would limit bandwidth and require an often impractically large explicit load capacitance at the output. Miller-compensation can be used to make other poles in the system dominant instead, but at high price in terms of bandwidth and efficiency. In the case of a ringamp with very small transistors used in gain stages 1 and 2 and dynamic biasing (i.e. just a basic inverter in this case), output pole stabilization becomes a realistic possibility. Poles p_1 and p_2 can be placed at very high frequencies, which allows us to place p_3 at a sufficiently stable location using a much more reasonably sized load capacitance.

This analysis describes the steady-state condition that a ringamp must reach in order to stabilize, but it does not tell us anything about *how* it does it. In many application scenarios, Fig. 1b is not the optimal implementation, and may not even be capable of meeting the required performance specs. At this point it is worth revisiting Fig. 1a and re-considering the "*how*" of ringamp stabilization. In the next section, we will take a closer look at a particular ringamp structure that uses time-domain feedback to dynamically adjust the output pole location and thereby relax stability constraints. Afterwards, in Sect. 4, we will look at other existing solutions, including techniques to extend ringamp operation into high accuracy applications.

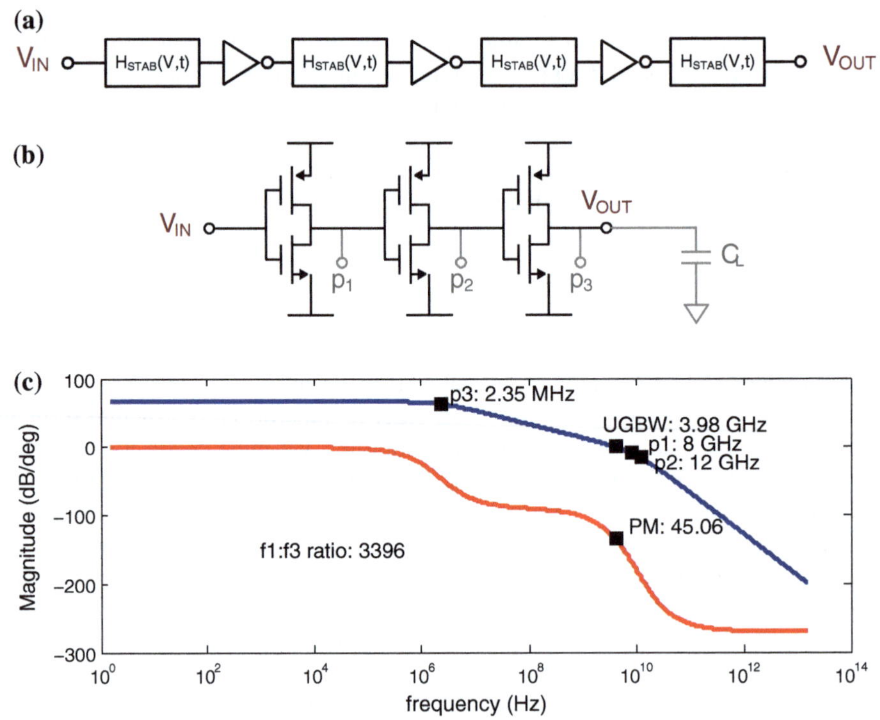

Fig. 1 A generalized three-stage ringamp in (**a**) is shown for the specific case of a three-inverter ring oscillator in (**b**). An example frequency response of (**b**) is given in (**c**)

3 Utilizing Time-Domain Feedback

We will now consider the specific case of the ringamp topology shown in Fig. 2 that utilizes time-domain feedback to enhance stabilization, and thereby improve efficiency. Fundamentally, the ringamp of Fig. 2 is a ring oscillator that has been split into two separate signal paths. A different offset is embedded into each signal path in order to create a range of input values for which neither output transistor M_{CN} nor M_{CP} of Fig. 2 will fully conduct. If this non-conduction "dead-zone" is sufficiently large, the ring amplifier will operate by slewing-to, stabilizing, and then locking into the dead-zone region. When placed in the example switched capacitor MDAC feedback structure also shown in Fig. 2, this charging and settling behavior results in the waveforms of Fig. 3b.

Before we examine how and why this occurs, it is useful to first understand some of the basic characteristics of the structure itself. To begin with, consider the capacitor C_1 of Fig. 2. C_1 is used to cancel the difference between the MDAC virtual-node sampling reference (V_{CMX}) and the trip-point of the first stage inverter. This ensures that the ideal settled value for V_{IN} will always be V_{CMX}, independent of the actual inverter threshold. Any sources of offset that are

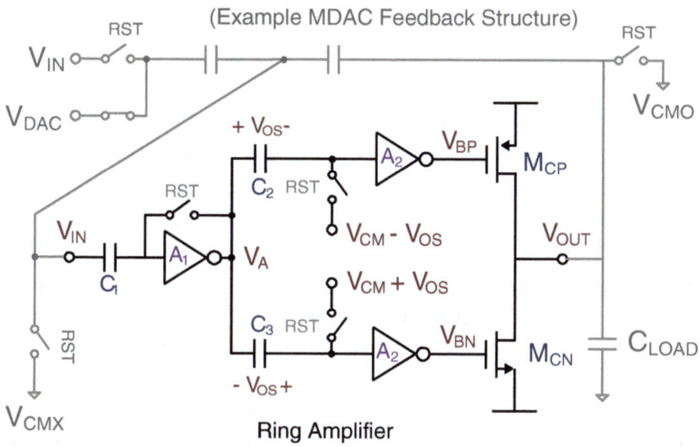

Fig. 2 The ringamp and basic switched-capacitor feedback network that we will primarily consider in this section. Devices and parameters that are referenced throughout the paper are labeled

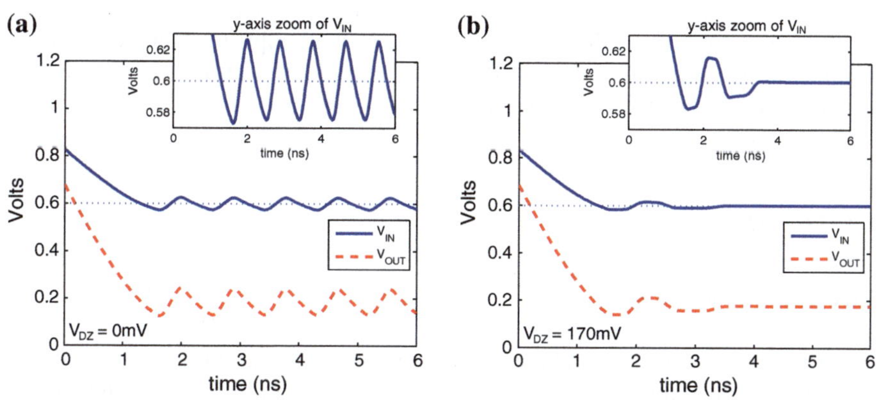

Fig. 3 Input and output charging waveforms of Fig. 2. In **a**, when $V_{DZ} = 0$ mV, the ringamp is functionally identical to a three-inverter ring oscillator. In **b**, the dead-zone is set large enough to generate stability ($V_{DZ} = 170$ mV) and the ringamp functions as an amplifier

generated after the first stage inverter will not be removed by C_1, but the input-referred value of such offsets will typically be negligibly small.

The dead-zone of the ringamp in Fig. 2 is embedded prior to the second stage inverters by storing a voltage offset across capacitors C_2 and C_3. Any value for V_{IN} within the dead-zone region is a viable steady-state solution for the ring amplifier, and the input-referred value of the dead-zone will determine the overall accuracy of the amplifier for most practical cases. In other words, the error at V_{IN} when the ringamp has stabilized and locked will be

$$-\left|\frac{V_{DZ}}{2 \cdot A_1}\right| \leq \epsilon_{V_{IN}} \leq \left|\frac{V_{DZ}}{2 \cdot A_1}\right| \tag{2}$$

where $V_{DZ} = 2V_{OS}$, A_1 is the final settled small-signal gain of the first stage inverter, and finite gain effects of the latter stages are ignored (revisited later).

Considering all this, we will now examine the interplay of small-signal, large-signal, and time-domain operation in a ringamp. Its behavior can be subdivided with respect to different modes of operation in time: slewing, stabilization, and steady-state. To illustrate key concepts, we will use the exaggerated charging waveform of Fig. 4 (taken from the ring amplifier of Fig. 2) that has been designed with relatively low bandwidth, excessive drive current, and a dead-zone size that biases the ringamp right at the edge of stability. Although one would never wish to make a real design in this way, as a teaching example it is quite useful. V_{CMX} is set to 0.6 V, and thus the ideal settled value of V_{IN} will also be 0.6 V. For the sake of simplicity and generality V_{OUT} is not shown (because it is simply a scaled, shifted, signal-dependent replica of V_{IN}). Unless otherwise stated, any mention of the amplitude of the fed-back signal will refer to the amplitude seen at V_{IN}.

In Fig. 4 we can clearly see three main phases of operation. Initially, from 0 to 2 ns, the ringamp rapidly charges toward the dead-zone. Then, from 2 ns to about 14 ns it oscillates around the dead-zone region as it attempts to stabilize. By 15 ns, with the output transistors M_{CP} and M_{CN} both completely cutoff, the ring amplifier reaches a steady-state solution within the dead-zone, and remains locked.

3.1 Operation

3.1.1 Initial Ramping

In the initial slew-charging phase of operation, the ring amplifier is functionally equivalent to the circuit of Fig. 5. The first two stages of the ring amplifier act like a pair of bi-directional continuous-time comparators that correctly select which output transistor (M_{CN} or M_{CP}) to use depending on the value of the input signal. The selected output transistor then operates as a maximally-biased current source and charges the output load with a ramp. In this initial charging phase the ringamp behaves similar to a zero-crossing based circuit [4, 5].

The ramping phase ends when the input signal crosses the threshold of the comparator and the current source turns off. Due to the finite time delay of the comparator, there will be some amount of overshoot beyond the comparator threshold, which will be given by:

$$\Delta V_{overshoot} = \frac{t_d \cdot I_{RAMP}}{C_{OUT}} \tag{3}$$

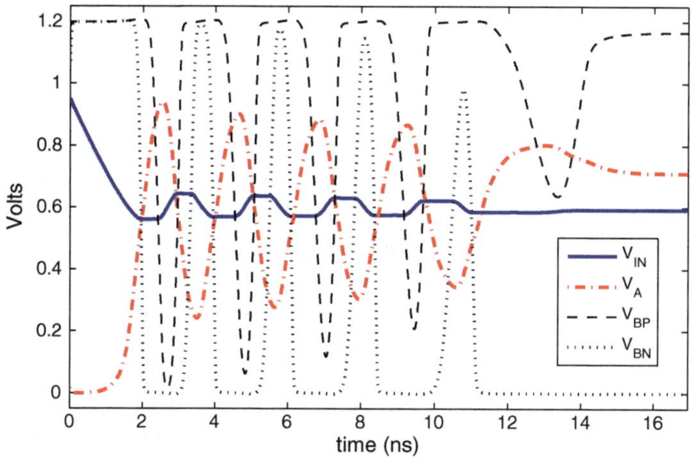

Fig. 4 Example ring amplifier operation for an exaggerated design biased at the edge of stability, showing the three key phases of operation: *1* initial ramping, *2* stabilization, and *3* steady-state

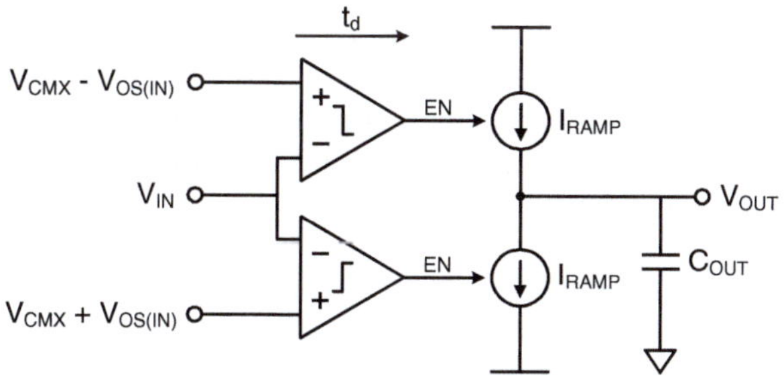

Fig. 5 Conceptual model of a ringamp during the initial slew-charging phase of operation. This model only applies to the initial charging phase and does not include the key ringamp stabilization mechanisms. $V_{OS(IN)}$ is the input-referred value of the dead-zone offset

where t_d is the time delay of the comparator decision, I_{RAMP} is the current supplied by the active current source, and C_{OUT} is the total loading capacitance seen at the output. This overshoot is with respect to the trip point, which will be on the boundary of the dead-zone. It will be more useful later on if we consider Eq. 2 as well, and express the input-referred overshoot with respect to the ideal settled value (the center of the dead-zone):

$$\Delta V_{init} = \frac{t_d \cdot I_{RAMP}}{\psi C_{OUT}} - \left| \frac{V_{DZ}}{2 \cdot \hat{A}_1} \right| \tag{4}$$

where ψ is the scaling factor that refers the output overshoot to the ringamp's input (and depends on feedback factor, parasitics, and feedback structure) and $\hat{A}_1$ is the effective gain of the first stage inverter at the end of the ramping operation (explained later).

3.1.2 Stabilization

After the initial charging ramp, the ring amplifier will begin to oscillate around the target settled value with amplitude ΔV_{init}. With no dead-zone, the structure is functionally identical to a three-inverter ring oscillator, and will continue to oscillate indefinitely (Fig. 3a). However, as the size of the dead-zone is increased, the ringamp will eventually reach an operating condition where it is able to self-stabilize, such as in Fig. 4. If the dead-zone size is increased further still, the time required to stabilize decreases substantially, and for most practical designs, a ringamp will stabilize in only one or two periods of oscillation (i.e. Fig. 3b).

The most fundamental mechanism in the process of stabilization is the progressive reduction in the peak overdrive voltage applied to the output transistors M_{CN} and M_{CP} on each successive period of oscillation. This effect is illustrated in Fig. 4 by the progressive decrease in amplitude of the signals V_{BP} and V_{BN}. When the following relation is true, the trough (minimum value) of V_{BP} will be limited by the finite-gain of the first two stages, and begin to de-saturate from rail-to-rail operation:

$$\hat{A}_2[\hat{A}_1(min(\tilde{V}_{IN}) - V_{CMX}) - V_{OS}] \geq V_{SS} - V_{CM} \tag{5}$$

(where $\tilde{V}_{IN}$ is the peak-to-peak amplitude, and $\hat{A}_1, \hat{A}_2$ are the negative-valued effective instantaneous inverter gains). A similar relation can also be expressed for the lower signal path and V_{BN}:

$$\hat{A}_2[\hat{A}_1(max(\tilde{V}_{IN}) - V_{CMX}) + V_{OS}] \leq V_{DD} - V_{CM} \tag{6}$$

The key point to notice in these expressions is that each signal path is being fed a different shifted replica of the oscillatory waveform generated at V_A. The upper path is given a replica where the *peaks* of the wave are *lowered* closer to the second stage inverter's threshold, and the lower path is given a replica where the *troughs* of the wave are *raised* closer to the threshold of the second stage inverter. For a sufficiently large shift in each path (V_{OS}), this creates the possibility that even for relatively large values of $\tilde{V}_{IN}$, finite gain effects will simultaneously limit the overdrive voltage that is applied to both M_{CP} and M_{CN}. This stands in stark contrast to the behavior of a three-inverter ring oscillator, where the decrease in V_{OV} of one output transistor necessarily means an increase in V_{OV} applied to the other.

When Eqs. 5 and 6 are true, the resulting reduction in V_{OV} applied to the output transistors M_{CN} and M_{CP} will reduce the magnitude of the output current I_{RAMP}. This decrease in output current will also cause a decrease in the amplitude of $\tilde{V}_{IN}$ by a proportional amount, due to Eq. 4. The left sides of Eqs. 5 and 6 are therefore reduced further, and the V_{OV}'s of M_{CN} and M_{CP} will decrease even more for the next oscillation cycle. This effect will continue to feedback until the input signal amplitude becomes smaller than the input-referred value of the dead-zone, at which point the ring amplifier will stabilize and lock into the dead-zone.

If we combine Eqs. 5 and 6 and rearrange, we see that in order to trigger this progressive overdrive reduction effect, the input signal must satisfy the following relation:

$$\tilde{V}_{IN} \leq \frac{1}{\hat{A}_1}\left(\frac{V_{DD} - V_{SS}}{\hat{A}_2} - V_{DZ}\right). \quad (7)$$

Furthermore, at the beginning of the stabilization phase:

$$\tilde{V}_{IN} = 2 \cdot \Delta V_{init} \quad (8)$$

Finally, using Eqs. 3, 4, 7, and 8, we can express the stability criterion in terms of the dead-zone (i.e. settled accuracy) and the initial slew rate (i.e. speed):

$$\frac{t_d \cdot I_{RAMP}}{\psi C_{OUT}} \leq \frac{1}{2 \cdot \hat{A}_1}\left(\frac{V_{DD} - V_{SS}}{\hat{A}_2} - 2 \cdot V_{DZ}\right) \quad (9)$$

Recall once again that $\hat{A}_1$ and $\hat{A}_2$ are negative valued gains.

From this relation we see that there is a clear design tradeoff between accuracy, speed, and power. Let's assume for a moment that only t_d, I_{RAMP}, and V_{DZ} can be adjusted. To increase speed, one can either increase the initial ramp rate or decrease the time required to stabilize. Both options require sacrificing either accuracy (by increasing V_{DZ}) or power (by decreasing t_d). Likewise, to increase accuracy (by decreasing V_{DZ}), one must either decrease I_{RAMP} or decrease t_d accordingly. While these simple tradeoffs serve as a good starting point, as we will soon discover, every parameter in Eq. 9 is variable to some extent.

The discussion thus far is only a first-order model, and there are additional bandwidth, slewing, and device biasing dynamics which are not represented. Let's take a moment to evaluate this model in the form of a practical example. Consider a pseudo-differential ringamp where $A_1 = A_2 = -25\frac{V}{V}$, $V_{DZ} = 100$ mV, $V_{DD} = 1.2$ V, and $V_{SS} = 0$ V. By Eq. 2, the input-referred size of the dead-zone will be about 4 mV, which for a 2 V pk-pk input signal would ideally be accurate enough to achieve an input-referred SNDR of 54 dB. By Eq. 7, the maximum allowable peak-to-peak amplitude of $\tilde{V}_{IN}$ is approximately 6 mV, and by Eq. 4, the maximum allowable input-referred overshoot at the end of the initial ramping phase must be less than 5 mV.

This isn't a very encouraging result, since such a small overshoot will place a tight constraint on the parameters in Eq. 3. However, if one were to simulate this same scenario, it will turn out that the peak-to-peak amplitude of oscillation can be significantly larger than the predicted 6 mV and still achieve stability. A closer look at Fig. 4 reveals an important contributor to this disparity between theory and practice. Although the AC small-signal gain of the first stage inverter, A_1, may be $-25\frac{V}{V}$, the effective instantaneous value

$$\hat{A}_1(t) = \frac{V_A(t)}{V_{IN}(t)} \tag{10}$$

in the actual transient waveform will be several times smaller at the beginning of stabilization. Thus, although the overall accuracy of the ringamp is determined by the final, settled, small-signal value of A_1, the stability criterion is determined by the initial, transient, large signal effective value of A_1. This reduction in A_1 occurs because the first stage inverter inherently operates around its trip point, where it will be slew limited. The maximum slewing current that the inverter can provide will be

$$I_{slew} = I_P - I_N \tag{11}$$

and for a square law MOSFET model, this will become:

$$I_{slew} = 2k'\frac{W}{L}\left(\frac{V_{DD} - V_{SS}}{2}\right)V_{IN} \tag{12}$$

Notice here that the slew current is linearly related (not quadratically) to the input voltage. Thus, for the first stage inverter, slew rate limiting (and finite bandwidth) has an important impact on determining the effective value of $\hat{A}_1$ during stabilization (and to a lesser extent, the value of $\hat{A}_2$). This dynamic adjustment of the effective inverter gain is a very attractive characteristic, and improves the design tradeoff between speed, accuracy, and power by a significant factor. Similar effects also influence the operation of the second stage inverters in an additional way: although Eqs. 5 and 6 assume rail-to-rail swing for the second stage inverters when $\tilde{V}_{IN}$ is large, in reality the output swing of the second stage inverters may never completely reach rail-to-rail, regardless of the value of $\tilde{V}_{IN}$ due to slew rate limiting, finite bandwidth, and triode device operation.

Relating the discussion of progressive overdrive reduction in this section back to the steady-state stability discussion of Sect. 2, this behavior can be conceptualized as a dynamic adjustment of the ringamp's output pole corner frequency. The decrease in output current due to V_{OV} reduction increases the output impedance (R_o) of the ringamp, and pushes the output pole (formed by R_o and C_{LOAD}) to lower frequency. As the V_{OV} reduction effect gains momentum on each successive oscillation half-period, the output pole progressively pushes to lower and lower frequency. By the time the ringamp is locked into the dead-zone and the output transistors are in cutoff, R_o is infinite and the output pole is at DC.

3.1.3 Steady State

Thus far, we have defined the steady-state condition for a ring amplifier as the complete cutoff of both output transistors, with the input signal lying solidly within the dead-zone, such as is the case in Fig. 4. However, considering the discussion about pole adjustment in the previous paragraph, it's clear that the ringamp can in

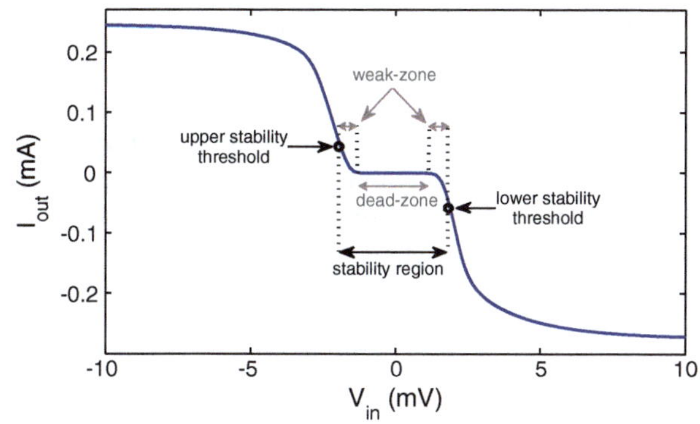

Fig. 6 DC sweep of V_{IN} versus I_{OUT} for a typical configuration of Fig. 2, illustrating the full input-referred characteristic near the dead-zone region. In addition to a true "dead-zone" where both output transistors are in cutoff, there is also a small boundary region "weak-zone" where the output pole location is low enough to create stability

fact be stable for a range of low frequency output pole locations down to DC. Such a situation will in practice occur often, even for a large dead-zone, since there is always a finite probability that the ring amplifier will happen to stabilize right at the edge of the dead-zone. If that happens, one of the output transistors will still conduct a small amount of current to the output, and may never fully shut off before the amplification period ends. The existence of this stable, boundary-region "weak-zone" is illustrated in the V_{IN} versus I_{OUT} plot of Fig. 6. The weak-zone isn't an inherent problem for ring amplification operation, since any low-bandwidth settling will only serve to further improve accuracy. However, there are sometimes higher-level structural considerations that make it advantageous to ensure that both output transistors are completely non-conducting once settled. We will see some designs where this is the case later on, in Sect. 4.

3.2 Key Advantages

Ring amplifiers are in many ways both structurally and functionally quite different from conventional opamps, and it is in these differences that the ringamp finds a unique advantage in the context of modern low-voltage CMOS process technologies. In this section, we will examine several of these important benefits in greater detail.

3.2.1 Output Compression Immunity

In low-voltage scaled environments, kT/C noise, SNR, and power constraints will typically be dictated by the usable signal range available [1], and any practical amplification solution for scaled CMOS must therefore utilize as much of the available voltage range as possible. As it turns out, ring amplifiers are almost entirely immune to output compression, and this enables them to amplify with rail-to-rail output swing.

To understand the basis of this output compression immunity, we must consider two scenarios. First, imagine a ringamp whose dead-zone is large enough that when the ringamp is locked into the center of the dead-zone, both M_{CN} and M_{CP} will be in cutoff. In other words, when:

$$V_{DZ} \geq \left| \frac{V_{DD} - V_{SS} - 2V_T}{A_2} \right| \tag{13}$$

As a rule of thumb, this relation will usually hold for low and medium accuracy ringamps up to about 60 dB. Under this scenario, M_{CN} and M_{CP} function as current sources whose linearity and small-signal gain has no appreciable effect on settled accuracy. The internal condition of the ringamp depends only on the signal at the input, and it will continue to steer toward the dead-zone until M_{CN} and M_{CP} are completely cut-off, regardless of whether they are in saturation or triode. Final settled accuracy will be governed by Eq. 2, independent of the characteristics at the output.

Now let's consider the condition where Eq. 13 does not hold. This will occur when the dead-zone is very small, and accuracies in the 60–90 dB range are desired. Although other practical issues in the ringamp structure of Fig. 2 may hinder such design targets, we will consider a ringamp structure later in Sect. 4 where it applies. In this scenario, the stability region of Fig. 6 is so small that the two weak-zones touch, and M_{CN} and M_{CP} will still conduct a small amount once settled. The ringamp's steady state condition will essentially be that of a three stage opamp, and the open loop gain will be the product of the three stage gains. With no true dead-zone, the distortion term of Eq. 2 becomes zero, and finite loop gain will become the fundamental limitation on accuracy. At first glance, generating sufficient loop gain appears to be a problem, since the gain of M_{CN} and M_{CP} will depend on output swing (which must be as large as possible in nanoscale CMOS). Consider the case where all three stages have a gain of 25 dB when operating in saturation. In the best case, the open loop gain will be 75 dB, and in the worst case perhaps 50 dB. Even in the best case, this seems to suggest that to build an 80 dB accurate ringamp, an additional gain stage is required.

Luckily, there is another effect at play here. In the ideal square-law MOSFET model M_{CN} and M_{CP} will be in saturation when $V_{OV} < V_{DS}$. Furthermore, the small signal output impedance, r_o, is inversely proportional to the drain current, I_D. In the context of the progressive overdrive voltage reduction that occurs in ringamp stabilization, both V_{OV} and I_D will in fact trend towards zero. This implies that during steady-state, M_{CN} and M_{CP} will remain in saturation or weak-inversion even for very small values of V_{DS}, and moreover, that their gain will be enhanced by a dynamic boost in r_o. Thus, even for a nominal open loop gain of 75 dB, with a wisely chosen topology it is possible to have an enhanced steady-state gain of at least 90 dB, even when swinging close to the rails.

Although output swing has little effect on ringamp accuracy, it will indeed affect speed, both with respect to slewing and settling. In the initial ramping phase, the selected current source transistor will be biased with the maximum possible V_{OV}, and this guarantees that for much of the possible output range it will initially be operating in triode. As seen in Fig. 7, for settled output values near mid-rail, I_{RAMP} will be the highest and the initial ramping will be faster, but more time will be required to stabilize for the reasons discussed in Sect. 3. Likewise, for values close to the rails, I_{RAMP} will be smaller, so the initial ramping will be slower but the stabilization time will be shorter. For the most part, this works out quite nicely, since the total time required to reach steady state in each case turns out to be approximately the same. However, for extreme cases very close to the rails, the large RC time constant of the output transistor in triode operation will require a comparatively long time to reach its target value. Ultimately, it is this RC settling limitation that will usually dictate the maximum output swing possible for a given speed of operation.

3.2.2 Slew-Based Charging

Whereas a conventional opamp charges its output load with some form of RC-based settling, the output transistors M_{CN} and M_{CP} in the ring amplifier behave like digitally switched current sources, and charge the output with slew-based ramping. This is a much more efficient way to charge, since only one of the current sources in Fig. 5 will be active at a time, and the only power dissipated will be dynamic. Furthermore, during the initial ramping operation, M_{CN} or M_{CP} (whichever is selected) will be biased with the maximum V_{OV} possible for the given supply voltage. This is a major benefit, because it means that even for large capacitive loads, small transistor sizes can still produce high slew rates, and with small output transistors, the second stage inverters will be negligibly loaded by M_{CN}/M_{CP}. This effectively decouples the internal power requirements from that of the output load size, and for typical load capacitances in the femto and pico-farad range, the internal power requirements are more-or-less independent of output capacitance. This unique property stands in stark contrast to the power-loading relationship for a conventional opamp, where settling speed is typically proportional to g_m/C_{LOAD}. Even for large load capacitances, where the size of M_{CN}/M_{CP} does have an appreciable effect on the internal power requirements, the ratio of static-to-dynamic power will scale very favorably.

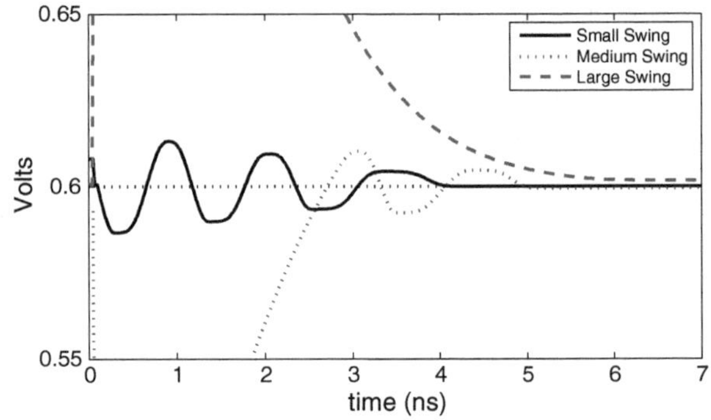

Fig. 7 Zoomed stabilization waveform of V_{IN} for three output swing cases: small (output near mid-rail), medium, and large (final output near the supply)

3.2.3 Performance Scaling with Process

In order for a technique to be truly scalable, it must meet two criteria. First, the given technique must operate efficiently *in* a scaled environment. This requirement has been our primary focus thus far. Second, the technique must inherently scale *with* advancing process technology, improving in performance simply by migrating into a newer technology. It is this second criteria that we will discuss now.

Intuitively, the ring amplifier seems like a prime candidate to benefit from process scaling, simply due to its structural similarity to a ring oscillator. The stability criterion of Eq. 9 suggests this to be true. As stated previously, the internal power consumption of a ringamp is governed much more by inverter power-delay product and internal parasitics than the size of the output load (in stark contrast to conventional opamps). Since the power-delay product of an inverter decreases approximately linearly in accordance with decreasing feature size [6], the ringamp's inverter chain propagation delay, t_d, can be expected to scale according to digital process performance as well. With the relationships in Eq. 9, this reduction in t_d can be directly traded for an improvement in any of the three main design specifications: speed, accuracy, and power.

The simple scaling experiment conducted in [7] suggests that this is indeed the case. The results of the test are shown in Fig. 8. The upper trend line represents predicted power efficiency with the power spent in charging the fixed (not scaled) load capacitance included. The lower trend line shows the power efficiency with the ideal power required to charge the load capacitance subtracted out. The result is a power efficiency trend that scales very well with advancing technology node. Although there are not yet enough measured ringamp designs to verify predicted trend, we do see a similar level of speed and efficiency scaling between the 0.18 μm CMOS design of [8] and the 65 nm CMOS design of [9]. Furthermore, recent investigations of ringamp structures in 28 nm CMOS have produced results that also support this hypothesis.

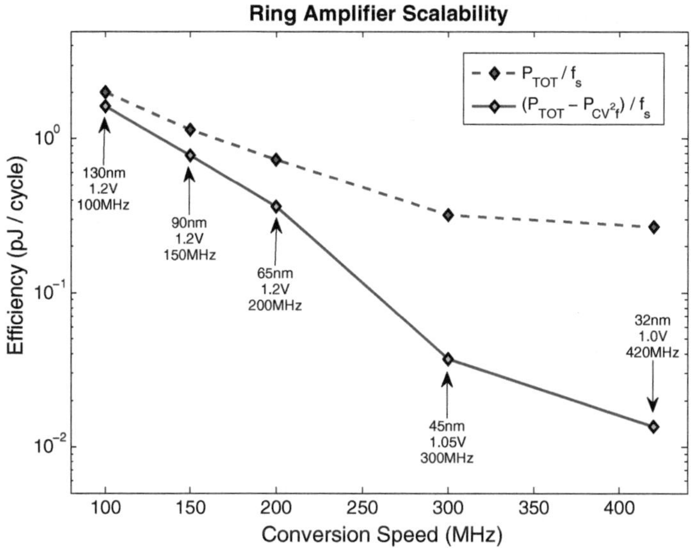

Fig. 8 Approximate power efficiency scaling trends predicted by [7]

4 A Diversity of Solutions

There are many different opamp structures available to designers to meet diverse needs. The same should also be true for ringamps. Accuracy, speed, power, and design effort are all important factors in choosing the best ringamp structure for the job. Although the exploration of possible ringamp techniques is still in its infancy, we can at least look at the comparative merits of those ringamp implementations which are already known and reported in the literature.

To begin with, for high-speed, medium-accuracy amplification, the ringamp circuits of Figs. 1b and 2 are both attractive candidates. In technologies where the gain of the ringamp is more than that required for accuracy specifications, the ringamp of Fig. 2 is a good choice. Technologies where this is the case include older processes such as 0.18 μm, 0.13 μm, and 90 nm, and possibly even some newer processes such as 14 nm FinFET, where inverter gains are improved [3]. The dead-zone embedding in the structure of Fig. 2 allows bandwidth to be decoupled from the small-signal gain of the inverters. Even if the small-signal gain of the inverter chain is very high, the dead-zone allows us to effectively reduce it and improve the gain-bandwidth product while still benefiting from the fast transient behavior associated with high gain inverters. This structure is used for the purposes of medium-accuracy amplification in the 10.5b pipelined ADC of [8], and as a sub-component of the 15b Split-CLS pipelined ADC of [10] and the Composite Ringamp Amplifier Block of [11].

Alternatively, if working in a technology with low intrinsic inverter gains such as 65, 45, 32, 28, and 22 nm planar CMOS, it may not be possible to set a true

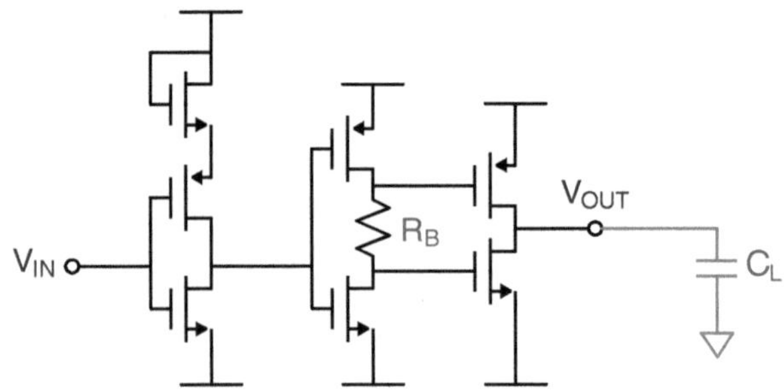

Fig. 9 Self-biased ringamp structure of [9]

dead-zone with the structure of Fig. 2. In this case, the simple inverter chain of Fig. 1b may actually be a realistic option to consider. The structure still scales excellently and retains many (although not all) of the benefits of a dead-zone stabilized ringamp. However, the ringamp structure introduced in [9], shown in Fig. 9, provides an even better solution for most scenarios. Like Fig. 1b it is also a single chain of inverters. However, it uses resistor R_B to embed a weak-zone offset for improved dynamic biasing, enhancing both efficiency and speed. The design of [9] also introduces the idea of using the time information contained in the internal nodes of a ringamp to perform quantization, and this time-domain information is used to build a 1.5b pipeline stage sub-ADC that results in significantly relaxed timing constraints.

For design scenarios requiring high accuracy amplification, a different approach to ringamp design is required. Fully differential operation is typically mandatory, and the amplifier must either produce very high effective gain or be assisted by calibration. In [10] the technique of Split-CLS [12] is used to combine a pseudo-differential ringamp structure with a fully differential telescopic opamp. Figure 10a shows the basic principle of this technique. The ringamps provide an initial fast and coarse charge of the output. Then, the opamp is coupled into the output via the level shifting capacitors (C_{CLS}) and proceeds to fine settle the output with greatly relaxed slew and swing requirements. At the end of an amplification period, the total accuracy of the output is the combined accuracy of the ringamp *and* the telescopic opamp. This structure is capable of very high accuracy, and allows a robust differential opamp to be the final determinant of accuracy.

Another approach to precision amplification is the structure of Fig. 10b. Introduced in [11], a Composite Ring Amplifier Block uses only ring amplifiers (no opamps) and consists of a coarse but fast and efficient pseudo-differential ringamp connected in parallel with a differential-input, single-output precision ringamp. When placed in this parallel configuration, the coarse ringamp will automatically and asynchronously cutoff and transfer control to the fine ringamp at the correct moment of operation. Initially, all ringamps are enabled, and contribute charge to the output. However, the coarse ringamps have a larger slewing capability and dominate the initial charging behavior, quickly settling both the differential and common-mode levels close to their final target values. As the coarse ringamps enter their dead-zone, they automatically disconnect from the output.

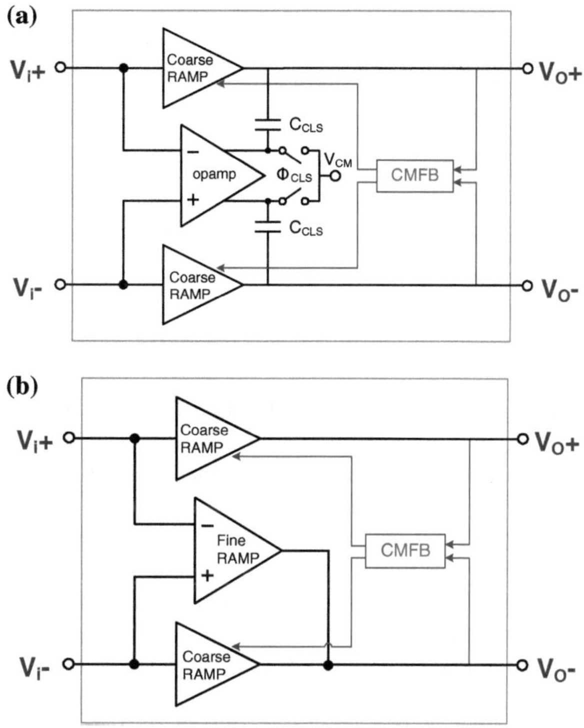

Fig. 10 Ringamp structures suitable for high-accuracy applications. **a** Split-CLS structure [10], **b** composite ring amplifier block [11]

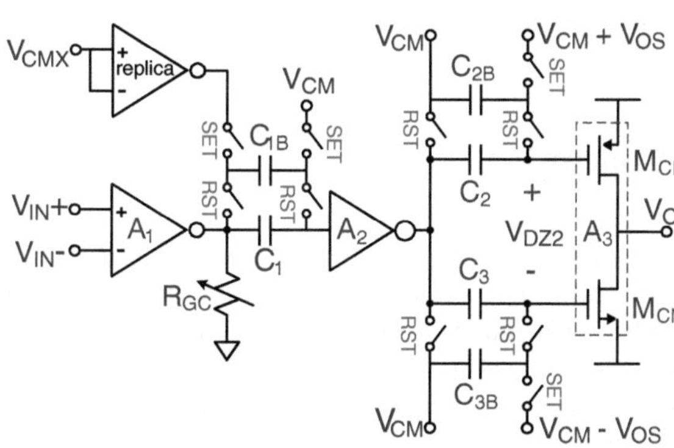

Fig. 11 Precision ringamp circuit used in the composite ring amplifier block (Fig. 10b) of [11]

The fine ringamp remains active, however, since its stability region (weak-zone) is necessarily quite small and thus completely enclosed by the larger stability region (dead-zone) of the coarse ringamps. With the common-mode level already settled to sufficient accuracy by the coarse ringamp's CMFB, and V_{O+} floating, the fine ringamp simply settles V_{O-} differentially around a stationary V_{O+}. Thus, in addition to providing speed enhancement this scheme removes the need for

Table 1 Summary of ringamp ADCs

	VLSI'12 [8]	ISSCC'12 [10]	VLSI'13 [11]	ISSCC'14 [9]
Technology	0.18 μm	0.18 μm	0.18 μm	65 nm
Supply voltage (V)	1.3	1.3	1.2	1.2
Resolution (b)	10.5	15	15	10.5
Input range (V pk-pk)	2.2	2.5	2.4	2
Sampling rate (Msps)	30	20	20	100
SNDR (dB)	61.5	76.8	75.9	56.3
SFDR (dB)	74.2	95.4	91.4	67.6
Total power (mW)	2.6	5.1	2.96	2.46
FoM (fJ/c-step)	90	45	29	46

common-mode feedback in the fine ringamp, permitting a single ended output to be used without loss of accuracy, thereby minimizing both complexity and power.

A simplified schematic of the fine ringamp is depicted in Fig. 11. Notably, the offset V_{DZ2} is embedded just prior to M_{CN}/M_{CP}, and allows the settled value of V_{OV} to be precisely set (and weak-zone operation to be guaranteed). However, the constraint that this places on the value of V_{DZ2} consequently limits its ability to tune stability. This is solved by observing that stability is actually determined by the input-referred value of V_{DZ2}, which can also be tuned by adjusting the gain of either the first or second stage inverter. Thus, a tunable gain-control resistor (R_{GC}) is used to set the size of the stability region. There is almost no linearity requirement for this tuned resistance, and it is implemented in [11] as a simple 3-bit DAC composed of tiny MOSFET resistive elements.

A summary of the measured performance of the ringamp-based pipelined ADCs discussed here is provided in Table 1.

5 Conclusion

Even in the initial design attempts listed in Table 1, we already see very promising performance numbers being achieved across a range of target accuracies and speeds. Performance will continue to improve in the future for a couple of reasons. First, simply scaling down into newer technology nodes should yield substantial benefits. Second, the potential for using time-domain properties to improve ring-amp efficiency is by no means exhausted. For example, the idea to exploit the time-domain behavior of a ringamp to perform quantization in [9] is a solution with attractive benefits. Some of the insight gained in the research of other emerging time-domain techniques such as VCO-based quantizers [13] may prove relevant to ringamp research as well.

Just how many niches in the circuit ecosystem ring amplification can enhance diversity in remains to be determined. It is already evident that they are useful in pipelined ADCs. Many other ADC architectures can benefit from ringamps as